Biological Oxidation Systems

Volume 2

Biological Oxidation Systems

Volume 2

edited by

C. Channa Reddy

College of Agriculture and
Environmental Resources Research Institute
The Pennsylvania State University
University Park, Pennsylvania

Gordon A. Hamilton

Department of Chemistry
The Pennsylvania State University
University Park, Pennsylvania

K. M. Madyastha

Department of Organic Chemistry
Indian Institute of Science
Bangalore, India

ACADEMIC PRESS, INC.

Harcourt Brace Jovanovich, Publishers

San Diego New York Boston London Sydney Tokyo Toronto

Academic Press Rapid Manuscript Reproduction

This book is printed on acid-free paper. ♾

Academic Press, Inc.
San Diego, California 92101

United Kingdom Edition published by
Academic Press Limited
24-28 Oval Road, London NW1 7DX

Library of Congress Catalog Card Number: 90-83444

ISBN 0-12-584552-9 (alk. paper)

Printed in the United States of America
90 91 92 93 9 8 7 6 5 4 3 2 1

CONTENTS VOLUME 1

II. Oxidases and Oxygenases Containing a Flavin/Pteridine

III. Copper Containing Oxidases and Oxygenases

IV. Nonheme Iron-Containing and Other Oxygenases

V. Peroxidases in Cell Protection and Metabolism

CONTENTS VOLUME 2

VIII. Lipid Peroxidation and Other Nonenzymatic Reactions of Oxygen

SPONSORS

Biomol Research Laboratories
Cayman Chemical Company
Chugai Pharmaceuticals
Council for Scientific and Industrial Research, India
Department of Atomic Energy, India
Department of Biotechnology, India
Department of Environment, India
Hoffmann La Roche
Indian National Science Academy
International Union of Biochemistry
Landmark Travel
Merck Frosst Canada
Merrell Dow Pharmaceuticals
Monsanto Company
National Science Foundation, USA
Norwich Eaton Pharmaceuticals
Schering Plough Corporation
Union Carbide Corporation
University Grants Commission, India

FOREWORD

The Bangalore Symposium on Oxygen Systems presented a rewarding opportunity for both young and established scientists to share a broad array of knowledge on the essential roles, and the dangers, of dioxygen reactions in biology.

This foreword, meanwhile, gives me the freedom to extol the overall excellence of my colleagues' orchestration and performance at the symposium. Allow me to specify what was special about this gathering and why.

John Buchan—scholar, author, diplomat, Governor General of Canada, and the first Lord Tweedsmuir—once observed that a liberal education should endow its recipients with three qualities: humility, humanity, and humor. Humility because "if we are educated men, with the treasures of the world's thought behind us, we shall not be inclined to overvalue ourselves or to claim too much for the work of our hands." To Buchan, you see, knowledge presupposed humility. Humanity because "we need a deeper respect for human nature. There can be no such respect in those who obliterate the personality and make beings mere featureless details." And humor because "in a time like the present...there is a tendency for popular leaders to exalt themselves into a kind of bogus deity and to think their shallow creeds a divine revelation. The answer to all such folly is laughter."

The symposium planning and implementation epitomized John Buchan's three H's. The continuing exponential growth of science certainly commands attention. From the hundred or so scientific journals in 1800, an increase of tenfold each fifty years has brought the total to above 100,000. Meanwhile, 80 or 90 percent of all the scientists who have ever lived are alive now. Clearly, these astonishing numbers compel some judicious sorting.

The scientific sessions focused attention on oxygen systems in biology, closely considered the positive and negative effects on living matter, and featured a freedom of access to, and transfer of, the current pool of knowledge within the Symposium's selection of the international participants.

The informal meetings between the sessions fostered conversation, understanding, friendship, and more learning. Tours introduced smaller groups of first-time and return visitors to India to the variety, flavor, customs, and long history of its civilization as well as to relaxed and restful extensions of friendships.

While the preceding remarks imply emphasis on scientific content as the main purpose of the papers and discussions, a few related observations on the early concepts of oxygen may be in order. The scientific road is but two centuries from Priestly's "different kinds of air" and Lavoisier on "oxygine" and one century from Pasteur with his "life without air" and McMunn with "myo- and histo-haematins." We are a bare half century past Hayaishi and Mason who validated "oxygenases" with dioxygen 18.

Reflecting on these two centuries of science we see a progression from a time when simple breathing and body warmth defined life— a time that served as a preamble to discovery along paths now well travelled, themselves leading at present to uncharted scientific regions. There has been a quickening in both formal and interpersonal dissemination of the knowledge base, manpower development, and advances in food, health and biosphere protection. Simultaneously, a growing awareness of common heritages has emerged, dependent upon equitable resource allocations and mandated by growing population pressures.

Biology now enjoys the precision and excitement to attract colleagues from physics and chemistry who can augment theory and provide the measurements to characterize dynamics and mechanisms precisely. Science and utility move forward rapidly together, addressing the varied roles of oxygen in biology— an advance certain to benefit all of mankind. Such an important advance was the result of the thorough exchanges that occurred among my colleagues last October in Bangalore.

I. C. Gunsalus

PREFACE

Aerobic life is a highwire act supported by oxygen chemistry, balanced precariously between the biosynthetic power provided by the rich energy supplies of respiration and the destructive effects of oxygen radicals formed during O_2 metabolism.

The myriad chemical reactions of a cell's metabolic activity produce a variety of free radicals. Superoxide radicals result from mitochondrial functions. Xanthine oxidase and aldehyde oxidase systems also generate O_2 radicals. Hydroxyl, alkylperoxyl, and alkoxyl radicals result from the oxidation of polyunsaturated fatty acid by the autooxidation processes. Thus, antioxidants (such as vitamin C, vitamin E, and glutathione) and antioxidant enzymes (such as superoxide dismutase, glutathione peroxidases, and catalase) all play crucial roles in protecting cells from deleterious reactions between their structural and functional components and the free radicals.

On the other side of this delicate balance, aerobic life depends on reactions of O_2 for much more than the mitochondrial respiration that supplies its energy requirements. For example, O_2 reacts directly with metabolites in numerous important anabolic and catabolic pathways, including aromatic amino acid metabolism, catecholamine biosynthesis and degradation, prostaglandin, leukotriene and steroid hormone biosynthesis, drug and xenobiotic metabolism, liginin degradation, and photorespiration. A bewildering variety of enzymes catalyze the reactions of O_2 with this array of metabolites. Mason and his co-workers and Hayaishi and his colleagues were the first to report, independently in 1955, the enzyme-catalyzed fixation of molecular oxygen into aromatic compounds. Since then, a voluminous literature has accumulated on the occurrence of such enzyme systems in microbes, higher plants, and animals.

Despite the fact that enzymes that catalyze O_2 reactions are so common and important in nature, the mechanisms of these reactions are known in considerably less molecular detail than the mechanisms of most other biological reactions. For example,

1. What are the exact structures of the important intermediates in the various enzymic oxidations?

2. What characteristics of individual enzymes permit the formation and stabilization of highly reactive intermediates?

3. What alterations in protein structure allow different enzymes, all having the same cofactor (for example, heme or non-heme iron), to carry out such different reactions?

4. What are the common structural and mechanistic features that permit enzymes with quite different requirements (for example, one requiring copper and another requiring iron) to carry out virtually identical reactions?

We have entered a new era in biology with the advent of molecular biology techniques, and these new tools already enable us to understand the structures and mechanisms of several enzymes in enough detail to begin answering these questions. For example, the elucidation of gene structures for enzymes has led to new insights and new approaches to the regulation of enzymes that utilize O_2 as a substrate.

These two volumes present the contributions of the scientists invited to address the October, 1989, Bangalore Symposium, the theme of which was the molecular mechanisms by which O_2 and its reduction products like superoxide and peroxides react in biological systems. It is hoped that these proceedings will give a current and representative picture of the research being conducted in a wide range of fields, all of which is contributing to an understanding of the biochemistry of O_2. Many major diseases and degenerative processes—including carcinogenesis, allergies, cardiovascular disorders, and aging—are associated in some still poorly understood way with imbalances of the oxidative and antioxidant systems. Ideally, these proceedings will suggest new ways and encourage new efforts to integrate the emerging techniques from the various biological oxidation and antioxidant systems so as to address the complex questions of human health.

The Symposium on Biological Oxidation Systems, held in Bangalore, India on October 22–26, 1989, was sponsored by the National Science Foundation. This Symposium was jointly organized by the College of Agriculture and the Environmental Resources Research Institute of Penn State University and the Indian Institute of Science in Bangalore. We gratefully acknowledge the support of Dr. Lamartine F. Hood, Dean of the College of Agriculture; Dr. Donald G. Simmons, Head of the Veterinary Science Department; Dr. Archie J. McDonnell, Director of the Environmental Resources Research Institute at Penn State University; and Prof. C. N. R. Rao, Director of the Indian Institute of Science, Bangalore. Our special thanks for their invaluable assistance go to Dr. George R. Hildenbrandt and Mr. Thomas McCawley. They assisted us not only with the planning and running of the Symposium but also with compiling and preparing the chapters for this publication. Their devoted, conscientious work and careful attention to detail made the organizing and editing tasks possible. We particularly want to thank our office staffs at Penn State and the Indian Institute of Science for their valuable assistance. Also, we would like to express our thanks for the assistance we received from the graduate students and other staff of the Institute during the conference. It was largely because of their dedicated and diligent work that this conference was a success. We would like to extend our appreciation to Dr. Pallu Reddanna for his assistance throughout the course of the Symposium preparations. Finally, the organizers of the Symposium would like to gratefully acknowledge the financial support for this conference from the agencies and corporations listed on page xvi.

THE EDITORS

CHAPTER VI

Oxidation Mechanisms in Carcinogenesis

Oxidations Catalyzed by Prostaglandin H Synthase

T.E. Eling, D.C. Thompson, G.L. Foureman, J.F. Curtis and M.F. Hughes
Eicosanoid Biochemistry Section, Laboratory of Molecular Biophysics
National Institute of Environmental Health Sciences, National Institutes of Health
Research Triangle, NC 27709 USA

Abstract

The metabolism of xenobiotics during arachidonic acid metabolism has been investigated as an oxidative system for the conversion of carcinogens to reactive metabolites. In this paper, we have briefly summarized the important findings of these investigations from this laboratory as well as other laboratories in this area of research. In general, oxidations that occur during arachidonic acid metabolism are catalyzed by the peroxidase activity of prostaglandin H synthase or by peroxyl radicals. Free radical metabolites are formed which can be mutagens, and some data suggest that prostaglandin H synthase may be of importance for the activation of chemicals in extrahepatic tissues.

Introduction

Exposure to environmental chemicals and other toxicants is generally recognized as an important cause of human cancer and disease. For carcinogenesis and other toxic effects, numerous investigations indicate that chemicals must be metabolized to exert their effects. Extensive literature testifies to the importance of the cytochrome P-450 (P-450) monooxygenases in the metabolic activation of chemicals to reactive electrophiles. However, in 1975 a report (1) showed that during the oxidation of arachidonic acid to prostaglandins by prostaglandin H synthase (PHS) chemicals present in the incubation which included carcinogens were oxidized. This was termed cooxidation. The dissimilarity between P-450 and PHS indicated the possibility that PHS could serve as an alternate enzyme for chemical metabolism, particularly in many extrahepatic tissues which are low in monooxygenase activity. Investigations indicate that multiple mechanisms exist for the metabolism of chemicals during arachidonic acid oxidation, and these events can play an important role in determining the eventual toxicity of a chemical.

Biochemistry of PHS

PHS is the initial enzyme in the pathway which commits arachidonic acid to the formation of prostaglandins. Two catalytic activities copurify with PHS - cyclooxygenase and peroxidase (2). The cyclooxygenase catalyzes the addition of two moles of oxygen to one mole of arachidonic acid forming a cyclic endoperoxide hydroperoxide, prostaglandin (PG) G_2. The peroxidase subsequently reduces the hydroperoxide to the corresponding alcohol, PGH_2. PGH_2 is then converted into a variety of other prostanoid metabolites, including thromboxanes, prostacyclin, and prostaglandins E_2 and $F_{2\alpha}$. PHS activity is found in almost every mammalian tissue that has been investigated. The highest concentrations of PHS, and therefore the primary source of PHS for investigative studies, are found in ram or bull seminal vesicles. Other cells and tissues which contain relatively high levels of PHS include kidney medulla, platelets, vascular endothelial cells, the alimentary tract, brain, lung and bladder (3-7).

PHS cyclooxygenase is the site of action of the nonsteroidal anti-inflammatory drugs (8). In contrast to PHS cyclooxygenase, there are no known inhibitors which are specific for PHS peroxidase. Compounds which inhibit PHS peroxidase-dependent cooxidation reactions, such as methimazole, probably do so by directly reducing cosubstrate free radicals rather than by inhibiting the enzyme (9).

Mechanism of Cooxidation

Peroxidase-Derived

Chemicals that are cooxidized during arachidonic acid metabolism by PHS can be separated into classes depending upon their ability to donate electrons to the peroxidase. Many aromatic amines and phenols are excellent electron donors to the peroxidase, serving as reducing cosubstrates during the reduction of PGG_2 to PGH_2. During the donation of electrons, these cosubstrates are oxidized to free radical metabolites. The classical reaction catalyzed by peroxidase is shown below.

Equation 1: $$ROOH + 2DH \xrightarrow{\text{Peroxidase}} ROH + 2D\cdot + H_2O$$

However, peroxidases, under some conditions, can also catalyze peroxide reduction by direct transfer of the peroxide oxygen to an acceptor molecule as shown below (10, 11).

Equation 2: $$RO^{18}O^{18}H + A \xrightarrow{\text{Peroxidase}} RO^{18}H + AO^{18}$$

The purification of PHS in sufficient quantities has made possible investigations into the mechanisms responsible for peroxide reduction at the expense of a cosubstrate. Early studies indicated that PHS peroxidase is very similar to horseradish peroxidase (HRP). Ruf and his colleagues (12, 13) have extensively investigated the higher oxidation states of PHS using rapid spectroscopic methods and low temperature ESR. Intermediate I was proposed as [(protoporphyrin IV)$^{\bullet}$ ($^{+}Fe^{IV}O$)] analogous to HRP compound I. Intermediate II resembles compound II of HRP but contains a tyrosyl radical with the proposed structure [(protoporphyrin IX)($Fe^{IV}O$)tyr•] which is formed by an electron transfer from a tyrosine residue to the porphyrin cation radical. Intermediate II resembles the ES complex of cytochrome *c* peroxidase. Thus, intermediate II of PHS is different from compound II of HRP. One of the unique features of PHS is that two enzymatic activities, the cyclooxygenase and the peroxidase, reside within a single protein. Ruf and coworkers (12, 13) have also proposed that the tyrosyl radical of intermediate II participates in the cyclooxygenase reaction. By this proposed mechanism, the tyrosyl radical initiates the cyclooxygenase by abstracting a hydrogen at C-13 of arachidonic acid to form a carbon centered radical. This radical rearranges and reacts with molecular oxygen to form a cyclic endoperoxide peroxyl radical which reoxidizes the tyrosine yielding the tyrosyl radical and PGG_2. Thus, the heme Fe maintains the same oxidation state during the cyclooxygenase reaction. This interaction between the peroxidase activity and cyclooxygenase activity may assist in understanding the self inactivation of PHS that occurs during arachidonic acid oxidation and may explain the protection that a reducing cosubstrate affords against this self inactivation.

Hydroperoxide-Derived Oxidation

There are several xenobiotics that are metabolized during oxidation of arachidonic acid catalyzed by PHS, yet are not reducing cosubstrates as described above. These chemicals are oxidized by a hydroperoxide-dependent mechanism that is not related to PHS turnover. Examples of xenobiotics that undergo this type of cooxidation include polycyclic aromatic hydrocarbons (PAH), aflatoxin, and diphenylisobenzofuran. Investigations into this mechanism of oxidation have primarily centered on the metabolism of the PAH and their metabolites (Figure 1).

A P-450-derived metabolite of benzo[a]pyrene (BP), 7,8-dihydroxy-7,8-dihydroBP (BP-7,8-diol), which is not a reducing cosubstrate for PHS, is epoxidized to the proposed ultimate carcinogen, anti-9,10-epoxy-7,8-dihydroxy-7,8,9,10-tetrahydroBP (anti-BPDE) with PHS, in

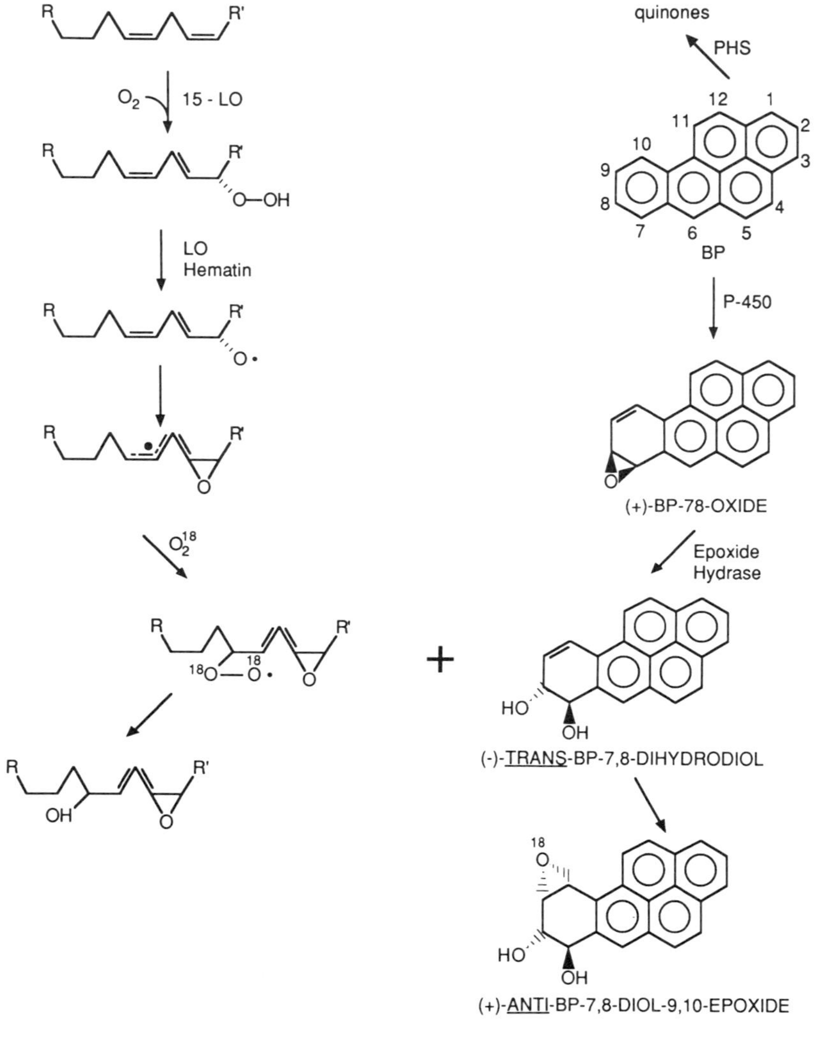

Figure 1. *The Metabolism of Benzo(a)pyrene and Its Metabolites by Peroxyl Radicals*

the presence of either arachidonic acid or 15-HPETE (14-16). Dix *et al.* (17, 18) proposed that this mechanism of epoxidation, as catalyzed by hematin, occurs by the donation of one electron from hematin to the hydroperoxide, forming an alkoxyl radical (Figure 1). The alkoxyl radical cyclizes to an epoxy allylic radical, which traps oxygen forming a peroxyl radical, a known epoxidizing agent. Thus, formation of a peroxyl radical is a requirement for epoxidation of BP-7,8-diol to occur. Epoxidation of BP-7,8-diol by peroxyl radicals to the anti BPDE occurs by a stereo-selective mechanism that is distinct from the epoxidation catalyzed by P-450. The stereoselectivity of the epoxidation mediated by peroxyl radicals and P-450 can be differentiated using the (+) enantiomer of BP-7,8-diol (19, 20-24). The (+)-BP-7,8-diol is epoxidized to (+)-syn-BPDE by P-450 and (-)-anti-BPDE by peroxyl radicals. Thus, (+)-BP-7,8-diol can be used as a probe to determine the pathway of epoxidation in different systems. Recent studies in our laboratory (25) indicate that 15-lipoxygenase catalyzes the epoxidation of BP-7,8-diol.

Chemical Reactions Initiated by PHS Synthase

Since PHS cooxidation reactions were first described, a large body of literature has been assembled which characterizes known cooxidation reactions. Although the list of reactions described in the literature is extensive, diverse, and ever expanding, one common trait is shared by all cooxidation reactions. PHS peroxidase cooxidations are initiated as one electron oxidations of the reducing cosubstrate to free radical metabolites. Any subsequent reactions of these free radicals will be determined by their physical and chemical properties. Once a reducing cosubstrate is oxidized, the resulting free radical can have several fates which will depend on the nature of the generated radical, i.e., its stability. In addition, its fate will also depend upon what other chemicals are present for the free radical to react in its environment.

Thus, the fate of the primary free radical is dependent on the stability of the radical, the site of generation and the environment in which it exists. The free radical can: 1) react with molecular oxygen as exemplified by carbon-centered free radicals which results in the formation of peroxyl radicals as described for phenylbutazone (58); 2) react with endogenous or exogenous thiols which reduces the free radical with the resultant formation of a thiyl free radical as exemplified by aminopyrine (29) or phenol (55); 3) undergo coupling reactions as shown for 2-aminofluorene (2-AF) (34, 36) resulting in the formation of oligomeric products; 4) undergo further oxidation to two electron oxidation products as shown for aminopyrine (29) or benzidine (27); 5) react with a secondary compound to generate a secondary metabolite of potential toxicological interest (58); 6) react with macromolecules to form potentially toxic adducts (42-44).

Biological Implications

Aromatic Amines

Metabolism of Aromatic Amines: Aromatic amines are metabolized by PHS peroxidase via one electron oxidation to nitrogen centered free radicals. The ease of oxidation is altered by substituents which affect the electron density of the nitrogen. For example, benzidine is an excellent cosubstrate for PHS peroxidase, while the less easily oxidized 2-aminofluorene (2-AF) (26) is a poorer cosubstrate. Acetylation of the amine reduces the ease of oxidation by PHS and other peroxidases. Acetylbenzidine, *e.g.*, is a much poorer cosubstrate than benzidine for PHS (27). The addition of electron withdrawing groups to the aromatic ring of N-methyl anilines decreases the metabolism of these compounds by PHS peroxidase (28).

Evidence for a one electron oxidation of aromatic amines by PHS peroxidase to free radical metabolites comes from a number of investigations. Aminopyrine and benzidine are oxidized by PHS peroxidase to cation free radicals that were detected by ESR (29-33). The carcinogen 2-AF, while a poor cosubstrate (34), is also oxidized by PHS peroxidase to free radical metabolites. Due to the instability of the free radicals, indirect evidence was used to support the hypothesis for a one electron oxidation. Recently, Petry *et al.* (35) investigated the metabolism of carcinogenic heterocyclic aromatic amines derived from the pyrolysis of amino acids and proteins. Free radicals were not observed by ESR but the detection of a glutathionyl radical (GS•) formed by reduction of a free radical by glutathione indicated that these amines are also undergoing a one electron oxidation.

The stable isolated metabolites produced by PHS peroxidase catalyzed cooxidation of aromatic amines is dictated by the chemical nature of the free radical. Tertiary and secondary aromatic amines undergo N-dealkylation. For primary aromatic amines, the free radicals undergo nitrogen-to-nitrogen or nitrogen-to-carbon coupling reactions. The carcinogen 2-AF is oxidized by PHS peroxidase and HRP to 2,2'-azobisfluorene, 2-aminodifluorenyl amine and polymeric material (36). Also detected was 2-nitrofluorene. Extensive studies with this aromatic amine indicate that N-hydroxylation is not catalyzed by PHS peroxidase since N-hydroxy-2-aminofluorene was not detected in incubation mixtures. Benzidine is also oxidized to azobenzidine by peroxidases but a complete characterization of the stable metabolites that are formed by PHS peroxidase or HRP is not published. The bladder carcinogen 2-naphthylamine (2-NA) (37) is oxidized by PHS peroxidase to 2-amino-1,4-naphthoquinone-N^4-naphthylimine presumably formed by the coupling of 2-NA to the oxygenated intermediate 2-imino-1-naphthoquinone (Figure 2). In contrast, HRP produces nitrogen-to-nitrogen and nitrogen-to-carbon coupling products indicative of free radical oxidation. The formation of oxygenated metabolites with PHS, but not with HRP, is an example of potentially important differences between the peroxidase mechanisms for these two enzymes.

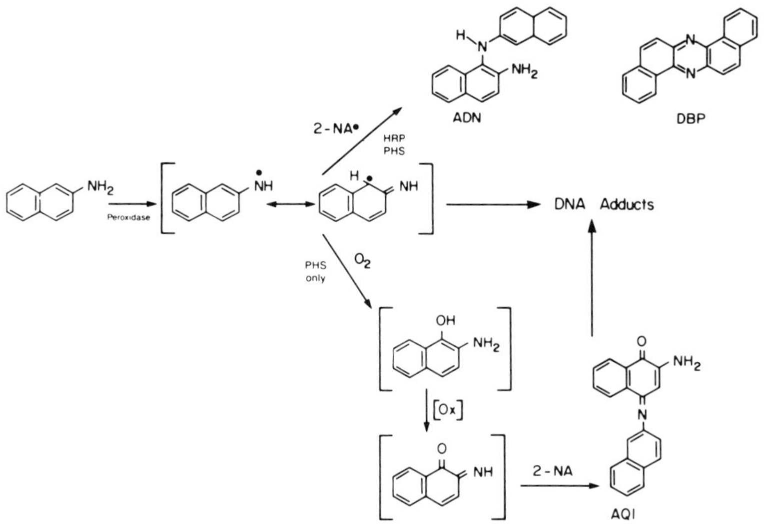

Figure 2. *The Oxidation of 2-Naphthylamine by PHS and HRP*

Formation of Mutagens: The formation of mutagenic metabolites of aromatic amines catalyzed by PHS peroxidase was extensively investigated by coupling ram seminal vesicle (RSV) microsomal fraction as an activating system to bacterial tester systems such as *Salmonella typhimurium* TA-98 (38, 39). As seen in Table 1, PHS peroxidase oxidizes many aromatic amines to products that are mutagenic. In addition to the classical aromatic amine carcinogens such as benzidine and 2-NA, heterocyclic amino acid pyrolysates such as IQ and MeIQ likewise are activated to mutagens by PHS peroxidase.

In other studies, we have used an acetyltransferase deficient bacterial tester strain, TA-98/1,8-DNP_6, which gave essentially no mutagenic response to benzidine or 2-AF. This implies that bacterial acetyltransferases play a role in the formation of the ultimate

Table 1. *Formation of Aromatic Amine Mutagens by PHS*

	Relative Mutagenicity	
Chemical	TA-98	pYG-121
2-Aminofluorene	++	++++
Benzidine	+	++
Acetylbenzidine	+++	+++++
2-Naphthylamine	++	NT
2,4-Diaminoanisole	+	NT
2,5-Diaminoanisole	++	NT
Aniline	-	NT
Acetylaminofluorne	-	-
Trp-1-P	-	-
Glu-P-1	-	+
IQ	++	++++
MeIQ	++	++++
3,3-Dimethoxybenzidine	-	-
o-Toludine	-	-
3,3',5,5'-Tetramethylbenzidine	-	-

NT = Not tested.

mutagen. Recently (40), we have used a bacterial strain developed by Watanabe from TA-1538 with high acetyltransferase activity. With this new strain, pYG-121, a number of aromatic amines, including benzidine, 2AF and the heterocyclic aromatic amino acids, exhibit enhanced mutagenic response catalyzed by PHS peroxidase (see Table 1). The role of bacterial acetyltransferases in mutagen formation is, at the present, an enigma and may indicate potential uncharacterized PHS-dependent metabolites.

Characterization of Aromatic Amine-DNA Adducts: Characterization of DNA adducts could provide unique biochemical markers to measure peroxidase-catalyzed activation of aromatic amines *in vivo*. However, characterization is exceptionally difficult due to problems in preparing the DNA adducts in sufficient quantities and purity for structural analysis. Kadlubar and coworkers characterized several DNA adducts of benzidine (41, 42) and 2-NA (43). We have also attempted to characterize the peroxidase-catalyzed DNA adducts of 2-AF (44). These studies indicate the formation of two adducts that are different from N-(deoxyguanosin-8-yl)-AF adducts formed by the reaction of N-hydroxy-2-AF with deoxyguanosine. An intensive effort to characterize the peroxidase-catalyzed adducts of 2-AF has not been successful primarily due to the inability to prepare an unstable intermediate for reacting with DNA *in vitro*. Only small amounts of these adducts could be prepared by catalyses with HRP or PHS peroxidase.

PHS Activation of Aromatic Amines in vivo: Evidence to support PHS activation of carcinogens in vivo has come from several studies measuring the DNA-adducts detected in tissues from dogs fed a dose of an aromatic amine carcinogen. Dogs are used in these studies since they develop bladder cancer as do humans upon exposure to aromatic amines. These studies are exceptionally difficult and costly. Animals must be given curie amounts of the ^{3}H-labeled aromatic amine which contributes to the high costs and causes a difficult decontamination problem. Despite the high amount of radioactivity administered, the level of ^{3}H bound to DNA is low. In spite of this difficulty, these studies detected peroxidase catalyzed aromatic amine adduct formation. Kadlubar observed peroxidase catalyzed benzidine (42) and 2-NA DNA adducts (43) in the urethelium but not the liver of dogs. Krauss *et al.* (44) recently detected peroxidase catalyzed 2-AF-DNA adducts in the kidney and bladder epithelia of dogs fed 2-AF. Zenser *et al.* (45, 46) characterized glutathione adducts of benzidine and ANFT catalyzed by PHS. They also observed this adduct in incubations with rabbit medullary slices. These data indicate that peroxidase-mediated activation of aromatic amines occurs *in vivo* but further studies are required to fully support the hypothesis that PHS may be important in the activation of carcinogenic aromatic amines *in vivo*.

Polycyclic Aromatic Hydrocarbons

BP7,8-diol is epoxidized to predominantly the anti-BPDE by peroxyl radicals formed during metabolism of arachidonic acid by microsomal PHS (15, 16). Arachidonic acid-dependent epoxidation of BP-7,8-diol is catalyzed by microsomes isolated from a number of sources, including the target organs of BP-induced tumorigenesis such as rat lung, mouse skin and human lung. BP-7,8-diol is oxidized in the presence of PHS to derivatives that bind to protein and nucleic acids (20). BP-7,8-diol and other PAH dihydrodiols are mutagenic in bacterial tester systems with PHS and arachidonic acid (47). PHS is more selective than P-450 in activation of PAH dihydrodiols to mutagenic products. Only those dihydrodiols with double bonds adjacent to the bay region area are mutagenic with PHS (47).

BP-7,8-diol oxidation to anti-BPDE by peroxyl radicals is observed in cellular systems. Mouse embryo fibroblasts, which contain PHS and P-450, metabolize BP-7,8-diol to anti and syn BPDE (48). Enriched fractions from Clara cells and alveolar type II cells isolated from rat lung also epoxidize BP-7,8-diol by an arachidonic acid- and P-450-dependent mechanism (49). Peroxyl radicals catalyze anti-BPDE formation and binding to DNA in explant tissue cultures prepared from hamster trachea and human bronchus (50). Indomethacin inhibits the enhancement of epoxidation in the hamster trachea, but has no effect in the human bronchus. Hamster trachea metabolizes arachidonic acid primarily to PHS metabolites, whereas human bronchus metabolizes arachidonic acid to lipoxygenase products. Thus, two different arachidonic acid-metabolizing systems may be supplying the peroxyl radical necessary for anti-BPDE formation. Recently, we observed that 15-lipoxygenase epoxidizes BP-7,8-diol by a mechanism analagous to the hematin/hydroperoxide system (25). Lipoxygenases are ubiquitous enzymes present in mammals and 15-lipoxygenase is present in human pulmonary epithelial cells in high concentrations. Lipoxygenases may have an important but unrecognized role in human carcinogenesis.

Mouse skin, a target organ of BP-induced carcinogenesis, possesses P-450 and PHS activity. Keratinocytes freshly isolated from hairless mice were used to investigate which system was epoxidizing (+)-BP-7,8-diol (23). In cells isolated from control animals, anti-BPDE was the predominant product, indicative of peroxyl radical-mediated epoxidation. In cells isolated from mice pretreated with β-naphthoflavone, an inducer of P-450 with high specific activity towards PAH, the syn-BPDE was the major product. Thus, peroxyl radicals are responsible for epoxidation of BP-7,8-diol in skin of untreated animals, while P-450 is responsible in the skin of induced animals. BP-7,8-diol is not the only PAH epoxidized during metabolism of arachidonic acid by PHS. Other bay region dihydrodiols of PAH are also epozidied by peroxyl radicals.

Miscellaneous Compounds

Many other compounds are cooxidized by PHS peroxidase to reactive intermediates which may play a role in toxicity or carcinogenicity *in vivo*. For example, acetaminophen, aflatoxin B_1, 5-nitrofurans, diethylstilbestrol, and sulfur dioxide are metabolized by PHS to potential toxic or carcinogenic metabolites. The reader should consult the orginal research articles or more detailed reviews for a discussion of the many chemicals for which PHS dependent activation may be important.

PHS and Sulfur Centered Radicals

Mechanisms of Formation of Thiyl Radicals

The formation of nitrogen-, carbon-, and oxygen-centered free radical metabolites by PHS peroxidase and in some cases the eventual formation of peroxyl radicals can ultimately lead to a toxic response. These reactions are considered as an activation process which has been the major focus of studies on cooxidation. The presence of the ubiquitous tripeptide, glutathione (GSH), in living systems acts to compromise or alter this activation process. Glutathione also participates in a number of reductive reactions and in many cases can reduce nitrogen- and oxygen-centered free radicals. As a consequence of this reduction, glutathione is oxidized to a glutathionyl free radical (GS•) (51). Other thiols are capable of reducing cosubstrate-derived free radicals back to the ground state as was shown to occur for a number of free radicals generated by PHS peroxidase (29, 51). This reducing ability of thiols served as the basis for a study of the mechanism by which methimazole inhibits PHS peroxidase-catalyzed reactions (9).

There are two mechanisms by which PHS peroxidase forms GS•. Glutathione serves as a cosubstrate for PHS peroxidase and is oxidized to GS• (53). In the second mechanism, the peroxidase first oxidizes a compound that is a more efficient reducing cosubstrate than GSH (*e.g.*, phenols, aromatic amines). The cosubstrate-derived radical produced by this oxidation is then reduced by GSH forming both GS• and the parent reducing cosubstrate. With efficient reducing cosubstrates such as aminopyrine, phenol and acetaminophen this process leads to markedly enhanced levels of GS•. Recent studies suggest that GS• may be formed intracellularly (54).

Formation of GSH Conjugates by Peroxidases

GS• may also add directly to compounds, thereby forming GSH conjugates. This new route for formation of GSH conjugates, mediated by peroxidases through the production of GS•, was first demonstrated by Stock *et al.* (55) and is compared to the glutathione transferase (GST)/monooxygenase pathway in Figure 3. In the presence of GSH, PHS, and arachidonic acid , styrene was directly converted to two isomeric GSH conjugates in contrast to the four isomeric GSH conjugates formed by GST-catalyzed addition of GSH to styrene oxide. GS• initiates the reaction by adding to a double bond proximal to a highly conjugated system such as an aromatic ring. This results in the formation of a carbon-centered radical which reacts with molecular oxygen to yield a peroxyl radical (Figure 4). Other studies (56) showed that an exocyclic double bond, as found in styrene, is required for a compound to undergo GS• addition. With the precocenes, antijuvenile hormones which possess this structural arrangement, up to 94% conversion to GSH-precocene conjugates was observed in the presence of PHS peroxidase when supplemented with a GS• enhancing agent, phenol (56).

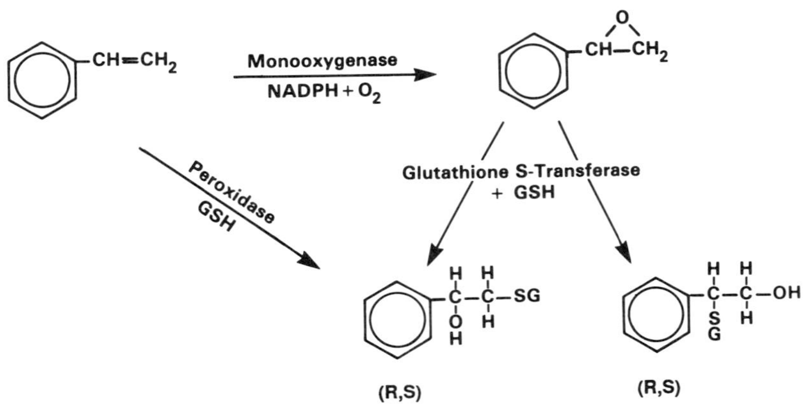

Figure 3. *The Mechanisms for Formation of Glutathione Conjugates*

An important point to note in this peroxidase mechanism is that formation of the electrophilic epoxide is not required for GSH conjugate formation as with transferases . This peroxidase-mediated pathway of glutathione conjugate formation could then be viewed as an "epoxide shunt". Such a shunt would be of considerable toxicological relevance in the case of compounds such as the PAH dihydrodiols whose epoxidation forms the ultimate carcinogen, the PAH diol epoxides. Peroxidase-mediated formation of PAH dihydrodiol-GSH conjugates would shunt the dihydrodiols away from further activation. Several PAH dihydrodiols, which possess structural similarity to styrene, undergo GS• addition catalyzed by PHS peroxidase (56). From these results, it appears that neither the position of the double bond nor the extent of the conjugating system (either 3 or 4 rings) prevents the conversion of the parent dihydrodiol to GSH conjugates by the PHS peroxidase-mediated pathway. Peroxidase-catalyzed formation of GSH conjugates is therefore general and may represent an alternative route to detoxication of a wide variety of compounds.

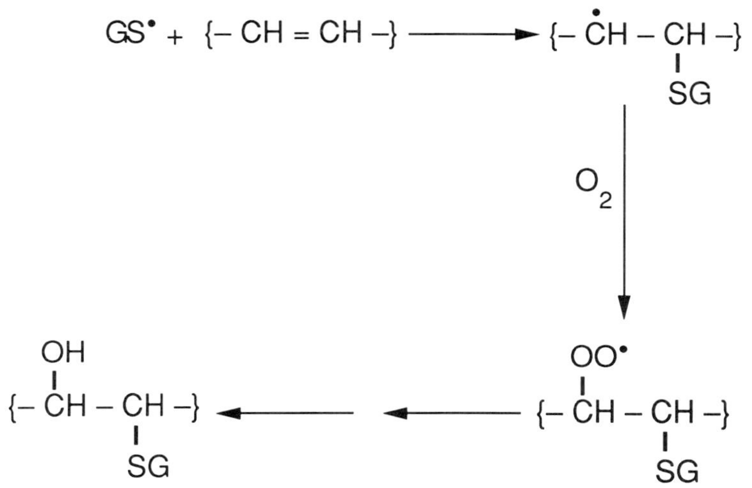

Figure 4. *Reaction Mechanism for Peroxidase-Catalyzed Glutathione Conjugate Formation*

GSH is not the only compound which can form peroxidase-catalzyed addition products from sulfur-centered free radicals. During investigation of the bisulfite-derived peroxyl radical and its interaction with BP-7,8-diol, Curtis *et al.* (57) observed the formation of a product which was characterized as a BP-7,8-diol-sulfonate. Formation of this sulfonate product may potentially be a detoxification reaction by removing the precursor of an ultimate carcinogen.

Summary

The cooxidation reactions that occur during arachidonic acid metabolism are dependent upon the peroxidase activity of PHS. For some chemicals that are not cosubstrates, the epoxidation reactions that occur are dependent upon the subsequent formation of peroxyl radicals. A large and diverse number of chemicals are metabolized by an equally large and diverse number of chemical reactions. The unifying theme is the free radical nature of these oxidations. The subsequent reactions that these chemicals undergo is dictated by the nature of the free radical and the environment in which the radical is generated.

We think that ample evidence exists that these free radical-mediated reactions can contribute to the formation of toxic metabolites, but in some cases can also participate in the detoxification of chemicals.

Acknowledgments

The authors wish to acknowledge Ms. Debbie Garner for her editorial assistance in preparing this manuscript.

References

1. Marnett, L.J., Wlodawer, P., & Samuelsson, B. (1975) *J. Biol. Chem.* **250**, 8510-8517
2. Miyamoto, T., Ogino, N., Yamamoto, S., & Hayaishi, O. (1976) *J. Biol. Chem.* **251**, 2629-2636
3. Christ, E.J., & Van Dorp, D.A. (1972) *Biochim. Biophys. Acta* **270**, 537-545
4. Yoshimoto, T., Magata, K., Ehara, H., Mizuno, K., & Yamamoto, S. (1986) *Biochim. Biophys. Acta* **877**, 141-150
5. Zenzer, T.V., & Davis, B.B. (1988) *Prostaglandins Leukotrienes and Essential Fatty Acids: Reviews* **31**, 199-207
6. Pace-Asciak, C.R., & Rangaraj, G. (1977) *Biochim. Biophys. Acta* **486**, 579-582
7. DeWitt, D.L., Day, J.S., Sonnenburg, W.K., & Smith, W.L. (1983) *J. Clin. Invest.* **72**, 1882-1888
8. Higgs, G.A., & Vane, J.R. (1983) *Br. Med. Bull.* **39**, 265-270
9. Petry, T.W., & Eling, T.E. (1987) *J. Biol. Chem.* **262**, 14112-14118
10. Egan, R.W., Gale, P.H., Vanden Heuvel, W.J. A., Baptista, E.M., & Kuehl, F.A., Jr. (1980) *J. Biol. Chem.* **255**, 323-326
11. Egan, R.W., Gale, P.H., Baptista, E.M., Kennicot, K.L., Vanden-Heuvel, W.J.A., Walker, R W., Fagerness, P.E., & Kuehl, F.A., Jr. (1981) *J. Biol. Chem.* **256**, 7352-7361
12. Karthein, R., Dietz, R., Nastainczyk, W., & Ruf, H.H. (1988) *Eur. J. Biochem.* **171**, 313-320
13. Dietz, R., Nastainczyk, W., & Ruf, H. H. (1988) *Eur. J. Biochem.* **171**, 321-328
14. Marnett, L.J., Johnson, J.T., & Bienkowski, M.J. (1979) *FEBS Letters* **106**, 13-16
15. Sivarajah, K., Mukhtar, H., & Eling, T. (1979) *FEBS Letters* **106**, 17-20
16. Marnett, L.J., & Bienkowski, M.J. (1980) *Biochem. Biophys. Res. Commun.* **96**, 639-647
17. Dix, T.A., & Marnett, L.J. (1981) *J. Am. Chem. Soc.* **103**, 6744-6746
18. Dix, T.A., Fontana, R., Panthani, A., & Marnett, L.J. (1985) *J. Biol. Chem.* **260**, 5358-5365

19. Panthananickal, A., Weller, P., & Marnett, L.J. (1983) *J. Biol. Chem.* **258**, 4411-4418
20. Panthananickal, A., & Marnett, L.J. (1981) *Chem.-Biol. Interact.* **33**, 239-252
21. Thakker, D.R., Yagi, H., Akagi, H., Koreeda, M., Lu, A.Y.H., Levin, W., Wood, A.W., Conney, A.H., & Jerina, D. M. (1977) *Chem.-Biol. Interact.* **16**, 181-300
22. Deutsch, J., Leutz, J.C., Yang, S. K., Gelboin, H.V., Chang, Y.L., Vatsis, K.P., & Coon, M.J. (1978) *Proc. Natl. Acad. Sci. USA* **75**, 3123-3127
23. Eling, T.E., Curtis, J., Battista, J., & Marnett, L.J. (1986) *Carcinogenesis* **7**, 1957-1963
24. Pruess-Schwartz, D., Nimesheim, A., & Marnett, L.J. (1989) *Cancer Res.* **49**, 1732-1737
25. Hughes, M.F., Chamulitrat, W., Mason, R.P., & Eling, T.E. (1989) *Carcinogenesis* **10**, 2075-2080, 1989.
26. Bull, A.W. (1987) *Carcinogenesis* **8**, 387-390
27. Josephy, P.D., Eling, T.E., & Mason, R.P. (1983) *J. Biol. Chem.* **258**, 5561-5569
28. Sivarajah, K., Lasker, J.M., Eling, T.E., & Abou-Donia, M.B. (1982) *Mol. Pharmacol.* **21**, 133-141
29. Eling, T.E., Mason, R.P., & Sivarajah, K. (1985) *J. Biol. Chem.* **260**, 1601-1607
30. Lasker, J.M., Sivarajah, K., Mason, R.P., Kalyanaraman, B., Abou-Donia, M.B., & Eling, T.E. (1981) *J. Biol. Chem.* **256**, 7764-7767
31. Josephy, P.D., Eling, T., & Mason, R.P. (1982) *J. Biol. Chem.* **257**, 3669-3675
32. Josephy, P.D., Mason, R.P., & Eling, T. (1982) *Cancer Res.* **42**, 2567-2570
33. Josephy, P.D., Eling, T.E., & Mason, R.P. (1983) *Mol. Pharmacol.* **23**, 766-770
34. Boyd, J.A., Harvan, D.J., & Eling, T.E. (1983) *J. Biol. Chem.* **258**, 8246-8254
35. Petry, T.W., Krauss, R.S., & Eling, T.E. (1986) *Carcinogenesis* **7**, 1397-1400
36. Boyd, J.A., & Eling, T.E. (1984) *J. Biol. Chem.* **259**, 13885-13896
37. Boyd, J.A., & Eling, T.E. (1987) *Cancer Res.* **47**, 4007-4014
38. Robertson, I.G.C., Sivarajah, K., Eling, T.E., & Zeiger, E. (1983) *Cancer Res.* **43**, 476-480
39. Wild, D., & Degen, G.H. (1987) *Carcinogenesis* **8**, 541-545
40. Josephy, P.D., Chiu, A.L.H., & Eling, T.E. (1989) *Cancer Res.* **49**, 853-856
41. Yamazoe, Y., Roth, R.W., & Kadlubar, F.F. (1986) *Carcinogenesis* **7**, 179-182
42. Yamazoe, Y., Beland, F.A., & Kadlubar, F.F. (1985) *Proc. Am. Assoc. Cancer Res.* **26**, 1335
43. Yamazoe, Y., Miller, D.W., Weiss, C.C., Dooley, K.L., Zenser, T.V., Beland, F.A., & Kadlubar, F.F. (1985) *Carcinogenesis* **6**, 1379-1387
44. Krauss, R.S., Angerman-Stewart, J., Dooley, K.L., Kadlubar, F.F., & Eling, T.E. (1989) *Biochem. Toxicol.* **4**, 111-117
45. Wise, R.W., Zenser, T.V., Rice, J. R., & Davis, B. B. (1986) *Carcinogenesis* **7**, 111-115.
46. Rice, J. R., Zenser, T.V., & Davis, B.B. (1985) *Carcinogenesis* **6**, 585-590

47. Guthrie, J., Robertson, I.G.C., Zeiger, E., Boyd, J.A., & Eling, T.E. (1982) *Cancer Res.* **42**, 1620-1623
48. Boyd, J.A., Barrett, J.C., & Eling, T.E. (1982) *Cancer Res.* **42**, 2628-2632
49. Sivarajah, K., Jones, K. G., Fouts, J.R., Devereux, T., Shirley, J.E., & Eling, T.E. (1983) *Cancer Res.* **43**, 2632-2636
50. Reed, G.A., Grafstrom, R.C., Krauss, R.S., Autrup, H., & Eling, T.E. (1984) *Carcinogenesis* **5**, 955-960
51. Ross, D., Albano, E., Nilsson, U., & Moldeus, P. (1984) *Biochem. Biophys. Res. Commun.* **125**, 109-115
52. Ross, D., Mehlhorn, R.J., Moldeus, P., & Smith, M.T. (1985) *J. Biol. Chem.* **260**, 16210-16214
53. Eling, T.E., Curtis, J. F., Harman, L.S., & Mason, R.P. (1986) *J. Biol. Chem.* **261**, 5023-5028
54. Schreiber, J., Foureman, G.L., Hughes, M.F., Mason, R.P., & Eling, T.E. (1989) *J. Biol. Chem.* **264**, 7936-7943
55. Stock, B.H., Schreiber, J., Guenat, C., Mason, R.P., Bend, J.R., & Eling, T.E. (1986) *J. Biol. Chem.* **261**, 15915-15922
56. Foureman, G.L., & Eling, T.E. (1989) *Arch. Biochem. Biophys.* **269**, 55-68
57. Curtis, J.F., Hughes, M.F., Mason, R. P., & Eling, T.E. (1988) *Carcinogenesis* **9**, 2015-2021
58. Reed, G.A., Brooks, E.A., & Eling, T.E. (1984) *J. Biol. Chem.* **259**, 5591-5595

Oxidation Reactions by Endoperoxide Synthase

Robert W. Egan
Schering-Plough Research, 60 Orange Street, Bloomfield, NJ 07003 USA

It has been recognized since the 1970's that two separate reactions were catalyzed by endoperoxide synthase, oxygenation and peroxidation. These reactions convert arachidonic acid to the hydroperoxide PGG_2 by the addition of molecular oxygen and the PGG_2 is then reduced to PGH_2 by a peroxidase. Stoichiometric considerations dictated that an oxidizing species had to be released into the medium and something other than arachidonic acid would also be oxidized. What it would be and where it would occur should be dictated by the redox potential and enzyme binding properties of the reductant. Since then a large body of literature has developed on various peroxidases and it has been demonstrated that the peroxidase actively associated with endoperoxide synthase is phenomenologically and mechanistically similar to other peroxidases such as cytochrome *c* peroxidase, horseradish peroxidase and hemoglobin.

The pioneering work of Smith and Lands had shown that endoperoxide synthase underwent irreversible catalytic inactivation (1) and Marnett and Samuelsson demonstrated that this enzyme could oxidize a variety of organic reductants (2). At about the same time, we demonstrated that these phenomena were both related to the peroxidase activity and proposed a common mechanism for these apparently unrelated events (3). This manuscript will review these concepts as they relate to inactivation of endoperoxide synthase and other redox-based enzymes and to the oxidation of organic reducing agents.

Figure 1 demonstrates the metabolism of several hydroperoxides by the hydroperoxidase (4). These reactions were monitored by incubating microsomes with ^{14}C-labeled substrate for about 30 sec, then extracting the products and remaining substrate into chilled ether, chromatrographing in an appropriate solvent system and observing the pattern of radioactivity either with a radiochromatogram scanner or by scintillation counting. The panel in the upper left shows PGG_2 metabolism to PGH_2 by the hydroperoxidase. In the absence of any additive, this reaction was not very substantial. However, this pattern was drastically different in the presence of 500 μM phenol where conversion was about 75% reduction of 100 μM 15-hydroperoxy PGE_1 (HPE_1) exclusively to PGE_1 is shown on the middle panel.

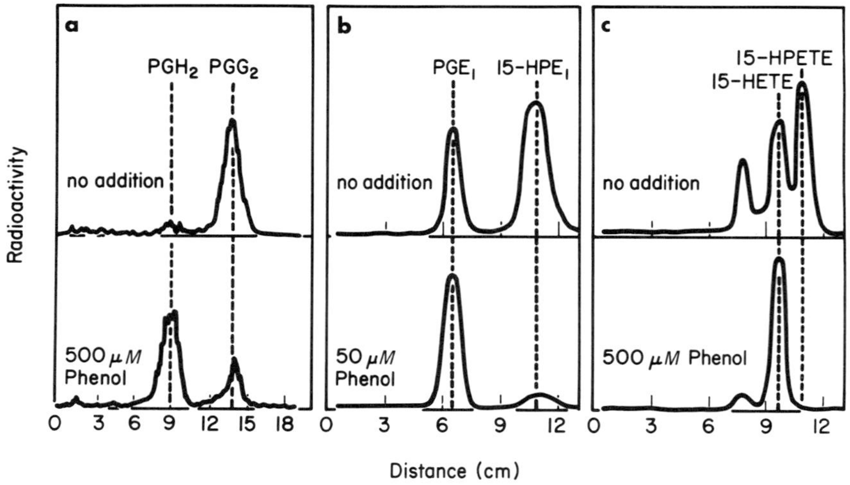

Figure 1. *Metabolism of Hydroperoxides.* In each trace, radioactivity is plotted vertically versus distance along the chromatogram. A, [1-^{14}C] PGG_2 (24 μM) was incubated with 0.05 mg of microsomal protein for 30 s; B, 15-[1-^{14}C]HPE_1, (71 μM) was incubated with 0.5 mg of protein for 60 s; C, 15-[1-^{14}C]HPETE (112 μM) was incubated with 0.5 mg of microsomal protein for 60 s. In each panel, the dotted lines represent the location of chromatographic standards detected by mass assay. The upper traces represent reaction with no additive while the lower curves depict the same reactions in the presence of phenol.

The last panel depicts the metabolism of 15-HPETE, the other hydroperoxide examined in detail. 15-Hydroxyeicosatetraenoic acid (15-HETE) was the major product along with an unknown, more polar material. The bottom panels show these reactions in the presence of phenol. In each instance, the extent of hydroperoxide reduction was increased. These comparative studies were performed with a host of other reducing agents with comparable results.

Arachidonic acid stimulates oxygen uptake when added to buffered enzyme (Figure 2, top panel) (5). During a two minute incubation of PGG_2 with the enzyme, the PGG_2 was converted into PGH_2. As a consequence, the arachidonic acid stimulated rate was retarded 90 percent. However, a second enzyme addition to this reaction mixture resulted in oxygen uptake comparable to the control. When enzyme was added last to the buffer containing PGG_2 and arachidonic acid, the curve coincided closely with the control, indicating that PGG_2 itself did not cause inhibition. These experiments suggest that the cyclooxygenase was inactivated by being present during the reductive breakdown of PGG_2 and not by PGG_2 itself or by the organic products of PGG_2 breakdown. Hence, the deactivation most likely occurred as a consequence of something released during the reduction of the PGG_2 to PGH_2, the stage at which oxidizing equivalents were generated. 15-HPE, and 15-HPETE gave qualitatively similar results.

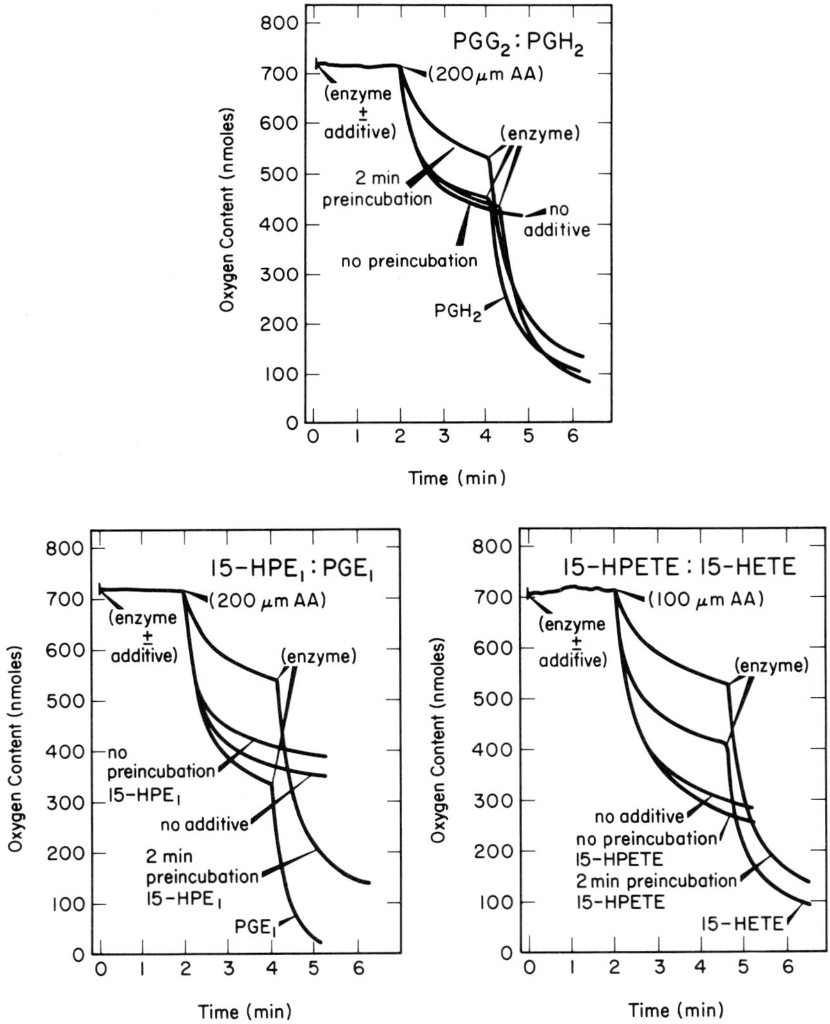

Figure 2. *Deactivation of Cyclooxygenase by Hydroperoxides.* Traces with PGG_2 and PGH_2 were at 3.6 and 40 μM, respectively, using 1.31 mg of microsomal protein. The 15-HPE_1 was 15 μM and PGE_2 was 1000 μM with 1.59 mg of microsomal protein. With 15-HPETE and 15- HETE, we used 2.5 and 100 μM, respectively, along with 2.1 mg of microsomal protein, AA, arachidonic acid.

Having established that the cyclooxygenase component of endoperoxide synthase was deactivated irreversibly by the activity of the hydroperoxidase, we examined whether the same applied to the hydroperoxidase itself (4). As opposed to the continuous oxygen monitor readout for oxygenase activity, these reactions were monitored using radioactive PGG_2, HPE_1 and 15-HPETE and measuring product formation at various times after initiating the reaction by adding substrate to microsomal enzyme. In each instance, the reaction proceeded rapidly then terminated and remained unchanged for ten minutes. To establish that this was due to irreversible deactivation of the enzyme, we added more substrate or reducing substrate with no effect. However, in each instance, when we added an equal aliquot of enzyme at point A there was renewed reaction that was, in all respects, similar to the initial reaction. The hydroperoxidase, therefore, suffered the same consequences as the oxygenase component of the endoperoxide synthase.

To better understand the cyclooxygenase mechanism, electron spin resonance (ESR) methods have been utilized to detect radical formation during the reaction (Figure 3) (3). In 1967, Nugteren and co-workers observed an ESR signal following addition of arachidonic acid to aerobic seminal vesicle microsomes (6). However, fatty acids

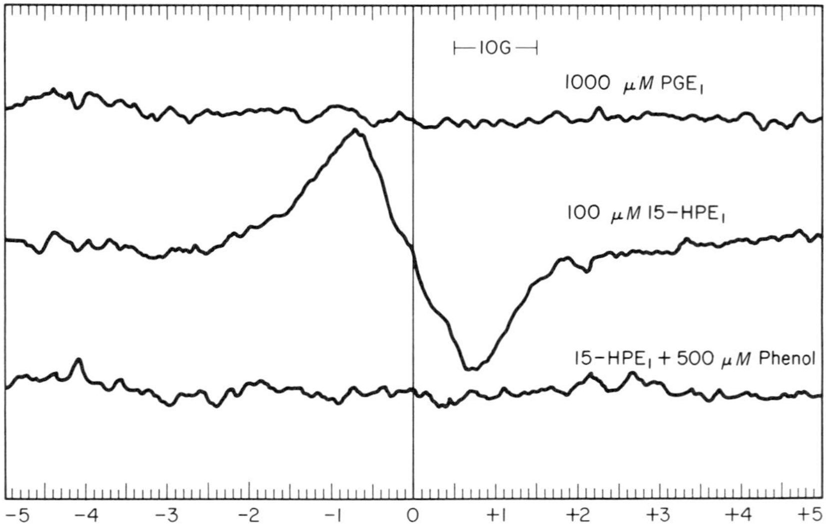

Figure 3. *Electron Spin Resonance Signal from Endoperoxide Synthase and 15-HPE_1.* After the reaction, the signal was recorded on a Varian E-109E spectrometer at - 195°C over a 200-G scan range in the g = 2 region, with a modulation amplitude of 3.2 G, a modulation frequency of 100 KHz, and a microwave power of 64 mW.

could have undergone several side reactions in the presence of microsomes, so we felt it important to extend this observation.

Reactions were conducted by incubating 100 μM 15-HPE$_1$ with the microsomes for 30 seconds. The mixture was then freeze-quenched at -195° and maintained at that temperature for observation in the spectrometer. The relative number of radicals present is roughly proportional to the peak-to-peak intensity of the differential signal. The signal generated by mixing 15-HPE$_1$ with microsomes was a broad line with no discernable hyperfine structure, centered at a g-value of 2.0. No signal was observed when the reaction was conducted with denatured enzyme. The reduction product of 15-HPE$_1$, PGE$_1$, did not cause a signal. Phenol also depressed this signal 82%, as shown in the lower trace. Based on these data, we can state with confidence that endoperoxide synthase was responsible for catalyzing the signal.

While we do not know the radical responsible for this signal, we do know that if we put in certain reductants such as aminopyrine and dimethylphenylene diamine we get signals that are characteristic of semi-stable intermediates in their oxidation that are well characterized from other studies (5). Both of these signals are relatively complex and have been seen by Eling and his associates (7). In the case of aminopyrine, the ESR signal is known to be the result of a blue colored radical intermediate in aminopyrine oxidation. Hence, the peroxidase can catalyze oxidation reactions that then proceed in solution off the surface of the enzyme, because this signal is characteristic of a well-known species that is detected in the solution-phase oxidation of aminopyrine.

When reducing agents are added to the medium, the cyclooxygenase functions for a large number of turnovers before it deactivates. This lead us to the conclusion that the protection of the enzyme by these compounds, which have very different structures (4), was the result of their redox properties and that the enzyme was probably oxidizing itself to death. This could have been occurring at the heme center or away from the iron at an essential amino acid, so we tested some amino acids to establish which ones were able to protect the oxygenase activity (Table 1). It turned out that tryptophan was the most effective, leading us to the conclusion that the oxidizing capacity of the endoperoxide synthase would be readily turned on itself in the absence of more potent reducing agents, and that it was oxidation, possibly at a tryptophan residue or at a heme center, that was leading to inactive enzyme after a certain number of turnovers.

Having characterized the disposition of the hydroperoxide during this reaction, our next concern was the fate of the cosubstrate for the peroxidase where we have focused on organic sulfides. Indeed, if these oxidizable materials were acting as radical scavengers and reducing agents then they should be metabolized in a peroxidase-dependent reaction.

Table 1. *Effects of Amino Acids on the ESR Signal*

Amino Acid	Concentration (mM)	Changes in ESR Intensity[a]
Methionine	1	+14
Histidine	1	-5
Cysteine	1	-9
Cystine	1	+13
Tyrosine	1	-10
Trytophan	1	-51

[a] plus is an increase and minus is a decrease

The structures of the sulfides of interest are shown in Figure 4 (5,8). Methylphenylsulfide (MPS) is the sulfide end of sulindac sulfide, each of which was converted by the hydroperoxidase to its analogous sulfoxide. MPS metabolism has been followed using [methyl ^{14}C] labeled MPS which we prepared from thiophenol and [^{14}C]methyl iodide. Following incubation with enzyme and hydroperoxide, methylphenylsulfoxide (MPSO) was extracted into chloroform, chromatographed on TLC plates then counted. [^{14}C]Sulindac sulfide was incubated similarly, but extracted into ether and chromatographed on silica-impregnated paper. Both reactions were protein dependent, destroyed by heat denaturation, and elicited by PGG_2, 15-HPE_1, 15-HPETE and H_2O_2.

The actual mechanism of oxygen transfer to the sulfide has been examined by mass spectral determination of the distribution of oxygen-18 in the products, with [^{18}O]15-HPE_2 as reactant (Figure 5) (5). Labeled 15-HPE_2 was prepared by the cyclooxygenase-catalyzed reaction of arachidonic acid with $^{18}O_2$. Because of the dioxygenase reaction mechanism, the isotopic composition of the 15-HPE_2 and PGE_2 would be identical and the PGE_2 was used to establish this parameter because it produced a more interpretable mass spectrum. Mass spectra for the tetratrimethylsilylated derivatives of both enzymatically-prepared (middle panel) and authentic (upper panel)

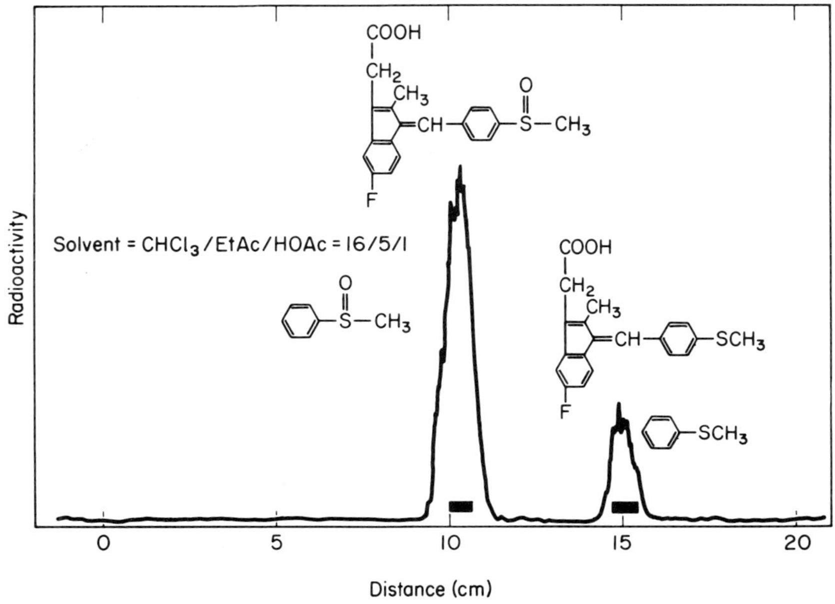

Figure 4. *Endoperoxide Synthase-Catalyzed Oxidation of Sulindac Sulfide and MPS.* The chromatogram shows the pattern of radioactive products in the ether extract from a 60-sec incubation of 100 μM sulfide with 0.56 μg of microsomal protein and 150 μM 15-HPE$_1$. The bars indicate the location of nonradioactive authentic sulindac (structure to the left of the large peak), and sulindac sulfide (structure over the small peak).

PGE$_2$ were the basis for comparison. The molecular ion for authentic silylated PGE$_2$ was 640, the largest peak in this region. Because the labeled PGE$_2$ could contain either zero, one, two or three ^{18}O-atoms, the ratios of 640, 642, 644 and 646 were compared to establish the isotopic distribution, which showed that 78% of the hydroxyl at carbon-15 contained ^{18}O-atoms. Consequently, 78% of the 15-HPE$_2$ would possess ^{18}O-atoms in the hydroperoxide.

Labeled 15-HPE$_2$ was used as substrate for the oxidation of MPS to MPSO with purified enzyme. The PGE$_2$ resulting from this reaction (lower panel) showed 74% oxygen-18 at carbon 15 compared with 78% in the starting 15-HPE$_2$. This excellent agreement confirmed that the hydroperoxide moiety was not exchanging in some unexpected fashion. The spectra of MPSO are on the right. On top is the spectrum of MPSO itself. The molecular ion at 140 is predominant with no 138 ion and only 5% the intensity at 142. Hence, the predominant ion of MPSO with ^{18}O in the sulfoxide would be 142. The product of the reaction with [^{18}O]15-HPE$_2$ and

MPS is shown on the bottom panel. This MPSO has a large 142 and a small 140 ion. By comparing these peak heights, it was determined that 86% of the oxygen in the sulfoxide was ^{18}O. Therefore, the oxygen atoms in this sulfoxide were transferred directly from the hydroperoxide.

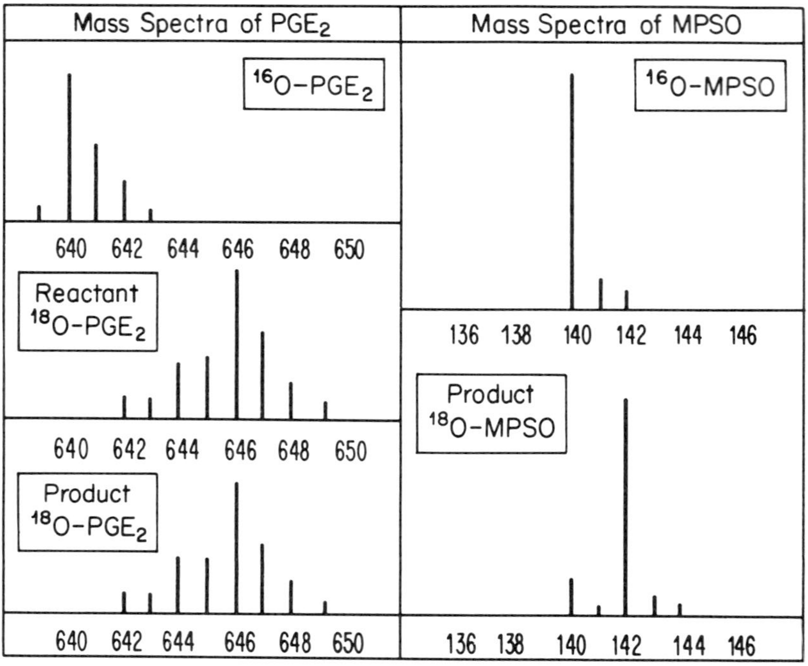

Figure 5. *Mass Spectrometric Analyses of MPS Reaction*: Mass spectra were obtained using an LKB 9000 gas chromatograph-mass spectrometer with 70 eV ionizing potential, 60 μA trap current, 3.5 kV accelerating potential and 250°C source temperature. Samples containing about 10 μg of PGE_2 were immediately subjected to trimethylsilylation with 20 μl of NO-bis (trimethyl-silyl) trifluoroacetamide/dimethylformamide (1:1) for 17 h at room temperature prior to examination. Aliquots of 1-2 μg of these samples were introduced through the direct inlet system and heated slowly at 100-150°C prior to collecting data. Samples of MPSO were studied under identical spectrometer conditions, but were introduced through the gas chromatograph. About 1 mg of MPSO was injected into a 5-foot column of 3% OV-17 at 148°C with 5-min retention time.

We have also examined the stoichiometry of these two reactions (5). With 15-HPE_2 as oxidizing substrate, the ratios of sulfide oxidized to HPE_2 reduced were calculated using MPS with purified enzyme and using sulindac sulfide with microsomal protein. In both instances, there was equimolar conversion between sulfide and hydroperoxide. This stoichiometry would strongly suggest that the oxidation reaction was occurring at the enzyme surface as opposed to proceeding in solution via an oxidizing species such as OH released by the enzyme. Were this reaction going on in solution, there would certainly have been less than a one-to-one stoichiometry.

With this knowledge of the location of the reducing substrate during oxidation, we decided to examine the chiral character of the MPSO formed by the hydroperoxidase. Although we used polarimetry to confirm our results and to tell us which enantiomer was in excess, our primary quantitative determination involved an NMR technique using a chiral shift reagent. Sulfoxides exhibit R and S optimal isomers due to the lone pair on the sulfur in addition to the three legitimate bonds.

The concept is that sulfoxides bind weakly to transition metal complexes. If that complex is chiral itself then the sulfoxide adducts are diasteriomers and, therefore, have different physical properties enabling separation. The shift reagent we used was an europium camphorato complex with optically active camphorato ligands. The NMR of MPSO itself has an unsplit absorption at 2.55 ppm for the 5 aromatic protons (Figure 6). When exposed to the chiral shift reagent, these resonances were shifted downfield and the aromatic protons were resolved.

Resonances at 10.4, 7.9 and 5.3 ppm are the ortho, meta and para, and methyl protons of MPSO, respectively. These assignments gave the expected 2:3:3 ratios, and were further confirmed by studies at varying $Eu(hfc)_3$ levels. All other peaks originated from the europium complex and are not important. The 0.126 ppm splitting of the methyl resonances was due to non-equal complex formation of the two MPSO diastereomers, with the complex of the S-MPSO enantiomer at higher field. In agreement with the polarimetry data, the average ratio of enantiomers in the NMR was 44:56, not a very large difference. This is shown more clearly in the inset, which is an expansion of the region at 5.3 ppm. The R-enantiomer, which is dextrorotatory in the polarimeter, is shifted further downfield than the S-enantiomer. Consequently, the S-MPSO is in excess by about 30%.

The low chirality of the enzymatically-formed MPSO indicates that the sulfide is not held rigidly within its binding site. As a result, electrophilic attack on the sulfur could involve either lone pair with almost equal frequency. Such a lack of stereospecificity has also been noted with serine hydroxymethylase (9) and certain bacterial sulfide oxidations (10,11). With other sulfides or different experimental conditions, the chirality of the reaction might even be increased.

However, it is clear that there is an optical preference of the hydroperoxidase, further suggesting that the sulfide is bound when oxidized.

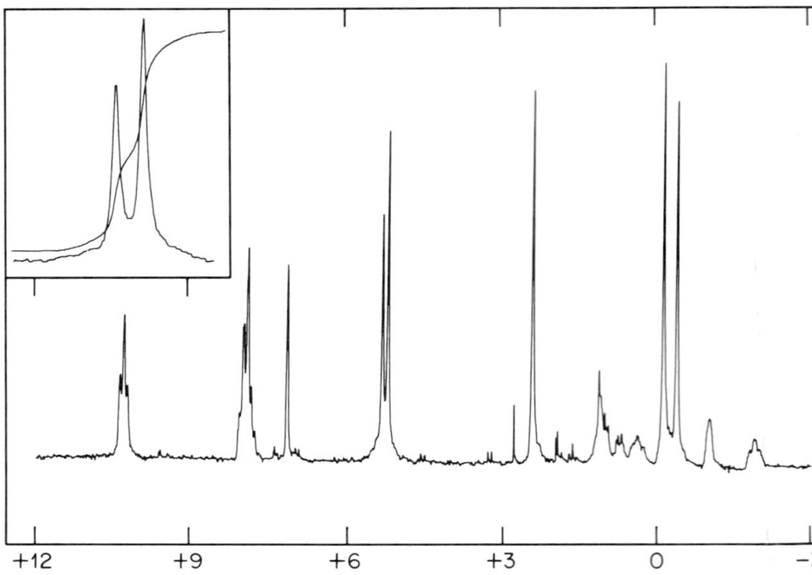

Figure 6. *Enantiomeric Composition of Enzymatically Prepared MPSO.* Enzymatically prepared MPSO (2.04 mg) and 5.79 mg of Eu(hfc)$_3$ were added to 0.6 ml of CCl_4. C_6D_6 (50 μl) was added for NMR field/frequency lock. The spectrum was obtained on a Varian XL- 100 spectrometer at 100.06 MHz in the Fourier transform mode. Operating conditions were 2000-Hz spectral width, 150 60°-pulses at 4 s repetition rate, 0.12-Hz line broadening, 5-μM outer diameter sample tube, and 28°C probe temperature. Inset, an expansion of the methyl resonances at 5.3 ppm, along with their integrals.

Of course, other reducing substrates for the enzyme have also been examined. Of particular interest are studies by Marnett *et al.* that showed with diphenylisobenzofuran (DPBF) and phenylbutazone (PB) that the ^{18}O in each product arose from dissolved molecular oxygen (12). With our knowledge that the sulfide was bound when oxidized, it seemed possible that we could distinguish the location of the primary oxidation of these molecules by looking for competition with the MPS (Table 2). Although it is clear that chain propagation proceeds in solution, it seemed possible that the primary initiating step involved enzyme-bound PB. The hydroperoxidase reaction was investigated at as much as a seven-fold molar excess of PB over MPS. Both MPS and PB stimulated the extent of 15-HPE_2 metabolism, 500 and 1000 μM PB more so than 150 μM MPS. In combination, they stimulated more than either one individually. Were PB competing with MPS for the oxidizing equivalents released by the hydroperoxidase, it should decrease the oxidation of MPS to MPSO. As shown in the last column, the extent of MPS conversion increased by 32% at 1000 μM PB. Hence, the primary locus of PB oxidation was different from MPS, very likely in solution via an enzyme-generated oxidant. Nevertheless, radical forming substrates such as PB protect the cyclooxygenase from self-destruction, indicating that the enzyme inactivation can result from a species present in solution. Such an oxidant possesses many characteristics akin to the hydroxyl radical.

Table 2. *Effect of Phenylbutazone on Methyl Phenyl Sulfide Oxidation*

MPS (μM)	Phenylbutazone (μM)	Product Formed: PGE_2 (nmol/ml)	Product Formed: MPSO (nmol/ml)
0	0	42	
150	0	69	66
0	500	91	
0	1000	96	
150	500	106	71
150	1000	125	87

Reactions measuring 15-HPE_2 reduction utilized [1-^{14}C]15-HPE_2 and nonradioactive MPS, whereas those measuring MPS oxidation used [methyl-^{14}C]MPS and nonradioactive 15-HPE_1. Purified enzyme was 150 μg/ml, hemin was 10 μM and 15 HPE_2 was 160 μM.

There is a stoichiometric equimolar reaction between sulfide and 15-HPE_1 (Figure 7). The functional groups involved in the reaction are placed in proximity and enclosed within a dashed box with the asterisks designating oxygen-18 atoms. The oxygen atom is transferred to the sulfide and, following the reaction, all the atoms are accounted for. This precludes indirect oxidizing species such as molecular or singlet oxygen as contributors to this cooxygenation.

Figure 8 shows a mechanism for cooxygenation that accommodates both modes of oxygen incorporation described previously. The enzyme contains heme with iron in a valance state of 3. As for many peroxidases, it is oxidized in a formal sense to Fe(V) upon reduction of the hydroperoxide. From this stage, there are two

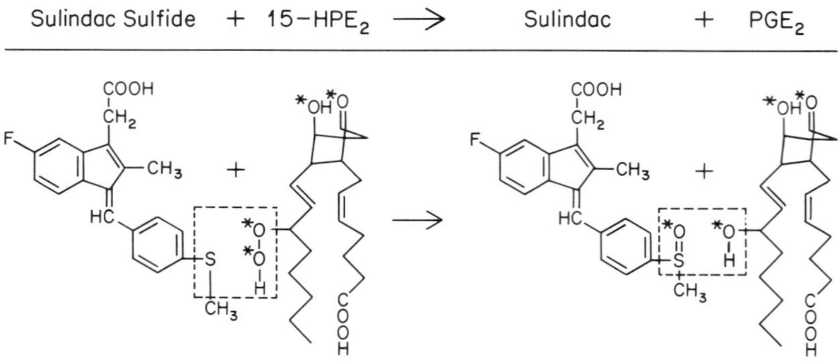

Figure 7. *Mechanism of Sulfide Oxidation by Endoperoxide Synthase*

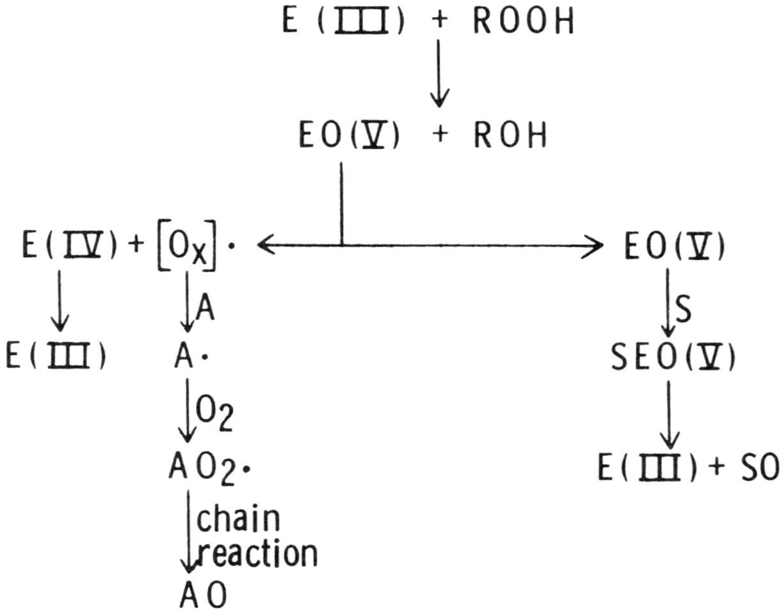

Figure 8. *Mechanisms of Hydroperoxidase-Catalyzed Oxidations.* E(III), enzyme with Fe(III) heme; EO(V), enzyme with bound oxygen atom and Fe(V) heme; E(IV), enzyme with Fe(IV) heme; SEO(V), EO(V) with S bound to the active site; [Ox]-oxidizing species released into solution; A, substrate oxidized in single electron steps; A•, radical form of A, AO•, peroxy radical of A, AO•, oxygen atom adduct of A•, S, substrate oxidized in two-electron steps; SO, oxygen atom adduct of S.

routes for discharge of this oxidizing potential, via bond formation as with the sulfides to give FeIII and sulfoxide or via release of an oxidant [Ox] into solution. This can then initiate a radical chain with substrate A to give oxidized product. In the former case, the oxygen atom arises from the hydroperoxide and you get an equimolar reaction with enzyme-bound cosubstrate. In fact, with sulfides the substrate may be bound during hydroperoxide reduction. In the latter case, you can get virtually any stoichiometry with the oxygen in the product arising from dissolved molecular oxygen. Because an oxygen atom is transferred directly from MPS to MPSO, the sulfur must be in proximity to the iron complex. Nevertheless, it is not bound in the inner sphere of the iron because one site is occupied by the hydroperoxide and the 6th coordination position is on the other site of the planar protoporphorin group. Hence, sulfide binding occurs on the same side of the heme as hydroperoxide binding, and must permit adequate motion to account for the low chiral character of the product.

In summary, I have discussed two ramifications of the oxidative behavior of endoperoxide synthase. This arose from the recognition that when you combine two molecules of oxygen with one of arachidonic acid to form PGH_2 oxidizing equivalents must be released and that they must react with something. This is not peculiar to endoperoxide synthase and some other enzymes that perform oxidation reactions also undergo oxidative deactivation. Although these two phenomena, self-inactivation and cosubstrate oxidation, appear dissimilar, they are not. With co-substrate oxidation, a small molecule reducing agent is oxidized and with enzyme deactivation the enzyme itself serves as the reducing agent. During cosubstrate oxidation, the enzyme initiates the reaction but the reducing agent such as aminopyrine, phenylbutazone and methylphenylsulfide dictate whether the reaction will proceed stoichiometrically on the enzyme or through a chain mechanism in solution.

References

1. Smith, W.L. and Lands, W.E.M. (1972) J. Biol. Chem. 247, 1038
2. Marnett, L.J., Wlodawer, P. and Samuelsson, B. (1975) J. Biol. Chem. 250, 8510
3. Egan, R.W., Paxton, J. and Kuehl, F.A. Jr. (1976) J. Biol. Chem. 251, 7329
4. Egan, R.W., Gale, P.H. and Kuehl, F.A. Jr. (1979) J. Biol. Chem. 254, 3295
5. Egan, R.W., Gale, P.H., Baptista, E.M., Kennicott, K.L., Vanden Heuvel, W.J.A., Walker, R.W., Fagerness, P.E. and Kuehl, F.A. Jr. (1981) J. Biol. Chem. 256, 7352

6. Nugteren, D.H., Beerthuis, R.K. and van Dorp, D.A. (1966) Recl. Trav. Chim. Pays-Bas 85, 405
7. Sivarajah, K., Lasker, J.M., Eling, T.E. and Abou-Donia, M.B. (1982) Mol. Pharmacol. 21, 133
8. Egan, R.W., Gale, P.H., Vanden Heuvel, W.J.A., Baptista, E.M. and Kuehl,F.A. Jr. (1980) *J. Biol. Chem.* **255**, 323
9. Tatum, C.M. Jr., Benkovic, P.A., Benkovic, S.J., Potts, R., Schleicher, E. and Floss, H.G. (1977) Biochemistry 16, 1093
10. Holmlund, C.E., Sax, K.J., Nielsen, B.E., Hartman, R.E., Evans, R.H. Jr. and Blank, R.H. (1962) J. Org. Chem. 27, 1468
11. Dodson, R.M., Newman, N. and Tsuchiya, H.M. (1962) J. Org. Chem. 27, 2707
12. Marnett, L.J., Bienkowski, M.J., Pagels, W.R. and Reed, G.A. (1980) in Advances in Prostaglandin and Thromboxane Research (Samuelsson, B., Ramwell, P.W. and Paoletti, R., eds.) Vol. 6, pp. 149-151. Raven Press, New York

Molecular Mechanisms for the Peroxidase Catalyzed Activation of Arylamine Carcinogens to Cytotoxic Metabolites

P.J. O'Brien, S. Khan, S. Jatoe, L.G. McGirr
Faculty of Pharmacy, University of Toronto
19 Russel Street, Toronto, Ontario M5S 2S2 Canada

Abstract

1. The cytotoxicity of the phenacetin metabolite N-hydroxyphenacetin (2mM) towards isolated hepatocytes is inhibited by the carboxyesterase inhibitor bis(-p-nitrophenyl)-phosphate. Nitrosophenetole was also thirteen fold more cytotoxic than N-hydroxyphenacetin. This suggests that N-hydroxyphenacetin is cytotoxic as a result of N-deacetylation to N-hydroxyphenetidine which autooxidizes to nitrosophenetole.
2. Lactoperoxidase, H_2O_2, bromide or hypochlorite alone, markedly enhanced N-hydroxyphenacetin cytotoxicity. With hypohalite activating systems, oxidative N-deacetylation occurred and nitrosophenetole was rapidly and stoichiometrically formed.
3. p-Phenetidine, the N-deacetylated metabolite of phenacetin was not toxic towards isolated hepatocytes at 20 mM. However, 0.3 mM p-phenetidine was cytotoxic if horseradish peroxidase and H_2O_2 were present. Cytotoxicity was probably due to the N-(4 ethoxy-phenyl)-p-benzoquinonediimine formed by dimerisation of phenetidine cation radicals.
4. p-Phenetidine was also cytotoxic if preincubated with chloroperoxidase and H_2O_2. A nearly stoichiometric conversion to nitrosophenetole was found.
5. p-Phenetidine (0.5 mM) was also cytotoxic if either hypochlorite or lactoperoxidase, H_2O_2, bromide was present. The hypohalite activating system oxidatively O-deethylated p-phenetidine to benzoquinoneimine and N-chlorobenzoquinoneimine. Similar products were formed when hypochlorite was added to p-aminophenol. The presence of halide therefore enables peroxidase-H_2O_2 to catalyze a two-electron oxidation of arylamine rather than a one-electron oxidation to radicals.

Biological Oxidation Systems, Volume 2

Introduction

The induction of cancer by carcinogenic xenobiotics is believed to be initiated by metabolism of the xenobiotic to electrophilitic metabolites which covalently link to nucleoplitic groups on informational cellular macromolecules such as DNA(1). Much attention has been focused on the N-oxidation of arylamines and arylamides by cytochrome P-450 dependent hepatic mixed function oxidases. However, peroxidases (donor-H_2O_2 oxidoreductases, EC.l.11.1.7) also catalyze the N-oxidation of arylamines. The latter oxidation usually differs in being a one-electron oxidation to free radicals which often form dimeric or polymeric products. However, peroxidases are also effective at catalyzing the irreversible binding of xenobiotics (particularly arylamines) to DNA and protein, and is therefore considered to be potentially genotoxic or cytotoxic. Comprehensive reviews have recently appeared on the activation of xenobiotics by peroxidases (2-5) or the peroxidase activity of prostaglandin H synthase (6,7). About 20% of the DNA adducts formed in the urothelium of dogs treated with the bladder carcinogen 2-naphthylamine are estimated to be of peroxidative origin (8). It is likely that the benzidine: DNA adducts formed with peroxidase (9,10) are also formed in dog bladders in vivo.

The physiological role of mammalian peroxidases is likely to (i) remove toxic peroxides, (ii) defend against microbes in saliva, milk, plasma or after phagocytosis by leukocytes, (iii) defend against tumor cells, parasites, worms, etc., (iv) hemolyse old erythrocytes in the spleen, and (v) inactivate thyroxine, oestradiol, chemotactic factors, leukotrienes. For all of these functions, halides increase the effectiveness of the peroxidase. The halide effect can partly be attributed to the peroxidase catalyzed formation of oxidant hypohalites. However, for thyroid peroxidase function or extrathyroidal peroxidase activity, iodide is essential for the iodination and coupling of the tyrosine residues involved in thyroxine formation.

One intriguing hypothesis is that target organ peroxidases or peroxidases in activated circulating polymorphonuclear leukocytes, monocytes or eosinophils can also metabolise drugs or carcinogens to cytotoxic or genotoxic products. Polymorphonuclear leukocytes, monocytes and eosinophils are the predominant cell types infiltrating the tissues of patients with a variety of infectious disorders, allergies, autoimmune disorders and tumors. These cells become activated by perturbation of the plasma membrane by bacterial bodies (*e.g.* the chemotactic N-formylmethionyl-leucyl-phenylalanine); chemotactic complement fragments; leukotrienes; platelet-activating factors;

immobilized immune complexes; or contact with opsonized bacteria. A cyanide resistant respiratory burst ensues in which oxygen is converted to hydrogen peroxide and degranulation resulting in peroxidase release occurs (11).

These circulating cells containing peroxidase migrate to any tissues associated with inflammatory disease where they become activated: *e.g.* rheumatoid arthritis, myocardial reperfusion injury, respiratory distress syndrome, ulcerative colitis, alcoholic liver cirrhosis. Eosinophils are not a striking feature of the acute inflammatory response but are involved in allergic and immune responses and accumulate in areas of helminthic infestations. Extensive degranulation of eosinophils and deposition of eosinophil peroxidase also occurs in certain pathologic conditions such as Hogkin's disease, chronic myelogenous leukemia, atopic dermatitis and endomyocardial fibrosis (12). Anti-inflammatory drugs may act by being metabolized by the peroxidase in these cells to products which prevent the tissue destructive events in inflammatory disease. However, these and other drugs may also be metabolized by these peroxidases to toxic products which cause bone marrow toxicity (leukopenia, agranulocytosis, neutropenia, etc.), lupus, hypothyroidism, thyroid cancer, etc. (13).

Chronic inflammation has also been associated with the subsequent development of cancer, and tumor formation has often been attributed to chronic irritation. Malignant tumors occur at a variety of inflammatory sites (14). Many tumor promoters are also powerful activators of leukocyte oxygen metabolism. Thus leukocytes could activate chemical carcinogens to genotoxic intermediates and contribute to carcinogenesis.

We have shown that leukocytes with the carcinogens ^{14}C-methylaminoazobenzene, benzidine, aminofluorene or phenol can result in its irreversible binding to nuclear DNA if the leukocytes are activated by the tumor promoter phorbol myristate acetate (10,15). The binding was dependent on oxygen and prevented by azide and cyanide indicating that myeloperoxidase catalyzed the oxidative activation of these xenobiotics resulting in DNA binding. Other investigators have shown that ^{14}C-chloraniline and 4-methylaniline bind to a lesser extent to DNA in activated leukocytes (16). However, binding to RNA occurred to a much greater extent. Activated leukocytes also catalyzed the binding of benzo(a)pyrene-7,8-dihydrodiol to exogenous DNA (17). 14C-Phenol and 1-naphthol were also found to be extensively bound to protein in activated leukocytes. Phenol binding was attributed to a myeloperoxidase mechanism but naphthol binding may also involve a superoxide anion radical catalyzed oxidation to 1,4-naphthoquinone (18).

The blood contains approximately 0.1M chloride, 50-100 μM bromide and < 1 μM iodide(19). Some peroxidases can utilize the halides as substrates and oxidize them to the corresponding hypohalous acids. Their order of effectiveness is iodide > bromide > chloride. However, only myeloperoxidase and chloroperoxidase are effective at utilizing chloride. Eosinophil and lactoperoxidase probably utilise bromide in vivo whereas most peroxidases can utilise iodide. Activated leukocytes have also been shown to form taurine monochloramine from taurine (19), 2 chloro-1,3,5-trimethoxybenzene from 1,3,5-trimethoxybenzene (20) and monochloramine from ammonium salts (21). It has also been estimated that 0.16 mM hypochlorite is generated in one hour by activated neutrophils (4 x 10^6 cells ml^{-1}) thereby making hypochlorite the primary oxygen metabolite (22).

$$H_2O_2 + X^- + H^+ \rightarrow HOX + H_2O$$

$$(X^- = Cl^-, Br^-, I^-, \text{and } SCN^-)$$

Recent evidence suggests that much of the antibacterial, antiparasitic, antiprotozoal, antitumor cell functions and proinflammatory activity of neutrophils *in vivo* could be largely due to the oxidant properties of hypochlorite (23). Much of this activity in eosinophils has also been attributed to the formation of hypochlorite (24). The role that hypohalites play in drug activation has yet to be investigated for most drugs (11).

Phenacetin was widely used as an antipyretic and analgesic until recently. Chronic hemolytic anemia, methemoglobinemia, analgesic nephropathy and papillary or solid tumors (transitional cell carcinomas) mainly in the renal pelvis (with some in the ureters and bladder) often resulted if phenacetin was ingested in daily therapeutic doses for a year (25-27). Phenacetin administered to rats long-term induces tumors in the kidneys, bladder, and nose (28). Analgesic nephrotoxicity can be induced in some selected strains of rats (29).

Phenacetin is metabolized *in vivo* by cytochrome P-450 dependent mixed function oxidase by O-deethylation to acetaminophen which could be responsible for the therapeutic effects of phenacetin. Ring hydroxylation to 2-hydroxyphenetidine and some N-hydroxylation to N-hydroxyphenacetin also occurs. Phenacetin is also N-deacetylated to p-phenetidine (30). Proximal tubular cell necrosis in rats has been produced after intravenous administration in rats of 2-hydroxyphenetidine and N-hydroxyphenacetin (31). Liver tumors in rats have also been produced after chronic oral administration of

N-hydroxyphenacetin (32). The ultimate carcinogenic electrophiles have been suggested to be nitrenium ions derived from the sulfate or glucuronide esters of N-hydroxyphenacetin (29). However the N-deacetylated metabolite N-hydroxy-p-phenetidine has been shown to bind to DNA and is a direct-acting mutagen (33). Recently kidney medullary prostaglandin synthetase and arachidonate have been shown to catalyze the irreversible binding of phenetidine to kidney microsomes and could form the basis of phenacetin induced kidney toxicity (34). The binding was attributed to p-phenetidine quinone-imine and quinone-dimine dimers which were shown previously to be products of the peroxidative oxidation of p-phenetidine and toxic to isolated hepatocytes (35).

In the "multistage process of carcinogenesis" it is probably the cytotoxicity caused by the chemical carcinogens that provides the stimuli for cell replication. Replication of cells with irreversible DNA damage which survive cell replication would give rise to transformed cells. Cell transformation could also arise from the death of initiated cells and the release of clastogenic or mitogenic factors. Selective pressure leading to transformed foci or colony formation would arise if the initiated cells are more resistant to the cytotoxic action of the carcinogens than the transformed cell. In the following, evidence for a peroxidase-halide-H_2O_2 mechanism for the oxidative activation of an arylamine carcinogen to cytotoxic products is presented for the first time. Isolated hepatocytes were used as a target and the phenacetin metabolites N-hydroxyphenacetin and p-phenetidine were activated by various peroxidases, H_2O_2 and bromide or chloride. A peroxidase - halide mechanism for phenacetin induced hepatotoxicity and kidney carcinogenicity is proposed.

Methods

Lactoperoxidase (bovine milk) (LPO), horseradish peroxidase (type VI) (HRP), chloroperoxidase (CPO), bis(p-nitrophenyl) phosphate (BNPP) sodium bromide, H_2O_2 and hypochlorite were purchased from Sigma Chem. Co. (St. Louis, MO). p-Phenetidine and p-aminophenol were obtained from Aldrich Chem. Co. (Milwaukee, WI). N-Hydroxyphenacetin (36), nitrosophenetole (37), benzoquinoneimine (38)and N-chlorobenzoquinoneimine (38) were synthesized as described.

Male Sprague-Dawley rats (200-250g body weight) were used throughout the experiments. Hepatocytes were isolated by collagenase perfusion of the liver as described (39). Isolated hepatocytes (10^6 cells per ml) were incubated in 10 ml Krebs-Henseleit buffer, pH

7.4, in rotating round-bottom flasks at 37°C under a carbogen atmosphere. Cell viability was estimated by the Trypan Blue exclusion method (39). Where indicated, the reaction mixture was preincubated before addition to the isolated hepatocytes. However, normally unless shown otherwise H_2O_2 (1mM) was added last to the incubation mixture containing hepatocytes (10^6 cells/ml), horseradish peroxidase type VI (1 μg/ml) or lactoperoxidase (1 unit/ml) or chloroperoxidase (1 unit/ml) bromide (0.1M) and the phenacetin metabolite (concentration as indicated). The spectral changes associated with a standard incubation mixture without hepatocytes were recorded (200-700nm) on a Shimadzu UV 240 spectrophotometer at various times.

Results

As shown in table 1 the cytotoxicity induced by the phenacetin metabolite N-hydroxyphenacetin was decreased if the isolated hepatocytes were preincubated for ten minutes with the N-deacetylase inhibitor bis p-nitrophenyl phosphate. This suggests that intracellular N-deacetylase activates NOH phenacetin. Because the N-deacetylated product N-hydroxyphenetidine rapidly autoxidizes, with a half-life in neutral buffer of about 2-5 mins (40), to p-nitrosophenetole the susceptibility of hepatocytes to p-nitrosophenetole was next investigated. Nitrosophenetole was found to be thirteen fold more cytotoxic than N-hydroxy phenacetin and could therefore be responsible for the cytotoxicity of N-hydroxyphenacetin.

Possible activation by various peroxidase-H_2O_2 systems was next investigated. Lactoperoxidase or horseradish peroxidase and H_2O_2 did not increase the toxicity of N-hydroxyphenacetin (2 mM) spectral analysis without hepatocytes present revealed that N-hydroxyphenacetin was oxidized very slowly (in contrast to NOH acetylaminofluorene). However, as shown in Fig 1, if bromide was present N-hydroxyphenacetin cytotoxicity was markedly increased. Spectral studies showed that N-hydroxyphenacetin was rapidly oxidized nearly stoichiometrically by lactoperoxidase, H_2O_2, bromide to nitrosophenetole. At the concentrations used lactoperoxidase, bromide and H_2O_2 by themselves were not cytotoxic towards isolated hepatocytes.

The cytotoxicity of p-phenetidine, the deacetylated metabolite of phenacetin was next investigated. As shown in table 2, p-phenetidine was not cytotoxic to isolated hepatocytes even at 1mM. However in the presence of horseradish peroxidase or lactoperoxidase and H_2O_2, p-phenetidine was markedly cytotoxic at 0.15 mM. As shown in fig 2,

Table 1. *Activation of NOH Phenacetin to Toxic Products*

Additions	Time (min):	Cytotoxicity (% hepatocytes taking up trypan blue) 60	120	180	240
none		18±2	20±3	21±2	25±3
NOH Phenacetin (2mM)		28±3	50±4	76±6	100
NOH Phenacetin (2mM)		32±3	41±3	58±6	82±6
Nitrosophenetole (0.15mM)		28±3	49±4	62±5	93±5
NOH Phenacetin (0.5mM)		19±2	21±3	22±2	24±3
NOH Phenacetin (2mM)[a] + LPO + H_2O_2		27±4	48±4	78±7	100
NOH Phenacetin (0.5mM)[a] + LPO + Br^- + H_2O_2		34±4	52±4	85±7	100
NOH Phenacetin (0.5mM)[a] + methionine (1mM) + LPO, Br^- + H_2O_2		23±2	26±3	26±3	31±3
NOH Phenacetin (0.5mM) + HOCl (1mM)		31±3	49±3	55±5	83±7
HOCl (1mM)		18±3	21±2	22±4	25±3

[a] Preincubated for 15 mins. before addition of hepatocytes.

a spectral analysis of the reaction mixture in the absence of hepatocytes showed that p-phenetidine underwent rapid oxidation initially to an orange product with maximal absorbance at 265, 306, 450nm and then to a red product with absorbance at 265, 306 and 497nm. If bromide was present in the reaction mixture however completely different non coloured products were formed absorbing at 253, 260 and 292 nm. These products were also cytotoxic. Hypochlorite also formed similar products with p-phenetidine. Furthermore as can be seen in table 2, hypochlorite markedly activated phenetidine to cytotoxic products. Chloroperoxidise and H_2O_2 also activated p-phenetidine to cytotoxic products. Other investigators have identified the products formed when phenetidine is oxidized by H_2O_2 and horseradish peroxidase (41). The 253, 260nm products formed when phenetidine is oxidized by hypochlorite had an identical spectra to benzoquinoneimine synthesised as previously described (38). As this indicated that an oxidative deethylation had occurred, reaction products formed when hypochlorite reacted with p-aminophenol was also investigated.

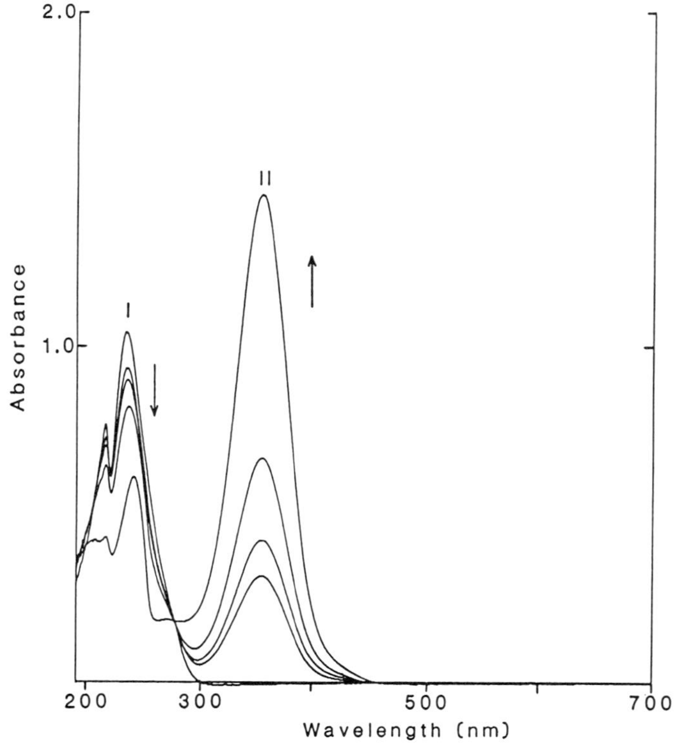

Figure 1. *N-hydroxyphenacetin (I) Oxidation by Lactoperoxidase to Nitrosophenetol (II).* Reaction conditions: N-hydroxyphenacetin (100 μM) was oxidized by lactoperoxidase (1 unit/ml) and H_2O_2 (100 μM) in the presence of KBr (0.1M) in 3.0 ml of 0.1M phosphate buffer, pH 7.0 at 20°C and scanned repetitively.

As shown in Fig. 3, the initial product formed with one molar equivalent of hypochlorite was also benzoquinoneimine but a product absorbing at 292nm was the major product formed with two equivalents of hypochlorite. This spectra was identical to N-chlorobenzoquinoneimine synthesised as previously described (38). As shown in Fig. 4, the product formed with chloroperoxidase and H_2O_2 absorbed at 353 mm. This spectra was identical to that observed with nitrosophenetole synthesised as previously described (37) indicating that nitrosophenetole was the major product formed by chloroperoxidase.

Table 2. *Oxidative Activation of Phenacetin to Toxic Products*

Additions	Time (min):	Cytotoxicity (% hepatocytes taking up trypan blue) 60	120	180	240
none		18±2	20±3	21±2	25±3
p-Phenetidine (10mM)		18±2	21±3	25±3	27±3
p-Phenetidine (0.15mM) + HRP + H_2O_2 (0.3mM)		36±4	52±4	58±4	100
p-Phenetidine (0.3mM)[a] + CPO + H_2O_2		29±3	38±3	42±4	68±5
p-Phenetidine (0.3mM) + LPO + Br^- + H_2O_2		34±4	45±4	79±4	74±6
p-Phenetidine (0.5mM) + HOCl (1mM)		30±3	41±4	45±5	100

[a] reaction mixture preincubated at pH6.5 for 15 mins before addition of hepatocytes.
HRP: horseradish peroxidase, CPO: chloroperoxidase, LPO: lactoperoxidase.

Discussion

It is widely assumed that cytochrome P-450 also catalyzes a two electron oxidation of phenetidine as well as phenacetin. However, although hamster liver microsomes catalyze phenacetin N hydroxylation (36), such activity in uninduced isolated hepatocytes is low as phenetidine (10mM) and phenacetin (20mM) were not toxic to hepatocytes (results not shown). This means that in order to study the different pathways for the metabolic activation of phenetidine it is necessary to add an exogenous activating system. Exogenous peroxidases markedly activated phenacetin metabolites to highly cytotoxic products. However, at least three different routes of peroxidase catalyzed metabolic activation exist depending on the peroxidase involved and whether a halide is present.

In the absence of an added activating system, the phenacetin metabolite N-hydroxyphenacetin was cytotoxic at 2mM. N-Hydroxyphenacetin was less cytotoxic in the presence of the N-deacetylase inhibitor, bis (p-nitrophenyl)phosphate. Nitrosophenetole was highly cytotoxic at 0.15mM. This suggests that the cytotoxic mechanism of N-hydroxyphenacetin involves N-deacetylation to N-hydroxyphenetidine which autoxidizes to nitrosophenetole. The N-deacetylation of N-hydroxyphenacetin was the rate limiting step in rat

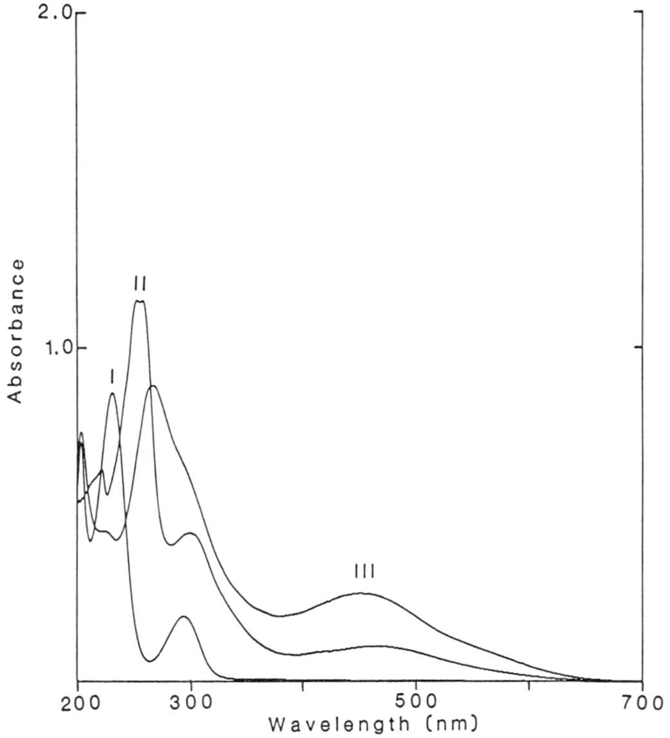

Figure 2. *p-Phenetidine Oxidation by Lactoperoxidase: H_2O_2 in the presence and absence of KBr.* Reaction conditions: p-Phenetidine (100 μM (I) was oxidized by lactoperoxidase (1 unit/ml) and H_2O_2 (100 μM) in the absence (II) or presence (III) of KBr (0.1M) in 3 ml 0.1M acetate buffer, pH 5.5 and the spectra recorded after 5 minutes.

hepatocytes. Rat liver microsomes however had twenty fold less specific activity than hamster microsomes at N-deacetylating N-hydroxyphenacetin (42). N-Hydroxy-acetylaminofluorene was much more cytotoxic in isolated hepatocytes (results not shown) probably because it was more readily N-deacetylated(40).

p-Phenetidine also became cytotoxic in the presence of chloroperoxidase and H_2O_2. Furthermore p-nitrosophenetole was the principal product formed which suggests that activation by chloroperoxidase involves a two electron oxidation of p-phenetidine to N-hydroxyphenetidine which autoxidizes to the highly cytotoxic

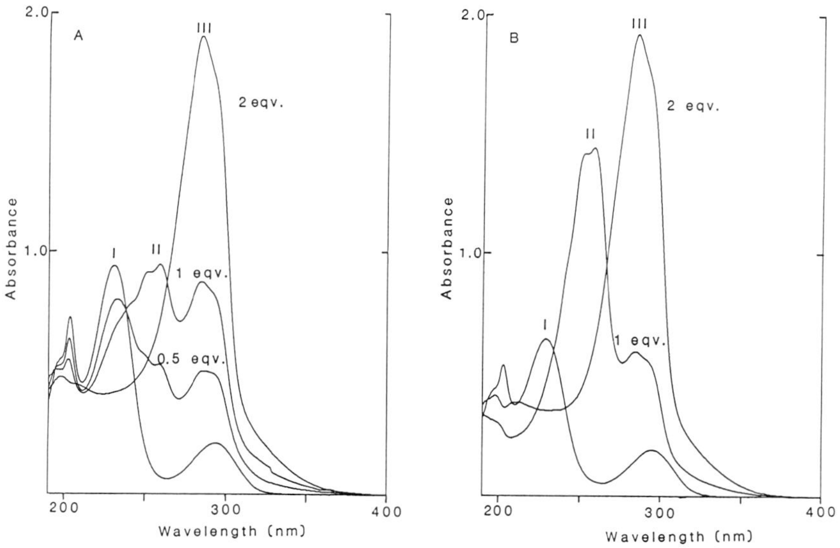

Figure 3. *Oxidation of p-Phenetidine (A) or p-Aminophenol (B) by Hypochlorite.* Reaction conditions: Various equivalents of hypochlorite were added to p-phenetidine or p-aminophenol (100 μM) in 0.1M phosphate buffer pH 7.4 at 20°C. Spectra were recorded after 1 min. I. p-Phenetidine or p-Aminophenol; II. p-Benzoquinoneimine; III. N-Chloro-p-benzoquinoneimine

nitrosophenetole. Although chloroperoxidase is located in fungi and not mammalian systems, its interest is because chloroperoxidase has many similarities to cytochrome P-450 particularly with regards to its ability to catalyze a two electron oxidation of substrates (43).

Most peroxidases and H_2O_2 catalyze a one electron oxidation of N-hydroxyarylamines to nitroxy radicals which then disproportionate. In the case of N-hydroxy-2-acetylaminofluorene the disproportionation products are nitrosofluorene and N-acetoxy-2-acetylaminofluorene. We have shown that nitrosofluorene is much more cytotoxic to isolated hepatocytes than N-acetoxy-acetylaminofluorene (44). Unfortunately it was found that N-hydroxyphenacetin unlike N-hydroxyacetylaminofluorene was a poor peroxidase substrate. However, when bromide was added to a lactoperoxidase/H_2O_2 reaction mixture, N-hydroxyphenacetin became highly toxic to hepatocytes. A spectral analysis of the reaction mixture showed a nearly stoichiometric conversion to nitrosophenetole. Similar results were obtained when chloride was added to a myeloperoxidase/H_2O_2 reaction mixture. As cytotoxicity could be prevented by methionine, a

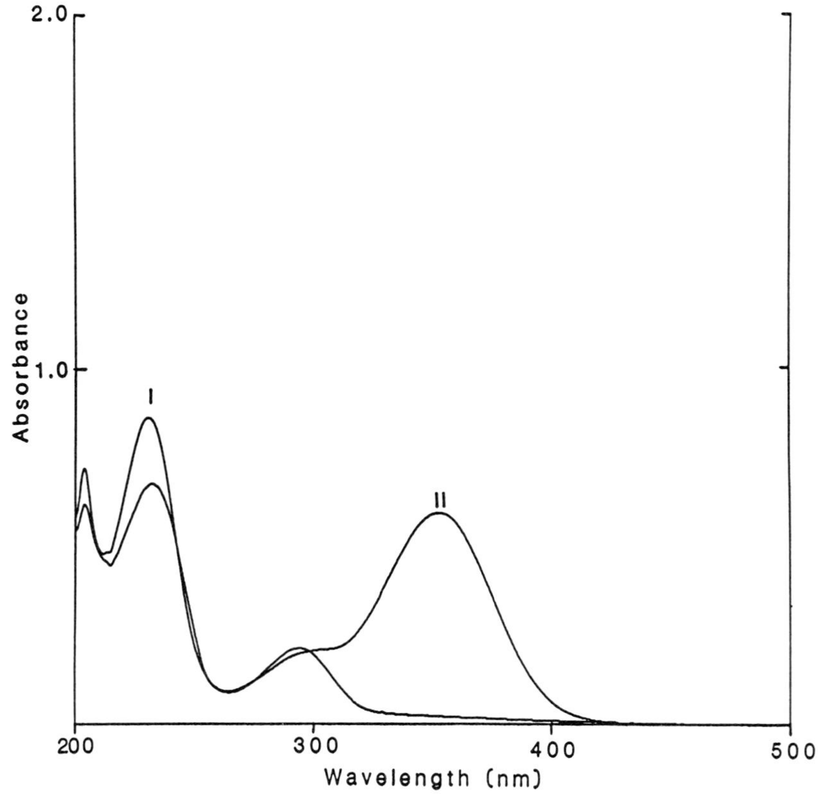

Figure 4. *p-Phenetidine (I) Oxidation by Chloroperoxidase.* H_2O_2 to nitrosophenetole (II). Reaction conditions: p-Phenetidine (100 μM), chloroperoxidase (3.0 units/ml) and H_2O_2 (500 μM) in 3.0 ml of 0.1M phosphate buffer, pH 7.4. Reaction mixture was scanned 5 minutes after the addition of H_2O_2.

scavenger of hypochlorite, it is likely that hypobromite or hypochlorite were the peroxidase products formed which oxidized the N-hydroxyphenacetin to nitrosophenetole. Other investigators have also reported that lactoperoxidase, H_2O_2 and bromide oxidizes N-hydroxyacetyl-aminofluorene to nitrosofluorene and have suggested this as a mechanism for the induction of mammary gland tumors by N-hydroxyacetylaminofluorene (45). N-hydroxyphenacetin also became cytotoxic to isolated hepatocytes if a non toxic- concentration of hypochlorite was added.

p-Phenetidine became markedly cytotoxic when activated by hydrogen peroxide and horseradish peroxidase or the mammalian peroxidases lactoperoxidase and myeloperoxidase. The dimer products formed indicate that these peroxidases catalyze a one electron oxidation to 4 ethoxyaniline cation free radical. Recently fast-flow electron spin resonance has been used to detect this radical in a lactoperoxidase system (46). The major metabolites identified by other investigators were the dimeric species N-(4 ethoxyphenyl) -p-benzoquinoneimine, N,N'- diethoxyazobenzene and the trimeric metabolite 2 amino 5 p ethoxybenzoquinone-di-p-ethoxyphenylimine. The intermediate N-(4 ethoxyphenyl)-p-benzoquinone diimine was also trapped with butylated hydroxyanisole (41). The latter intermediate seemed to alkylate GSH to unidentified mono and di GSH conjugates. The diimine of the dimer and/or trimer is probably responsible for the cytotoxicity as cytotoxicity was prevented by the dimine trapping agent 2,6(CH_3)2 phenol (results not shown).

p-Phenetidine was still cytotoxic if bromide was added to the lactoperoxidase, H_2O_2 reaction system. However, no coloured products were formed. A spectral analysis of the reaction mixture revealed a stoichiometric formation of benzoquinoneimine (253nm, 260nm) and N-chloroquinoneimine (290nm). As no dimers were formed the presence of bromide changed the lactoperoxidase mechanism from a one electron oxidation to a two electron oxidation. Methionine prevented the cytotoxicity indicating that the latter oxidation was probably mediated by the hypobromite formed by the lactoperoxidase. p-Phenetidine was also cytotoxic in the presence of nontoxic concentrations of hypochlorite and a spectral analysis of the reaction mixture revealed the formation of benzoquinoneimine and N-chloroquinonemine. Because oxidation by hypochlorite involved O-deethylation of the p-phenetidine, the oxidation of p-aminophenol by hypochlorite was also investigated. Identical spectra changes occurred indicating that similar products were formed. Benzoquinoneimine and N-chloroquinoneimine were synthesised and tested for cytotoxicity towards isolated hepatocytes. The concentration required for 50% cytotoxicity at 2 hours was 0.15mM for benzoquinoneimine and 0.25 mM for chlorobenzo-quinoneimine. The cytotoxicity of p-phenetidine activated by H_2O_2, lactoperoxidase and bromide is therefore largely due to benzoquinoneimine.

A chemical mechanism for the oxidation of N-hydroxyphenacetin by hypohalites is outlined in figure 5. Although N-hydroxy-phenetidine rapidly autoxidizes to nitrosophenetole, a mechanism for the oxidative N-deacetylation of N-hydroxylamines by hypohalite to nitroso metabolites is given for comparison. By contrast cytochrome P-450 catalyzes the oxidative O-deethylation of N-hydroxphenacetin to N-hydroxyacetaminophen (47).

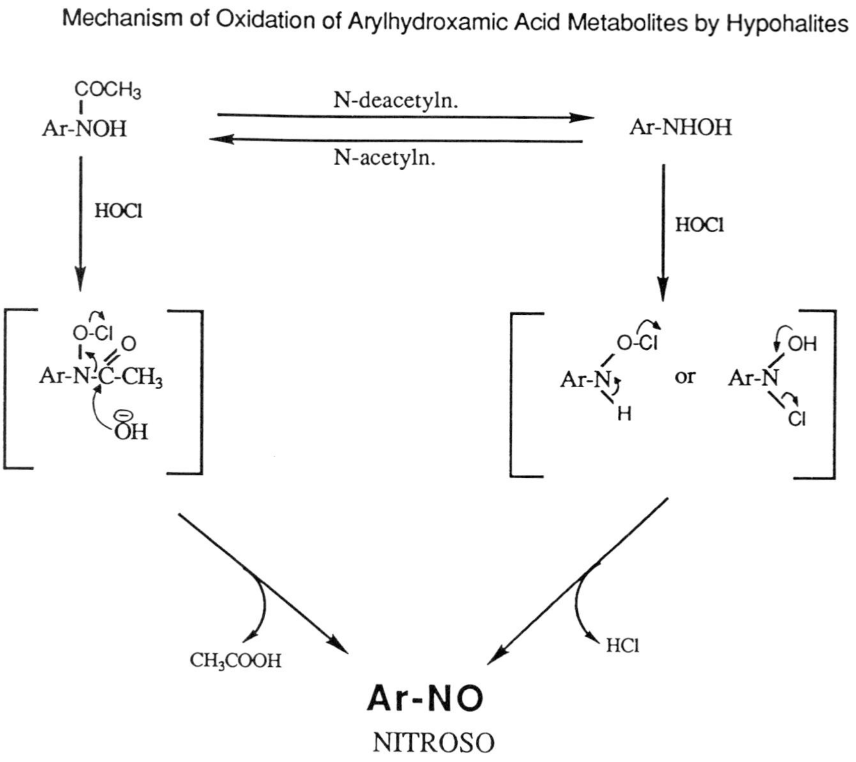

Figure 5. *Mechanism of Oxidation of Arylhydroxamic Acid Metabolism by Hypohalides*

A chemical mechanism for the oxidative O-deethylation of p-phenetidine or oxidation of p-aminophenol by hypohalites is outlined in figure 6. By contrast as shown above chloroperoxidase catalyzes the N-oxidation of p-phenetidine to N-hydroxy-phenetidine and nitrosophenetole. Microsomes and NADPH also catalyze the covalent binding of p-[^{3}H]phenetidine to microsomal proteins possibly as a result of a cytochrome P-450 catalyzed oxidation to nitrosophenetole (48). However the major pathway for cytochrome P-450 is believed to involve the O-deethylation of phenetidine to p-aminophenol (42).

Phenacetin or acetaminophen causes centrilobular hepatic necrosis in rodents (49) and hamsters are particularly susceptible. In rats, phenacetin is more toxic than acetaminophen (50). N-Hydroxyphenacetin fed to rats induces liver tumors in most animals and kidney tumors in some animals (32). Peroxidases in the liver seem to be located in the endoplasmic reticulum and nuclear

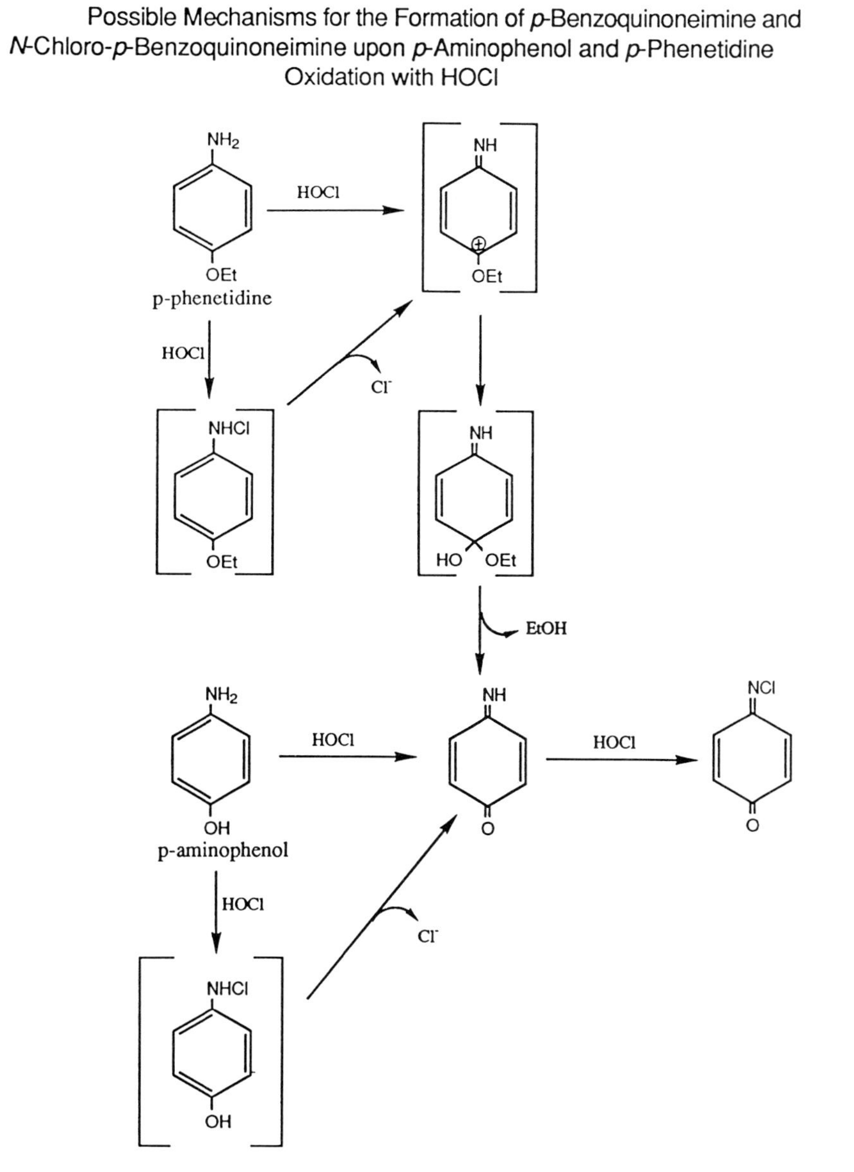

Figure 6. *Possible Mechanisms for the Formation of p-Benzoquinoneimine and N-Chloro-p-Benzoquinoneimine Upon Oxidation of p-Aminophenol or p-Phenetidine with HOCl*

envelope of the phagocytic reticuloendothelial cells (Kupffer cells) which line the bile ducts in the liver and form as much as 18% of all liver cells (51). Hepatocytes however contain 30 times lower peroxidase specific activity than the non-parenchymal cells (52). Kupffer cells peroxidases could therefore play a role in the activation of phenacetin metabolites. The reduced spectra of Kupffer cell peroxidase is similar to myeloperoxidase (51) which could also indicate that this peroxidase can form hypochlorite. The invasion of activated neutrophils or eosinophils into the liver infected by hepatitis or damaged by alcoholism could also activate phenacetin metabolites as a result of the high peroxidase content of the invading cells. In humans phenacetin abuse is associated with interstitial nephritis and papillary necrosis in the kidney. Which of the numerous cell types in the kidney contain peroxidase is not known although prostaglandin hydroperoxidase has been located in the kidney medulla and can activate p-phenetidine (33) It is also interesting that the hypochlorite products of p-aminophenol and phenetidine were found to be similar. p-Aminophenol and phenacetin were previously found to cause extensive renal tubular necrosis and interstitial nephritis(52).

In conclusion, the target tissues for phenacetin probably contain endogenous peroxidases which together with physiological concentrations of halides could activate phenacetin metabolites to highly cytotoxic products. Alternatively activation could also occur by the peroxidases of white blood cells which invade these tissues as a result of infection or inflammation.

Acknowledgements

This research was supported by the Natural Science and Engineering Council of Canada. We wish to thank Dr. A.D. Rahimtula for synthesising N-hydroxyphenacetin and p-nitrosophenetol and Dr. J. Uetrecht for research discussions.

References

1. Miller, J.A. (1970) *Cancer Res.* **30**, 559-570
2. O'Brien, P.J. (1984) *Free Radicals in Biology VI* (Pryor, W.A., ed), pp. 289-322, Academic Press
3. O'Brien, P.J. (1988) *Free Radicals Biol. Med.* **4**, 169-183
4. O'Brien, P.J. (1985) *Env. Health Perspect.* **64**, 219-232
5. Meunier, B. (1987) *Biochemie* **69**, 3-9
6. Boyd, J.A., and Eling, T.E. (1985) *Environ. Health Perspect.* 64, 45-51
7. Josephy, D. (1988) *Free Rad. Biol. Med.* 6, 533-542

8. Yamazoe, Y., Miller, D.W., Weiss, C.C., Dooley, K.L., Zenser, T.V., Beland, F.A., and Kadulbar, F.F. (1985) *Carcinogenesis* **6**, 1379-1387
9. Yamazoe, Y., Zenser, T.V., Miller, D.W., and Kadlubar, F.F. (1989) *Carcinogenesis* **9**, 1635-1641
10. O'Brien, P.J. Gregory, B., Fanney, R., Davison, W., Rahintula, A.D., and Tsuouta, Y. (1985) *Microsomes and Drug Oxidations* (Boobis, A.R., Caldwell, J., de Matteis, F., and Elcombe, C.R., eds) pp. 100-112, Taylor and Francis Ltd., London
11. O'Brien, P.J. (1988) *The Respiratory Burst and Its Physiological Significance* (Sbarra, A.J., and Strauss, R.R., eds) pp. 203-232, Plenum Press
12. Samoszuk, M.K., Gidanian, F., and Rietveld, C. (1988) *Am. J. Pathol.* **132**, 455-460
13. Uetrecht, J.P. (1988) *The Respiratory Burst and Its Physiological Significance* (Sbarra, A.J., and Strauss, R.R., eds) pp. 233-244, Plenum Press
14. Gordon, L.I., and Weitzman, S.A. (1988) *The Respiratory System and Its Physiological Significance* (Sbarra, A.J., and Strauss, R.R., eds) pp. 277-298, Plenum Press
15. Tsuruta, Y., Subrahmanyam, V.V., Marshall, W., and O'Brien, P.J. (1985) *Chem.-Biol. Interacn.* **53**, 25-35
16. Corbett, M.D., and Corbett, B.R. (1988) *Chem. Res. Toxicol.* **1**, 356-363
17. Twerdok, L.E., and Trush, M.A. (1988) *Chem. Biol. Interacn.* **65**, 261-273
18. Eastmond, D.A., French, R.C., Ross, D.. and Smith, M.T. (1987) *Chem.-Biol. Interacn.* **63**, 47-62
19. Weiss, S.J., Klein, R., Slivka, A., and Wei, M. (1982) *J. Clin. Invest.* **70**, 598-607
20. Foote, C.S., Goyne, T.E., and Lehrer, R.I. (1983) *Nature* **301**, 715-716
21. Thomas, E.L., Grisham, M.B., and Jefferson, M.M. (1983) *J. Clin. Invest.* **72**, 441-453
22. Grisham, M.B., Jefferson, M.M., Melton, D.F., and Thomas, E.L. (1984) *J. Biol. Chem.* **259**, 10404-10412
23. Weiss, S.J. (1986) *Adv. in Free Radical Biology and Medicine* **2**, 91-106
24. Weiss, S.J., Test, S.T., Eckmann, C.M., Roos, D., and Regiane, D. (1986) *Science* **234**, 200-203
25. Raaflaub, J., and Dubach, U.C. (1969) *Klin. Wsch.* **23**, 1286-1287
26. Johansson, S., and Wahlqvist, L. (1977) *Acta Path. Microbiol. Scand.* Sect A. **85**, 768-774
27. McCredie, M., Stewart, J.H., Ford, J.M., and MacLennan, R.A. (1983) *Br. J. Urol.* **55**, 220-224
28. McLean, S., Davies, N.W., Watson, H., Favretto, W.A., and Bignall, J.C. (1981) *Drug Metab. Dispos.* **9**, 255-260
29. Vaught, J.B., McGarvey, P.B., Lee, M.S., Garner, C.D., Wang, C.Y., Linsmaier-Bednar, E.M., and King, C.M. (1981) *Cancer Res.* **41**, 3424-2429
30. Smith, R.L., and Timbrell, J.A. (1974) *Xenobiotica* **4**, 489-501

31. Calder, I.C., Creek, M.J., Williams, P.J., Funder, C.C., Green, C.R., Ham, K.N., and Tange, J.D. (1973) *J. Med. Chem.* **16**, 499-506
32. Calder, I.C., Goss, D.E., Williams, P.J., Funder, C.C., Green, C.R., Ham, K.N., and Tange, J.D. (1976) *Pathology*, **8**, 1-6
33. Larsson, R., Ross, D., Berlin, T., Olsson, L.I., and Moldeus, P. (1985) *J. Pharmacol. Exp. Therap.* **235**, 475-480
34. Mulder, G.J., Kadlubar, F.F., Mays, J.B., and Hinson, J.A. (1984) *Mol. Pharmacol.* **26**, 342-347
35. Larsson, R. Linqvist, T., Lindeke, B., and Moldeus, P. (1986) *Chem.-Biol. Interacn.* **60**, 317-330
36. Hinson, J.A., and Mitchell, J.R. (1976) *Drug Metab. Disp.* **4**, 430-435
37. Hays, J.T., De Butts, E.H., and Young, H.L. (1967) *Org. Chem.* **32**, 153-158
38. Hodnett, E.M., Prakash, G., and Amirmoazzami, J. (1978) *J. Med. Chem.* **21**, 11-16
39. Moldeus, P., Hogberg, J., and Orrenius, S. (1978) *Methods Enzymol.* **52**, 60-66
40. Lenk, W., and Scharmer, U. (1980) *Xenobiotica* **10**, 582-591
41. Ross, D., Larsson, R., Norbeck, K., Ryhage, R., and Moldeus, P. (1985) *Mol. Pharmacol.* **27**, 277-286
42. Nohmi, T., Mizokami, K., Kawano, S., Fukuhara, M.,, and Ishidate, M. (1987) *Jpn. J. Cancer Res.* (Gann) **78**, 153-161
43. Dawson, J.H. (1988) *Science* **240**, 433-439
44. Jatoe, S., and O'Brien, P.J. (1990) *Progr. in Pharmacol. and Clin Pharmacol.* (in press)
45. Malejka-Giganti, D., Ritter, C.L., Decker, R.W., and Suilman, J.M. (1986) *Cancer Res.* **46**, 6200-6206
46. Fischer, V., Harman, L.S., West, P.R., and Mason, R.P. (1986) *Chem. Biol. Interacn.* **60**, 115-127
47. Hinson, J.A., Nelson, S.D., and Gillette, J.R. (1979) *Mol. Pharmacol.* **15**, 419-427
48. Hinson, J.A., and Mays, J.B. (1986) *J. Pharmacol. Exp. Therap.* **238**, 106-112
49. Boyd, E.M.. (1960) *Toxicol. Appl. Pharmacol.* **1**, 240-249
50. Boyd, E.M., and Bereczky, G.M. (1966) *Br. J. Pharmacol.* **26**, 606-610
51. Fahimi, H.D., Gray, B.A., and Herzog, V.K. (1976) *Laboratory Investigation* **34**, 192-201
52. Calder, I.C., Funder, C.C., and Green, C.R. (1971) *Brit. Med. J.* 4, 518-521
53. Van Berkel, T.J. (1974) *Biochem. biophys. Res. Comm.* **61**, 204-209

Inhibitors of Carcinogenesis

Bandaru S. Reddy
Division of Nutrition and Endocrinology, American Health Foundation, Valhalla, New York, 10595, USA

Abstract

During the last 20 years, a substantial amount of progress has been made in the understanding of the relationship between the dietary constituents and development of cancer in man. Laboratory animal model studies have provided useful data for evaluating the role of dietary constituents and synthetic compounds as inhibitors of carcinogenesis. Current research in laboratory animal models also indicate that these inhibitors protect against carcinogenesis by minimizing carcinogen formation or by inhibiting events at the stages of cancer initiation and post initiation (promotion). Much research still remains to be done in this important area of cancer prevention.

Introduction

It is becoming increasingly clear that dietary habits and nutritional factors play an important role in the causation and development of a number of major human cancers worldwide (1-2); this conclusion is amply supported by evidence from experiments conducted in laboratory animal models (3). For example, populations consuming diets high in certain fats and low in fiber have an increased risk of cancer of the colon, breast, prostate, and possibly other organs (3, 4,). On the other side of the spectrum, several foodstuffs (vegetables, fruits, and grains) are also a source of naturally occurring anticarcinogenic (anti-initiating and antipromoting) agents that inhibit the formation of carcinogens from precursors or tumor promotion (5, 6). Several synthetic compounds with tumor inhibiting activity have been identified (Table 1) and tested for their anticarcinogenic properties (5-7). These chemical substances both synthetic and naturally occurring are termed as chemopreventive agents (5). The chemical diversity of these chemopreventive suggests that inhibition of carcinogenesis is not a highly selective phenomenon and that multiple strategies exist for bringing about this desired effect (5).

Biological Oxidation Systems, Volume 2

Table 1 *Inhibitiors of Carcinogenesis*

Inhibitors	Source
A. Naturally occurring inhibitors	
Aromatic isothiocyanates (benzylisothiocynate, phenylethylisothiocyanate, and phenylisothiocyanate)	Brussels sprouts, cauliflower, and cabbage
Tocopherols	Vegetable oils
Ascorbic acid	Citrus fruits and leafy vegetables
Fibers	Cereal products, grains, fruits, and vegetables
Coumarins and Lactones (coumarin, limettin, lactone, etc.)	Citrus fruits and vegetables
Flavones and flavonoids (quercetin, myricetin, kaempferol, chrysin, α-napthoflavone, and pentamethylether, etc.)	Grains, fruits and vegetables
Indoles (indole-3-acetonitrile, indole-3-carbinol, 3,3-diindoylmethane, etc.)	Brussels sprouts, cabbage and cauliflower
Phenols (caffeic acid, ellagic acid, ferulic acid chlorogenic acid, etc.)	Coffee, tea, soybeans, oats, appples, etc.
Plant sterols (β-sitosterol)	Oils
Carotenoids (β-carotene)	Carrots, yams, etc.
Calcium	Milk
Protease inhibitors	Soybeans, seeds, nuts, and legumes
B. Synthetic inhibitors	
Phenols	Butylated hydroxyanisole, and butylated hydroxytoluene
Non-steriodal antiinflammatory drugs	Indomethacin and piroxicam
ODC inhibitor	Difluoromethylornithine
Steroids	Dehydroepiandrosterone

Chemopreventive agents can be divided into 2 broad categories based on their mode of action in that some of them are effective against complete carcinogens (initiation phase) and that some are effective against tumor promoters (post initiation phase). In addition, some of chemopreventive agents act both during the initiation and post initiation phases of carcinogenesis (5-7). Briefly, the mechanism of action of agents that act during the initiation phase of carcinogenesis is by inhibiting enzymes involed in activation of carcinogen, by increasing levels of the enzymes and cofactors involved in carcinogen detoxification, or by scavenging the ultimate carcinogens (5-7). With regard to mechanism of action of agents during the post initiation phase, these agents can hinder the progression of changes leading to form a neoplasm (5-7).

Inhibition at Initiation Phase of Carcinogenesis

Based on mode of action, inhibitors of carcinogenesis acting at the stage of initiation can be divided into 2 broad categories. They include the compounds (a) that prevent the formation of carcinogens from precursor or procarcinogens and (b) that inhibit carcinogenic process by increasing the detoxification and/or decreasing the activation of carcinogens.

A vast majority of N-nitrosamines and N-nitrosamides produce tumors in a variety of tissues of experimental animals (8). N-nitroso compounds are formed from the reactions of a nitrosating species such as nitrogen trioxide or dinitrogen tetroxide with primary, secondary, or tertiary amines, or secondary amides (9). The formation of potentially carcinogenic N-nitroso compounds from commonly occurring precursors accounts for their ubiquitous presence in the environment. Endogenous formation of N-nitroso compounds in the human as well as in animal models from nitrite and amines or amides have been documented (10, 11). Ascorbic acid was the first naturally occurring substance found to effectively block the formation of N-nitroso compounds (9, 12, 13). Animals administered with appropriate precursor compounds form nitroso carcinogens and subsequently develop tumors. Under these conditions, administration of ascorbic acid in the diet to the animals will prevent the formation of nitroso compounds and occurrance of tumors. Thus, treatment of pregnant rats or hamsters with ascorbate together with nitrite and ethylurea (precursor of ethylnitrosourea) inhibited the induction of tumors in the offspring (14, 15). High levels of dietary sodium ascorbate reduced the incidence of lung adenomas in Swiss or Strain A mice treated with morpholine, piperazine, N-methylaniline, methylurea, or ethylurea along with nitrite (16). In humans, inhibition of N-nitroso compound formation by ascorbic acid, using urinary levels of nitrosoproline as an indicator, was also demonstrated (17). α-Tocopherols and phenols also have been shown to inhibit the formation of N-nitroso compounds. α-Tocopherol, a lipid-soluble, water-insoluble vitamin, is an effective inhibitor of N-nitrosation reactions in dispersons, and thus could be effective in inhibiting the formation of N-nitroso compounds in the GI tract.

The critical step in the metabolic activation of N-nitrosamines to ultimate carcinogenic forms is the hydroxylation of the α-carbon atom; this reaction is catalyzed by cytochrome P-450-dependent mixed function oxidases. Modulation of the enzyme activities responsible for α-hydroxylation of N-nitrosamines can determine the carcinogenicity of these compounds. A large and diverse group of compounds both naturally occurring and synthetic falls into the category of inhibitors of carcinogen activation and enhancers of carcinogen detoxification (Table 1). An example of this type of inhibition is the prevention of 1,2-dimethylhydrazine (DMH)-induced colon carcinogenesis by disulfiram which inhibits the metabolism of procarcinogen to its carcinogenic metabolites (18). Isothiocyanates which are naturally occurring constituents of cruceferous vegetables have also been found to be effective inhibitors of tumorigenesis in several different animal models (5, 6). Phenyl isothiocyanate (PITC), benzyl isothiocyanate (BITC) and phenylethyl isothiocyanate (PEITC) inhibited mammary tumor formation in Sprague-Dawley rats when administered orally prior to 7,12-dimethyl-benz(a)anthracene (DMBA) administration (19). In the same study, dietary BITC and PEITC also inhibited DMBA-induced neoplasia of the forestomach and lung of ICR/Ha mice. BITC pretreatment also inhibited benzo[a] pyrene (BP)-induced lung and forestomach tumors and N-nitrosodiethylamine-induced forestomach tumors in A/J mice (20). Also, PEITC inhibited 4-(methylnitrosamino)-1-(3-pyridyl)-1-butanone (NNK)-induced lung tumorigenesis in A/J mice and F344 rats (21, 22). The inhibitory activity of PEITC appears to be related to its inhibitory effects on NNK metabolism in lung microsomes.

The cellular levels of glutathione (GSH), a powerful nucleophile, together with the enzymes GHS transferases and GSH peroxidases are critical in maintaining defense against carcinogens. Because several carcinogenes of diverse types are either being electrophilic per se or being metabolically activated to electrophilic species (23), it is evident that anticarcinogenic agents acting through enhancement of the GSH-GSH transferase system could offer the possibility of being especially effective in the human setting. Diet and dietary constituents have been found to markedly influence the levels of GSH transferases. Sparnins *et al.* (24) reported that mice fed an unrefined diet (Purina Lab Chow) had higher liver and intestinal GSH transferase activities than did mice fed a semipurified diet. The inclusion of chopped dried green coffee beans, Brussels sprouts, instant tea, or powdered tea leaves in the semipurified diet significantly increased GSH transferases in both organs. Coumarin, α-angelicalactone, indole-3-carbinol and indole-3-acetonitrile, which are constituents of Brussels sprouts and cabbage also significantly increased GSH transferase in liver and small intestine. Inclusion of dried green coffee beans in the diet of Sprague-Dawley rats prior to the administration of DMBA significantly reduced the incidence of mammary tumors (25). The active components of green coffee beans responsible for the induction of GSH-transferase were isolated and identified as palmitate esters of the dieterpenes kahweol and cafestol.

Inhibition at Post Initiation Phase of Carcinogenesis

These agents inhibit carcinogenesis when administered subsequent to a course of carcinogen administration that result in the occurrence of cancer. In Table 1, several classes of compounds both synthetic and naturally occurring that inhibit the tumor promotion are listed.

Quercetin, applied topically to the skin of CD-1 mice at a dose level of 30 μmol per mouse significantly inhibited the promoting effect of 12-0-tetradecanoylphorbol-13-acetate (TPA) on DMBA-induced skin tumors (26). Similar effects on the inhibition of TPA actions were observed with morin, kaempferol, and fisetin (27). Quercetin strongly inhibits mouse skin lipoxygenase, an observation in agreement with previous studies on the inhibition, by various flavonoids, of lipoxygenase and prostaglandin synthetase (28), and its antipromotional activity could well be due to an effect at this locus. In studies with rats, administration of β-sitosterol, a plant sterol, in the diet at 200ppm to F344 rats significantly reduced both the incidence and multiplicity of colon tumors due to the intrarectal administration of methylnitrosourea (MNU), a direct acting carcinogen that does not require metabolic activation (29).

Burkitt (30) observed that populations consuming a diet rich in fiber have a low incidence of colon cancer, whereas those eating refined carbohydrates with little fiber have a higher incidence of the disease. In several populations consuming diets high in total fat, dietary fiber acts as a protective factor for colon cancer risk. Recent studies comparing rural and urban populations in Finland, Denmark, and Sweden, and urban populations in New York, indicated that one of the factors contributing to the low risk of colon cancer in rural Scandinavia, in contrast to New York, appears to be high dietary fiber intake that is mainly in the form of whole-grain cereals, even though all of these populations have a high fat intake (31, 32). These studies demonstrate a protective role of fiber in colon cancer.

Studies examining the role of various types of dietary fiber in animal models have confirmed the human evidence. The effect of a semipurified diet containing 15% pectin or wheat bran and 20% fat on colon carcinogenesis induced by AOM was studied in F344 rats by Watanable *et al.* (33). The addition of pectin or wheat bran to the diet greatly inhibited colon tumor incidence induced by AOM. Wheat bran also protected rats against DMH-induced colon cancer, regardless of whether the carcinogen is administered orally or subcutaneously (34, 35). A recent study by Kroes *et al.* (36) suggested a protective effect of wheat bran on DMH-induced colon carcinogenesis in rats. The protective effect occurred at low, medium, and high-fat levels and was dose-related.

The effect of dietary wheat bran and dehydrated citrus fiber at 15% level and 5% dietary fat on intestinal carcinogenesis induced by AOM and 3,2-dimethyl-4-aminobiphenyl(DMAB) was studied in male F344 rats (37, 38). Animals fed the wheat bran or citrus fiber and treated with AOM had a lower incidence (number of animals with tumors) and multiplicity (number of tumors/tumor-bearing rat) of colon tumors and tumors of the small intestine than did those fed the control diet and treated with AOM, whereas animals fed the wheat bran and treated with DMBA had a lower incidence and multiplicity of colon tumors. In another study, Reddy *et al.* (39) tested the effect of dietary lignin on DMBA-induced colon carcinogenesis in rats. Animals fed a semisynthetic diet containing 7.5% lignin inhibited the multiplicity of colon adenocarcinomas compared to those fed control diet containing 5% alphacel. Several studies have shown an inhibitory effect of dietary cellulose on DMH-induced colon carcinogenesis (40).

Dietary calcium has been shown to act as a protective factor on colon cancer development in humans (41). There are studies to indicate that dietary calcium in excess protects against chemically-induced colon cancer in laboratory animals. Wargovich *et al.* (42) demonstrated that the incidence of AOM-induced colon adenocarcinomas was significantly inhibited in F344 rats fed the high-fat diets containing 0.32% calcium as calcium lactate as compared to those fed the high-fat diet containing 0.04% calcium or high-fat diet alone. Pence and Budding (43) showed that supplemental dietary calcium in the form of calcium gluconate at 1% in a high-fat diet inhibited colon carcinogenesis in F344 rats induced by DMH whereas supplemental calcium had no inhibitory effect in animals fed the low-fat diet. Supplemental vitamin D_3 when given alone but not when fed in combination with calcium inhibited the DMH-induced colon carcinogenesis. Thus, the combination of both supplemental calcium and vitamin D_3 appears to antagonize the inhibitory effect of each supplement when fed alone. Other investigators demonstrated that supplemental calcium at 1% in a high-fat diet increased the DMH-induced colon carcinogenesis in F344 rats (44). In another study supplemental calcium in the form of calcium gluconate at 0.2-2% in a low-fat diet had no effect on DMH-induced colon carcinogenesis in F344 rats.

Recently, interest in the marine oils emerged from the findings that cancer incidence rates are generally low in Eskimos in Alaska and Greenland compared to North Americans and other western population groups who consume food containing high dietary fat. Although fish oils are rarely found in western diets, food consumed by Eskimos contains large amounts of oils derived from fish and seals, and other marine animals. These studies resulted in an interest in the role of fish oils in health and disease. The high levels of highly polyunsaturated omega-3 fatty acids such as eicosapentaenoic acid (c20:5, n-3) and docosahexaenoic acid (c22:6, n-3) present in the marine oils make them unique dietary fats. Several studies in laboratory animal models demonstrated that high levels of dietary Menhaden fish oil had no enhancing effect on DMBA-induced and MNU-induced mammary tumors when compared to high corn oil diet (46, 47). Dietary intake of 20% Menhaden oil when compared to 20% corn oil produced an inhibition of preneoplastic lesions in rat pancreas (48). Our recent studies demonstrated that high-fat diets containing 22.5% Menhaden oil + 1% corn oil and increasing levels of Menhaden fish oil produced a significant inhibition of AOM-induced colon tumors in rats when compared to a 23.5% corn oil diet. (49, 50). Eicosapentaenoic acid and docosahexaenoic acid present in Menhaden oil have been shown to inhibit the oxidative metabolism of arachidonic acid by the cyclooxygenase pathway that is involved in the synthesis of prostaglandins. It is therefore possible that the lack of a colon tumor-promoting effect of Menhaden oil might be due to its inhibitory effect on prostaglandin synthesis.

Compounds which inhibit inflammation can be expected to be antipromotional. A large number of experimental animal tumors and human cancers have been shown to contain and/or produce large quantities of prostaglandins, particularly those of E series. Therefore, the possibility exists that lowering of colonic prostaglandin synthetic activity by specific substances may inhibit tumorigenesis. Prostaglandin synthesis inhibitors which include non-steroidal antiinflammatory drugs such as piroxicam (4-hydroxy-2-methyl-N-2-pyridinyl-2H-1,2-benzothiozine-3carbamide-1,1-dioxide) and indomethacin have been tested as chemopreventive agents in colon carcinogenesis. Indomethacin given in drinking water or administered i.p. resulted in the inhibition of chemically-induced colon tumors in rodents (51, 52). Also, indomethacin reduced ornithine decarboxylase (ODC) induction, the first and rate limiting enzyme in polyamine biosynthesis, in the colon of rats (53). Dietary piroxicam has been shown to inhibit MNU, MAM acetate and AOM-induced colon carcinogenesis in rats (54, 55). Studies from our laboratory demonstrated that increasing levels of piroxicam in the diet, when fed 1 or 13 weeks after carcinogen treatment inhibit colon tumor incidence suggesting the inhibitory action of piroxicam during the development and progression of colon tumors (55).

Intracellular polyamine levels and polyaminesynthetic enzyme activities have been shown to be high in neoplastic cells and cells undergoing neoplastic transformation by carcinogens (56). ODC, a rate-limiting enzyme of the polyamine pathway, converts ornithine to putrescine and plays a role in normal and neoplastic cell proliferation in the intestinal mucosa and other organs (57, 58). Studies in the mouse skin tumor promotion model, (58) and in the colon tumor promotion model (50) have demonstrated that tumor promoters induce ODC activity suggesting a relationship between the tumor promotion and the induction of ODC activity in their respective target organs. Therefore, suppression of colonic polyamine levels by the inhibitors of ODC activity may prevent the tumor promotion. Difluoromethylornithine (DFMO) reduces intracellular polyamines by specific and irreversible inhibition of ODC. Studies using DFMO as a chemopreventive agent have shown that this compound can suppress the growth of human tumor xenografts (59), and inhibit the tumor formation in athymic mice implanted with human small cell lung carcinoma cells (60). DFMO administered in drinking water as a 1% solution significantly reduced the MNU-induced mammary carcinogenesis in female Sprague-Dawley rats (62). Studies have also demonstrated that DFMO administered in drinking water or in diet can inhibit chemically-induced colon carcinogenesis in rodent models (62).

Conclusion

In this review, attempts have been made to describe the compounds occurring naturally in fruits, vegetables and grains and synthetic compounds that can protect against carcinogenesis by minimizing carcinogen formation, or by inhibiting events at the stages of cancer initiation and promotion. Attemps have also been made in this review to discuss the mechanisms of action of various inhibitors of carcinogenesis so that the understanding of mechanism of action will lead to the identification and/or development of more effective inhibitors of carcinogenesis (6).

References

1. Doll, R., and Peto. R. (1981)) *J. Natl. Cancer Inst.* **66**, 1191-1308
2. National Academy of Sciences National Research Council Committee on Diet, Nutrition and Cancer (1982) *Diet, Nutrition and Cancer*. Washington, DC: National Academy Press
3. Reddy, B.S. (1986) in *Diet, Nutrition and Cancer: A Critical Evaluation* (Reddy, B.S., and Cohen, L.A., eds) Vol. 1, pp. 46-65. C.R.C. Press, Boca Raton
4. Wynder, E.L., Weisburger, J.H., and Horn, C. (1983) *Cancer Surveys* **2**, 557-576
5. Wattenberg, L.W. (1985) *Cancer Res.* **45**, 1-8
6. Fiala, E.S., Reddy, B.S., and Weisburger, J.H. (1985) *Ann. Rev. Nutr.* **5**, 295-321
7. Boone, C.W., Kelloff, G.J., and Malone, W.E. (1990) *Cancer Res.* **50**, 2-9
8. Druckrey, H., Preussmann, R., Ivankovic, S., Schmahl, D., and Afkham, J. (1967) *Z. Krebsforsch.* **69**, 103-201
9. Mirvish, S.S. (1975) *Toxicol. Appl. Pharmacol.* **31**, 325-351
10. Ishiwata, H., Boriboon, P., Nakamura, Y., Harada, M., Tanimura, A., and Ishidate, M. (1975) *J. Food Hyg. Soc. Jpn.* **16**, 19-24
11. Spiegelhalder, B., Eisenbrand, G., and Preussmann, R. (1976) *Food Cosmet. Toxicol.* **14**, 545-548
12. Dahn, H., Loewe, L., and Burton, C.A. (1960) *Helv. Chim.* Acta **43**, 320-333
13. Mirvish, S.S. (1983) *J. Natl. Cancer Inst.* **71**, 631-647
14. Ivankovic, S., Preussmann, R., Schmahl, D., and Zeller, J. (1973) *Z. Krebsforsch* **79**, 145-147
15. Rustia, M. (1975) *J. Natl. Cancer Inst.* **55**, 1389-1393
16. Mirvish, S.S., Cardesa, A., Wallcave, L., and Shubik, P. (1975) *J. Natl. Cancer Inst.* **55**, 633-636
17. Ohshima, H., and Bartsch, H. (1981) *Cancer Res.* **41**, 3658-3662

18. Fiala, E.S., Bobotas, G., Kulakis, C., Wattenburg, L.W., and Weisburger, J.H. (1977) *Biochem. Pharmacol.* **26**, 1763-1768
19. Wattenberg, L.W. (1978) *J. Natl. Cancer Inst.* **58**, 395-398
20. Wattenberg, L.W. (1981) *Cancer Res.* **41**, 2991-2994
21. Morse, M.A., Wang, C.-X., Stoner, G.D., Mandal, S., Conran, P.B., Amin, S.G., Hecht, S.S., and Chung, F.-L. (1989) *Cancer Res.* **49**, 549-553
22. Morse, M.A., Eklind, K.I., Amin, S.G., Hecht, S.S., and Chung, F.L. (1989) *Carcinogenesis* **9**, 1757-1759
23. Miller, E.C., and Miller, J.A. (1981) *Cancer* **47**, 1055-1064
24. Sparnins, V.L., Venegas, P.L., and Wattenberg, L.W. (1982) *J. Natl. Cancer Inst.* **68**, 493-496
25. Wattenberg, L.W. (1983) *Cancer Res.* **43**, 2448S-2453S
26. Kato, R., Nakadate, T., Yamamoto, S., and Sugimura, T. (1983) *Carcinogenesis* **4**, 1301-1305
27. Nakadate, T., Yamamoto, S., Aizu, E., and Kato, R. (1984) *Gann* **75**, 214-222
28. Baumann, J., von Bruchhausen, F., and Wurm, G. (1980) *Prostaglandins* **20**, 627-639
29. Cohen, B.I., and Raicht, R.F. (1981) in *Inhibition of Tumor Formation and Development* (Zedeck, M.S., and Lipkin, M., eds.) pp. 189-201. Plenium, New York
30. Burkitt, D.P. (1971) *Cancer* **28**, 3-13
31. Reddy, B.S., Hedges, A.T., Laakso, K., and Wynder, E.L. (1978) *Cancer* **42**, 2832-2838
32. Jensen, O.M., MacLennan, R., and Wahrendorf, J. (1982) *Nutr. Cancer* **4**, 5-19
33. Watanabe, K., Reddy, B.S., Weisburger, J.H., and Kritchevsky, D., (1979) *J. Natl. Cancer Inst.* **63**, 141-145
34. Chen, W.F., Patchefsky, A.S., and Goldsmith, H.S. (1978) *Surg. Gynecol. Obstet.* **147**, 503-506
35. Wilson, R.B., Hutcheson, D.P., and Wideman, L. (1977) *Am. J. Clin. Nutr.* **30**, 176-181
36. Kroes, R., Beems, R.B., Bosland, M.C., Bunnik, G.S.J., and Sinkeldam, E.J. (1986) *Federation Proc.* **45**, 136-142
37. Reddy, B.S., Mori, H., and Nicolais, M. (1981) *J. Natl. Cancer Inst.* **66**, 553-557
38. Reddy, B.S., and Mori, H. (1981) *Carcinogenesis* **2**, 21-25
39. Reddy, B.S., Maeura, Y., and Wayman, M. (1983) *J. Natl. Cancer Inst.* **71**, 419-424
40. Freeman, H.J., Spiller, G.A., and Kim, Y.S. (1984) *Carcinogensis* **5**, 261-264
41. Slattery, M.L., Sorenson, A.W., and Ford, M.H. (1988) *Am. J. Epidemiol.* **128**, 504-514
42. Wargovich, M.J., Stephens, L.C., and Gray, K.N. (1989) *Proc. Am. Assn. Cancer Res.* **30**, 196

43. Pence, B.C., and Buddingh, F. (1988) *Carcinogenesis* **9**, 187-190
44. Kaup, S.M., Behling, A.R., Choquette, L.L., and Grager, J.L. (1989) *FASEB J.* **3**, A469
45. Karkare, M., Comer, P.F., and Glauert, H.P. (1989) *FASEB J.* **3**, A472
46. Carroll, K.K., and Braden, L.M. (1984) *Nutr. Cancer* **6**, 254-259
47. Jurkowski, J.J., and Cave, W.J., Jr. (1985) *J. Natl. Cancer Inst.* **74**, 1145-1150
48. O'Connor, T.P., Roebuck, B.D., Peterson, F., and Campbell, T.C. (1985) *J. Natl. Cancer Inst.* **75**, 959-962
49. Reddy, B.S., and Maruyama, H. (1986) *Cancer Res.* **46**, 3367-3370
50. Reddy, B.S., and Sugie, S. (1988) *Cancer Res.* **48**, 6642-6647
51. Pollard, M., and Luckert, P.H. (1983) *J. Natl. Cancer Inst.* **70**, 1103-1105
52. Narasawa, T., Sato, M., Tani, M., Kudo, T., Takahashi, T., and Goto, A. (1981) *Cancer Res.* **41**, 1954-1957
53. Verma, A.K., Ashendel, C.L., and Boutwell, R.K. (1980) *Cancer Res.* **40**, 308-315
54. Pollard, M., and Luckert, P.H. (1984) *Cancer Lett.* **25**, 117-121
55. Reddy, B.S., Maruyama, H., and Kelloff, G. (1987) *Cancer Res.* **47**, 5340-5346
56. Pegg, A.E. (1988) *Cancer Res.* **48**, 759-774
57. Luk, G.D., Marton, L.J., and Baylin, S.B. (1980) *Science* **210**, 195-198
58. Boutwell, R.K. (1978) *Carcinogenesis* **2**, 49-58
59. Takigawa, M., Verma, A.K., and Simsiman, R.C. (1982) *Biochem. Biophys. Res. Commun.* **105**, 969-976
60. Luk, G.D., Abeloff, M.D., Griffin, C.A., and Baylin, S.B. (1983) *Cancer Res.* **43**, 4239-4243
61. Thompson, H.J., Herbst, E.J., Meeker, L.D., Minoeha, R., Ronan, A.M., and Fite, R. (1984) *Carcinogenesis* **5**, 1649-1651
62. Nigro, N.D., Bull, A.W., and Boyd, M.E. (1986) *J. Natl. Cancer Inst.* **77**, 1309-1313

The Role of Iron in Asbestos-Catalyzed Damage to Lipids and DNA

Ann E. Aust and Loren G. Lund
Department of Chemistry and Biochemistry, Utah State University
Logan, UT 84322-0300 USA

Abstract

Asbestos exposure can induce bronchogenic carcinoma, mesothelioma, and pulmonary fibrosis. The mechanism by which asbestos causes cancer is unknown. Asbestos catalyzes lipid peroxidation, strand breaks and formation of 8-hydroxydeoxyguanosine in DNA, and generation of oxygen radicals, all reactions known to be catalyzed by iron. Through the use of iron chelators, such as desferrioxamine B, evidence is accumulating that iron, present at levels as high as 36% by weight in asbestos, is responsible for these reactions. Iron can be mobilized from asbestos *in vitro* by physiological chelators, such as citrate and ADP. If iron is mobilized *in vivo* by these low-molecular-weight chelators, it will remove iron from the normal control by cellular transport and storage proteins which may result in deleterious oxidation reactions with biomolecules, such as DNA.

Introduction

Asbestos is a group of fibrous hydrated silicates which have desirable properties for industrial use. Exposure to asbestos can induce fibrosis of the lungs and carcinoma of the lungs, esophagus, or stomach (1-3) or mesothelioma of the pleura or peritoneum (4). The molecular mechanism responsible for the toxicity and carcinogenicity of asbestos remains unknown. However, one important observation is that asbestos contains iron to levels as high as 36% by weight. The amphibole varieties, crocidolite and amosite, are strongly carcinogenic while the serpentine type, chrysotile, is less dangerous. Crocidolite is a hydrated silicate of iron and sodium (see 5 for review). Amosite is a hydrated silicate of iron and magnesium (5). Chrysotile is a silicate of magnesium with Fe^{2+} substituted to varying extents for Mg^{2+} and with monocrystalline magnetite (iron oxide) interspersed (5). The chemical composition of each type of asbestos can vary depending upon where the asbestos was mined and how it was milled (6).

Biological Oxidation Systems, Volume 2

Crocidolite and amosite have been reported to have iron contents which vary from 26%-36%. Chrysotile always contains much less iron with contents varying from 0-3%.

Iron has been implicated as a causative agent in a variety of disease states, including cancer. Iron ore miners experience an increased incidence of lung cancer (7,8). Individuals with hereditary hemochromatosis (9,10) and porphyria cutanea tarda (11,12), iron overload conditions, experience an increased incidence of liver cancer. Individuals with increased body iron stores have an increased risk of all types of cancer (13-15). Iron may cause cancer by damaging DNA. Iron is known to catalyze deleterious oxidation reactions which cause DNA strand breaks, formation of 8-hydroxydeoxyguanosine (8-OH-dG), lipid peroxidation, and protein damage (16). Asbestos catalyzes many of the same reactions leading to production of lipid oxidation products (17-25), oxygen radicals (26-29), DNA strand breaks (30-32), and 8-OH-dG (30,33).

The unusual presence of low-molecular-weight iron chelates in cells is thought to lead to the production of the highly reactive •OH via the iron-catalyzed Haber-Weiss reaction. It is the production of •OH that is generally considered to be dangerous to the cell. The •OH reacts with biomolecules at diffusion-controlled rates and therefore must react near the site of its formation. Another mechanism by which iron and oxygen can initiate oxidation of lipids (and perhaps other molecules) has been proposed by Aust and co-workers (34). Their results showed an absolute requirement for Fe^{2+}, Fe^{3+}, and O_2 to initiate lipid peroxidation. Thus, they have proposed that a Fe^{2+}:Fe^{3+}:O_2 complex is the initiating species. Both of these mechanisms require iron and oxygen within the immediate proximity of the target molecule. Therefore, the location and reactivity of the iron becomes very important when considering damage to biomolecules, such as DNA. If iron from asbestos is responsible for the toxicity and carcinogenicity attributed to asbestos, then it becomes important to determine whether the iron remains associated with the asbestos and catalyzes oxidative damage, whether it must be mobilized to catalyze damage, or whether it is redox active both on and off asbestos. If the iron remains associated with the asbestos, then the oxidation reactions would occur only in the immediate vicinity of the asbestos fibers. However, if the iron can be mobilized by low-molecular-weight chelators, the potential for damage may be expanded to essentially all areas of the cell and even to surrounding cells by passage of the iron chelates through gap junctions.

This review will discuss the literature and recent data from our laboratory which provides evidence that iron is responsible for asbestos-catalyzed lipid peroxidation, oxygen radical generation, and DNA damage.

Asbestos-Catalyzed Lipid Peroxidation

Asbestos has been shown to catalyze lipid peroxidation in treated red blood cells (17), macrophages (18,19), and in rat liver microsomes in the absence (20) or presence (21) of an NADPH-generating system. Kamal and co-workers (22) found that workers exposed to asbestos had significantly higher malondialdehyde levels in blood plasma than did age matched controls. This suggested that exposure to asbestos dusts results in lipid peroxidation in humans.

Since lipid peroxidation is known to be catalyzed by iron, the involvement of iron in asbestos-catalyzed lipid peroxidation has been investigated. Gulumian and Kilroe-Smith (23) studied crocidolite-catalyzed lipid peroxidation in rat lung microsomes. They determined that Fe, Mg, Ca, and Mn were the only metals present in the crocidolite sample being used for the studies. When the metals were added independently to the microsomes, Fe was the only metal capable of catalyzing lipid peroxidation. They concluded that iron in asbestos was responsible for catalyzing lipid peroxidation. Weitzman and Weitberg (24) showed that prewashing of three different forms of asbestos with desferrioxamine B, an iron chelator, inhibited the asbestos-catalyzed lipid peroxidation. They concluded from this that the iron from asbestos was catalyzing the lipid peroxidation. Pretreatment of asbestos with desferrioxamine B also inhibited lipid peroxidation in asbestos-treated A549 lung cell line (18) and C3H10T1/2 cell line (25) in culture, further supporting the role of iron in asbestos-catalyzed lipid peroxidation. Interestingly, magnetite, a solid iron oxide which contaminates some forms of asbestos, also catalyzes rat liver microsomal lipid peroxidation (21).

Generation of Oxygen Radicals by Asbestos

Weitzman and Graceffa (26) and Eberhart and co-workers (27) showed that asbestos could catalyze the generation of $\cdot OH$ and $O_2^{\cdot -}$ when it was incubated with H_2O_2. Although direct participation of iron was not demonstrated, it was proposed that generation of the radical species was the result of the iron-catalyzed Haber-Weiss reaction. Zalma and co-workers (28) studied the generation of oxygen radicals in the absence of H_2O_2, since the bioavailability of

H_2O_2 is usually low. Using ESR spin trapping, they were able to demonstrate the generation of •OH radicals by crocidolite and chrysotile asbestos. Although no direct evidence for participation of iron was presented, they surmised that the ability to generate the •OH depended upon the availability of Fe^{2+}. Kennedy and co-workers (29) studied various forms of asbestos, known to cause pneumoconiosis, for their ability to catalyze the production of •OH by measuring the release of methane from DMSO. They observed methane production by amosite, crocidolite, and two forms of chrysotile only when a reducing agent was included. Methane production was completely inhibited by transferrin or the iron chelator desferrioxamine. Thus, the authors concluded that iron from asbestos was involved in the generation of •OH, probably via the Fenton reaction, and that generation of •OH might be responsible for the tissue damage in the lung leading to pneumoconiosis.

Asbestos-Catalyzed DNA Damage

Two types of DNA damage have been observed after treatment of DNA with asbestos *in vitro*, DNA strand breaks in the presence of H_2O_2 (30) and formation of 8-OH-dG in the presence (30) or absence (33) of H_2O_2. Using the sensitive technique of *in situ* nick translation, Libbus and co-workers (31) were able to demonstrate DNA strand breaks in the DNA of crocidolite-treated rat embryo cells. From their microscopy studies, they concluded that physical contact between the nucleus and asbestos was not required for DNA damage and speculated that oxygen free radicals generated by the asbestos might be responsible for the DNA strand breaks. Turver and Brown (25) showed that crocidolite asbestos was capable of inducing S_1 nuclease sensitive sites in the DNA of C3H10T1/2 cells. Preincubation of the asbestos fibers with desferrioxamine B before treatment of the cells reduced the number of S_1 nuclease sensitive sites. Thus, the authors concluded that asbestos can cause DNA damage in treated cells by a mechanism that depends upon the presence of iron.

We have studied the induction of DNA single strand breaks in closed-circular, supercoiled ϕX174 DNA by crocidolite asbestos. As a control $FeSO_4$ was incubated with the DNA and we observed a decrease in closed circular DNA with an increasing $FeSO_4$ concentration (Table 1.). Ascorbate or crocidolite incubated alone with the DNA did not induce strand breaks. However, when ascorbate and crocidolite were incubated together with DNA, only 70% of the closed-circular DNA remained, indicating that strand breaks had occurred. When desferrioxamine B, an iron chelator, was preincubated with the crocidolite for 15 min or 24 hr before addition to the DNA and ascorbate, complete inhibition of DNA strand break activity was observed. This strongly suggests that the iron from asbestos was totally responsible for the induction of DNA strand breaks.

Table 1. *Single Strand Breaks Induced in φX174 DNA by $FeSO_4$ or Asbestos*[a]

Additions (Concentrations) (% of control)	Closed Circular DNA
None	100
$FeSO_4$ (50 μM)	89
(100 μM)	69
(200 μM)	48
Ascorbate (1 mM)	102
Crocidolite (1 mg/ml)	102
Ascorbate (1mM) plus:	
Crocidolite (1 mg/ml)	70
Desferrioxamine B (1 mM)[b]	98
Desferrioxamine B (1 mM)[c]	99
EDTA (1 mM)[b]	30
EDTA (1 mM)[c]	3

[a] Assays contained 0.5 μg closed-circular, supercoiled φX174 DNA incubated for 30 min under the indicated conditions. Samples were subjected to electrophoresis on 0.6% agarose mini-gels and the resulting DNA bands were visualized after staining with ethidium bromide. Gels, transilluminated with UV light, were photographed and the negatives were analyzed by integrated scanning densitometry. The amount of remaining closed-circular, supercoiled DNA was calculated as a percent of the total DNA in the sample. All samples were compared by expressing each as a percent of the untreated control DNA.
[b] 15 min preincubation with asbestos.
[c] 24 hr preincubation with asbestos.

Synergistic Effects of Cigarette Smoke and Asbestos

Asbestos exposed workers who also smoke have a 10 fold higher incidence of lung cancer than those who do not smoke (35). Jackson and co-workers (32) showed that cigarette smoke and crocidolite induced strand breaks in 78% of the PM2 DNA treated, while less than 10% developed breaks when exposed to either agent alone.

They also showed that •OH was generated only in the presence of both agents. The iron chelators 1,10-phenanthroline and desferrithiocin decreased both •OH formation and DNA strand breaks. They concluded that the iron contained in asbestos was catalyzing the formation of •OH from H_2O_2 generated by cigarette smoke which led to induction of DNA strand breaks.

Recent work by Qian and Eaton (36) showed that organic condensates of cigarette smoke could leach iron from amosite or crocidolite asbestos. The chelation process required reduction of the iron by reducing agent(s) present in the smoke extracts. Once the iron was chelated, it was demonstrated that it could be transported into red blood cells where the iron might catalyze deleterious oxidation reactions. The authors concluded that the presence of iron chelators in cigarette smoke may explain the synergistic effect on asbestos carcinogenicity.

Iron Mobilization from Asbestos

Presented here have been a variety of experimental data that show that the iron associated with asbestos is capable of catalyzing oxidation reactions in lipids and DNA. If the iron were mobilized from the asbestos, the potential for damage to all parts of the cell would be much greater. Holmes and Morgan (37) have shown the leaching of minerals from the asbestos in treated animals. They treated rats with Rhodesian chrysotile which had been irradiated with a high flux of thermal neutrons to produce radionuclides. They observed high levels of ^{59}Fe in blood and liver of the treated animals indicating that iron was leached from the asbestos *in vivo*.

We have been studying the mobilization of iron from asbestos to determine what chelators are capable of mobilizing iron and how mobilization affects its redox activity. We have found that mobilization requires the presence of a chelator (38). Thus, the report of iron mobilization from asbestos *in vivo* (37) must be the result of chelation. The physiological chelators, citrate, ADP, or ascorbate, all mobilized iron (Fe^{2+} and/or Fe^{3+}) from crocidolite at respective rates of 4.2, 0.3, or 0.5 nmol iron/mg asbestos/hr (38). The chelator EDTA, which binds both Fe^{2+} and Fe^{3+}, showed the highest rate of iron mobilization, 30 nmol/mg asbestos/hr (38). Preincubation of crocidolite with EDTA before addition to ϕX174 DNA greatly enhanced the asbestos-catalyzed DNA strand break activity (Table 1). This suggests that mobilization of iron from asbestos by EDTA enhanced its ability to catalyze DNA strand breaks. This may be true for other chelators as well and will be the subject of future investigations.

Conclusions

The literature and results presented here strongly suggest that the iron in asbestos is responsible for catalyzing the formation of oxygen radicals and subsequent oxidation reactions that have been attributed to asbestos. The uncontrolled entry of iron into the cell by internalization of asbestos fibers may present a challenge for which the cell is unprepared. Reactive iron on the fibers could lead to damage to biomolecules in the immediate vicinity. Mobilization of reactive iron within the cell by low-molecular-weight chelators could lead to greater reactivity of the iron and more widespread damage throughout the cell and perhaps even into neighboring cells by translocation of these chélates through gap junctions. Through generation of the highly reactive •OH, or a similarly reactive oxygen species, the targets of this damage could be carbohydrates, lipids, proteins, and DNA. This could result in the toxicity and/or carcinogenicity that has been attributed to asbestos.

Acknowledgements

This work was supported by a Faculty Research Grant from Utah State University.

References

1. Seidman, H., Selikoff, I.J., and Hammond, E.C. (1979) *Ann. NY Acad. Sci.* **330**, 61-89
2. Warnock, M..L., and Churg, A.M. (1975) *Cancer* **35**, 1236-1242
3. Hammond, E.C. (1966) *Natl. Cancer Inst. Monogr.* **19**, 127-204
4. Selikoff, I.J., Lilis, R., and Nicholson, W.J. (1980) *Ann. NY Acad. Sci.* **330**, 295-311
5. Harrington, S., Allison, A.C., and Badami, D.V. (1975) *Adv. Pharm. and Chemotherapy* **12**, 291-402
6. Cralley, L.J., Keenan, R.G., and Lynch, J.R. (1967) *Amer. Ind. Hygiene Assoc. J.*, September-October
7. Kinlen, L.J., and Willows, A.N. (1988) *Br. J. Ind. Med.* **45**, 219-224
8. Boyd, J.T., Doll, R., Faulds, J.S., and Leiper, J. (1970) *Br. J. Ind. Med.* **27**, 97-105

9. Niederau, C., Fischer, R., Sonnenberg, A., Stremmel, W., Trampisch, H.J., and Strohmeyer, G., (1985) *N. Engl. J. Med.* **313**, 1256-1262
10. Bradbeir, R.A., Bain, C., Siskind, V., Schofield, F.D., Webb, S., Axelsen, E.M., Halliday, J.W., Bassett, M.L., and Powell, L.W. (1985) *J. Natl. Canc. Inst.* **75**, 81-84
11. Okuda, K. (1986) *Hepatology* **6**, 1054-1056
12. Salata, H., Cortes, J.M., Enriquez, De Salamanca, R., Oliva, H., Castro, A., Kusak, E., Carreno, V., and Hernandez Guio, C. (1985) *J. Hepatol.* **5**, 477-481
13. Stevens, R.G., Beasley, R.P., and Blumberg, B.S. (1986) *J. Natl. Canc. Inst.* **76**, 605-610
14. Selby, J.V., and Friedman, G.D. (1988) *Int. J. Cancer* **41**, 677-682
15. Stevens, R.G., Jones, D.Y., Micozzi, M.S., and Taylor, P.R. (1988) *N. Engl. J. Med.* **319**, 1047-1052
16. Aust, S.D., Morehouse, L.A., and Thomas, C.E. (1989) *J. Free Rad. Biol. Med.* **1**, 3-25
17. Gabor, S., and Anca, Z. (1975) *Br. J. Ind. Med.* **32**, 39-41
18. Turver, C.J., Poole, A., and Brown, R.C. (1985) *In Vitro Effects of Mineral Dusts*, (Beck, E.G., and Bignon, J., eds) NATO ASI series G3, pp. 267-273. Springer-Verlag, Berlin
19. Goodlick, L.A., Pietras, L.A., and Kane, A.B. (1989) *Am. Rev. Respir. Dis.* **139**, 1265-1273
20. Gulumian, M., Sardianos, F., Kilroe-Smith, T., and Ockerse, G. (1983) *Chem.-Biol. Interac.* **44**, 111-118
21. Fontecave, M., Mansuy, D., Jaouen, M., Pezerat, H. (1987) *Biochem. J.* **241**, 561-565
22. Kamal, A.-A.M., Gomaa, A., Khafif, M.E., and Hammad, A.S. (1989) *Environ. Res.* **49**, 173-180
23. Gulumian, M., and Kilroe-Smith, T.A. (1987) *Environ. Res.* **43**, 267-273
24. Weitzman, S.A., and Weitberg, A.B. (1985) *Biochem. J.* **225**, 259-262
25. Turver, C.J., and Brown, R.C. (1987) *Br. J. Cancer* **56**, 133-136
26. Weitzman, S.A., and Graceffa, P. (1984) *Arch. Biochem. Biophys.* **228**, 373-376
27. Eberhardt, M.K., Raman-Franco, A.A., and Quiles, M.R. (1985) *Environ. Res.* **37**, 287-292
28. Zalma, R., Bonneau, L., Guignard, J., Pezerat, H., Jaurand, M.-C. (1987) *Can. J. Chem.* **65**, 2338-2341
29. Kennedy, T.P., Dodson, R., Rao, N.V., Ky, H., Hopkins, C., Baser, M., Tolley, E., and Hoidal, J.R. (1989) *Arch. Biochem. Biophy.* **269**, 359-364
30. Kasai, H., and Nishimura (1984) *Gann* **75**, 565-566
31. Libbus, B.L., Illenye, S.A., and Craighead, J.E. (1989) *Cancer Res.* **49**, 5713-5718
32. Jackson, J.H., Schrafstatter, I.U., Hyslop, PlA., Vosbeck, K., Sauerheber, R., Weitzman, S.A., and Cochran, C.G. (1987) *J. Clin. Invest.* **80**, 1090-1095

33. Leanderson, P., Soderkvist, P., Tagesson, C., and Axelson, O. (1988) *Br. J. Ind. Med.* **45**, 309-311
34. Aust, S.D., and Svingen, B. (1982) *Free Radicals in Biology* (Pryor, W.A., ed.) Vol. 5, pp. 1-28. Academic Press, Orlando
35. Hammond, E.C., Selikoff, I.J., and Seidman, H. (1979) *Ann NY Acad. Sci.* **330**, 474-490
36. Qian, M.., and Eaton, J.W. (1989) *Arch. Biochem. Biophys.* **275**, 280-288
37. Holmes, A., and Morgan, A. (1967) *Nature* **215**, 441-442
38. Lund, L.G., and Aust, A.E. (1990) *Arch. Biochem. Biophys.* (in press)

Selective Enhancement of Free Radicals by *cis*-Unsaturated Fatty Acids in Tumor Cells as a Strategy to Kill Tumor Cells *In Vitro* and *In Vivo*

U.N. Das

Department of Medicine, The Nizams Institute of Medical Sciences
Punjagutta, Hyderabad 500 482 India

Abstract

γ-linolenic acid, arachidonic acid, and eicosapentaenoic acid (GLA, AA, and EPA respectively) and to a limited extent linoleic acid and α-linolenic acid (LA and ALA respectively) were found to have selective tumoricidal action without harming the normal cells at the concentrations tested (10-40 μg/ml/1 x 10^4 cells). This tumoricidal action of GLA, AA and EPA appeared to be mediated by free radicals and lipid peroxides but not by prostaglandins and leukotrienes possibly by damaging DNA. Intratumoral injection of GLA in patients with gliomas showed substantial decrease/lysis of tumor as evidenced by serial CT scans without significant side-effects. Similar benefit was observed on oral supplementation of LA and GLA in 6 patients with stage IV Hodgkin's lymphoma with no significant side-effects. It was also observed that GLA can prevent genetic damage induced by radiation, benzo[a]pyrene and cis-platinum *in vivo*. These results indicate that GLA has both antimutagenic and tumoricidal actions and can be exploited as a potential anticancer drug.

Introduction

It is desired that agents attack only tumor cells without exerting any adverse effect on normal cells in the chemotherapy of malignant tumors. But, this is rarely achieved with the currently available drugs and radiation. Recent studies have indicated that tumor necrosis factor (TNF), lymphotoxin (1-3) *etc.*, may have such selective cytotoxic action on tumor cells but produce significant side-effects (4). Some eicosanoids are capable of inhibiting the proliferation of some types of tumor cells (5-7). Essential fatty acids (EFAs) are precursors of eicosanoids and are important structural components of cell membranes. EFAs are essential nutrients. All EFAs are polyunsaturated fatty acids (PUFAs) but many PUFAs are not EFAs (8,9).

Thus, linoleic acid (LA) and α-linolenic acid (ALA) are EFAs where as γ-linolenic acid (GLA), dihomo-γ-linolenic acid (DGLA), eicosapentenoic acid (EPA) and docosahexaenoic acid (DHA) are PUFAs and hence, are non-essential fatty acids. But, both GLA and DGLA can perform all the functions of the parent compounds on nutritional testing and so are also called as derived EFAs (9). There are two types of EFAs, the n-6 series derived from the dietary cis-linoleic acid (LA) and the n-3 series derived from ALA. The number indicates the position of the first double bond from the omega end of the molecule. The two series are not interchangeable in animals. Dietary LA and ALA have limited biological actions, are oxidized to provide energy, and are needed to maintain integrity of the skin to prevent fluid loss. Since, all the EFAs and PUFAs are cis-unsaturated fatty acids (c-UFAs) they are referred to as c-UFAs through out the length of this article.

Metabolism of EFAs

Both the n-6 and n-3 series of EFAs seem to be metabolized by the same enzymes and is given in figure 1. The Δ-6-desaturase (Δ-6-d) enzyme is important since its activity determines ultimately the tissue levels of GLA, DGLA and 1 series PGs. AA, the precursor of 2 series PGs can also be obtained from food which is sufficient to account for daily 2 series PGs, LTs and TXs production. There is now extensive evidence that in adult humans the activity of Δ-6-d is limited and can be further interfered with by a variety of factors which is summarized in reference 9. Both Δ-6-d and Δ-5-d are present in almost all tissues of the body except RBC and epidermal cells of the skin.

Tumor Cells and c-UFAs

Some EFAs and their metabolites have been shown to inhibit both tumor and normal cell growth *in vitro* in a concentration-dependent manner (10,11) and under some well defined culture conditions GLA was reported to show a marked differential cytostatic effect against tumor cells compared to its effects on normal cells (12). Tolnai and Morgan (13) reported antitumor activity of c-UFAs *in vitro* in 1962. These results may have potential clinical significance since, tumor cells derive most of their lipids including c-UFAs from the environment (14).

Tumor cells are known to be deficient in Δ-6-d activity (15), have low amounts of PUFAs especially that of AA, a substantial decrease

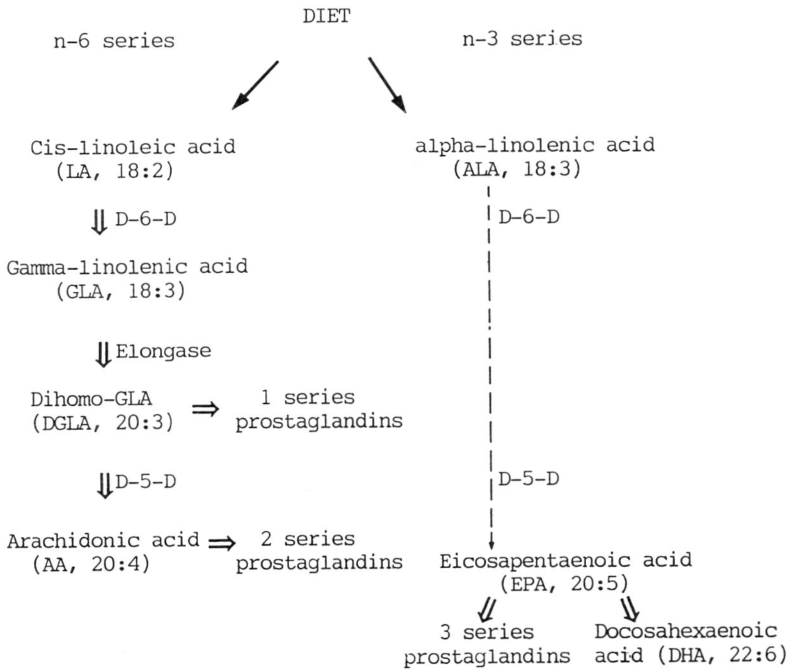

Figure 1. *Metabolism of Essential Fatty Acids (EFAs)*

in the P-450 system, and an increase in total antioxidant capacity compared to the normal cells(16-21). Studies performed by Benedetti *et al.* (17) and Player (18) suggested that lipid peroxidation is decreased in preneoplastic foci and rapidly growing neoplastic cells, where as mature cells readily show lipid peroxidation. Products formed as a result of lipid peroxidation and in particular 4-hydroxy-alkenals, have an inhibitory effect on DNA synthesis, cell division, and tumor growth (24). Thus, a decrease in lipid peroxidation may be one of the features of cancer. This is supported by the observation that mitochondrial or microsomal suspensions prepared from cancer cells peroxidize slowly (25-27). These studies coupled with the observation that tumor cells are highly susceptible to free radical-induced toxicity (16,17) suggest that if methods are designed whereby free radical generation in the tumor cells are increased, it is likely that selective tumor cell lysis can occur.

In this context, it is interesting to note that c-UFAs augment free radical generation in human neutrophils (28,29) and that this is a calmodulin dependent process (29). Further, it is well documented that Eskimos, whose diet is rich in EPA and DHA, are relatively free from cancer (30) and that the tumorigenicity of cancer cells is related to their AA content (31) *viz*: the higher the AA content the lower the invasive capacity and lower the AA content higher the invasive capacity. In view of these evidences, we studied the effect of various c-UFAs on the survival of tumor cells both *in vitro* and *in vivo* and the mechanism of their tumoricidal action. In addition, studies were also performed to look at the effect of c-UFAs on the mutagenicity of radiation, benzo[a]pyrene and cis-platinum. Some of these results have already been published (32-46) and here I have presented an over review of these studies.

Materials and Methods

Cells and Culture Conditions

Normal monkey kidney (CV-1), normal human fibroblasts (CCD-41-SK), human promyelocytic leukemia (HL-60), and human breast cancer (ZR-75 1) cells were used for these studies. The cells were seeded at 5×10^4 or 1×10^4 cells per plate or well depending on the experimental protocol and further details are given in reference 32-36. In all the experiments cell viability was tested by trypan blue dye exclusion method. The cells were checked for possible mycoplasma contamination and confirmed that there was no contamination during the period of the study.

Studies with Antioxidants and Metals

These studies were performed in 24-well culture plates. One day after seeding, 10 μg/ml or 20 μg/ml of GLA (which was used as representative of the fatty acids having tumoricidal action) was added to the cells. To these cultures antioxidants and metal salts were added as per the experimental protocol. The cultures were observed and cell viability was determined every day till day 7. These studies were done with ZR-75-1 cells as a representative of the tumor cells.

TBA-Reaction

ZR-75-1 and CV-1 cells were used in these studies. Cells were seeded at 5×10^4 cells/ml in 35 mm petri dishes and were supplemented with 20 μg/ml GLA, AA, DHA and EPA. At the end of day seven, medium and cells were harvested separately and assayed for TBA reaction as a measure of lipid peroxidation. TBA-reaction was done as described by Gavino *et al.* (47) and as describe earlier (48). The absorbance values obtained were converted to picomoles of

MDA-equivalent (MDA-eq) from a standard curve obtained with 1,1,3,3 tetramethoxypropane.

Experiments with Thymidine Labelled Cells

For these studies, CV-1, CCD-41-SK, HL-60, and ZR-75-1 cells were used. The cells were seeded at 1 x 10^4 cells/ml in 24-well culture plates. 24-hours after seeding, cells were labelled with thymidine (Specific activity 20.0 μCi/mmol) and then were supplemented with 20 μg/ml of GLA for 24 hrs. At the end of the incubation period, the supernatant was collected and counted in a counter to measure the amount of labelled thymidine released by the cells as a measure of damage to DNA.

Clinical Studies

Hodgkin's Lymphoma

Stage IV Hodgkin's disease is considered as an advanced disease which shows poor response to chemotherapy. Five patients of stage IV Hodgkin's lymphoma were selected for the study with their consent. After establishing the diagnosis and appropriate investigations all were given 6 X 500 mg capsules of evening primrose oil each containing linoleic acid 350 mg and GLA 8 mg. Further details of the study are given else where (40). All the patients were treated as outpatients and were seen at least once in 2 weeks.

Gliomas

Patients with highly malignant gliomas or who had relapse after taking radiotherapy, surgery and chemotherapy were selected for the study. After obtaining the consent of the patients they received 1 mg of GLA intratumorally every day for 10 days. CT scans were done both before and after GLA treatment.

Studies on Genetic Damage

Swiss male mice aged 7-8 weeks and weighing 24-27 gm were used for the study. One hour before and after giving the genotoxic agents such as γ-radiation, benzo[a]pyrene (BP) and cis-platinum, GLA was given intraperitoneally in saline. 5 to 6 animals were used in each group. All the animals were killed 24 to 36 hours after the treatment schedule. The bone marrow smears were prepared, stained with May-Grunwald-Giemsa and mounted in Euparol as described earlier (41,44,45). The presence of micronuclei in polychromatic erythrocytes (PCE) is an indication of genetic damage (41,47). The ratio of polychromatic to normochromatic erythrocytes (P/N) was used to estimate the effect of the treatments on the bone marrow. A P/N ratio of 1 is normal. Any increase in the number of normochromatic erythrocytes is an indication of bone marrow suppression.

Results

c-UFAs and Tumor Cells: In Vitro Studies

The effect of 20 μg/ml of various c-UFAs on the survival of human breast (ZR-75-1), normal monkey kidney (CV-1), and normal human fibroblast (CCD-41-SK) cells are given in figure 2. It is evident from these results that GLA, AA and EPA have tumoricidal action, where as DHA is without any significant tumoricidal action. Thus, AA = GLA > EPA > DHA in terms of their cytotoxic action on tumor cells. Both LA and ALA were also found to have potent cytotoxic action on ZR-75-1 cells (data not shown). But when all these fatty acids were tested on a series of tumor cells it was observed that GLA, AA, and EPA were the most potent and effective tumoricidal agents compared to ALA, LA and DHA. Lymphoma, leukemia and breast tumor cells were killed by almost all the c-UFAs (LA, ALA, GLA, DGLA, AA and EPA) except by DHA, where as other cell lines such as colon cancer cells were found to be sensitive only to GLA, AA and EPA (data not shown). Generally, the extent of cytotoxicity by c-UFAs was found to be influenced by the cell number, cell density (whether the cells were confluent or not), the type of cell line and fatty acid tested (32,33). In general, for a given sensitive cell line the results suggest that efficient cytotoxicity is obtained with c-UFA supplemented at a concentration of about 0.5 ng/cell (20 μg/ml 1 x 10^4 cells/well) relative to the seeding cell number (32,33). It is also seen that the maximum tumoricidal action of c-UFAs is seen by day 6 or 7.

The results shown in figure 3 indicate that both AA and EPA but not GLA are toxic to normal cells (CCD-41-SK) at doses 2 or 3 times the dose effective in killing the tumor cells. This is especially so when normal cells were supplemented with 60 μg/ml of AA or EPA. This suggests that GLA has more selective tumoricidal action compared to AA and EPA. Thus GLA > AA > EPA in their capacity to kill tumor cells selectively with least toxicity to normal cells.

The effect of various inhibitors or enhancers of lipid peroxidation and/or free radical generation on the tumoricidal action of GLA on human breast tumor cells is given in tables 1 and 2. It is clear from these results that antioxidants and superoxide dismutase (SOD) can effectively block the cytotoxic action of GLA (table 1). Similar results were obtained with AA and EPA (data not shown). The inhibitory

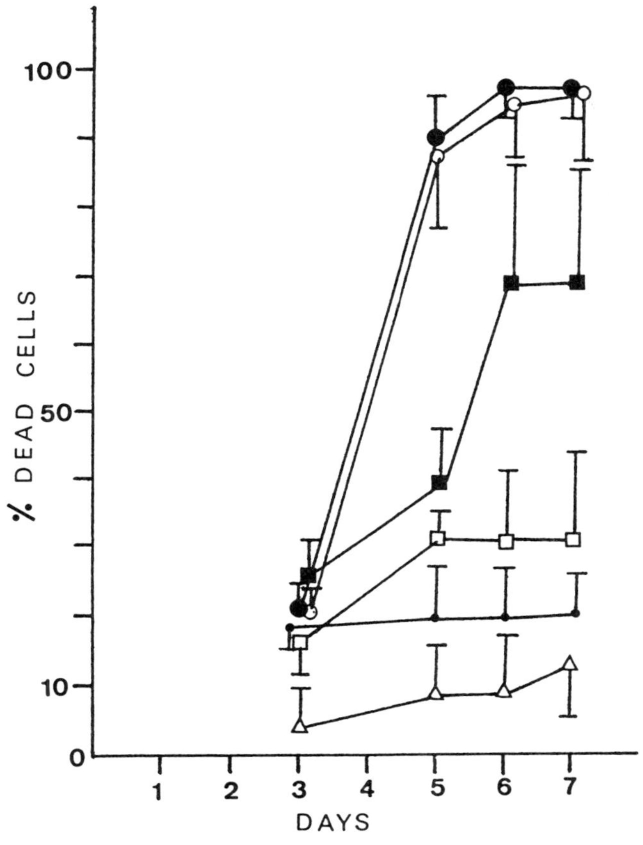

Figure 2. *Effect of 20 μg/ml of c-UFAs on the Survival of ZR-75-1, CV-1 and CCD-41 SK Cells in vitro*. Another 2 tumor cells also showed similar behavior compared to ZR-75-1 in the presence of c-UFAs.

●−● AA; ○−○ GLA; ■−■ EPA; □−□ DHA;

•-• unsupplemented ZR-75-1 cells

△−△ unsupplemented and c-UFA supplemented CV-1 and CCD-41 SK normal cells. These results are presented together as they were similar.

action of ATP is rather surprising. Both catalase and heat inactivated catalase enhanced the cytotoxic action of GLA suggesting that H_2O_2 does not participate in the tumoricidal action of c-UFAs. Both iron and copper salts, augmenters of lipid peroxidation and free radical generation (especially superoxide) (48) enhanced the cytotoxicity of GLA (table 2).

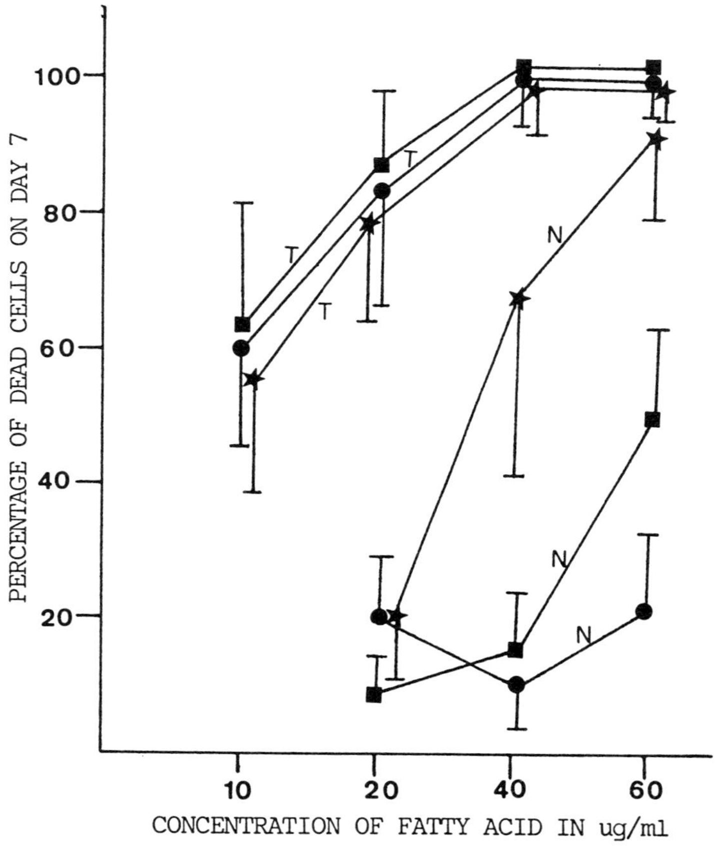

Figure 3. *Effect of GLA, AA, and EPA on the Survival of Normal (41-SK) and Tumor (ZR-75-1) Cells in vitro*

● = GLA; ■ = AA; ★ = EPA; N = Normal Cells; T = Tumor Cells

Table 3 depicts the MDA-eq detected in the medium and cells on day 7 after supplementation with 20 μg/ml of various c-UFAs. Both AA and GLA supplemented tumor cells (ZR-75-1) produced large amounts of MDA-eq (approximately 4-fold increase) compared to normal cells (CV-1) tested and controls. Both normal and tumor cells supplemented with EPA produced significantly increased amounts of

MDA-eq compared to controls. This may explain, at least in part, why normal cells were more susceptible to the cytotoxic action of EPA compared to that of GLA and AA (see figure 3).

Table 1. *Effect of Anti-Oxidants and Other Chemicals on the Cytoxicity of GLA to Human Breast Cancer Cells in vitro*

S. No.	Chemical/Enzyme	Dose Used	% of Dead Cells
1.	Control	---	20.0 ± 5.0
2.	GLA	20 μg/ml	95.0 ± 5.0
3.	Vitamin A	10 μg/ml	30.0 ± 7.0
4.	Vitamin E	10 μg/ml	10.0 ± 5.0
5.	Uric acid	1×10^{-4} M/ml	40.0 ± 10.0
6.	Cis-retinoic acid	1×10^{-5} M/ml	65.0 ± 18.0
7.	Superoxide dismutase	3×10^{-6} M/ml	20.0 ± 7.0
8.	ATP	200 μM/ml	22.0 ± 9.0
9.	Glutathione peroxidase	0.01 μg/ml	28.0 ± 8.0

All chemicals were tested along with 20 μg/ml of GLA on ZR-75-1 cells (1×10^4 cells/ml). All cultures were harvested on day 7 of the addition of GLA with and without the inhibitors.

Table 2. *Effect of Iron and Copper Salts and Catalase on the Cytotoxicity Induced by GLA on Human Breast Tumor Cells*

S. No.	Chemical/Enzyme	Dose Tested	% of Dead Cells
1.	Control	---	20.0 ± 5.0
2.	GLA	10 μg/ml	60.0 ± 15.0
3.	Catalase	1000 U/ml	80.0 ± 12.0
4.	Heat-inactivated Catalase	1000 U/ml	77.0 ± 8.0
5.	$FeCl_2$	20 μg/ml	98.0 ± 2.0
6.	$FeCl_3$	20 μg/ml	95.0 ± 5.0
7.	$CuSO_4$	4 μg/ml	78.0 ± 14.0

All chemicals were tested along with 10 μg/ml of GLA on ZR-75-1 cells (1×10^4 cells/ml). All the cultures were harvested on day 7 of addition of GLA with and without other chemicals.

Table 3. *Effect of Various c-UFAs on MDA-eq Formation in Normal and Tumor Cells on Day 7*

Cell line	Treatment	pmoles of MDA-eq	
		In the Cells	In the Medium
CV-1	Control	0.5 ± 0.3	0.3 ± 0.1
(Normal monkey	GLA	6.2 ± 1.1	1.1 ± 0.3
kidney cells)	AA	7.1 ± 0.4	2.8 ± 1.0
	EPA	15.0 ± 2.7	3.3 ± 0.5
	DHA	11.4 ± 3.0	10.1 ± 0.3
ZR-75-1	Control	0.2 ± 0.3	0.9 ± 0.2
(Human breast	GLA	26.5 ± 2.7	10.6 ± 2.4
cancer cells)	AA	35.9 ± 10.4	3.5 ± 1.6
	EPA	16.7 ± 1.7	10.2 ± 0.6
	DHA	11.3 ± 3.9	14.4 ± 0.4

All values are expressed as Mean ± S.D. of three seperate estimations. For other details see *Materials and Methods*.

The results given in table 4 indicate that GLA-treated tumor cells but not normal cells released significant amounts of thymidine compared to controls. This suggests that GLA can, possibly, damage DNA of tumor cells. But, this needs to be confirmed by other methods of assessing DNA damage.

Table 4. *Effect of 20 μg/ml of GLA on the Release of Thymidine from Radio-labelled Normal and Tumor Cells in vitro*

S.No.	Cell Line	Control	GLA Treatment
1.	CV-1	887.6 ± 82.5	916.5 ± 68.6
2.	CCD-41-SK	546.2 ± 41.2	564.7 ± 59.4
3.	HL-60	2903.5 ± 396.5	3446.3 ± 302.6*
4.	ZR-75-1	1579.4 ± 211.6	1995.2 ± 291.1*

*$p < 0.05$ compared to the control value. All values are in cpm. The details of the experiments are given in *Materials and Methods*.

Clinical Studies

The results of the clinical trial in Hodgkin's disease given in table 5 indicates that oral supplementation of LA and GLA can induce partial to complete remission of the disease (40). Further studies are needed to confirm these preliminary results.

Gliomas

Patients who had histological and radiological evidence of a malignant glioma, clinical and/or radiological evidence of tumor progression at least 1 month after the patient's last course of chemotherapy or radiotherapy and a tumor of sufficient size to warrant further therapy were selected for the study. Patients then underwent a neurosurgical procedure as indicated by their tumor. Patients were treated with a craniotomy for removal of the tumor (to the extent possible) followed by the placement of a catheter into the postoperative tumor cavity. Two weeks after surgery a CT scan was obtained which was compared with subsequent studies. Each patient received 1 mg of GLA in saline on alternative days/or daily to a maximum of 10 mg. The next day of the last injection, repeat CT scan was performed.

So far 6 patients have been treated by this regimen. All 6 have shown considerable regression in the size of the tumor and a typical case is given in figure 4. There were no side-effects during the therapy.

GLA and genetic damage

The results shown in figure 5 indicate that GLA is also capable of preventing genetic damage induced by cis-platinum to the bone marrow cells of mice. Similar results were obtained with BP and radiation (41,45). This genoprotective action of GLA was not due to a delay or arrest of mitosis of the bone marrow cells of mice (data not shown) and at the same time GLA did not interfere with the tumoricidal action of cis-platinum (data not shown).

Discussion

It is evident from the results presented here and elsewhere (32-36) that GLA, AA and EPA and to a limited extent LA and ALA have selective tumoricidal action and that free radicals and lipid peroxidation process has a role in this action. The enhancing effect of copper and iron lends support to this contention. The demonstration of an increase in the intracellular MDA-eq in tumor cells supplemented with GLA, AA and EPA adds strength to this concept. The increased amounts of MDA-eq observed in normals cells

Table 5. Details of the Clinical Data and Response to Oral Supplementation of LA and GLA in Hodgkin's Diseases.*

Patient	Age, Sex and Date of Hospitalization	Clinical Summary	State of Disease and Histology	Time Taken for Remission and Side Effects
1.	55, Male, February 15, 1984	Generalized pruritis, generalized lymphadenopathy, hepato-spleenomegaly	IV A, lymphocyte predominance	8 weeks, urticaria and diarrhoea
2.	62, Male, March 7, 1984 and August 12, 1984	H/O recurrent fever, loss of weight, generalized pruritis, generalized lymphadenopathy, hepato-spleenomegaly and mediastinal lymphadenopathy.	IV B, mixed cellularity	12 weeks, diarrhoea
3.	50, Male, June 9, 1984	Generalized lymphadenopathy, hepato-spleenomegaly	IV A, lymphocyte predominance	partial remission, no side-effects
4.	60, Male, June 18, 1984	Generalized lymphadenopathy, multiple subcutaneous nodules on the medial side of both arms and back (2 cm in diameter)	IV A, nodular sclerosis	partial remission, no side-effects
5.	50, Female, August 18, 1984	Generalized pruritis, generalized lymphadenopathy and hepato-spleenomegaly	IV A, lymphocyte predominance	6 weeks, no side-effects

* Further details can be seen in reference 40.

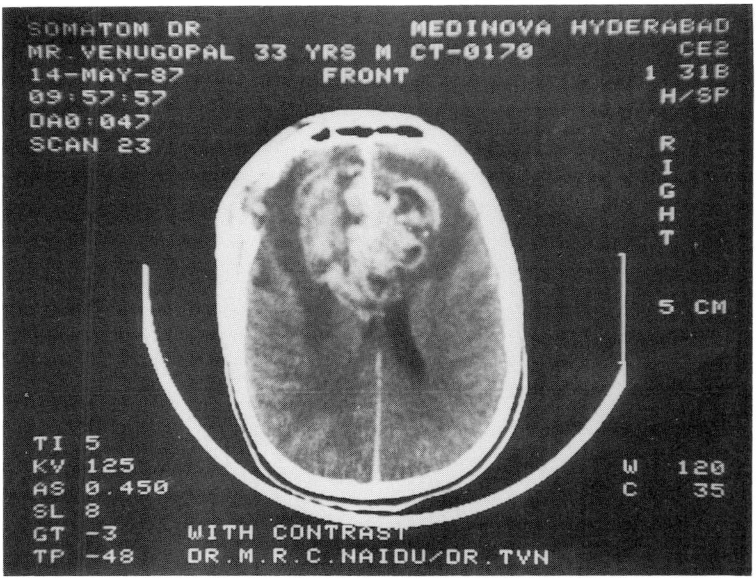

Figure 4a. *CT Scan of Head Showing Size of the Brain Tumor Before the Intra-tumoral Injection of GLA*

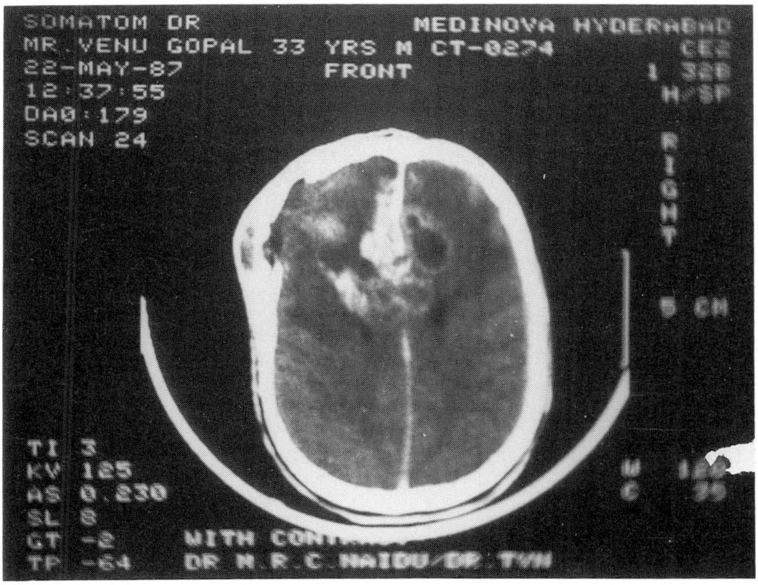

Figure 4b. *CT Scan of Head Showing Brain Tumor After the Intra-tumoral Injection of GLA.* The tumor has shown considerable reduction in the size. The hypodense area of the tumor suggests necrosis of the tumor.

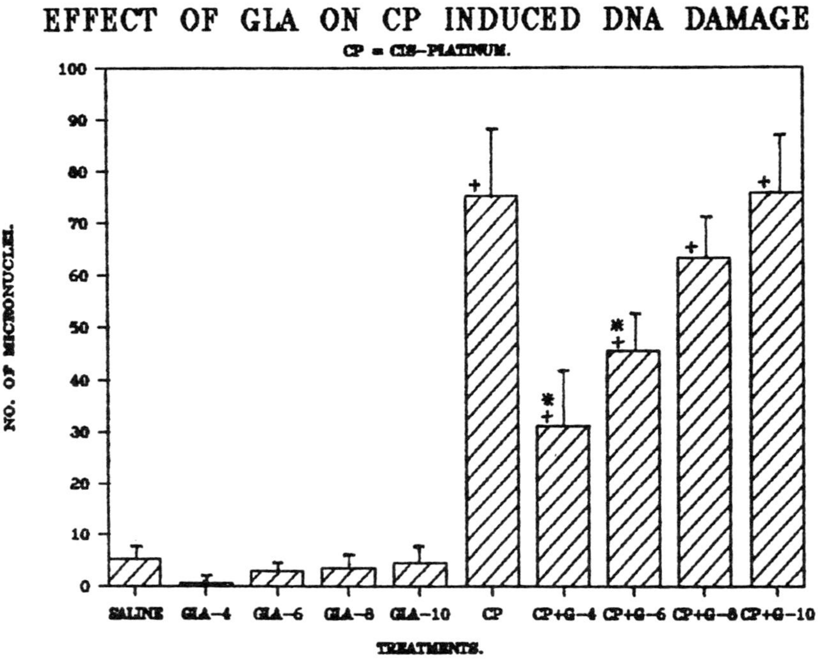

Figure 5. *Effect of GLA on Genetic Damage Induced by Cis-platinum to the Bone Marrow Cells of Mice as Assessed by Micronucleus Test.* Details of the test system is given in references 41, 44 & 45. GLA was given intraperitoneally one hour before CP. Cis-platinum was given intraperitoneally 25 μg/animal.

+ $P < 0.005$ Compared to Saline Group
* $P < 0.005$ Compared to Cis-Platinum Group
CP = Cis-Platinum
G = GLA, G-4, G-6, G-8, G-10 refers to GLA 10^{-4}, 10^{-6}, 10^{-8}, 10^{-10} M

supplemented with EPA suggests that increased lipid peroxidation and/or free radical generation in the cells beyond a limit can induce cell death since, EPA is toxic to normal cells at higher concentrations (figure 3). The inability of DHA to kill tumor cells (figure 2) in spite of its high degree of unsaturation and ability to under peroxidation readily may be due to its unstable nature and/or its ineffectiveness to increase lipid peroxidation in the cells (table 3).

It is known that free radicals can damage DNA, and/or cause induction of protein and nucleic acid cross-linking by lipid degradation products. The increase in the release of labelled thymiding by GLA treated tumor cells indicates that there could be damage to DNA.

The observed cytotoxicity of c-UFAs to tumor cells is not just a laboratory curiosity. Experiments performed in AKR mice (data not shown here) and the limited clinical studies depicted here suggest that GLA and possibly, other c-UFAs can indeed kill tumor cells even in an *in vivo* situation. The clinical studies are particularly interesting since no side-effects were observed. As the patients selected for the clinical studies had advanced malignancy and already took radiotherapy and/or chemotherapy but still had a tumor of sufficient size to warrant further therapy and had evidence of tumor progression, their response to GLA can be considered as significant. It is possible that if GLA is administered at a much early stage of the disease perhaps the response may be more impressive.

In addition, GLA was found to be capable of preventing genetic damage-induced by γ-radiation, BP and cis-platinum *in vivo* (41, 44,45). Both dose-response and time-course studies revealed that GLA is effective from 10^{-4} M to 10^{-8} M when given intraperitoneally and that a single dose of GLA can prevent DNA damage up to 48 to 72 hours after its administration (data not shown, 44). Further, GLA seems to be able to protect the bone marrow cells from the damaging action of γ-radiation and prevent bone marrow suppression. If this is true, GLA may be tried as a bone-marrow protective agent in patients receiving chemotherapy and radiation.

It was observed that GLA does not interfere with the tumoricidal action of cis-platinum *in vitro* (unpublished data). Thus, it looks as though GLA is a selective tumoricidal agent with ability to prevent bone marrow suppression induced by radiation and other agents. Though both AA and EPA also have selective tumoricidal action they do not show the genoprotective property of GLA. In fact, AA was found to be genotoxic by itself (43).

The results of these studies are interesting since, it is known that several anticancer drugs and lymphokines have the ability to enhance free radical generation and lipid peroxidation in the cells (39,49) and in the patients following chemotherapy (49). This indicates that free radicals and free radical-dependent lipid peroxidation process may be a common pathway by which tumor cells are eliminated both by the natural products of the immune cells and drugs. Thus, methods designed to specifically enhance free radical generation in the tumor cells may form a novel approach to cancer therapy. One such method seems to be c-UFA delivery to tumor cells. Though the possible interactions between c-UFAs, prostaglandins, lymphokines, immune system, free radicals, oncogenes and cell proliferation and cell death have not been discussed here, there seems to be a close link or interaction between these biologicals and biological processes and has already been discussed in detail elsewhere (16,37,51).

Acknowledgements

This work was supported by grants from the Department of Science and Technology, Indian Council of Medical Research, Indian National Science Academy, and Department of Atomic Energy of India to the author. During the tenure of this study the author was in receipt of INSA Research Fellowship.

References

1. Pennica, D., Nedwin, G.E., Hayflick, J.S., *et al.* (1984) *Nature* **312**, 724-729
2. Gray, P.W., Aggarwal, B.B., Benton,C.V., *et al.* (1985) *Nature* **312**, 721-724
3. Ransom, J.H., Evans, C.H., McCabe,R.P., *et al.* (1985) *Cancer. Res.* **45**, 851-862
4. Barba, D., Saris, S.C., Holder, C., *et al.* (1989) *J. Neuro. Surg.* **70**, 175-182
5. Tanaka, H., Yamamuro, T., Matsumoto, M., *et al.* (1985) *Prostaglandins* **30**, 167-173
6. Sakai, T., Yamaguchi, N., Shiroko, Y., *et al.*. (1984) *Prostaglandins* **27**, 17-26
7. Kobayashi, H. Ojima, M., Satoh, H. (1985) *Prostaglandins* **29**, 255-268
8. Sinclair, H.M. (1980) *Lancet.* **i**, 414
9 Das, U.N., Horrobin, D.F., Begin, M.E., *et al.*. (1988) *Nutrition* **4**, 337-341
10. Cornwell, D.G., Morisaki, N. (1984) in *Prostaglandins, Lipid Peroxides and Co-Oxidation Reactions* (Pryor, W.A., ed) Vol.5, pp. 95-148. Academic press, New York
11. Smith, D.L., Willis, A.L., Mahmud, I. (1984) *Prostaglandins Leukotrienes. Med.* **16**, 1-10
12. Dippenaar, N., Booyens,J., Fabbri, D., *et al.* (1982) *S. Afr. Med. J.* **62**, 505-509
13. Tolnai, S., Morgan, J.F. (1962) *Can. J. Biochem. Physiol.* **40**, 869-875
14. Reitz, R.c., Thompson, J.A., Morris, H.P. (1977) *Cancer. Res.* **37**, 561-567
15. Dunbar, L.M., and Bailey, J.M. (1975) *J. Biol. Chem.* **250**, 1152-1153
16. Das, U.N. (1987) *Med. Sci. Res.* **15**, 1485-1490
17. Galeotti, T., Bartoli, G.M., Bartoli, S., and Bertoli, E. (1982) in *Biological and Clinical Aspects of Superoxide and Superoxide Dismutase* (Bannister , J.V., ed) pp. 106-117. *Proc. Eur. Biochem. Soc. Symposium No.* **62**, Rome
18. Cheeseman, K.H., Buston, G.W., Ingold, K.U., Slater, T.F (1984). *Toxicol. Pathol.* **12**, 235-239
19. Cheeseman, K.H., Collins, M., Proudfoot, K., *et al.* (1986) *Biochem. J.* **235**, 507-514
20. Slater, T.F., Benedetto, C., Burton, G.W., *et al.* (1984) in *Eicosanoids and Cancer* (Thaler. Dao, H., *et al..*, eds) pp. 21-29. Raven Press, New York

21. Dianzani, M.U., Canuto, R.A., Rossi, M.A., *et al.*. (1984) *Toxicol. Pathol.* **12**, 189-199
22. Benedetti, A., Malvadi, G., Fulceri, R., Comporti, M.C.,(1984) *Cancer. Res.* **44**, 5712-5717
23. Player, T.J. (1982) in *Free radicals, Lipid Peroxidation, and Cancer* (McBrien, D.C.H. and Slater, T.F., eds) pp. 173-195, Academic Press, New York
24. Slater T.F., Cheeseman, K.H., and Proudfoot, K. (1984) in *Free Radicals in Moleculr Biology, Ageing, and Disease* (Armstrong,D., Sohal,R.S., Cutler,R.G., and Slater, T.F, eds) pp. 293-305. Raven Press, New York
25. Ahmed, S.M. and Slater, T.F. (1981) in *Recent Advances in Lipid peroxidation and Tissue Injury* (Slater, T.F. and Garnet, A., eds) pp. 177-194, Brunel University, Uxbridge
26. Burlakova, E.B. (1975) *Russian. Chem. Rev.* **44**, 871-880
27. Cheeseman, K.H. (1982) Ph.D thesis, Brunel Univerisity, Uxbridge
28. McPhail, L.C., and Synderman, L. (1984) *Contemp. Topics Immunobiol.* **14**, 247-281
29. Sangeetha, P., Das, U.N., Koratkar, R. (1989) *Prostaglandins Leukotrienes. Essential Fatty Acids*, in press
30. Editorial (1983) *Lancet* **i**, 1139-1141
31. Roos, D,S., and Choppin, P.W. (1984) *Proc. Natl. Acad. Sci., USA,* **81**, 7622-7626
32. Begin, M.E., Das, U.N., Ells, G., Horrobin, D.F. (1985) *Prostaglandins Leukotrines. Med.* **19**, 177-186
33. Begin, M.E., Ells, G., Das, U.N., Horrobin, D.F. (1986) *J. Natl. Cancer Inst.* **77**, 1053-1062
34. Begin, M.E., Ells, G., Das, U.N. (1986) *Anti Cancer Res.* **6**, 291-294
35. Das U.N., Huang, Y.S., Begin,M.E., Horrobin, D.F. (1987) *Free Radical Biol. Med.* **3**, 9-14
36. Das,U.N., Begin, M.E., Ells, G., Huang, Y.S., Horrobin, D.F. (1987) *Biochem. Biophys. Res. Commun.* **145**, 15-24
37. Das, U.N., Huang, Y.S., Begin, M.E., Horrobin, D.F. (1986) *IRCS. Med. Sci.* **4**, 1069-1074
38. Das, U.N., Begin, M.E. (1988) *Speculations. Sci. Tech.* **11**, 111-114
39. Das, U.N., Ells, G., Begin, M.E., Horrobin, D.F. (1986) *J. Free Radicals. Biol. Med.* **2**, 183-188
40. Singh, N.K., Das, U.N., Srivastava, P.K. (1987) *J. Assoc. Physicians India.* **35**, 137-138
41. Das, U.N., Devi, G.R., Rao, K.P., Rao, M.S. (1985) *Prostaglandins* **29**, 911-920
42. Das, U.N., Devi, G.R., Rao, K.P., Rao, M.S. (1985) *IRCS. Med.Sci.* **13**, 316-317
43. Das,U.N., Ells, G., Begin, M.E.(1987) *Nutrition Rep. Int.* **36**, 1267-1270
44. Das, U.N., Devi, G.R., Rao, K.P., Rao, M.S. (1989) *Prostaglandins*, in press

45. Das, U.N., Devi, G.R., Rao, K.P., Rao, M.S. (1985) *Nutrition Res.* !, 101-106
46. Das, U.N. (1989) *Nutrition* **5**, 106-110
47. Eisen, V., Walker, D.I. (1976) *Br. J. Pharmacol.* **57**, 527-532
48. Halliwell, B., Gutterridge J.M.C. (1986) *Biochem.J.* **219**, 1-14
49. Sangeetha, P., Das,U.N., Koratkar, R. (1989) *Free Radical. Biol. Med.,* in press
50. Berton, G., Zeni, L., Cassatella, M.A., Rossi, F. (1986) *Biochem. Biophys. Res. Commun.* **138**, 1276-1282
51. Das, U.N., Devi, G.R. (1988) in *Eicosanoids and Radiation* (Polgar,P., ed) pp. 107-118. Kluwer Acadamic Publisher, Boston

Inhibition of Binding of Aflatoxin B_1 to Renal DNA by Dehydroepiandrosterone

H.R. Prasanna, M.H. Lu, K.D. Nakamura, and R.W. Hart
Department of Health and Human Services, Food and Drug Administration, National Center for Toxicological Research, Jefferson, AR 72079 USA

Abstract

Dehydroepiandrosterone (DHEA), a naturally occurring adrenal steroid, has been shown to inhibit both spontaneous and chemically induced tumors in several species. In our earlier studies, we found that DHEA had a profound influence on the cytochrome P-450 mediated biological oxidation of the hepatocarcinogens dimethyl-nitrosamine (NDMA) and aflatoxin B_1 (AFB_1), as well as the mammary carcinogen 7,12-dimethylbenz(a)anthracene (DMBA), in the liver, resulting in their increased interaction with hepatic proteins followed by a concomitant decrease in their binding to DNA. In the present study, we have investigated the influence of DHEA on the metabolism of AFB_1 in kidney, lung, and brain tissues, the known recipients of this carcinogen in the body. The weights of kidney and lung increased ($p<0.05$) and that of brain decreased significantly ($p<0.001$) in DHEA-fed young male Fischer 344 rats. Total tissue, microsomal and cytosolic proteins, DNA content and cytosolic glutathione transferase activity of these tissues in DHEA-fed rats were not different from that of the controls. DHEA feeding decreased the total amount of AFB_1 bound to renal DNA and protein, but did not have any effect on the binding of AFB_1 to either tissue proteins or nucleic acids in lung and brain. In comparison with our earlier studies with liver, where DHEA protected hepatic DNA from damage by AFB_1 due to increased activation of AFB_1, the results of this study suggest that a similar effect of DHEA in the protection of kidney DNA might occur by a different mechanism.

Introduction

Chemoprevention of cancer by various natural and synthetic compounds has been a subject persued with great interest by many investigators (1). Dehydroepiandrosterone (DHEA), a naturally occurring adrenal steroid, has gained prominence in this area of study by

Biological Oxidation Systems, Volume 2

virtue of its ability to inhibit both chemically induced and spontaneous tumors (2). The levels of both DHEA and its sulfate in plasma and urine of humans have been inversely correlated with susceptibility to breast cancer (2) and cardiovascular diseases (3). DHEA is known to inhibit chemically induced tumors in lung and colon of mice and block phorbol ester promotion of skin tumors in mice (4). The effects of DHEA in protecting various species against chemically induced tumors are believed to be due to the non-competitive inhibition of glucose- 6-phosphate dehydrogenase (2), resulting in the depletion of NADPH, an obligate co-factor required for the metabolic activation of several carcinogens.

Recent studies from our laboratories have shown that DHEA has a profound influence on the cytochrome P-450 mediated oxidation of the hepatocarcinogens like dimethylnitrosamine and aflatoxin B_1, as well as the mammary carcinogen, 7,12-dimethylbenz(a)anthracene in the liver, resulting in their increased interaction with hepatic proteins followed by a concomitant decrease in their binding to DNA (5-9). These studies also indicated that there was increased total cytochrome P-450 content in the livers of DHEA-fed rats. This observation was subsequently confirmed by the elegant work of Dr. Estabrook and his co-workers (10). *In vitro* studies from our laboratories have suggested that DHEA interacts with the hepatic microsomes from rat resulting in the enhancement of the metabolic activation of AFB_1 (11).

In order to examine whether DHEA affords protection to non-hepatic tissues from the hepatocarcinogen AFB_1 (which also affects non-hepatic tissues) (12), in the present study we have investigated the influence of DHEA on the metabolism of AFB_1 in kidney, lung, and brain, of male Fischer 344 rats.

Materials and Methods

Chemicals

DHEA, NADPH, glutathione (GSH), and ribonuclease A were purchased from Sigma Chemical Co., St. Louis, MO. 1-Chloro-2,4-dinitrobenzene (CDNB) was acquired from Eastman Kodak Co., Rochester, NY. AFB_1 (>98% pure) and [^{3}H]AFB_1 (24 Ci/mmol) were obtained from Moravek Biochemical Inc., Brea, CA. Reverse phase HPLC of AFB_1 and [^{3}H]AFB_1 indicated high purity of those compounds (>98%).

Treatment of Animals

Male 2-month-old Fischer 344 rats were obtained from NCTR breeding colony and housed individually under a 12 hr light/dark cycle at 25 ± 2° and 50 ± 4% relative humidity. The rats were fed either a control diet (#0822) or 0.8% DHEA-containing diet prepared by Bioserve Inc., Frenchtown, NJ. After two weeks, each animal received a single intraperitoneal injection of $[^3H]AFB_1$ (600 μg/kg, 200 mCi/mmol) that had been dissolved in 200 μl dimethylsulfoxide. After three hours, the rats were sacrificed by decapitation and the kidneys, lung, and brain were removed, washed with 250 mM sucrose, frozen in liquid nitrogen and stored at -80°C.

Preparation of Subcellular Fractions

The tissues were thawed, minced, and mixed with eight volumes of ice cold 250 mM sucrose, 10 mM Tris-HCl, pH 7.4. All subsequent operations were conducted at 4°C. The tissue suspensions were homogenized with a Potter-Elvehjem homogenizer and centrifuged at 600 x g for 10 min to sediment a crude nuclear fraction. The supernatant was centrifuged at 12,000 x g for 15 min to separate mitochondria. The post-mitochrondrial supernatant was then centrifuged at 105,000 x g for 60 min to sediment microsomes. The resultant supernatant was designated cytosol.

Nucleic Acid and Protein Isolation

DNA was isolated from a protein of pooled kidney homogenates by a phenolic extraction procedure (13) and was quantified by the method of Burton (14). Protein was precipitated from sample homogenates with 10% trichloroacetic acid (TCA), and was made free of DNA, RNA, and lipid as described by Munro and Fleck (15). The residue was dissolved in 0.1 N NaOH and the protein content was determined by the method of Lowry *et al.* (16).

The amount of AFB_1 bound to DNA and protein was determined by the radiometric method. In the case of lung and brain, the perchloric acid extract of the TCA precipitates of the homogenates represented total nucleic acids and AFB_1 associated with them was estimated.

GST Assay

Cytosolic glutathione S-transferases (GST), which are involved in the modulation of binding of AFB_1 to DNA (17-20), were assayed as previously described (21) using CDNB as the substrate.

Statistical Analyses

Statistical analyses were conducted using Students t test.

Results

Dietary DHEA caused significant increases in the weights of kidney and lung ($p<0.02$ and $p<0.05$, respectively) but a decrease in brain weight ($p<0.001$). However, there was no significant change in either the protein or DNA content of these tissues, nor of the protein contents of the microsomal or cytosolic fractions. The cytosolic glutathione transferase activity of these tissues was also not different from those of control and DHEA-fed animals (Tables 1-3).

Table 1. *Effect of DHEA on Kidney Weight and Some of Its Constituents*

	Control	DHEA-fed
Kidney weight per 100 g body weight (g)	0.80 ± 0.04[a]	0.86 ± 0.08[b]
Protein mg/g tissue		
Homogenate	176.0 ± 27	164.0 ± 23
Microsomal	9.6 ± 1.5	7.5 ± 1.3
Cytosolic	68.0 ± 11	58.0 ± 13
DNA mg/g tissue	1.6 ± 0.1	1.3 ± 0.2
Cytosolic GST activity/mg protein	29.0 ± 4	31.0 ± 0.9

[a] Control rats were pair fed. The total amount of DHEA consumed per animal during 14 days was 0.88 g. Kidneys from 2 animals were pooled to generate one sample for biochemical analysis. Tissue weight, and protein and DNA were obtained from 8 animals. DNA and glutathione transferase assays were from 4 samples in each group. Analysis were conducted in duplicate for each sample. The results are presented as the mean ± standard deviation. GST activity is expressed as nmol of CDNB conjugated per minute at 25°C.

[b] $p < 0.02$

However, since these results do not indicate the influence of DHEA on the efficacy of the metabolic activation of AFB_1 in these tissues, the effect of dietary DHEA on the *in vivo* binding of AFB_1 to nucleic acids and proteins of these tissues was followed. The results are presented in Tables 4 and 5. The results show that the binding of AFB_1 to renal DNA is significantly reduced (40%) either per mg of DNA or for total DNA in the tissue. This trend is further reflected in the lower binding of AFB_1 to the renal total and cytosolic protein (31 and 38% inhibition, respectively). These parameters were not significantly different in lung and brain tissues of control and DHEA-fed animals.

Table 2. *Effect of DHEA on Lung Weight and Some of Its Constituents*

	Control	DHEA-Fed
Lung weight per 100 g body weight (g)	0.55 ± 0.05[a]	0.59 ± 0.05[b]
Protein (mg/g tissue)		
Homogenate	143.0 ± 35	130.0 ± 19
Microsomes	2.8 ± 0.9	2.9 ± 0.6
Cytosolic	55.0 ± 22	58.0 ± 13.7
DNA (mg/tissue)	2.7 ± 0.4	2.4 ± 0.2
Cytosolic GST activity/mg protein	22.0 ± 3.1	23.0 ± 2.2

[a] Experimental details were the same as described in *Materials and Methods* and in Table 1.

[b] $p < 0.05$

Table 3. *Effect of DHEA on Brain Weight and Some of Its Constituents*

	Control	DHEA-fed
Brain weight per 100 g body weight (g)	1.06 ± 0.04[a]	0.93 ± 0.03[b]
Protein (mg/g tissue)		
Homogenate	155.0 ± 46	148.0 ± 27
Microsomal	3.4 ± 0.5	2.6 ± 0.3
Cytosolic	65.0 ± 13	59.0 ± 12
DNA (mg/g tissue)	0.42 ± 0.08	0.5 ± 0.06
Cytosolic GST activity/mg protein	48.0 ± 10	45.0 ± 12

[a] Experimental details were the same as described in *Materials and Methods* and in Table 1.

[b] $p < 0.001$

Table 4. *Effect of DHEA on the Binding Of AFB_1 to DNA*

	pmol AFB_1 Bound per mg DNA		pmol AFB_1 Bound per Whole Tissue DNA or Total Nucleic Acids	
	Control	DHEA-fed[a]	Control	DHEA-fed
Kidney	12.3 ± 1.3	7.3 ± 1.8[b]	19.7 ± 2.1	9.5 ± 2.3[c]
	total pmol AFB_1 Bound to PCA Extractable Nucleic Acids		**pmol AFB_1 Bound per Whole Tissue DNA or Total Nucleic Acids**	
	Control	DHEA-fed	Control	DHEA-fed
Lung	49 ± 12	40 ± 3	113 ± 21	111 ± 47
Brain	35 ± 17	28 ± 5	42 ± 10	49 ± 9

[a] The results are presented as the mean ± standard deviation. Analyses were conducted on duplicate samples from individual samples. Each group consisted of 4 samples in each group.

[b] $p<0.01$

[c] $p<0.001$

Table 5. *Effect of DHEA on the Binding Of AFB_1 to TCA Precipitable Macromolecules*[a]

	pmols Bound to Total Tissue Protein			
	Homogenate		Cytosol	
Tissue	Control	DHEA-fed[b]	Control	DHEA-fed
Kidney	356. ± 21	247. ± 31[c]	164. ± 17	101. ± 8[c]
Lung	561. ± 66	585. ± 143	264. ± 44	191. ± 29
Brain	45. ± 8	33. ± 9	31. ± 7	26. ± 4

[a] The results are presented as the mean ± standard deviation. Analyses were conducted on duplicate samples from individual animals. Each group consisted of 4 samples (tissues from 2 animals per sample in each group).

[b] Bound refers to radioactivity associated with TCA-precipitable proteins which were freed from DNA, RNA, and lipids.

[c] $p < 0.005$, as compared with the controls.

Discussion

Following our earlier studies on the effects of DHEA on the biochemistry of livers in Sprague Dawley and Fischer 344 rats, and its profound influence on the cytochrome P-450 mediated activation of various carcinogens and the protection it afforded to hepatic DNA from NDMA, DMBA, and AFB_1-induced damage (5-9), in the present study we have chosen to examine the effect of the steroid on the bioactivation and subcellular effects of AFB_1 in kidney, lung, and brain. Additionally, we have measured the binding of AFB_1 to DNA in kidneys as this is considered to be an important step in the initiation of carcinogenesis. The results presented in this study show that the effect of the steroid on these tissues is not as dramatic as in liver (5). Of the three non-hepatic tissues studied, only in kidney was the binding of AFB_1 to both DNA and protein significantly inhibited (Tables 4 and 5). Unlike liver, where a reduction of AFB_1 binding to DNA was partly due to increased protein binding of AFB_1 (which was increased in the tissue due to dietary DHEA) as well as increased detoxication of AFB_1 due to increased amounts of GST in the tissue (which modulate the binding of AFB_1 to DNA) (17-20), the absence of such changes in the kidneys of DHEA-fed rats suggest that the reduced binding of AFB_1 to both renal DNA and protein may be due to an overall decrease in the metabolic activation of AFB_1.

Several possibilities exist for the decreased metabolic activation of AFB_1 in kidney by DHEA:

(a) The inhibition of glucose-6 phosphate dehydrogenase (G6PDH) by DHEA causing less generation of NADPH, an obligate cofactor for the mixed function oxidases for the activation of carcinogens, resulting in the inhibition of the activation of carcinogens (2).
(b) Alteration in the efficacy of the isozyme of cytochrome P-450 mediating the activation of AFB_1 in kidney.
(c) Status of detoxification of AFB_1 in the tissue.
(d) Limited transport of the carcinogen to the extrahepatic tissues.
(e) Alteration in renal DNA repair.
(f) Or a combination of any of the above mentioned possibilities.

With the data at hand, it is difficult to rule out any of these possibilities. However, certain general observations can be made. The absence of information regarding the inhibition of G6PDH by DHEA in non-hepatic tissues and the invalidity of such an inhibition in the liver as a viable mode of action of DHEA in that tissue (22-23), so far as carcinogen activation is concerned (5-9), all cast a doubt regarding DHEA action through that step in kidney. However, this possibility cannot be ruled out. The interaction of DHEA with microsomal activation of AFB_1 as demonstrated earlier in our laboratories (11) could operate in kidneys also, which could result in the alteration of the efficacy of cytochrome P-450 function. This, along with a possible modulation in the levels of the isozyme of cytochrome P-450 mediating the activation of AFB_1, could result in decreased activation of AFB_1, which could protect renal DNA from AFB_1 induced damage. However, quantitation of cytochrome P-450 in the renal tissue should resolve this tissue. Although the GST activity in the kidney cytosols of DHEA-fed rats are not different from the pair-fed controls, the measurement of efficacy of modulation AFB_1 binding to DNA by renal cytosols along with the glutathione levels is needed to understand the status of this important detoxication system in DHEA-fed rats.

In a related study, we have observed that the amount of the AFB_1 in these tissues was the same between control and DHEA-fed rats (data not shown), indicating that the limitation of transport of AFB_1 to kidneys of DHEA-fed rats is not a factor in causing reduced AFB_1 binding to renal DNA. Additionally, since the animals in the present study were exposed to AFB_1 for only a short period of time (3 h), it is unlikely that reduction of binding of AFB_1 to DNA is due to alteration in DNA repair. However, this possibility cannot be ruled out. The observation that DHEA protects both hepatic and renal DNA damage by AFB_1 with no significant differential effect on the binding of AFB_1 to nucleic acids and protein in lung and brain tissues suggest the absence of any possibility of an organotopic effect of AFB_1 by dietary DHEA. The results of this study lend further support to the role of DHEA, an endogenous adrenal steroid, as a chemopreventive agent.

Acknowledgments

We thank Pat Rutherford for help in the preparation of this manuscript.

References

1. Wattenberg, L.W. (1985) *Cancer Res.* **45**, 1-8
2. Schwartz, A.G., Whitecomb, J.M., Nyce, J.W., Lewbart, M.L., and Pashko, L.L. (1988) *Adv. Cancer Res.* **51**, 391-424
3. Bulbrook, R.D., Hayward, J.L., and Spicer, C.C. (1971) *Lancet* **ii**, 395-398

4. Barret-Conner, E., Khaw, K.T., and Yen, S.S. (1986) *New Engl. J. Med.* **315**, 1519-1524
5. Prasanna, H.R., Lu, M.H., Beland, F.A., and Hart, R.W. (1989) *Carcinogenesis* **10**, 2197-2200
6. Prasanna, H.R., Hart, R.W., and Magee, P.N. (1989) *Biochem. J.* **262**, 985-988
7. Prasanna, H.R., Magee, P.N., Harrington, G.W., and Hart, R.W. (1989) *J. Toxicol. & Env. Hlth.* **27**, 467-476
8. Prasanna, H.R., Hart, R.W., and Magee, P.N. (1989) *Carcinogenesis* **10**, 953-955
9. Prasanna, H.R., Hart, R.W., and Magee, P.N. (1989) *Toxin Reviews* 8(1), in press
10. Wu, H.W., Masset-Brown, J., Tweedie, D.J., Mulewich, L., Frenkel, R.A., Martin-Wixtrom, C., Eastabrook, R.W., and Baugh, R.A. (1989) *Cancer Res.* **49**, 2337-2343
11. Prasanna, H.R., Heflich, R.H., Minor, T.Y., Lu, M.H., and Hart, R.W. (1989) *Proc. Am. Assoc. Cancer Res.* **30**, 356
12. Busby, W.F. Jr. and Wogan, G.N. (1984) *Aflatoxins in Chemical Carcinogens* (Searle, C.E., ed) ACS Monograph 182, pp. 945-1136
13. Swann, P.F. and Magee, P.N. (1968) *Biochem. J.* **110**, 39-47
14. Burton, K. (1956) *Biochem. J.* **62**, 315-322
15. Munro, H.N. and Fleck, A. (1966) *Methods in Biochemical Anal.* **14**, 113-176
16. Lowry, O.H., Rosebrough, N.M., Farr, A.L., and Randall, R.J. (1951) *J. Biol. Chem.* **193**, 265-275
17. Prasanna, H.R., Lotlikar, P.D., Hacobian, N., and Magee, P.N. (1986) *Cancer Letts.* **33**, 259-267
18. Lotlikar, P.D., Insetta, S.M., Lyons, P.R., and Jhee, E.C. (1980) *Cancer Lett.* **9**, 143-149
19. Coles, B., Meyer, D.J., Ketterer, B., Stanton, C.A., and Garner, R.C. (1985) *Carcinogenesis* **6**, 693-697
20. Neal, G.E. and Green, J.A. (1983) *Chem. Biol. Inter.* **45**, 259-275
21. Habig, W.H., Pabst, M.J., and Jacoby, W. (1974) *J. Biol. Chem.* **149**, 7130-7139
22. Casazza, J.P., Schaffer, W.T., and Veech, R.L. (1986) *J. Nutr.* **116**, 304-310
23. Cleary, M.P., Fox, N., Lazin, B., and Billheimer, J.T. (1985) *Nutr. Res.* **5**, 1247-1257

CHAPTER VII

Enzymes in Prostaglandin and Leukotriene Metabolism

Prostaglandin H Synthase

Lawrence J. Marnett
Departments of Biochemistry and Chemistry, Center in Molecular Toxicology
Vanderbilt University, School of Medicine, Nashville, TN 37232 USA

Introduction

Prostaglandin H (PGH) synthase catalyzes the first two steps in prostaglandin, thromboxane, and prostacyclin biosynthesis (1-6). The first step is oxygenation of polyunsaturated fatty acids to hydroperoxy endoperoxides (cyclooxygenase activity) and the second step is reduction of the hydroperoxy endoperoxides to hydroxy endoperoxides (peroxidase activity) (2,7). The most common substrate for PGH synthase is arachidonic acid which is oxygenated to PGG_2 and PGH_2 (eq 1). The oxygenation step catalyzed by the cyclooxygenase represents the committed step in the biosynthesis of prostaglandins and thromboxanes but it is not rate-limiting in intact cells because release of substrate fatty acid by lipases must precede oxygenation (8,9). The peroxidase step is important for maximizing the flow of arachidonic acid to final metabolites. The K_m's of the isomerases that convert the cyclic peroxide to prostaglandins are lower for PGH_2 than for PGG_2, which implies PGH_2 is the preferred substrate (10,11). In addition, the peroxidase minimizes irreversible inactivation of PGH synthase by PGG_2 and other hydroperoxides (12).

Equation 1

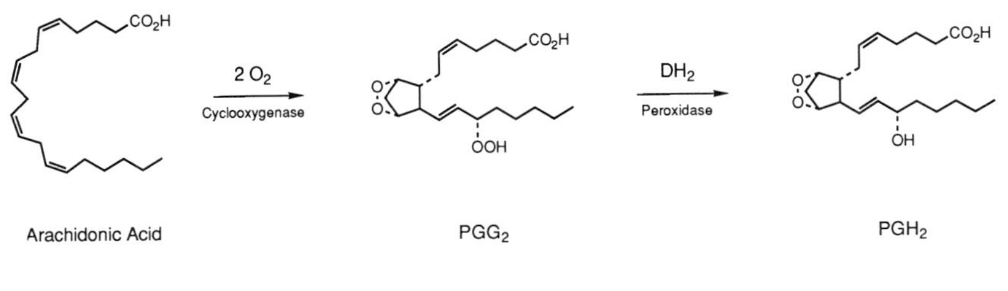

The cyclooxygenase activity of PGH synthase appears to be the *in vivo* target for non-steroidal anti-inflammatory agents such as aspirin, ibuprofen, indomethacin, acetaminophen, and phenylbutazone (13-15). A variety of mechanisms are responsible for inhibition. Aspirin acetylates a single serine hydroxyl group which apparently blocks binding of arachidonic acid to the oxygenase active site (16,17). Ibuprofen binds stereospecifically, but reversibly, to the protein, thereby inhibiting its activity (18). Indomethacin also binds tightly to the enzyme but, thereafter, causes time-dependent, irreversible loss of activity (19). The irreversible loss of activity may be due to a conformational change although the details are unclear. Phenylbutazone is oxygenated to a hydroperoxide by the peroxidase and then the hydroperoxide irreversibly inactivates the protein (20).

Steroids such as dexamethasone and cortisol are potent anti-inflammatory agents but do not inhibit cyclooxygenase activity *in vitro* (21). Recent work by Bailey and associates indicates that in certain cell types (e.g., aortic smooth muscle cells) steroids significantly depress synthesis of cyclooxygenase protein (22). Raz *et al.* have found that the glucocorticoids dexamethasone and triamcinolone acetonide prevent induction of PGH synthase in human dermal fibroblasts treated with interleukin-1 (23). The time course of the inhibitory effects suggests steroids act by inhibition of translation of PGH synthase mRNA. If this effect of glucocorticoids is general, it provides a mechanism for limitation of eicosanoid synthesis that complements the enzyme inhibition exhibited by non-steroidal anti-inflammatory compounds.

The peroxidase activity of PGH synthase oxidizes xenobiotics to reactive derivatives (24). These derivatives bind covalently to protein or nucleic acid and exert deleterious biological effects. Some of the compounds activated to mutagens or toxins are dihydrodiols of polycyclic hydrocarbons, aromatic amines, heterocyclic amines, aflatoxins, synthetic estrogens, trimethadone, and nitrofurans (25-36). The central role of PGH synthase in generating bioactive eicosanoids, the chemical transformations it catalyzes, its modulation by numerous pharmacological agents, and its potential involvement in toxicity and carcinogenesis makes it an important and challenging enzyme for study.

Occurence and Induction

PGH synthase is widely distributed in animal tissue (37). It exists at reasonably high levels in cells of the vasculature (platelets, endothelial cells) which accounts for a varying percentage of the activity detected in a tissue homogenate. The percentage of all cells that actually contain the enzyme is estimated to be around 20% (38). Sheep seminal vesicles are the preferred tissue for preparative biochemistry. In these glands, PGH synthase comprises approximately 5% of the total microsomal protein.

Emerging evidence suggests that PGH synthase levels fluctuate considerably in response to cell stimuli. Hormonal and membrane-active agents increase production of various eicosanoids by stimulation of the activity of phospholipases that release substrate fatty acids from membrane stores (39). As synthesis proceeds, PGH synthase undergoes self-catalyzed inactivation due to reaction of the protein and its heme prosthetic group with oxidants generated by cyclooxygenase and peroxidase catalytic action (12,40,41). In vascular smooth muscle cells in culture, serum effects a rapid synthesis of new PGH synthase protein following self-catalyzed inactivation (42). The molecule primarily responsible for this response appears to be epidermal growth factor and the epidermal growth factor response seems to be synergistically enhanced by transforming growth factor β (42). Epidermal growth factor also stimulates new PGH synthase protein synthesis in cultured human amnion cells (43). Other agents shown to stimulate PGH synthase protein synthesis include interleukin-1, the tumor promoter tetradecanoyl phorbol acetate, and tumor necrosis factor (23,44-46).

Immunohistochemical analysis of mouse 3T3 cells reveals intense staining for PGH synthase in the endoplasmic reticulum and nuclear membranes (47). Analysis of other cell types occasionally places the enzyme in the plasma membrane (48). PGH synthase is considered to be an integral membrane protein because it requires non-ionic detergent for solubilization. Topological models propose the protein is inserted in the endoplasmic reticulum with its amino terminus directed toward the lumen and its carboxyl terminus directed toward the cytoplasm (49).

Structural Properties

PGH synthase is readily purified from bovine or sheep seminal vesicles by a combination of detergent solubilization and ion-exchange chromatography (1-3,5-6,50-53). The enzyme from sheep seminal vesicles is isolated as a homodimer of 70 kDa subunits (2). It contains 3.5% carbohydrate of the high mannose type (54). It also contains variable amounts of ferric protoporphyrin IX (Fe^{3+}-PPIX) (55). The heme group is essential for both cyclooxygenase and peroxidase activities but is readily lost during the solubilization and chromatography steps used in purification. Cyclooxygenase and peroxidase activities are reconstituted by addition of Fe^{3+}-PPIX (4,56). Reconstitution with Mn^{3+}-PPIX produces an enzyme with high cyclooxygenase activity but very low peroxidase activity (<3% of Fe^{3+}-PPIX-reconstituted enzyme) (4).

The stoichiometry of heme binding to PGH synthase has been difficult to establish but recent estimates place it in the range of 0.5-1.0 heme/subunit (57,58). The iron is in the ferric oxidation state in resting enzyme; it is approximately 80% high-spin (59). Electronic spectra suggest the fifth ligand is histidine whereas electron paramagnetic resonance spectra have been alternatively interpreted to indicate the presence of histidine or tyrosine as the fifth ligand (59,60). During cyclooxygenase and peroxidase catalytic turnover, the heme center cycles from ferric to ferryl (Fe^{4+}) (59,61,62). There is no evidence for the production of the ferrous oxidation state during turnover although it is readily produced by chemical reductants (56).

The protein sequence of sheep seminal vesicle PGH synthase is available from its cDNA sequence (63-65). The open reading frame codes for 600 amino acids of which the first 24 represent a signal sequence. The mature protein (M_r = 66,175 omitting glycosyl) contains four consensus glycosylation sites and preliminary studies suggest at least three of the four sites are glycosylated (54,63). Application of computer algorithms to the derived protein sequence infers the existence of several hydrophobic regions. The two strongest are the signal sequence of the immature protein and Val^{267}-Leu^{282} of the mature protein (63,64). Although the requirement for detergent solubilization suggests PGH synthase is an integral membrane protein, application of the algorithm of Klein *et al.* predicts PGH synthase is a peripheral membrane protein (64).

The PGH synthase sequence exhibits no homology to globin proteins and only weak homology to several cytochromes P-450 (63). Its sequence is 22% identical over a stretch of 254 amino acids to a region of myeloperoxidase and 50% identical from Thr^{279} to Ile^{311} (63). This may indicate the existence of a heme-binding sequence in this region. DeWitt *et al.* have proposed that His^{309} is the proximal ligand to iron and H^{202} is the distal ligand (108).

Treatment of apoPGH synthase with trypsin (1% w/w) under non-denaturing conditions cleaves the protein into 33 and 38 kDa fragments that are separable by reversed phase HPLC (67). The 33 kDa fragment contains the amino-terminus of the intact protein (67). The amino-terminal sequence of the 38 kDa fragment places the trypsin cut between Arg^{253}-Gly^{254} (67). Reconstitution of apoPGH synthase with heme prior to trypsin treatment prevents cleavage of the protein (67-69). A similar protective effect against tryptic cleavage is provided by Mn^{3+}-PPIX (70). This indicates that iron and manganese porphyrins exert a significant conformational effect on the Arg^{253}-Gly^{254} peptide bond. ApoPGH synthase cleaved at Arg^{253}-Gly^{254} by trypsin appears to have lost one binding site for heme (66).

The ability of porphyrins to protect against proteolysis parallels their ability to stimulate incorporation of acetyl residues from acetylsalicylic acid into Ser^{506} (70). Acetylation of holoPGH synthase is approximately 100-fold faster than acetylation of apoPGH synthase (71). Heme-dependent alterations in protein conformation may account for the increased rate of acetylation. The apparent correspondence of heme effects on tryptic cleavage and aspirin acetylation may indicate the heme-binding site and Ser^{506} are close to each other in three-dimensional structure even though they are separated by over 250 residues in linear sequence.

Binding of the non-steroidal anti-inflammatory agents indomethacin, ibuprofen, flufenamic acid, and flurbiprofen to apoPGH synthase also protects against cleavage by trypsin (71). A close correspondence was observed between the enantioselectivity of protection against tryptic cleavage and inhibition of cyclooxygenase activity for the (R)- and (S)-enantiomers of flurbiprofen. It was assumed in this study that the site of trypsin cleavage of apoPGH synthase was Arg^{253} although it was not verified by sequencing (71). If the assumption is correct, it indicates that certain cyclooxygenase inhibitors, as well as the heme prosthetic group, alter the conformation of the Arg^{253}-Gly^{254} peptide bond so that it is not cleavable by trypsin.

Catalytic Properties

In 1967 Hamberg and Samuelsson proposed that PGH synthase oxygenates unsaturated fatty acids by a mechanism that is essentially enzyme-catalyzed lipid peroxidation (72). The *pro-S* hydrogen atom at C-13 is stereospecifically removed generating a carbon radical that couples to O_2 at C-11. Serial cyclization and trapping of a second molecule of O_2 followed by hydrogen donation from another molecule of fatty acid produces the hydroperoxy endoperoxide PGG_2. Extensive model studies and spin trapping experiments have validated the Hamberg and Samuelsson model, which is reproduced in Figure 1 (73-81). Although the series of molecular transformations appears complex, the O_2 trapping and cyclization steps proceed precisely as predicted from chemical model studies with the exception that the stereochemistry of the side chain attachment to carbons 8 and 12 is *trans* in the naturally occurring PGG_2 whereas model studies yield *cis* (79,80,82). This stereochemical alteration can be accomodated by postulating that the enzyme holds the fatty acid substrate molecule in the kinked conformation shown in the Figure. If the *pro*-S hydrogen is removed as shown, O_2 trapping and cyclization will occur to produce PGG_2 with the correct stereochemistry. The identity of the oxidizing agent on the protein that removes the hydrogen is uncertain although it has recently been proposed that it is a tyrosyl radical produced by oxidation of a side chain tyrosine by a higher oxidation state of the peroxidase (*vide infra*) (62).

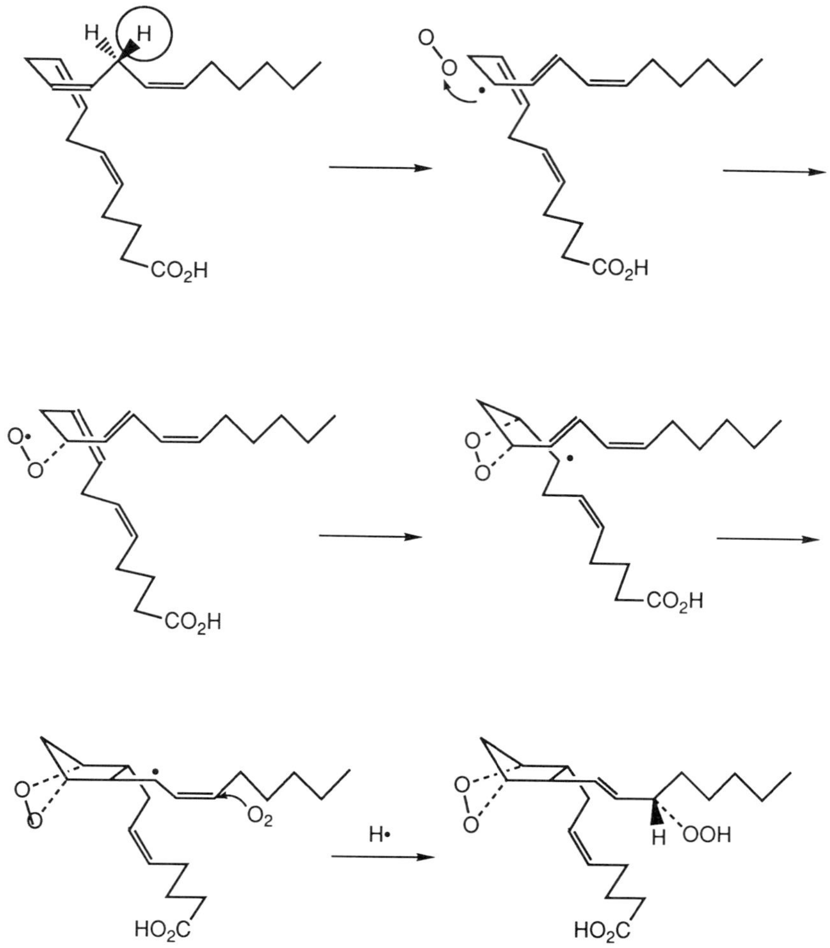

Figure 1. *Mechanism of Oxygenation of Arachidonic Acid by Cyclooxygenase*

The second reaction catalyzed by PGH synthase is the reduction of the hydroperoxy endoperoxide, PGG_2, to the hydroxy endoperoxide, PGH_2. Theoretically, the hydroperoxide could be reduced by a heme peroxidase or glutathione peroxidase. A peroxidase activity was detected in ram seminal vesicle microsomes that oxidizes organic molecules concomitant with arachidonic acid oxygenation (24,83). The peroxidase activity copurifies with the cyclooxygenase activity from sheep and bovine seminal vesicular gland microsomes and is an integral part of PGH synthase as shown by biochemical and immunochemical methods (7,84,85). The heme prosthetic group is essential for peroxidase activity. As stated earlier, apoprotein reconstituted with Fe^{3+}-PPIX exhibits peroxidase activity but apoprotein reconstituted with Mn^{3+}-PPIX does not (4).

The peroxidase of PGH synthase preferentially reduces fatty acid hydroperoxides (7). It is much less active toward H_2O_2 and almost inactive toward tertiary hydroperoxides such as cumene hydroperoxide and t-butyl-hydroperoxide (7). Reduction of the diagnostic substrates 5-phenyl-4-pentenyl-1-hydroperoxide (PPHP) and 10-hydroperoxy-8,12-octadecadienoic acid (10-OOH-18:2) demonstrates that PGH synthase peroxidase catalyzes concerted two-electron hydroperoxide reduction (12,66). This corresponds to the finding that addition of PPHP to purified PGH synthase generates a spectral intermediate that appears to be a two-electron oxidized form of the enzyme (61). Sequential one-electron reduction of this intermediate regenerates resting enzyme. The catalytic cycle is summarized in Figure 2. The overall process of turnover is analogous to that of the plant enzyme horseradish peroxidase, which proceeds via the intermediacy of higher oxidation states called Compound I and Compound II (86). These represent forms of the enzyme in which the heme prosthetic group is oxidized by two and one electrons relative to resting enzyme. The high turnover numbers of the peroxidase of PGH synthase and its mechanistic similarity to other peroxidases argue strongly that this activity evolved to fulfill a specific function and is not an adventitious activity resulting from the presence of heme on the cyclooxygenase.

When apoPGH synthase is treated with trypsin, the protein is cut at a single position producing 33 and 38 kDa fragments (*vide supra*) (67). The fragments remain tightly associated and addition of heme reconstitutes cyclooxygenase but not peroxidase activity (66). In addition to the absence of peroxidase activity, the trypsin-cleaved enzyme has one less heme binding site than the native protein (66). Incubation of trypsin-cleaved enzyme with diagnostic hydroperoxide substrates reveals an alteration in mechanism of hydroperoxide reduction from heterolytic to homolytic (66). This experiment demonstrates the close association between peroxidase catalytic efficiency and heterolytic peroxide bond scission.

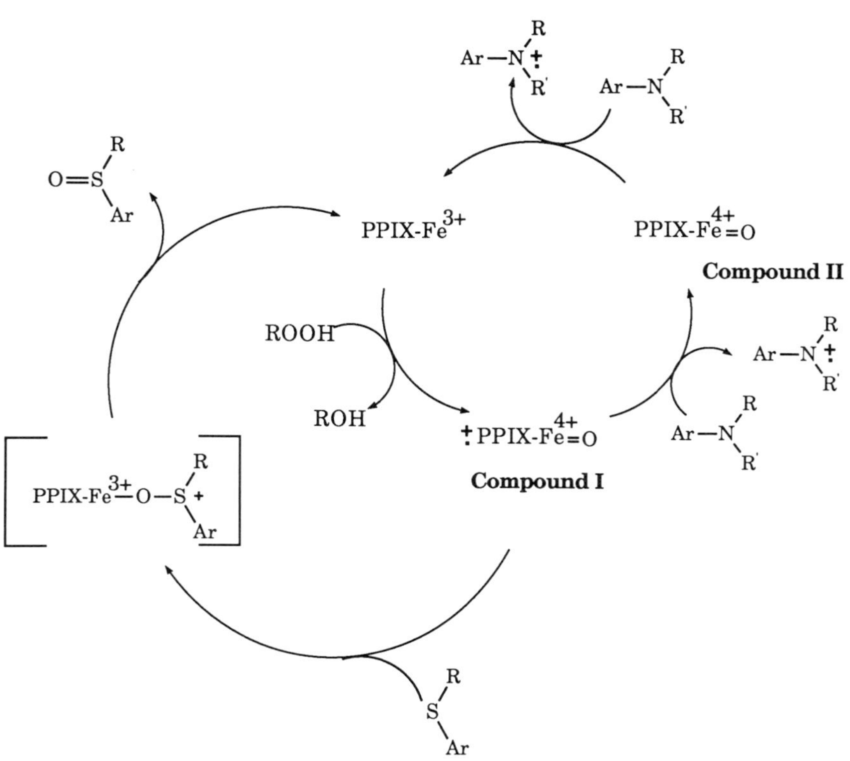

Figure 2. *Peroxidase Catalytic Cycle*

For catalytic turnover of PGH synthase peroxidase to occur, the higher oxidation states of the enzyme must be reduced. As with most peroxidases, a wide variety of compounds serve as reducing substrates. The two general mechanisms of reduction of the enzyme are electron transfer and oxygen transfer. Electron transfer usually occurs by two sequential one-electron transfers, although there is evidence for simultaneous operation of one- and two-electron reduction by acetaminophen (87). One-electron oxidized derivatives of several reducing substrates have been detected by EPR spectroscopy (88-90). Although electron transfer occurs at a site on the protein, the electron-deficient forms of the reducing substrates appear to diffuse into the bulk phase where they react to form stable products.

A detailed kinetic picture of reducing substrate specificity is not available. The extent of N-dealkylation of N-alkyl-aromatic amines suggests electron-releasing substituents enhance reduction and electron-withdrawing substituents retard reduction (91). This is supported by a study of the ability of a variety of compounds to support the catalytic reduction of PPHP (12). Electron-rich molecules are the best reductants but the presence of polar functionalities in the molecules affects their reducing substrate activity (12). For example, aromatic amines are relatively good substrates for horseradish peroxidase but very poor substrates for PGH synthase (92, 93). Some of the best electron transfer substrates for PGH synthase are polyhydroxylated compounds (hydroquinone, epinephrine) and aromatic diamines (benzidine, tetramethylphenylenediamine) (12). Among a series of naturally-occurring compounds, the best substrates are epinephrine and uric acid. The physiological reducing substrate may vary with tissue but speculation has centered on uric acid (94, 95). It is only a fair reducing substrate but is present in high concentrations (~300 μM) in human plasma (96, 97). Likewise, glutathione is a relatively poor reducing substrate but is present in very high concentrations in most cells (~6 mM) (98). Glutathione is also capable of reducing electron-deficient derivatives of highly active reducing substrates for PGH synthase which may provide a novel mechanism for reduction involving catalytic oxidation and reduction of a reducing substrate (27). The ultimate source of electrons would be glutathione but it would not be directly oxidized by the peroxidase.

The stable products of reducing substrates oxidized by electron transfer are determined by the chemistry of their electron-deficient derivatives. Most of the molecules donate a single electron and then the one-electron oxidized product diffuses from the enzyme. In the case of aromatic amines, some of the radical cations are stable and can be used to quantitate peroxidase turnover (83). In other cases, the radical cations undergo disproportionation to produce one molecule of the parent aromatic amine and one molecule of a two-electron oxidized derivative (eq 2) (99). The overall oxidation of the parent amine appears to be a two-electron oxidation but it actually occurs by one-electron oxidation followed by disproportionation. Phenols and catechols are oxidized to phenoxyl radicals which couple or disproportionate (eqs 3 and 4). Quinone oxidation products can add H_2O to form polyphenolic compounds that are also oxidized by the peroxidase (eq 5). This can lead to complex products (e.g., pyrogallol to purpurogallin). If the one-electron oxidized derivative of the reducing substrate contains a carbon-centered radical, it can couple to O_2 to form a peroxyl radical (eq 6). Peroxyl radicals abstract hydrogen from reactive donors or transfer oxygen to non-aromatic double bonds. The latter reaction may play a role in metabolic activation of precarcinogens to ultimate carcinogens (100).

Equation 2

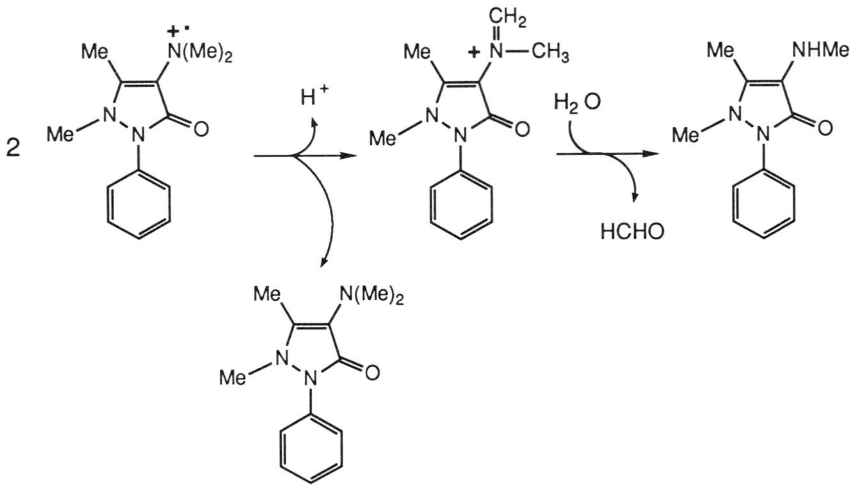

Equation 3

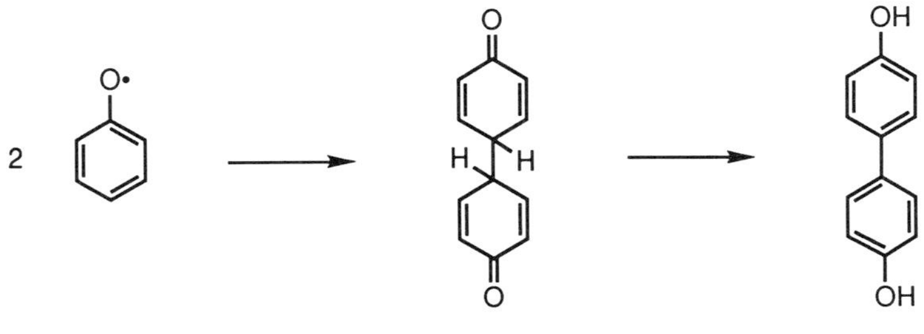

The peroxidase of PGH synthase catalyzes oxygen transfer from hydroperoxides to alkyl aryl sulfides (eq 7) (101,102). The results of experiments with isotopically-labeled hydroperoxides reveal the hydroperoxide oxygen is quantitatively incorporated into the sulfoxide oxygen (101). As with electron transfer reductants, the ability of sulfides to serve as substrates for the peroxidase is a function of electron density, steric hindrance, polarity, etc. (103).

Equation 4

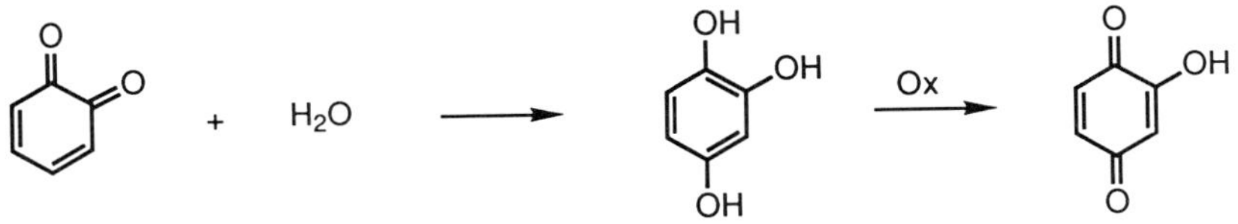

Equation 5

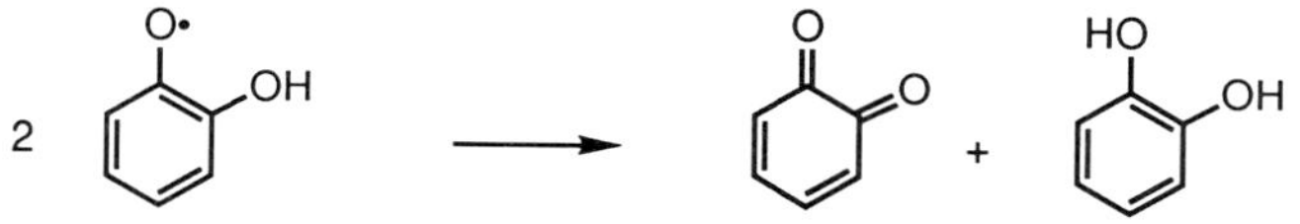

Equation 6

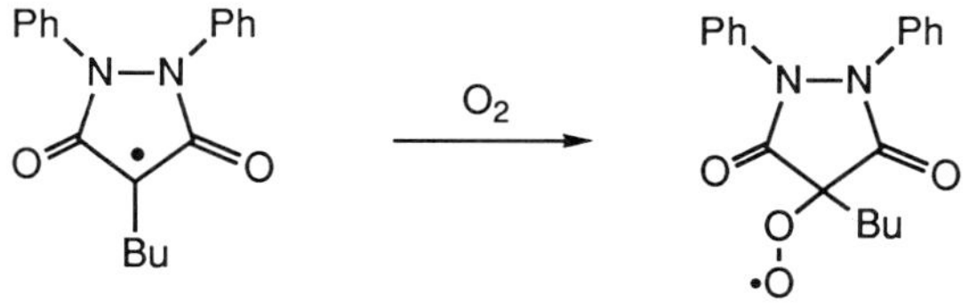

Equation 7

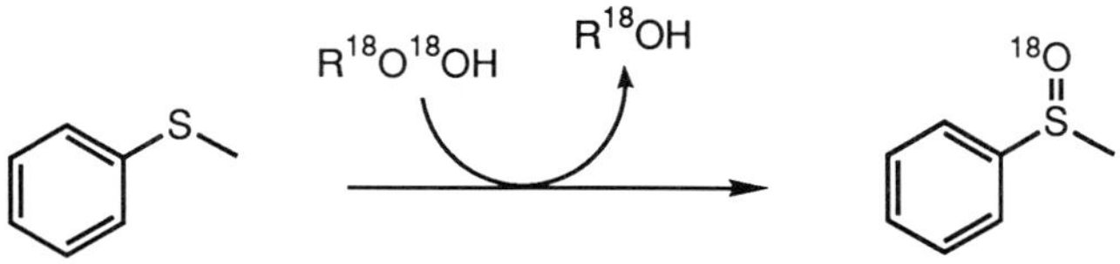

Thioanisoles substituted in the *para* position with amino and dimethylamino are much less effective reductants than anticipated, especially when compared to their half-wave potential for electrochemical oxidation (103). This is because the mechanism changes from oxygen transfer to electron transfer. For example, when *p*-(dimethylamino)-thioanisole is incubated with PGH synthase and PPHP at 37°C a bright blue color develops quickly that persists for about 3 min. The color is due to the formation of a radical cation centered on nitrogen. This radical is an intermediate in N-demethylation. *p*-(Methylamino)thioanisole and formaldehyde are produced as the sole products of oxidation; no sulfoxide is detected. The difference in the atom oxidized (N vs S) is consistent with the greater ease of N-oxidation.

Only a small difference in reducing substrate activity is observed when the methyl group of thioansiole is replaced by an ethyl group but introduction of an isopropyl group on the sulfur dramatically reduces the ability of the molecule to act as a reducing substrate (103). This suggests that the isopropyl group sterically hinders the attack of the ferryl-oxo compound I on the adjacent sulfur. The sensitivity of PGH synthase to steric effects in alkyl aryl sulfide oxidation seems to reflect a moderately restricted heme pocket or channel, intermediate in size between those of horseradish peroxidase and cytochrome P-450.

Relation of Cyclooxygenase to Peroxidase

PGH synthase is a hemeprotein with no more than one heme group per subunit that catalyzes oxygenation of polyunsaturated fatty acids and reduction of fatty acid hydroperoxides. The heme prosthetic group is required for both activities. Although we have a reasonable understanding of the mechanisms of both the cyclooxygenase and peroxidase activities, their structural and functional interrelation is uncertain. Docosahexaenoic acid, a competitive inhibitor of arachidonic acid binding to the cyclooxygenase, does not inhibit binding of lipid hydroperoxides to the peroxidase (104). This implies that arachidonic acid and PGG_2 bind to different sites on the protein. Markey *et al.* reported that several peroxidase reducing substrates which display enantiospecific inhibition of cyclooxygenase activity do not exhibit enantiospecific reduction of the peroxidase (12). For example, the L-enantiomer of etodolac is a poor inhibitor of cyclooxygenase (I_{50} > 300 μM) compared to the D- enantiomer (I_{50} = 2 μM), whereas the reducing substrate efficiency of the L-enantiomer in the PPHP assay is about

10% higher than that of the D-enantiomer. This implies that the site on the protein where oxidation of arachidonic acid occurs is different from the site where oxidation of peroxidase reducing substrates occurs.

Substrate oxidation in the cyclooxygenase reaction requires homolytic scission of the C-H bond at C-13 to produce a carbon-centered radical. Despite resonance stabilization of the ensuing pentadienyl radical, C-H cleavage is a high energy process requiring about 72 kcal/mol (105). Even if subsequent oxidations of arachidonic acid occur by an enzyme-regulated autoxidation (*vide supra*), PGH synthase must generate a powerful oxidizing agent to oxidize the first arachidonic acid molecule. The presence of a peroxidase on the protein capable of generating a ferryl-oxo derivative of the iron porphyrin has made the higher oxidation states of the peroxidase obvious candidates for initiators of the cylooxygenase reaction. It appears PGH synthase needs to be activated by hydroperoxides to function as a cyclooxygenase (106). PGG_2 is a very effective activator and presumably fulfills this role during normal catalytic turnover.

Karthein *et al.* detected a tyrosyl radical during the reduction of PGG_2 by PGH synthase (62). The same radical appears during the incubation of arachidonic acid and the time course of its formation and disappearance appears related to the oxygenation of the fatty acid to PGG_2 (62). These workers proposed the tyrosyl radical is the oxidizing agent that removes the pro-S hydrogen from C-13 in the first step of the cyclooxygenase reaction (62). According to their hypothesis, the tyrosyl radical is catalytically consumed and regenerated as each molecule of PGG_2 is produced (Figure 3).

There are several attractive features to this novel mechanism. It explains the requirement for hydroperoxide to maximally activate cyclooxygenase activity and the ability of glutathione and glutathione peroxidase to inhibit cyclooxygenase turnover. Hydroperoxide molecules are required to generate Compound I of the peroxidase which then generates the tyrosyl radical by intramolecular electron transfer. Removal of the hydroperoxide would prevent formation of Compound I. The proposed catalytic role of the tyrosyl radical explains how an enzyme could exist with very low peroxidase activity but relatively high cyclooxygenase activity, such as in the Mn^{3+}-PPIX-reconstituted enzyme or the trypsin-cleaved Fe^{3+}-PPIX-reconstituted enzyme. Since several hundred turnovers of the cyclooxygenase could occur for each tyrosine oxidized by the peroxidase, even a poorly efficient peroxidase should suffice to oxidize the tyrosine. In fact, although the original studies indicate the Mn^{3+}-PPIX-reconstituted enzyme has no peroxidase activity, studies in our laboratory with the PPHP assay indicate it possesses 3% of the peroxidase activity of the Fe^{3+}-PPIX-reconstituted enzyme (107).

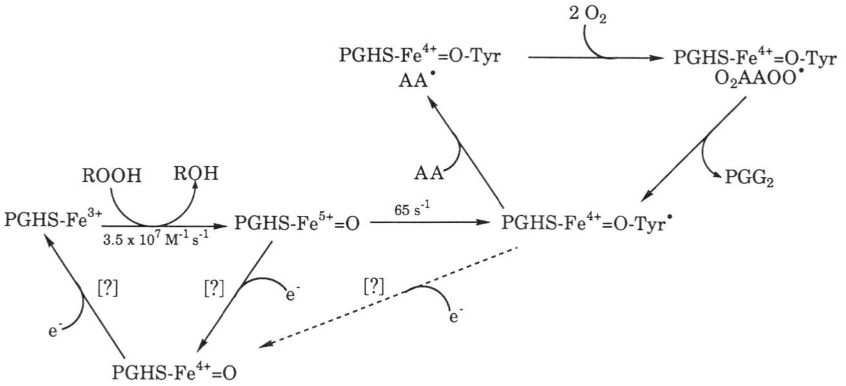

Figure 3. *Tyrosine-Mediated Transduction of Oxidants between Cyclooxygenase and Peroxidase*

The involvement of an essential tyrosine as a transducer of the oxidizing agent between the peroxidase and cyclooxygenase active sites would also explain how two different active sites could be present on one subunit but utilize a single heme prosthetic group for both reactions. The tyrosyl radical would be produced near the peroxidase active site (*i.e.* the heme group) but oxidize arachidonic acid at its binding site. The fatty acid substrate binding site appears to be distinct from the PGG_2 binding site, which is presumably adjacent to the heme group. The distance between the fatty acid and hydroperoxide binding sites may be relatively small and the two sites may be separated by the polypeptide backbone containing the tyrosine residue. EPR relaxation measurements estimate the distance from the tyrosyl radical to the heme group as 7-12 Å (62).

Conclusion

PGH synthase is one of only a few enzymes in all of nature that make hydroperoxides as metabolic end products or intermediates. Its oxygenase activity requires a heme prosthetic group but utilizes a substrate activation mechanism analogous to the non-heme iron-containing enzyme lipoxygenase. Its peroxidase activity is a classic heme peroxidase that reduces hydroperoxides heterolytically. Its heme-oxo derivative may be the oxidizing agent responsible for initiating fatty acid oxidation via a tyrosyl radical transducer. Studies of the structure and function of PGH synthase should provide important insights into this novel and important catalyst of biological oxidation.

Acknowledgements

Work in the author's laboratory was supported by research grants from the National Institutes of Health (GM23642 and CA47479).

References

1. Hemler, M., Lands, W.E.M., and Smith, W.L. (1976) *J. Biol. Chem.* **251**, 5575-5579
2. Van Der Ouderaa, F.J., Buytenhek, M., Nugteren, D.H., and Van Dorp, D.A. (1977) *Biochim. Biophys. Acta* **487**, 315-331
3. Miyamoto, T., Ogino, N., Yamamoto, S., and Hayaishi, O. (1976) *J. Biol. Chem.* **251**, 2629-2636
4. Ogino, N., Ohki, S., Yamamoto, S., and Hayaishi, O. (1978) *J. Biol. Chem.* **253**, 5061-5068
5. Yamamoto, S. (1982) *Methods. Enzymol.* **86**, 55-60
6. van der Ouderaa, F.J.G., and Buytenhek, M. (1982) *Methods. Enzymol.* **86**, 60-68
7. Ohki, S., Ogino, N., Yamamoto, S., and Hayaishi, O. (1979) *J. Biol. Chem.* **254**, 829-836
8. Lands, W.E.M., and Samuelsson, B. (1968) *Biochim. Biophys. Acta* **164**, 426-429
9. Vonkeman, H., and Van Dorp, D.A. (1968) *Biochim. Biophys. Acta* **164**, 430-432
10. Ogino, N., Miyamoto, T., Yamamoto, S., and Hayaishi, O. (1977) *J. Biol. Chem.* **252**, 890-895
11. Shimizu, T., Yamamoto, S., and Hayaishi, O. (1979) *J. Biol. Chem.* **254**, 5222-5228
12. Markey, C.M., Alward, A., Weller, P.E., and Marnett, L.J. (1987) *J. Biol. Chem.* **262**, 6266-6279
13. Vane, J.R. (1971) *Nature New Biol.* **231**, 232-235
14. Vane, J., and Botting, R. (1988) *FASEB J.* **2**, 89-96
15. Flower, R.J. (1974) *Pharmacol. Rev.* **26**, 33-67
16. Roth, G.J., and Majerus, P.W. (1975) *J. Clin. Invest.* **56**, 624-632
17. Roth, G.J., Machuga, E.T., and Ozols, J. (1983) *Biochemistry* **22**, 4672-4675
18. Kulmacz, R.J., and Lands, W.E.M. (1985) *J. Biol. Chem.* **260**, 12572-12578
19. Rome, L.H., and Lands, W.E.M. (1975) *Proc. Natl. Acad. Sci. USA* **72**, 4863-4865

20. Hughes, M.F., Mason, R.P., and Eling, T.E. (1988) *Mol. Pharmacol.* **34**, 186-193
21. Flower, R.J. (1988) *Br. J. Pharmacol.* **94**, 987-1015
22. Bailey, J.M., Makheja, A.N., Pash, J., and Verma, M. (1988) *Biochem. Biophys. Res. Comm.* **157**, 1159-1163
23. Raz, A., Wyche, A., and Needleman, P. (1989) *Proc. Natl. Acad. Sci. USA* **86**, 1657-1661
24. Marnett, L.J., Wlodawer, P., and Samuelsson, B. (1975) *J. Biol. Chem.* **250**, 8510-8517
25. Eling, T.E., and Krauss, R.S. (1985) in *Arachidonic Acid Metabolism and Tumor Initiation* (Marnett, L.J., ed) pp. 83-124, Martinus Nijhoff, Boston
26. Rice, J.R., Zenser, T.V., and Davis, B.B. (1985) in *Arachidonic Acid Metabolism and Tumor Initiation* (Marnett, L.J., ed) pp. 125-169, Martinus Nijhoff, Boston
27. Moldeus, P., Larsson, R., and Ross, D. (1985) in *Arachidonic Acid Metabolism and Tumor Initiation* (Marnett, L.J., ed) pp. 171-198, Martinus Nijhoff, Boston
28. Marnett, L.J., and Eling, T.E. (1983) in *Reviews in biochemical toxicology* (Hodgson, E., Bend, J.R., and Philpot, R.M., eds) pp. 135-172, Elsevier/North Holland, New York
29. Marnett, L.J. (1985) in *Arachidonic Acid Metabolism and Tumor Initiation* (Marnett, L.J., ed) pp. 39-82, Martinus Nijhoff, Boston
30. Marnett, L.J., Reed, G.A., and Dennison, D.J. (1978) *Biochem. Biophys. Res. Comm.* **82**, 210-216
31. Robertson, I.G.C., Sivarajah, K., Eling, T.E., and Zeiger, E. (1983) *Cancer Res.* **43**, 476-480
32. Wells, P.G., Nagai, M.K., and Spano Greco, G. (1989) *Toxicol. Appl. Pharmacol.* **97**, 406-414
33. Degen, G.H., Eling, T.E., and McClachlan, J.A. (1982) *Cancer Res.* **257**, 6957-6964
34. Wild, D., and Degen, G.H. (1987) *Carcinogenesis* **8**, 541-545
35. Battista, J.R., and Marnett, L.J. (1985) *Carcinogenesis* **6**, 1227-1229
36. Zenser, T.V., Palmier, M.O., Mattammal, M.B., Bolla, R.I., and Davis, B.B. (1983) *J. Pharmacol. Exp. Ther.* **227**, 139-143
37. Christ, E.J., and Van Dorp, D.A. (1972) *Biochim. Biophys. Acta* **270**, 537-541
38. Smith, W.L. (1985) in *Biochemistry of Arachidonic Acid Metabolism* (Lands, W.E.M., ed) pp. 77-93, Martinus Nijhoff, Boston
39. Galli, C., Galli, G., and Porcellati, G. (1978) *Phospholipases and Prostaglandins. Advances in Prostaglandin and Thromboxane Research. Volume 3*, Raven, New York
40. Smith, W.L., and Lands, W.E.M. (1972) *Biochemistry* **11**, 3276-3285
41. Kulmacz, R.J. (1986) *Arch. Biochem. Biophys.* **249**, 273-285
42. Bailey, J.M., Muza, B., Hla, T., and Salata, K. (1985) *J. Lipid Res.* **26**, 54-61

43. Casey, M.L., Korte, K., and MacDonald, P.C. (1988) *J. Biol. Chem.* **263**, 7846-7854
44. Raz, A., Wyche, A., Siegel, N., and Needleman, P. (1988) *J. Biol. Chem.* **263**, 3022-3028
45. Pfeilschifter, J., Pignat, W., Vosbeck, K., and Marki, F. (1989) *Biochem. Biophys. Res. Comm.* **159**, 385-394
46. Wu, K.K., Hatzakis, H., Lo, S.S., Seong, D.C., Sanduja, S.K., and Tai, H.H. (1988) *J. Biol. Chem.* **263**, 19043-19047
47. Rollins, T.E., and Smith, W.L. (1980) *J. Biol. Chem.* **255**, 4872-4875
48. Smith, W.L., DeWitt, D.L., and Allen, M.A. (1983) *J. Biol. Chem.* **258**, 5922-5926
49. DeWitt, D.L., Rollins, T.E., Day, J.S., Gauger, J.S., and Smith, W.L. (1981) *J. Biol. Chem.* **256**, 10375-10382
50. Kulmacz, R.J., and Lands, W.E.M. (1987) in *Prostaglandins and Related Substances: a Practical Approach* (Benedetto, C., McDonald-Gibson, R.G., Nigam, S., and Slater, T.F., eds) pp. 209-227, IRL Press, Washington,D.C.
51. Roth, G.J., Machuga, E.T., and Strittmatter, P. (1981) *J. Biol. Chem.* **256**, 10018-10022
52. Marnett, L.J., Siedlik, P.H., Ochs, R.C., Pagels, W.D., Das, M., Honn, K.V., Warnock, R.H., Tainer, B.E., and Eling, T.E. (1984) *Mol. Pharmacol.* **26**, 328-335
53. Mevkh, A.T., Sud'ina, G.F., Golub, N.B., and Varfolomeev, S.D. (1985) *Anal. Biochem.* **150**, 91-96
54. Mutsaers, J.H.G.M., Van Halbeek, H., Kamerling, J.P., and Vliegenthart, J.F.G. (1985) *Eur. J. Biochem.* **147**, 569-574
55. Yoshimoto, A., Ito, H., and Tomita, K. (1970) *J. Biochem.* **68**, 487-499
56. Van Der Ouderaa, F.J., Buytenhek, M., Slikkerveer, F.J., and Van Dorp, D.A. (1979) *Biochim. Biophys. Acta* **572**, 29-42
57. Ruf, H.H., Schuhn, D., and Nastainczyk, W. (1984) *FEBS Letts.* **165**, 293-296
58. Kulmacz, R.J., and Lands, W.E.M. (1984) *J. Biol. Chem.* **259**, 6358-6363
59. Kulmacz, R.J., Tsai, A.-L., and Palmer, G. (1987) *J. Biol. Chem.* **262**, 10524-10531
60. Karthein, R., Nastainczyk, W., and Ruf, H.H. (1987) *Eur. J. Biochem.* **166**, 173-180
61. Lambeir, A.M., Markey, C.M., Dunford, H.B., and Marnett, L.J. (1985) *J. Biol. Chem.* **260**, 14894-14896
62. Karthein, R., Dietz, R., Nastainczyk, W., and Ruf, H.H. (1988) *Eur. J. Biochem.* **171**, 313-320
63. Merlie, J.P., Fagan, D., Mudd, J., and Needleman, P. (1988) *J. Biol. Chem.* **263**, 3550-3553
64. DeWitt, D.L., and Smith, W.L. (1988) *Proc. Natl. Acad. Sci. USA* **85**, 1412-1416
65. Yokoyama, C., Takai, T., and Tanabe, T. (1988) *FEBS Letts.* **231**, 347-351

66. Marnett, L.J., Chen, Y.-N.P., Maddipati, K.R., Plé, P., and Labeque, R. (1988) *J. Biol. Chem.* **263**, 16532-16535
67. Chen, Y.-N.P., Bienkowski, M.J., and Marnett, L.J. (1987) *J. Biol. Chem.* **252**, 16892-16899
68. Kulmacz, R.J., and Lands, W.E.M. (1982) *Biochem. Biophys. Res. Comm.* **104**, 758-764
69. Kulmacz, R.J., and Wu, K.K. (1989) *Arch. Biochem. Biophys.* **268**, 502-515
70. Chen, Y.-N.P., and Marnett, L.J. (1989) *FASEB J.* **3**, 2294-2297
71. Kulmacz, R.J. (1989) *J. Biol. Chem.* **264**, 14136-14144
72. Hamberg, M., and Samuelsson, B. (1967) *J. Biol. Chem.* **242**, 5336-5343
73. Porter, N.A. (1986) *Acc. Chem. res.* **19**, 262-268
74. Nugteren, D.H., Vonkemann, H., and Van Dorp, D.A. (1967) *Rec. Trav. Chim* **86**, 1237-1245
75. Mason, R.P., Kalyanaraman, B., Tainer, B.E., and Eling, T.E. (1980) *J. Biol. Chem.* **255**, 5019-5022
76. Schreiber, J., Eling, T.E., and Mason, R.P. (1986) *Arch. Biochem. Biophys.* **249**, 126-136
77. Porter, N.A., and Funk, M.O. (1975) *J. Organic. Chem.* **40**, 3614-3615
78. O'Connor, D.E., Mihelich, E.D., and Coleman, M.C. (1984) *J. Am. Chem. Soc.* **106**, 3577-3584
79. Mihelich, E.D. (1980) *J. Am. Chem. Soc.* **102**, 7141-7143
80. O'Connor, D.E., Mihelich, E.D., and Coleman, M.C. (1981) *J. Am. Chem. Soc.* **103**, 223-224
81. Pryor, W.A., and Stanley, J.P. (1975) *J. Organic. Chem.* **40**, 3615-3617
82. Hemler, M.E., and Lands, W.E.M. (1980) *J. Biol. Chem.* **255**, 6253-6261
83. O'Brien, P.J., and Rahimtula, A. (1976) *Biochem. Biophys. Res. Comm.* **70**, 832-838
84. Roth, G.J., Siok.C.J., and Ozols, J. (1980) *J. Biol. Chem.* **255**, 1301-1304
85. Pagels, W.R., Sachs, R.J., Marnett, L.J., DeWitt, D.L., Day, J.S., and Smith, W.L. (1983) *J. Biol. Chem.* **258**, 6517-6523
86. Dunford, H.B., and Stillman, J.S. (1976) *Coord. Chem. Rev.* **19**, 187-251
87. Potter, D.W., and Hinson, J.A. (1987) *J. Biol. Chem.* **262**, 974-980
88. Lasker, J.M., Sivarajah, K., Mason, R.P., Kalyanaraman, B., Abou-Donia, M.B., and Eling, T.E. (1981) *J. Biol. Chem.* **256**, 7764-7767
89. Eling, T.E., Mason, R.P., and Sivarajah, K. (1985) *J. Biol. Chem.* **260**, 1601-1607
90. Marnett, L.J., Siedlik, P.H., and Fung, L.W.M. (1982) *J. Biol. Chem.* **257**, 6957-6964
91. Sivarajah, K., Lasker, J.M., Eling, T.E., and Abou-Donia, M.B. (1982) *Mol. Pharmacol.* **21**, 133-141
92. Job, D., and Dunford, H.B. (1976) *Eur. J. Biochem.* **66**, 607-614
93. Bull, A.W. (1987) *Carcinogenesis* **8**, 387-390

94. Ogino, N., Yamamoto, S., Hayaishi, O., and Tokuyama, T. (1979) *Biochem. Biophys. Res. Comm.* **87**, 184-191
95. Ohkuchi, M., Tsukamoto, M., Yokoyama, T., Shiratsuchi, M., and Uchida, Y. (1988) *Prostaglandins* **36**, 787-795
96. Proctor, P. (1970) *Nature* **228**, 868
97. Ames, B.N., Cathcart, R., Schwiers, E., and Hochstein, P. (1981) *Proc. Natl. Acad. Sci. USA* **78**, 6858-6862
98. Reed, D.J. (1985) in *Bioactivation of Foreign Compounds* (Anders, M.W., ed) pp. 71-108, Academic Press, New York
99. Josephy, P.D., and Iwaniw, D.C. (1985) *Carcinogenesis* **6**, 155-158
100. Marnett, L.J. (1987) *Carcinogenesis* **8**, 1365-1373
101. Egan, R.W., Gale, P.H., Vandenheuvel, W.J., Baptista, E.M., and Kuehl, F.A. (1980) *J. Biol. Chem.* **255**, 323-326
102. Egan, R.W., Gale, P.H., Baptista, E.M., Kennicott, K.L., Vandenheuvel, W.J., Walker, R.W., Fargeness, P.E., and Kuehl, F.A. (1981) *J. Biol. Chem.* **256**, 7352-7361
103. Ple, P., and Marnett, L.J. (1989) *J. Biol. Chem.* **264**, 13983-13993
104. Marshall, P.J., and Kulmacz, R.J. (1988) *Arch. Biochem. Biophys.* **266**, 162-170
105. Kerr, J.A. (1966) *Chem. Rev.* **66**, 465-500
106. Lands, W.E.M., Kulmacz, R.J., and Marshall, P.J. (1984) in *Free Radicals in Biology, Vol. VI* (Pryor, W.A., ed) pp. 39-61, Academic Press, Inc., New York
107. Labeque, R. unpublished results
108. DeWitt, D.L., El-Harith, E.A., Kraemer, S.A., Andrews, M.J., Yao, E.F., Armstrong, R.L., and Smith, W.L. (1990) *J. Biol. Chem.* **265**, 5192-5198

Peroxide Tone and Its Consequences

William E.M. Lands and Richard R. Keen
Departments of Biochemistry and Biochemistry (M/C536), University of Illinois, A-312 College of Medicine West, 1853 West Polk Street, Chicago, IL 60612 USA

Abstract

An average steady-state concentration of hydroperoxide (peroxide tone) can be envisioned, although it may differ appreciably among cells at different times. Hydroperoxides are required activators of the fatty acid oxygenase activities that catalyze the first committed step in the biosynthesis of eicosanoids (prostaglandins and leukotrienes), and these oxygenases act as amplifiers as they generate hydroperoxides as their reaction products. Thus biosynthesis of eicosanoids mediates a potent cellular response to hydroperoxide signals. The ability of picomole amounts of fatty acid hydroperoxide to activate cyclo-oxygenase activity provides a sensitive method to determine ambient levels of hydroperoxide. Little glutathione peroxidase-sensitive hydroperoxide was detected in antecubital venous plasma of normal volunteers. However, blood draining septic regions had significantly elevated amounts. Also, during normal recovery from thoracic surgery, hydroperoxides were detected in arterial (but not mixed venous) plasma, whereas abdominal surgery led to elevated levels in mixed venous (but not arterial) plasma. The small amounts of circulating hydroperoxide seem useful indicators of previously unrecognized tissue responses.

The Physiological Need for Hydroperoxides and Radicals

The occurrence of free radicals in cellular systems is often regarded to be a preamble to destruction and disaster, but some normal cellular events (e.g., activation of ribonucleotide reductase for DNA formation and maintenance of fatty acid oxygenase activity for eicosanoid formation) actually require free radicals and hydroperoxides to promote normal physiological processes. This apparent requirement raises the interesting question of how much oxidant is needed and whether or not the presence of such amounts is antithetical to a continued smooth operation of normal cellular processes. Apparently the synthesis of deoxyribonucleotides for the production of new DNA can continue once a transitory oxidative activation of the

Biological Oxidation Systems, Volume 2

reductase has been achieved (1, 2). The activated enzyme radical is very stable, and it can last for a very long time. Thus, even a minor transient burst of radical formation at one stage of the cell cycle may suffice to provide the stable activated reductase needed for S phase events, and a prolonged formation of new radicals seems to be not necessary for DNA synthesis.

In contrast, the fatty acid oxygenases that produce the precursors of eicosanoid autacoids seem to require a continual presence of hydroperoxides to maintain rapid rates of biosynthesis (3,4). The hydroperoxide requirement first became evident when we attempted to trap hydroperoxide intermediates as the hydroxide derivatives by using added glutathione peroxidase. Instead of promoting the accumulation of intermediates as the hydroxide(s), the added peroxidase activity prevented oxygenation from proceeding at a significant rate (3), and no reaction intermediates were detected. The overall reaction mechanism for prostaglandin endoperoxide synthase appears to be a cooperative interaction of the two catalytic activities expressed by the enzyme (5), with the hydroperoxide-produced peroxidase intermediate initiating the radical chain oxygenation reaction of fatty acid oxygenase. Thus, prostaglandin biosynthesis is a potent regulatory physiological system through which cells respond to the presence of HOOH or lipid hydroperoxide and cause major modulations in cell behavior.

One can envision a steady state level of hydroperoxide ("peroxide tone") that could sustain the activity of the fatty acid oxygenases, but there is little evidence that eicosanoid biosynthesis occurs in a steady, sustained manner. Rather, much evidence suggests that eicosanoid biosynthesis is normally suppressed in normal tissues, and it proceeds only in response to some stimulus. Several experiments have demonstrated pulsatile formation of circulating prostaglandins in plasma (6,7) which would indicate frequent changes in intra-cellular peroxide tone. Much latent prostaglandin endoperoxide synthase appears to be present in most tissues (8,9), and the suppression of its action is attributable to a limited supply of the non-esterified substrate and the hydroperoxide activators (10-12). An intermittent and discontinuous biosynthesis of eicosanoids may suffice for those intermittent physiological responses that require eicosanoid modulation, and the biosynthesis may remain suppressed most of the time. Thus, cells may have only transient requirements for the hydroperoxide activators of fatty acid oxygenases. Alternatively, we might consider that the causes and consequences of transient hydroperoxide generation may require only a transient compensatory biosynthesis of eicosanoids.

Amplification of Hydroperoxides

Unusual kinetic features of the prostaglandin-forming cyclooxygenase include an irreversible, self-catalyzed inactivation of the enzyme and a self-catalyzed activation associated with accumulation of the resultant hydroperoxide product. To check the validity of these unexpected features in our early studies, we used soybean lipoxygenase as a control oxygenase system and confirmed that both types of fatty acid oxygenase catalyze both autoactivating and autoinactivating reactions (13,14). Subsequently, these bizarre mechanistic features have been consistently noted for all fatty acid oxygenases studied (although not all fatty acid "suicide" substrates promote the same degree of inactivation). The relative instability of the activated fatty acid oxygenases probably underlies the reaction inactivation ("suicide substrate") phenomenon(4,13,14), and the presence of an "activated ferryl species" in the reaction mechanism of the cyclooxygenase may explain its instability(2). The self-catalyzed inactivation provides a limitation on the overall amount of eicosanoid that can be formed by a given amount of oxygenase, and this type of negative feedback occurs without the need for a diffusible end product.

The fact that the fatty acid oxygenases synthesize a hydroperoxide that can activate an oxygenase makes the initiation of the fatty acid oxygenase reaction an explosive event as the reaction generates the required activator. Also, there is a possibility that the hydroperoxide produced from one oxygenase might stimulate the activity of different oxygenases (or even the ribonucleotide reductase?). Thus, triggering eicosanoid biosynthesis creates an inescapable amplification of intracellular peroxide tone. The prostaglandin-forming cyclooxygenase activity has the most thoroughly studied and the most clearly defined requirement for hydroperoxide (0.02 nmole/ml (15)). This low concentration of hydroperoxide is below that normally detected by usual analytical procedures, and also below that usually employed in toxicity studies with cells and tissues. In fact, that level is far below the 1 nmole/ml level where most cellular peroxidases are effective. Thus the biosynthesis of prostaglandins can be effectively triggered by very small amounts of fatty acid hydroperoxide, and then produce more and more lipid hydroperoxide as the biosynthesis proceeds faster and faster.

Three regions of peroxide tone were proposed to characterize the responsive ranges of hydroperoxides (16): from 10^{-9} to 10^{-8} M where triggering of cyclooxygenase activity probably occurs; 10^{-8} to 10^{-7} M where amplification of the hydroperoxide levels occur as the reaction proceeds; and 10^{-7} to 10^{-6} M where the tendency for excessive (or pathological) accumulation of hydroperoxide may be antagonized by intra-cellular peroxidases.

The rapid amplification of peroxide tone by cyclooxygenase using arachidonic acid (20:4n-6) as substrate permits the enzyme to respond explosively to triggering amounts of hydroperoxide even when the system is being suppressed by significant amounts of peroxidase. The vigor of this response contrasts with the slower oxygenation rate with eicosapentaenoic acid (20:5n-3) which keeps the amplification low (R.B. Pendleton, PhD thesis, Univ.of Illinois, 1989), especially in the presence of small additional amounts of hydroperoxide-scavenging peroxidase activity. The kinetic difference between these two acids in forming prostaglandins, apparently causes the n-3 type of polyunsaturated fatty acid to be more effective as an antagonist of the rapid formation and function of n-6 eicosanoids than as a facile precursor of active eicosanoid (17), and it provides a biochemical rationale for some beneficial roles for fish in human health (18).

Quantitating Hydroperoxides

With three functional ranges of intra-cellular hydroperoxide to be examined in studying the control of eicosanoid biosynthesis, it became important to develop analytical procedures to confirm the actual levels that occur in biological systems. This confirmation has proved to be difficult to achieve because of various limitations in all of the analytical methods currently available for quantitating lipid hydroperoxides. After considerable effort, our lab group concluded that there is NO good, convenient and reliable method for quantitating hydroperoxides in the submicromolar range(19). Nevertheless, efforts continue in many of our labs to determine the amounts of hydroperoxides in biological systems.

Our search for a sensitive reagent led us to examine the cyclooxygenase because of its obviously strong response to very small (10^{-8} M) amounts of hydroperoxide. The rapid explosive self-activation of the cyclooxygenase created short lag times that were difficult to measure accurately, but when the reaction was inhibited with cyanide and phenol (4), the longer lag times were reproducibly measured. This led to a standardized enzymatic micromethod for determining fatty acid hydroperoxides (20) in the range of 10 to 200 picomoles per assay. Assays were reproducible with standard hydroperoxide in buffer, but the presence of proteins in biological samples interfered with the assay. Thus, we employed 50% ethanol to denature much of the interfering protein and remove it by centrifugation. Unfortunately, greater variability was noted for these treated plasma supernatants.

For plasma from different individuals, the recoveries of added hydroperoxides were different, ranging from 70 to 100%. Thus different individuals may have different amounts (or types) of unknown inhibitory materials that interfere with reproducibility, but an estimate of the hydroperoxide content is still possible.

The amount of activator measured in the ethanolic supernatants of plasma from the antecubital vein of apparently normal volunteers was equivalent to approximately 0.5 μM fatty acid hydroperoxide (20). At that time, preliminary unpublished results (P.J.M.) suggested that fatty acid hydroperoxides stimulated the oxygenase whether or not they were esterified. Now we know that the hydroperoxides esterified in phospholipid, triacylglycerols, or cholesterol esters are essentially ineffective as activators relative to the non-esterified fatty acid hydroperoxides(R.B. Pendleton, PhD thesis, University of Illinois, 1989). Also, the fatty acid hydroperoxides are nearly one thousand-fold more effective than HOOH on a molar basis. Another important correction regarding the enzymatic assay is that the reported removal of activator from human venous plasma by treatment with glutathione peroxidase (21) has not been confirmed by subsequent research. Extensive study with hundreds of assays has indicated that most of the activator detected in the antecubital venous plasma of volunteers is resistant to removal by treatment with glutathione peroxidase and therefore is probably not a fatty acid hydroperoxide (R. Keen and W. Lands, unpublished results). These important revisions in the evidence for hydroperoxides in human plasma certainly illustrate the need for caution in interpreting results with biological samples, but they do not contradict the evidence that in pure buffer systems the cyclooxygenase activation assay is a very sensitive detector of fatty acid hydroperoxides.

Results with the thiobarbituric acid (TBA) method have been described extensively, but it is an indirect assay, and many investigators have particularly noted its lack of validity as well as its low sensitivity. When measuring fatty acid hydroperoxides in buffer systems, Asakawa(22) noted that optimal conditions permitted only approximately one nmol-equivalent of malonaldehyde to be detected for 16 nmols of fatty acid hydroperoxide. Thus the reported levels of TBA-positive material in 1 ml of human plasma (3-5 neq. MDA) correspond to approximately 50-70 nmoles of lipid hydroperoxide. The possible interfering chromogenic materials in human plasma make it very difficult to assign all of the absorbance to either malonaldehyde or lipid hydroperoxide, and the high calculated levels of lipid hydroperoxide in human plasma seem unlikely to be valid. Chromatographic purification can resolve the TBA-malonaldehyde derivative from interfering materials, but the method still cannot unequivocally indicate the amount of hydroperoxide in the sample being assayed.

Other assays for hydroperoxide include iodometric methods sensitive to redox potential and chemiluminescent methods based upon peroxidase action on hydroperoxides (e.g., 23). However, at this stage of development, each method has serious deficiencies in providing convincing evidence on the presence or absence of sub-micromolar amounts of hydroperoxide in biological systems.

Intra-cellular Peroxide Tone and Anti-inflammatroy Drug Efficacy

Because so many pathophysiological events are associated with too rapid a biosynthesis of prostaglandins, much effort has been placed upon finding effective inhibitors of this biosynthesis. The inhibitors may be classified among three types based upon their mechanism of action (24, 25). The action of the time-dependent irreversible inactivators of prostaglandin endoperoxide synthase (*e.g.* aspirin and indomethacin (26, 27)) differs appreciably from the reversible action of the simple competitive inhibitors (*e.g.* ibuprofen (28)) and the antagonists of hydroperoxide activation (*e.g.* acetamido-phenol (11, 29)). On the other hand, both of the first two types of inhibitor act independently of the peroxide tone, whereas the effectiveness of the latter type of agent appears to be markedly enhanced by decreasing the ambient hydroperoxide tone. This enhancement was demonstrated with purified cyclooxygenase in vitro by adding increasing amounts of glutathione peroxidase and observing increased inhibitory capacity of the agents (30). Apparently only when the peroxide tone is below a critical range do the radical-trapping or hydroperoxide-antagonizing agents cause significant inhibition of prostaglandin synthesis. This concept led to the further concept that tissues experiencing general hyperalgesic conditions may have lower levels of hydroperoxide than are found in inflammatory conditions (29). Such a mechanistic concept helps interpret how acetamido-phenol can be effective as an analgesic agent but not as an anti-inflammatory agent, and it may lead to the development of new analgesic agents with more selective actions.

Extra-cellular Hydroperoxides and Trauma

Macrophages and neurophils produce large amounts of HOOH and O_2^-, and in the presence of iron salts, these "aggressive" oxygen derivatives can generate significant amounts of extra-cellular hydroperoxides (31). The increased hydroperoxides in supernatants from guinea pig granulocytes activated by phorbol myristyl acetate were detected by the cyclooxygenase activation assay (31). In 30 min, the accumulated activator was equivalent to 20-50 pmol-eq. of fatty acid hydroperoxide per 10^6 cells. Catalase removed about one-half of the phorbol-induced activator, indicating that a considerable amount of HOOH had accumulated. Although HOOH is about 1000-fold weaker activator than fatty acid hydroperoxide, it was accumulated about 1000-fold more in the culture medium. These extra-cellular hydroperoxides may stimulate (31,32) the intra-cellular fatty acid oxygenases (that would, in turn, amplify intra-cellular peroxide tone).

Sepsis is the major single cause of death following major surgery and trauma (33), and early diagnosis of localized infection before systemic responses are activated can significantly improve patient outcome. Exploratory surgery is often the most definitive test for diagnosis of sepsis (34,35) , but it can be associated with mortality rates as high as 83%. Because our preliminary studies showed hydroperoxides produced in inflammatory conditions (31), and we found elevated levels of hydroperoxide in the plasma of septic patients (36), we have examined arterial and mixed venous plasma from twelve additional septic patients in the intensive care unit. The difference between glutathione peroxidase-sensitive fatty acid hydroperoxides in plasma from involved and uninvolved regions of the same individual had a mean value of 0.45 ± 0.14 μM ($p<0.005$). The results indicate that the enzymatic assay for hydroperoxides may be a useful indicator of oxidative events associated with severe infection. They also indicate that the fatty acid hydroperoxides are rapidly cleared from the blood upon passing through capillary beds.

In developing control data for the level of hydroperoxides in arterial and mixed venous plasma, we have studied an additional set of eighteen patients in the intensive care unit. These patients had undergone recent major thoracic or abdominal operations and showed no signs or symptoms of sepsis. The plasma hydroperoxides of these control patients showed some surprising features: following thoracic procedures, the arterial plasma had 0.58 ± 0.13 μM ($p<0.01$) more peroxidase-sensitive hydroperoxide than the mixed venous plasma, whereas following abdominal procedures, the arterial plasma had 0.65 ± 0.17 μM ($p<0.01$) less than the mixed venous plasma! In response to the surgical intervention, the pulmonary and intra-abdominal tissues were generating significant amounts of previously unrecognized fatty acid hydroperoxides that were rapidly cleared from the circulation by tissue capillary beds. Such a rapid clearance would explain the inability to detect significant amounts of fatty acid hydroperoxides in the antecubital venous samples.

Summary

Much data suggests that HOOH can serve as an inter-cellular signal that can stimulate non-enzymatic formation of extra-cellular lipid hydroperoxides and stimulate fatty acid oxygenases to form intra-cellular hydroperoxides. Transient pulses of formation of eicosanoids by fatty acid oxygenases are probably associated with transient increases in peroxide tone. Technical problems associated with detecting the small transient amounts of lipid hydroperoxides in biological systems remain a serious challenge.

References

1. Reichard, P. (1988) *Ann. Rev. Biochem.* **57**, 349-374
2. Stubbe, J.A. (1989) *Ann. Rev. Biochem.* **58**, 257-285
3. Lands, W.E.M., Lee, R.E., and Smith, W.L. (1971) *Ann. NY Acad. Sci.* **180**, 107-122
4. Hemler, M.E., and Lands, W.E.M. (1980) *J. Biol. Chem.* **255**, 6253-6261
5. Kulmacz, R.J., Miller, J.F., Jr. and Lands, W.E.M. (1985) *Biochem. Biophys. Res. Comm.* **130**, 918-923
6. Granstrom, E., and Samuelsson, B. (1978) *Adv. Prostaglandin and Thromboxane Res.* (ed. J. Frolich) Raven Press, NY, Vol. 5, pp. 1-8
7. Granstrom, E., and Kindahl, H. (1978) *Adv. Prostaglandin and Thromboxane Research* (ed. J. Frolich) Raven Press, NY, Vol. 5, pp119-195
8. Crowshaw, K. (1969) *Federation Proc.* **28**, 845
9. Jouvenaz, G.H., Nugteren, D.H., Beerthuis, R.K., and vanDorp, D.A. (1970) *Biochim. Biophys. Acta* **202**, 231-234
10. Cook, H.W., and Lands, W.E.M. (1976) *Nature* **260**, 630-632
11. Lands, W.E.M., Cook, H.W., Rome, L. H. (1976) *Adv. Prostaglandin and Thromboxane Research* (eds. B. Samuelsson and R. Paoletti) Raven Press, NY, Vol. 1, pp. 7-17
12. Marshall, P.J., Kulmacz, R.J., and Lands, W.E.M. (1987) *J. Biol. Chem.* **262**, 3510-3517
13. Smith, W.L. and Lands, W.E.M. (1972) *Biochem.* **11**, 3276-3285
14. Smith, W.L. and Lands, W.E.M. (1972) *J. Biol. Chem.* **247**, 1038-1042
15. Kulmacz, R.J., and Lands, W.E.M. (1983) *Prostaglandins* **25**, 531-540
16. Lands, W.E.M., and Marshall, P. J. (1985) *Rheumatology-85* (eds. P.M. Brooks and J.R. York) Elsevier Science Publishers B. V. (Biomedical Division) Amsterdam pp. 47-50
17. Lands, W.E.M. (1989) *J. Internal Med.* **225**: Suppl. 1, 11-20

18. Lands, W.E.M. (1986) *Fish and Human Health*, Academic Press, Orlando
19. Lands, W.E.M. (1988) *Free Radicals: Methodology and Concepts* (eds. C. Rice-Evans and B. Halliwell) Richelieu Press, London Vol. 4, pp. 213-224
20. Marshall, P.J., Warso, M.A., and Lands, W.E.M. (1985) *Anal. Biochem.* **145**, 192-199
21. Warso, M.A., and Lands, W.E.M. (1985) *J. Clin. Invest.* **75**, 667-671
22. Asakawa, T., and Matsushita, S., (1980) *Lipids* **15**, 137-140
23. Frei, B., Yamamoto, Y., Niclas, D., and Ames, B.N. (1988) *Anal. Biochem.* **175**, 120-130
24. Lands, W.E.M. (1981) *TIPS* (March) 78-80
25. Lands, W.E.M. (1985) *Adv. Drug. Res.* **14**, 147-164
26. Smith, W.L., and Lands, W.E.M. (1971) *J. Biol. Chem.* **246**, 6700-6704
27. Lands, W.E.M., LeTellier, P.R., Rome, L.H., and Vanderhoek, J.Y. (1973) *Adv. Biosci.* **9**, 15. 27
28. Rome, L.H., and Lands, W.E.M. (1975) *Proc. Nat. Acad. Sci. USA* **72**, 4863-4865
29. Lands, W.E.M., and Hanel, A.M. (1982) *Prostaglandins* **24**, 271-277
30. Hanel, A.M., and Lands, W.E.M. (1982) *Biochem. Pharmacol.* **31**, 3307-3311
31. Marshall, P.J., and Lands, W.E.M. (1986) *J. Lab. Clin. Med.* **108**, 525-534
32. Baud, L., Nivez, M.-P., Chansel, D., and Ardaillou, R. (1981) *Kidney Intl.* **20**, 332-339
33. Macheido, G.W., LoVerme, P.J., McGovern, P.J., Jr. and Blackwood, J.M. (1981) *Surg. Gyn. Obst.* **152**, 757-759
34. Macheido, G.W., Tikellis, J. Suval, W., Lee, B.C., Blackwood, J.M., and Rush, B.F., Jr. (1985) *Amer. Surg.* **51**, 149-154
35. Macheido, G.W., and Suval, W.D. (1988) *Surg. Clinics No. Am.* **68**, 215
36. Lands, W.E.M., and Pendleton, R.B. (1987) *Anticarcinogenesis and Radiation Protection* (eds. P. A. Cerutti, O. F. Nygaard and M.G. Simic) Plenum Press, NY, pp81-85

Involvement of Glutathione Peroxidases in Prostaglandin Biosynthesis

John R. Burgess, Yang Hong, Mei Chang, George Hildenbrandt
R.W. Scholz and *C. Channa Reddy*
Department of Veterinary Science, 226 Fenske Lab, Penn State University
University Park, PA 16802 USA

Abstract

With our increasing knowledge of the arachidonic acid cascade, it is becoming clear that the glutathione peroxidases (GSH-Pxs) play an important role both in the biosynthesis of eicosanoids, and in the regulation of key enzymes associated with the cascade. However, the precise nature of GSH-Px involvement in eicosanoid metabolism is not completely understood. We have determined the catalytic efficiencies of selenium-dependent GSH-Px (Se-GSH-Px) and selenium-independent GSH-Px (nonSe-GSH-Px) toward hydroperoxyeicosatetraenoic acids (HPETEs) as well as prostaglandins (PG) G_2 and H_2. All HPETEs, including 5-,8-,9-,11-,12-, and 15-HPETE, were determined to be good substrates for Se-GSH-Px with a K_m of ~10 μM and V_{max} of ~100 μmoles•min^{-1}•mg^{-1}, whereas nonSe-GSH-Px was found to be 50 times less efficient than Se-GSH-Px with V_{max} of ~ 2 μmoles•min^{-1}•mg^{-1}. Similarly, Se-GSH-Px was found to be very efficient in the reduction of PGG_2 to PGH_2 as indicated by K_m and V_{max} values of 12 μM and 78 μmoles•min^{-1}•mg^{-1}, respectively. These kinetic constants are comparable to those determined for 15-hydroperoxyeicosatetraenoic acid (15-HPETE), H_2O_2, and cumene hydroperoxide with this enzyme. On the other hand, PGH_2 was determined to be a poor substrate indicating that Se-GSH-Px reduces the hydroperoxide moiety more efficiently than the endoperoxide moiety. Thus, Se-GSH-Px is a good candidate for converting PGG_2 to PGH_2 without the free radical-mediated cooxidation associated with the prostaglandin endoperoxide synthase or other heme peroxidases. Selenium-independent GSH-Px associated with the glutathione S-transferase (YaYa) from rat liver cytosol was very active in the conversion of PGH_2 to $PGF_{2\alpha}$ with a V_{max} of 6 μmoles•min^{-1}•mg^{-1}. When PGG_2 was used as a substrate for the nonSe-GSH-Px $PGF_{2\alpha}$ was the ultimate product; however, a novel intermediate compound appeared which was later identified by several methods of structural analysis as 15-hydroperoxy $PGF_{2\alpha}$. Thus nonSe-GSH-Px catalyzes the reductive cleavage of the endoperoxide faster than the reduction of the 15-hydroperoxy moiety, enabling the enzyme to compete with other PGH_2 metabolizing enzymes both by diverting PGG_2 from the PGH_2 pool and by direct conversion of PGH_2 to $PGF_{2\alpha}$. A survey of the catalytic specificities of GSTs from several different tissues and species revealed that the maximum $PGF_{2\alpha}$ forming activity is associated with the GST isozymes consisting of Ya size subunits. Furthermore, we have shown that in rat liver, Ya subunit was induced 2.5 fold in the Se-deficient state with a concomitant increase in mRNA coding for this subunit. A corresponding increase in the capacity for transformation of PGH_2 to $PGF_{2\alpha}$ was also observed in these Se-deficient rat livers. This specificity of GST isozymes in PG biosynthesis, coupled with their tissue specific expression, may be a mechanism by which organisms modulate the types of PGs produced in various tissues.

Introduction

Arachidonic acid is the most abundant polyenoic fatty acids found in the phospholipids of mammalian tissues. It is a biosynthetic precursor of several classes of biologically active eicosanoids. Once released from phospholipids by the action of phospholipases, arachidonic acid is metabolized by one of the two pathways shown in Figure 1. The cyclooxygenase pathway produces prostaglandins (PGs), thromboxane (TX) and prostacyclin (PGI), whereas the lipoxygenase(s) pathway leads to the formation of leukotrienes (LTs) and lipoxins. The immediate oxygenation products of arachidonic acid by way of the cyclooxygenase and lipoxygenase pathways include PGG_2 and 5-hydroperoxyeicosatetraenoic acid (5-HPETE), respectively. Not only do these products serve as precursors for the biosynthesis of PGs and LTs, but they also have significant influence on the activities of various enzymes associated with the arachidonic acid cascade, including cyclooxygenase, lipoxygenases and prostacyclin synthase (1-10). Thus, an enzyme system that can reduce these reactive hydroperoxides has the potential for modulating the arachidonic acid cascade.

Glutathione peroxidases (GSH-PXs), both selenium-dependent GSH-Px (Se-GSH-Px) and selenium-independent GSH-Px (nonSe-GSH-Px), have the capacity for reducing potentially reactive hydroperoxides to less reactive alcohols (1-3). Thus, these enzymes can influence the formation of biologically active eicosanoids in at least three different ways. First, they can play a biosynthetic role in catalyzing the reduction of hydroperoxy intermediates, 5-HPETE and PGG_2, to their corresponding hydroxy derivatives (2-4) (Fig. 1); second, GSH-Pxs may play a regulatory role in controlling the level of fatty acid hydroperoxides necessary for the activation of lipoxygenase (LO) and cyclooxygenase (CO) enzymes (4-8); and third, GSH-Pxs may protect peroxide-sensitive enzymes, such as PGI_2 synthase, from inactivation by hydroperoxides (9,10). In the following, we review the experimental evidence for the possible involvement of GSH-Pxs in the biosynthesis of eicosanoids.

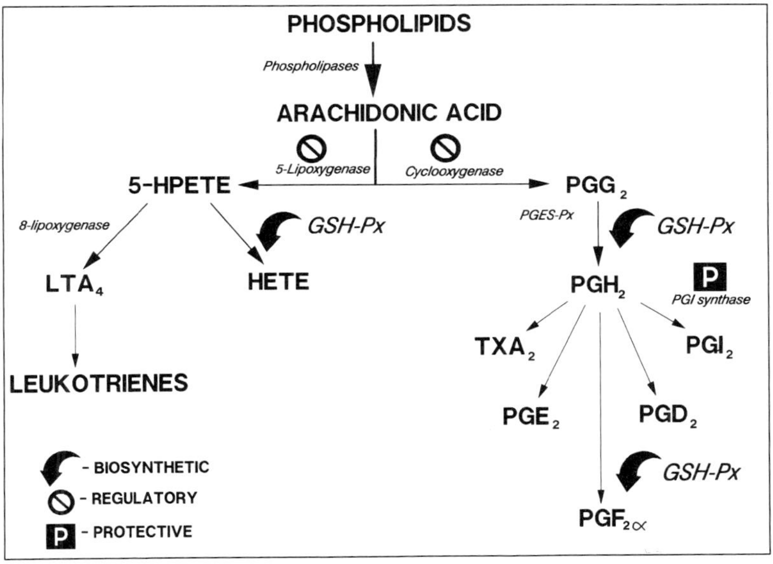

Figure 1. *Proposed Roles for GSH Peroxidases in Eicosanoid Biosynthesis*

Glutathione Peroxidases

Two distinct types of GSH-Pxs have been characterized to date. One class of GSH-Pxs is Se-dependent and the other is Se-independent. Originally GSH-Px (EC 1.11.1.9) was thought to be exclusively associated with an 84,000 molecular weight protein present in mammalian cells. This enzyme is a tetramer with four identical subunits, each containing 1 g-atom of selenium in the form of selenocysteine (11,12). A second Se-GSH-Px has been recently isolated from human plasma which is slightly different from the cellular enzyme in physical and kinetic properties (13,14). Se-GSH-

Pxs catalyze the reduction of both organic hydroperoxides and H_2O_2 and thus differ from a group of selenium-independent GSH-Pxs (nonSe-GSH-Px) associated with certain forms of the glutathione S-transferases (GSTs) (EC 2.5.1.18) (15,16) that can use only organic hydroperoxides as substrates. The functional relationship between the two classes of GSH-Pxs is not entirely clear. Both types of GSH-Pxs are ubiquotously distributed in animals. The cellular Se-GSH-Px is located primarily in the cytosolic and mitochondrial compartments, while the nonSe-GSH-Px has been identified in both cytosolic and microsomal cell fractions (17,18). The relative abundance of these two types of GSH-Pxs within a given tissue is species specific. In rat liver, for example, Se-GSH-Px has been reported to comprise more than 75 percent of the total GSH-Px activity, while in sheep lung and sheep liver the nonSe-GSH-Px predominates (19-21). The distribution of these two types of GSH-Pxs may play an important role in determining the type and amount of eicosaniod produced in a given tissue.

Catalytic Efficiency of Glutathione Peroxidases Towards Hydroperoxy-eicosatetraenoic acids

As mentioned earlier, in the divergent pathways of eicosanoid formation leading to various classes of biologically active compounds, arachidonic acid is initially oxygenated either by cyclooxygenase or by lipoxygenase to form cyclic endoperoxide or HPETE, respectively. In the lipoxygenase(s) pathway, one molecule of oxygen is incorporated into arachidonic acid to form HPETEs (Fig. 1). These HPETEs can be further transformed into LTs or reduced to their corresponding hydroxy eicosatetraenoic acids (HETEs). Selenium-dependent GSH-Px is most likely the enzyme responsible for this reduction in cells, due to its wide distribution and catalytic efficiency. This is supported by our study on the catalytic efficiency of the purified Se-GSH-Pxs from rat liver and bovine erythrocytes towards 5-, 8-, 9-, 11-, 12-, and 15-HPETEs which revealed that all HPETEs are equally good substrates with an apparent K_m of $\sim 10\,\mu M$ and an apparent V_{max} of $\sim 100\,\mu moles \cdot min^{-1} \cdot mg^{-1}$. These kinetic constants are comparable to those determined for H_2O_2, a classical substrate for Se-GSH-Px, whereas the catalytic efficiency of nonSe-GSH-Px toward HPETEs is ~ 50 times less than that determined for Se-GSH-Px.

Catalytic Efficiency of Glutathione Peroxidases Towards PGG_2

During cylooxygenase-catalyzed arachidonic acid oxidation, two molecules of oxygen are incorporated to produce PGG_2 which contains both a bicyclic endoperoxide and a hydroperoxide group. It is generally believed that the intrinsic hydroperoxidase activity associated with the PG endoperoxide synthase (PGES) accounts for the transformation of PGG_2 to PGH_2. However, during the PGES catalyzed reduction of PGG_2 to PGH_2, highly reactive oxidants are generated that have been implicated in the pathogenic mechanisms underlying various disease processes including carcinogenesis (22,23). Alternatively, Se-GSH-Px is well suited to reduce the 15-OOH of PGG_2 to form PGH_2 (2). A K_m value of 12 μM and a V_{max} of 78 μmoles•min^{-1}•mg^{-1} were determined for PGG_2 with the purified Se-GSH-Px from rat liver cytosol in a coupled assay system (2). We obtained similar results with the commercially available bovine erythrocyte GSH-Px (23). The kinetic parameters estimated for Se-GSH-Px with PGG_2 as a substrate are far more favorable than the kinetic values reported for the peroxidase activity of PGES (6,24,25). In contrast, PGG_2 was found to be a comparatively poor substrate for the nonSe-GSH-Pxs isolated from both rat liver and sheep lung (2,26). For example, the Michaelis constant for the nonSe-GSH-Px associated with the major cationic GST isozyme of sheep lung was determined to be 100 μM with PGG_2 as substrate, and the maximum velocity was 2.1 μmoles min^{-1} mg^{-1} (2).

Since Se-GSH-Px is relatively abundant and ubiquitously distributed in animal tissues, it is an excellent candidate to replace the heme-mediated peroxidase activity of PGES as the major pathway of PGG_2 reduction to PGH_2 *in vivo*. An argument against the involvement of cytosolic Se-GSH-Px in this reduction is inaccessibility of the membrane-associated substrate. However, PGG_2 is sufficiently hydrophilic to partition into the aqueous cytosol and can be released from PGES without reduction. We have observed this release in incubations of ram seminal vesicle microsomes with arachidonic acid (2). Furthermore, no physiological reductant has been identified for the PGES peroxidation reaction. Additionally, and very importantly, Se-GSH-Px-catalyzed reduction of PGG_2 generates no free radicals and, thus, the hazard of xenobiotic cooxidation is greatly minimized (C.C. Reddy and R.A. Floyd, unpublished results).

Catalytic Efficiency of Glutathione Peroxidases in $PGF_{2\alpha}$ Formation

In many cases the metabolic fate of PGH_2 is determined by the expression of a particular metabolizing enzyme, resulting in the formation of products charactersitic of a given tissue. For example, in platelets PGH_2 is exclusively converted to TXA_2 catalyzed by TX synthase (27); whereas in vascular endothelium, PGH_2 is transformed mostly into PGI_2 by PGI synthase (28). The classcial PGs, such as $PGF_{2\alpha}$, PGE_2, and PGD_2, are formed in most tissues, but to differing degrees. Tissue-specific biosynthesis of these PGs seems to result from the differential distribution of the individual enzymes responsible for their formation. This tissue specificity may be an important factor in the regulation of physiological functions.

While the enzymes responsible for the isomerization of PGH_2 to PGE_2 and PGD_2 are reasonably well understood, the enzyme(s) responsible for $PGF_{2\alpha}$ formation in a given tissue is less clearly defined. $PGF_{2\alpha}$ is a biological mediator in a variety of physiological functions including uterine muscle contraction during parturition and dissolving of the corpus leuteum during the menstraul cycle (29). Its biosynthesis follows one of three pathways: (i) 9-keto-reduction of PGE_2 (30,31); (ii) 11-ketoreduction of PGD_2 (32,33); and (iii) a direct reduction of the endoperoxide of PGH_2 (2,26,34-36). Recently, the ketoreduction of PGD_2 has been shown to result in the formation of a PGF product with C-9 and C-11 hydroxy groups in the anti-configuration which is sterochemically different from authentic $PGF_{2\alpha}$ (37,38). This product appears to possess some biological properties. It remains to be seen whether PGE_2 ketoreduction results in a stereochemically distinct $PGF_{2\alpha}$, and whether this product is biologically active. Considering these recent discoveries, the direct reduction of PGH_2 is likely the primary route of formation of $PGF_{2\alpha}$ with established physiological properties. Two distinct enzymes have been shown to catalyze the direct reduction of PGH_2 to $PGF_{2\alpha}$ *in vitro*. We have recently reported the GSH-dependent conversion of PGH_2 to $PGF_{2\alpha}$ by the GSTs (2,26,36). Additionally, a ketoreductase for PGD_2 isolated from bovine lung has been shown to utilize PGH_2 as a substrate forming $PGF_{2\alpha}$ as the primary product (39). It remains to be established, however, which of these two enzyme systems is responsible for the formation of $PGF_{2\alpha}$ in tissues.

The catalytic efficiency of the Se-dependent and Se-independent GSH-Pxs towards PGH_2 was determined by product analysis on TLC. In these experiments [1-^{14}C]PGH_2 was used as substrate and the formation of radiolabeled products was determined after extraction and separation by TLC. Interestingly, PGH_2 was not a substrate for the purified Se-GSH-Px from either bovine erythrocytes or rat liver cytosol (2,36). The product profile of this reaction was identical to the nonenzymatic transformation of PGH_2. On the other hand, a significant amount of $PGF_{2\alpha}$ formation occurred when PGH_2 was incubated with nonSe-GSH-Px and GSH, and a linear relationship emerged between $PGF_{2\alpha}$ formation and the concentration of enzyme protein. An estimate of the kinetic constants for the nonSe-GSH-Px associated with one of the rat liver GSTs (YaYa) revealed that the PGH_2 concentration giving half maximal velocity was 0.17 mM, and that the V_{max} for $PGF_{2\alpha}$ synthesis was 6.0 μmole•min^{-1}•mg^{-1} (2,26). For the nonSe-GSH-Px from sheep lung cytosol, a K_m of 77 μM and a V_{max} of 960 nmoles•min^{-1}•mg^{-1} were calculated (2).

Since the endoperoxide of PGH_2 was found to be a good substrate for the nonSe-GSH-Px, [1-^{14}C]PGG_2 was also tested as a substrate for this enzyme and the products extracted and separated by TLC (2). The nonSe-GSH-Px-catalyzed reaction with PGG_2 produced a slightly different product profile from that observed with PGH_2. A band that did not correspond to any of the classical PGs was detected on TLC (2). This new product, referred to as a novel compound, migrated between PGE_2 and $PGF_{2\alpha}$ on the TLC plate. Spraying the plate with N,N-dimethyl-*p*-phenylenediamine produced a pink color in the area of the novel compound, indicating the presence of a hydroperoxide moiety in the molecule. A significant amount of $PGF_{2\alpha}$ was also formed during this short time period, as well as a minute amount of 15-hydroperoxy PGE_2. When the reaction was allowed to proceed for 10 min in the presence of the nonSe-GSH-Px, $PGF_{2\alpha}$ was the major product formed and the amount of the novel compound was reduced drastically. To investigate the nature of the unknown hydroperoxide compound, Se-GSH-Px was added to the reaction mixture one minute after initiating the reaction with PGG_2 to reduce the hydroperoxy group. The band corresponding to the novel compound completely disappeared, and the amount of $PGF_{2\alpha}$ increased correspondingly. The structure of the novel compound was verified as 15-hydroperoxy-$PGF_{2\alpha}$ (15-$HPPGF_{2\alpha}$) by $Pb(OAc)_4$ dehydration followed by co-chromatography with authentic 15-keto-$PGF_{2\alpha}$, as well as, by mass spectral analysis of the compound after reduction with Se-GSH-Px (2). A time-course of the reaction with PGG_2 revealed that the endoperoxidase activity of nonSe-GSH-Px is sufficiently more active than the hydroperoxidase activity to result in a significant accumulation of 15-$HPPGF_{2\alpha}$ (Fig. 2) (2). $PGF_{2\alpha}$ was the major product on longer incubations with the nonSe-GSH-Px indicating that the enzyme can reduce both the endoperoxide and hydroperoxide groups of the molecule.

Based on these results an alternative pathway for $PGF_{2\alpha}$ synthesis was proposed (2) and is presented schematically in Fig. 3. In the presence of active Se-GSH-Px, or in the event that PGES-associate peroxidase activity is more dominant, PGG_2 would be converted to PGH_2 and subsequently transformed by a specific enzyme to $PGF_{2\alpha}$, PGE_2, PGD_2, TXA_2, or PGI_2 depending on the tissue specificity (Fig. 3A). If the Se-GSH-Px is not very active, as might occur in some tissues in the Se-deficient state, abundant nonSe-GSH-Px could convert a substantial portion of the PGG_2 to $PGF_{2\alpha}$ via 15-HPPGF$_{2\alpha}$ (Fig. 3B). Even if PGH_2 is formed as the predominant intermediate, GST isozymes with nonSe-GSH-Px activity would convert the endoperoxide to $PGF_{2\alpha}$, thus reducing the amount of other eicosanoid products by competing for the PGH_2 substrate.

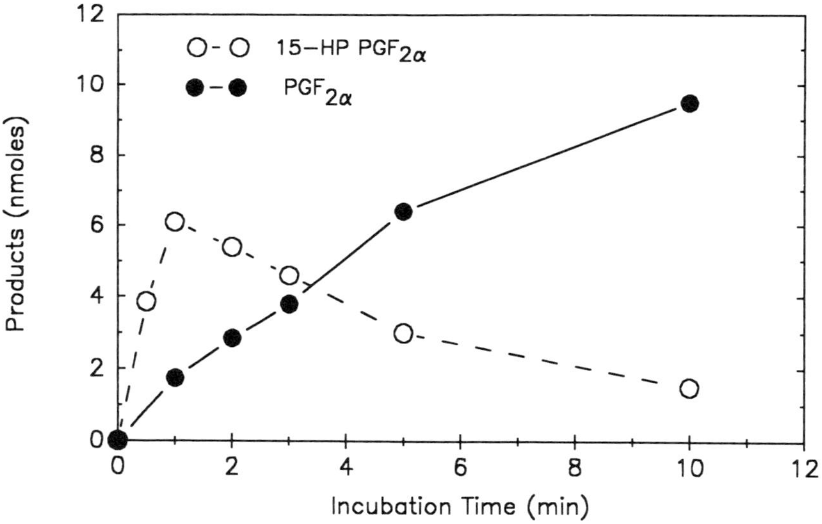

Figure 2. *Time Course of Sheep Lung GSTCII-Catalyzed Conversion of PGG_2 to $PGF_{2\alpha}$.* The conversion of PGG_2 to $PGF_{2\alpha}$ was carried out for varying periods of time as indicated. Figure taken from Hong *et al.* (2), and the procedure for standard assays and product analysis is described therein.

Specificity of nonSe-GSH-Px Isozymes in the Transformation of PGH_2 to $PGF_{2\alpha}$

The GSTs are a large family of multifunctional proteins which exist as distinct isomeric forms in most mammalian tissues. They are dimeric proteins composed of subunits with molecular weights

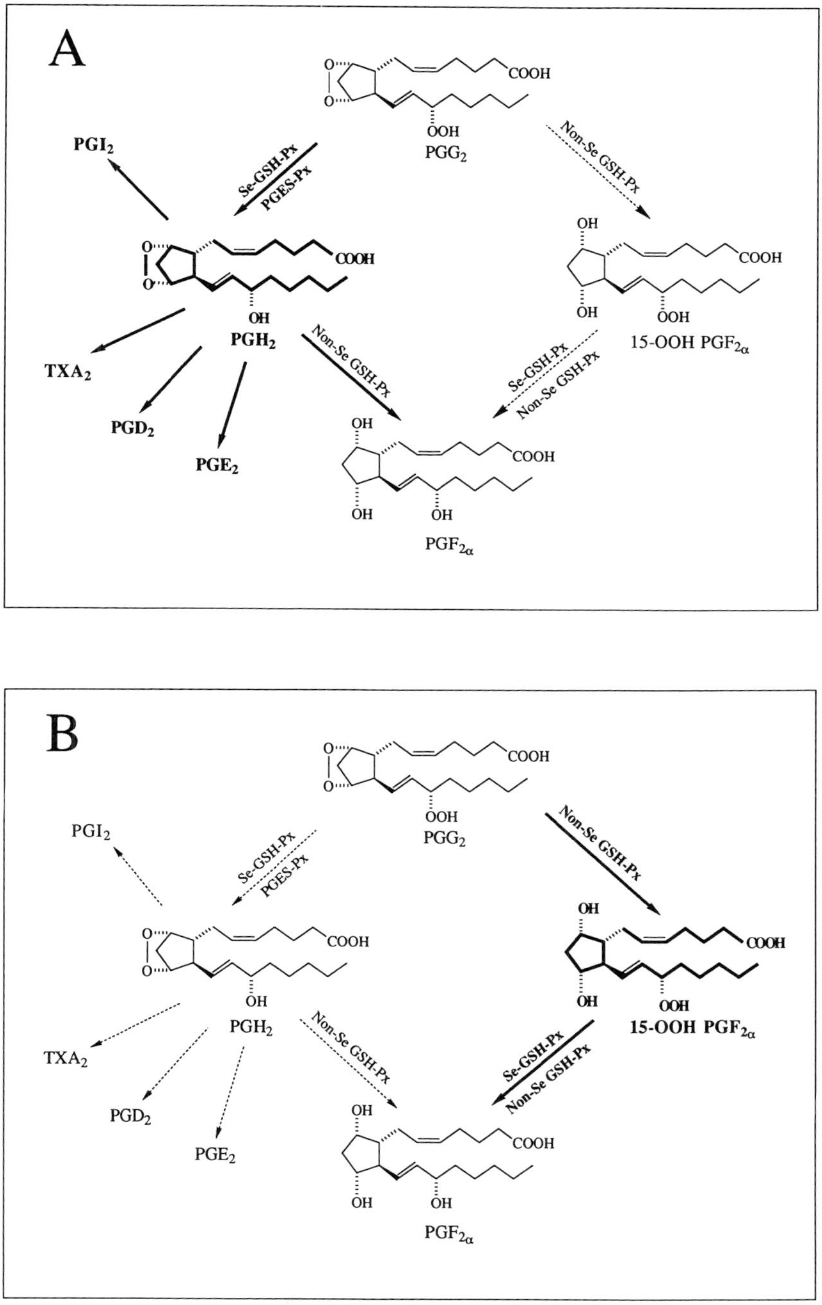

Figure 3. *Proposed Pathway of $PGF_{2\alpha}$ Biosynthesis Involving Se-GSH-Px and nonSe-GSH-Px.* **A.** Se-GSH-Px dominant; **B.** nonSe-GSH-Px dominant.

ranging from 23,000 to 28,000 (17). In rat liver the majority of the GSTs are either homo- or heterodimers of three subunit types: Ya (M_r=25,600), Yb (M_r=27,000), and Yc (M_r=28,000) (40,41). Different GST isozymes within a given tissue can be distinguished by their catalytic, physical, and immunological properties and by their primary sequence (17,40-45). They do, however, share a common feature in that all of them catalyze the nucleophilic attack of GSH on electrophilic centers of a wide spectrum of organic compounds, including organic peroxides (17,46). Even with this common property, isozymes exhibit different rates toward various electrophilic substrates. This latter phenomenon makes it more interesting to understand the tissue specific expression of GSTs as well as to explore, heretofore unidentified, physiological functions in addition to their role in the detoxication process. One such function would involve GSTs in the conversion of PGH_2 to $PGF_{2\alpha}$. The group of GSTs exhibiting nonSe-GSH-Px activity are basic proteins and belong to the alpha group of isozymes which are composed of both the Ya and Yc subunit types (2,17,19,26,41).

To determine which form of the GSTs are most active in catalyzing the reduction of PG endoperoxides, we screened the purified GST isozymes from several species and tissues for $PGF_{2\alpha}$ forming activity (2,26). The specific activities of individual GST isozymes isolated from sheep lung and seminal vesicles, and rat and human liver are presented in Table 1. $PGF_{2\alpha}$ was the major product formed enzymatically from PGH_2 in reactions of the three cationic GSTs from sheep lung. The seven anionic isozymes, A_I-A_{VII}, catalyzed the isomerization of PGH_2 primarily to PGE_2 and PGD_2, with very little $PGF_{2\alpha}$ formation. Similarly, in sheep seminal vesicles only the cationic isozyme was capable of converting PGH_2 to $PGF_{2\alpha}$. Also, we have previously shown that the highest PG endoperoxidase activity of rat liver GSTs is associated with the cationic isozymes (26). Human liver cationic GSTs were also found to be the most active in the conversion of PGH_2 to $PGF_{2\alpha}$ (2). Furthermore, the cationic GSTs with the most $PGF_{2\alpha}$ forming activity from each of these tissues contained only Ya size subunits (2,26). This association of PG endoperoxide reduction with Ya size subunits is consistent with the fact that rat brain cytosolic GSTs which lack this subunit type (47) showed very little $PGF_{2\alpha}$ forming activity. In general, the isozymes possessing high nonSe-GSH-Px activity toward cumene hydroperoxide also showed the greater endoperoxidase activity with PGH_2 as substrate.

Table 1. *Comparison of Individual GST Isozymes from Several Tissues for the Conversion of PGH_2 to $PGF_{2\alpha}$*[a,b]

Rat Liver GSTs		Sheep lung GSTs		Sheep SV GSTs		Human liver GSTs	
Isozyme[c] (subunit)	$PGF_{2\alpha}$ formation	Isozyme	$PGF_{2\alpha}$ formation	Isozyme	$PGF_{2\alpha}$ formation	Isozyme	$PGF_{2\alpha}$
AI (Yb)	36.0	AI	12.0	AI	13.0	AI	30.0
AII (Yb)	100.0	AII	4.0	AII	N.D.[d]	AII	28.0
CIV (Ya)	1936.0	AIII	1.2	AIII	2.0	CI	131.0
CV (Yb)	168.0	AIV	3.2	N	4.4	CII	609.0
CVI (YaYc)	427.0	AV	1.2	C (Ya)	1280.0	CIII	679.0
CVIII (Yb)	90.0	CI (Ya)	260.0			CIV	710.0
CIX (Yc)	200.0	CII (Ya)	650.0			CV	703.0
CX (Yc)	196.0	CIII (Ya)	826.0				

[a] Table adapted from Hong *et al.* (2).
[b] All values are expressed as specific activities (nmoles · min^{-1} · mg^{-1}).
[c] The numbering system for the GST isozymes is based on the increasing order of their pI values. A, anionic; C, cationic; N, neutral.
[d] N.D. = No detectable activity.

Effect of Dietary Selenium Status on the GSH S-Transferase and $PGF_{2\alpha}$ Forming Activities in Rat Liver

The relationship between Se-GSH-Px and nonSe-GSH-Px in eicosanoid metabolism and the relevance of the alternative pathway for $PGF_{2\alpha}$ formation were also investigated utilizing Se-deficient rat livers (48). It is well established that dietary deficiency of Se results in a drastic decrease in Se-GSH-Px activity in a variety of rat tissues, as well as, an increase in GST with nonSe-GSH-Px activity (49,50). This alteration of the activities of the two types of GSH-Px enzymes was used to provide insight into the relative importance of these two enzymes in prostanoid formation in a given tissue. We examined the change in GST activity, nonSe-GSH-Px activity, and both protein content and mRNA levels for individual GST subunit types. These parameters were correlated with the PG endoperoxide reductase activities of cytosol samples obtained from Se-deficient or Se-adequate rat livers (48).

We found the increased GST activity toward 1-Cl-2,4-dinitrobenzene in Se-deficient livers to be associated with an increase in the

Table 2. *The Effect of Dietary Status of Selenium on GSH-Peroxidases, GSH-S-Transferase, and PG-Endoperoxidase in Rat Liver Cytosol*

Analysis	Dietary Status		p value[g]
	+Se	-Se	
Se-GSH-Px Activity[a]	387.7±46	5.5±2	<0.01
nonSe-GSH-Px Activity[b]	175.4±46	231.8±11	<0.05
GST Activity[c]	854.6±138	1532.0±149	<0.01
GST Protein Content[d]			
Ya	1.0±0.29	2.53±0.85	<0.05
Yb	1.0±0.40	1.49±0.61	NS
Yc	1.0±0.27	1.79±0.49	<0.05
GST Poly(A)RNA levels[e]			
Ya	1.0	2.1	
Yc	1.0	1.7	
Total PG endoperoxidase[f]			
Total Activity	2.89±1.7	6.51±2.3	<0.05
plus S-decylGSH (250 μM)	1.68±1.7	2.40±2.6	NS

[a] MEAN±SD, n=6, activity expressed as μmoles H_2O_2 reduced $\cdot$ min^{-1} $\cdot$ mg^{-1}.
[b] MEAN±SD, n=6, activity expressed as μmoles cumene hydroperoxide reduced $\cdot$ min^{-1} $\cdot$ mg^{-1} after subtracting Se-GSH-Px activity.
[c] MEAN±SD, n=6, activity expressed as μmoles 1-Cl-2,4-dinitrobenzene conjugated $\cdot$ min^{-1} $\cdot$ mg^{-1}.
[d] MEAN±SD, n=4, protein content expressed relative to Se controls, values from (48).
[e] MEAN, duplicate analysis, mRNA levels expressed relative to Se controls, values from (48).
[f] MEAN±SD, n=6, activity expressed as nmoles PGH_2 converted to $PGF_{2\alpha}$ $\cdot$ min^{-1} $\cdot$ mg^{-1}.
[g] Level of significance between means confirmed by Students two-tailed t-test after determining dietary treatment effect by analysis of variance.

protein content of specific subunit types (Table 2.). Western blot analysis using antisera raised against specific GST isozymes revealed that Se-deficiency resulted in a 2.5-fold increase in the amount of the Ya subunit and a 1.8-fold increase in the amount of the Yc subunit, but no change in the amount of the Yb subunit (48). The results of Northern blot analysis of poly(A)RNA isolated from the Se-deficient and Se-adequate livers mirrored the observed changes in protein content. Using specific probes prepared from cDNA clones of individual GST subunits, we observed a 2.1-fold increase in the relative levels of mRNA coding for the Ya subunit, and a 1.7-fold increase in the relative mRNA levels coding for the Yc subunit in the Se-deficient rat liver (48). We concluded that Se deficiency leads to the induction of specific rat liver GST isozymes.

When PG endoperoxidase activity was assessed using [1-^{14}C]-PGH_2 as substrate (48), we observed a greater than 2-fold increase in the total conversion of PGH_2 to $PGF_{2\alpha}$ in the Se-deficient rat liver cytosol (Table 2). In the presence of S-decylGSH, an efficient GST inhibitor (51), the $PGF_{2\alpha}$ formation was inhibited by ~ 40% in control samples and >60% in the Se-deficient samples. No significant change in the GST-independent $PGF_{2\alpha}$ formation was observed between the two treatments. These results suggest that the increase in the conversion of PGH_2 to $PGF_{2\alpha}$ observed in Se-deficient rat liver was due entirely to the induction of specific GST isozymes responsible for PG endoperoxidase activity in that tissue. In rat lung, however, dietary Se status had no effect on the conversion of PGH_2 to $PGF_{2\alpha}$. This result can be explained by the observation that rat lung contains unusually low amounts of nonSe-GSH-Px activity (49,50).

Several situations other than Se deficiency have been shown to result in the induction of specific GST isozymes. Rats treated with phenobarbital and 3-methylcholanthrene, chemical inducers of GSTs, also show an elevation in their liver levels of mRNA coding for the Ya and Yb size subunits (44,52,53). Elevation in transcriptional activity of the GST genes, triggered by drugs, may account for this increase in GST mRNA levels (54). Although the induction pattern is not the same, a similar process may occur in Se deficiency where accumulation of substrates for nonSe-GSH-Px could activate the transcription of the specific GST genes which would result in higher nonSe-GSH-Px activity. This further implies that environmental factors and pathological conditions which result in increased lipid peroxidation or otherwise increased peroxide levels might also induce changes in GST expression.

In summary, the GSH-Pxs have an important biosynthetic function in the formation of eicosanoids. Se-GSH-Px is very efficient in catalyzing the reduction of fatty acid hydroperoxides. Since this enzyme is widely distributed in mammalian tissues, it is probably responsible for the conversion of HPETEs to HETEs. The Se-GSH-Px is also an excellent candidate for the reduction of PGG_2 to PGH_2, especially in tissues where this enzyme activity is very high. The nonSe-GSH-Px enzymes are more active in catalyzing the reduction of the PG endoperoxide moiety, and are capable of forming $PGF_{2\alpha}$ from either PGG_2 or PGH_2. They are also induced in response to Se deficiency, possibly due to accumulation of peroxides. Further investigation using Se deprivation or oxidative stress to alter the GST isozyme profiles of other species and tissues coupled with studies of S-decylGSH effects on PG formation should contribute substantially to our understanding of the roles of nonSe-GSH-Px in prostanoid metabolism.

References

1. O'Brien, P. J. (1988) in *Cellular Antioxidant Defense Mechanisms* **Vol. 1**, 73-87 (Chow, C. K., ed.) CRC Press, Boca Raton
2. Hong, Y., Li, C.-H., Burgess, J. R., Chang, M., and Reddy, C. C. (1989) *J. Biol. Chem.* **264**, 13793-13800
3. Flohe, L. (1982) in *Free Radicals in Biology* (edited by Pryor, W. A.) Academic Press, New York, **Vol. 5**, 223-254
4. Reddanna, P., Whelan, J., Burgess, J. R., Eskew, M. L., Hildenbrandt, G., Zarkower, A., Scholz, R. W., and Reddy, C. C. (1989) Ann. N.Y. Acad. Sci., **570**, 136-145
5. Whelan. J., Reddanna, P., and Reddy, C. C. (1986) Fed. Proc. **45**, 1935
6. Lands, W. E. M., Kulmacz, R. J., and Marshall, P. J. (1984) in *Free Radicals in Biology* (Pryor, W. A., ed.) **Vol. VI**, 39-61, Academic Press, New York, NY
7. Hemler, M. E., and Lands, W. E. M. (1980) *J. Biol. Chem.* **255**, 6253-6260
8. Marshall, P. J., Kulmacz, R. J., and Lands, W. E. M. (1987) *J. Biol. Chem.* **262**, 3510-3517
9. Moncada, S., Gryglewski, R. J., Bunting, S., and Vane, J. R. (1976) *Prostaglandins* **12**, 715-737
10. Schiavon, R., Freeman, Q. E., Guide, G. C., Perona, G., Zatti, M., and Kakkar, V. V. (1984) *Thromb. Res.* **34**, 389-396
11. Flohe, L., Gunzler, W. A., and Schock, H. H. (1973) *FEBS Letters* **32**, 132-134
12. Rotruck, J. T., Pope, A. L., Ganther, H. E., Swanson, A. B., Hafeman, D., and Hoekstra, W. G. (1973) *Science* **179**, 588-590
13. Takahashi, K., Avissar, N., Whitin, J. C., and Cohen, H. J. (1987) Arch. Biochem. Biophys. **256**, 672-686
14. Maddipati, K. R., and Marnett, L. J. (1987) J. Biol. Chem. **262**, 17398-17403
15. Lawrence, R. A., and Burke, R. F. (1976) *Biochem. Biophys. Res. Commun.* **71**, 952-958
16. Prohaska, J. R., and Ganther, H. E. (1976) *Fed. Proc.* **36**, 1094
17. Mannervik, B. (1985) *Adv. Enzymol. Related Areas Mol. Biol.* **57**, 357-427
18. Reddy, C. C., Tu, C.-P. D., Burgess, J. R., Ho, C. Y., Scholz, R. W., and Massaro, E. J. (1981) *Biochem. Biophys. Res. Commun.* **101**, 970-978
19. Reddy, C. C., Burgess, J. R., Gong, Z. Z., Massaro, E. J., and Tu, C.-P. D. (1983) *Arch. Biochem. Biophys.* **224**, 87-101

20. Li, C.-H., Srikumar, K., Burgess, J. R., Hong, Y., Hildenbrandt, G., Rao, M. K., Reddanna, P., Li, N., Tu, C.-P. D., and Reddy, C. C. (1986) *Contemp. Themes in Biochem.* **6**, 662-663
21. Lawrence, R. A., and Burk, R. F. (1978) J. Nutr. **108**, 211-215
22. Gale, P. H., and Egan, R. W. (1984) in *Free Radicals in Biology* (Pryor, W. A., ed.) **Vol. VI**, pp. 1-38, Academic Press, New York, NY
23. Marnett, L. J. (1984) in *Free Radicals in Biology* (Pryor, W. A., ed.) **Vol. VI**, pp. 63-94, Academic Press, New York, NY
24. Kulmacz, R. J., and Lands, W. E. M. (1983) *Prostaglandins* **81**,1227-1234
25. Ohki, S., Ogino, N., Yamamoto, S., and Hayaishi, O. (1979) *J. Biol. Chem.* **254**, 5061-5068
26. Chang, M., Hong, Y., Burgess, J. R., Tu, C.-P. D., and Reddy, C. C. (1987) *Arch. Biochem. Biophys.* **259**, 548-557
27. Hamberg, M., Svensson, J., and Samuelsson, B. (1975) *Proc. Natl. Acad. Sci. USA* **72**, 2994-2998
28. Moncada, S., Gryglewski, R. S., Bunting, S., and Vane, J. R. (1976) *Nature* **263**, 663-665
29. Oliw, E., Granstrom. E., and Angaard, E. (1983) in *Prostaglandins and Related Substances* (Pace-Asciak, C., and Granstrom, E., eds) pp. 1-44, Elsevier, Amerstdam
30. Lin, Y.-M., and Jabarak, J. (1978) *Biochem. Biophys. Res. Commun.* **81**, 1227-1234
31. Kruger, S., and Schlegal, W. (1986) *FEBS Lett.* **157**, 481-485
32. Watanabe, K., Shimizu, T., and Haiyaishi, O. (1981) *Biochem. Int.* **2**, 603-610
33. Wong, P. Y.-K. (1981) *Biochim. Biophys. Acta.* **659**, 169-178
34. Hamberg, M., and Samuelsson, B. (1967) *J. Biol. Chem.* **242**, 5336-5343
35. Qureshi, Z., and Cagen, L. M. (1982) *Biochem. Biophys. Res. Commun.* **104**, 1255-1263
36. Burgess, J. R., Hong, Y., Chang, M., Rao, M. K., Tu, C.-P. D., and Reddy, C. C. (1985) *Biochem. Biophys. Res. Commun.* **142**, 441-447
37. Watanabe, K., Iguchi, Y., Iguchi, S., Arai, Y., Hayaishi, O., and Roberts, L. J., II, (1986) *Proc. Natl. Acad. Sci. USA* **83**, 1583-1587
38. Liston, T. E., and Roberts, J. L., II (1985) *Proc. Natl. Acad. Sci. USA* **82**, 6030-6034
39. Watanabe, K., Yoshida, R., Shimizu, T., and Haiyaishi, O. (1985) *J. Biol. Chem.* **260**, 7035-7041
40. Mannervik, B., and Jensson, H. (1982) *J. Biol. Chem.* **257**, 9909-9912
41. Tu, C.-P. D., and Reddy, C. C. (1985) *J. Biol. Chem.* **260**, 9961-9964
42. Vander Jagt, D. L., Hunsaker, L. A., Garcia, K. B., and Royer, R. E. (1985) *J. Biol. Chem.* **260**, 11603-11610
43. Tu, C.-P. D., Lai, H.-C. J., Li, N., Weiss, M. J., and Reddy, C. C. (1984) *J. Biol. Chem.* **259**, 9434-9439

44. Telakowski-Hopkins, C. A., Rodkey, J. A., Bennett, C. D., Lu, A. Y. H., and Pickett, C. B. (1985) *J. Biol. Chem.* **260**, 5820-5825
45. Lai, H.-C. J., Grove, G., and Tu, C.-P. D. (1986) *Nucleic Acid Res.* **14**, 6101-6114
46. Jakoby, W. B., and Habig, W. H. (1980) in *Enzymatic Basis of Detoxification* (Jakoby, W. B., ed) **Vol.II**, pp.63-94, Academic Press, New York
47. Thyagaraju, K., Burgess, J. R., Reddanna, P., Hong, Y., Tu, C.-P. D., and Reddy, C. C. (1986) *Fed. Proc.* **45**, 1652
48. Chang, M., Burgess, J.R., Scholz, R. W., and Reddy, C. C. (1990) *J. Biol. Chem.* **265**, 5418-5423
49. Reddy, C. C., Grasso, T. L., Scholz, R. W., Labosh, T. J., Thomas, C. E., and Massaro, E. J. (1982) in *Prostaglandins and Cancer: First International Conference* (Powells, T. J., Ramwell, P. W., Honn, K. V., and Bockman, R. S., eds.) pp. 149-154, A. R. Liss, New York, NY
50. Reddy, C. C., Thomas, C. E., Scholz, R. W., Labosh, T. J., and Massaro, E. J. (1983) in *Advanced Modern Environmental Toxicology* (Mehlman, A., ed.) **Vol. 5**, pp. 395-410, Princenton Press, Princeton, New Jersey
51. Hildenbrandt, G., Tu,, C.-P. D., and Reddy, C. C. (1986) *Fed. Proc.* **45**, 1651
52. Pickett, C. B., Donohue, A. M., Lu, A. Y. H., and Hales, B. F. (1982) *Archives of Biochem. Biophys.* **215**, 539-543
53. Telakowski-Hopkins, C. A., Rodkey, J. A., Bennett, C. D., Lu, A. Y. H., and Pickett, C. B. (1985) *J. Biol. Chem.* **259**, 5182-5188
54. Ding, V. D.-H., and Pickett, C. B. (1985) *Arch. Biochem. Biophys.* **240**, 553-559

Non-Cyclooxygenase Prostaglandin Biosynthesis: Formation of PGA_3 Isomers Via an Allene Oxide

Alan R. Brash, Steven W. Baertschi, and Thomas M. Harris
Departments of Clinical Pharmacology, Vanderbilt University School of Medicine
Nashville, TN 37232 USA

Abstract

Allene oxides are unstable epoxides that have been detected recently as intermediates in the enzymic transformation of lipid hydroperoxides to ketols and cyclopentenones. In plants, this metabolic pathway is implicated in the biosynthesis of the growth hormone jasmonic acid. In certain types of coral, there is evidence for conversion of arachidonic acid via an allene oxide to prostaglandins of the E and A series, and to the prostaglandin analogues named clavulones. The precise sequence of reactions in the lipoxygenase/ allene oxide pathway to the prostaglandins remains to be elucidated. This paper includes a summary of the background to present work on the biochemistry of allene oxides. Also discussed are procedures we developed which have allowed the isolation and direct characterization of the very unstable allene oxides. Finally we describe our investigations of the conversion of arachidonic acid in the coral *Plexaura homomalla*, and some potential allene oxide intermediates in the non-cyclooxygenase metabolism of arachidonic acid to the prostaglandins.

Allene Oxides in Chemistry and Biochemistry

Allene oxides are unstable epoxides with a versatile chemistry. The epoxyene moiety in the allene oxide has the characteristics of an epoxide, a double bond, and an enol ether (1). The name allene oxide is itself a contraction of the terms "allene" and "epoxide", and comes from the fact that a simple synthesis of these derivatives involves epoxidation of an allene. The first report of isolation of an allene oxide came in 1968 (2). The molecule was protected from hydrolysis by bulky hydrophobic groups flanking the highly reactive epoxyene moiety (Scheme 1). Subsequent work has described the preparation of other allene oxides with varying degrees of stability (reviewed in ref. 1). In many other studies the allene oxide has been proposed as the key reactive intermediate, although usually isolation of the epoxide was not attempted.

Scheme 1

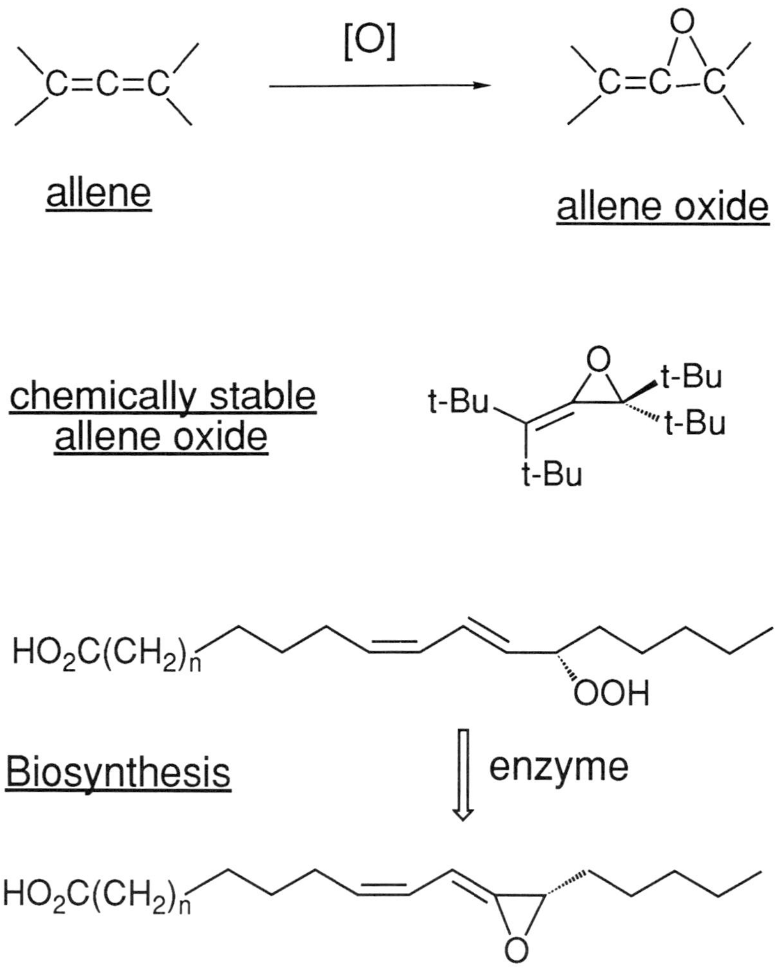

A well recognized chemical reaction of allene oxides is their potential to rearrange to cyclopentenones (see refs 1, and 3-7 for an introduction to the literature). Such cyclization products are related in structure to the PGA series prostaglandins. In the past two or three years it has been recognized that allene oxides are intermediates in *biochemical* pathways leading to the formation of this type of derivative. The immediate precursors in the biosynthetic schemes are the hydroperoxides derived via lipoxygenase metabolism of polyunsaturated fatty acids. One class of organism in which this biological transformation is now known to occur includes the sea whip and sea fan corals.

Prostaglandins in Coral

In 1969, Weinheimer and Spraggins reported the finding of 15R-prostaglandin A_2 (methyl ester 15-acetate derivative) in the sea whip coral *Plexaura homomalla* (8). *P. homomalla* is a member of the gorgonian class of corals and it is abundant in most areas of the Caribbean. Typical estimates of the prostaglandin content were 2-3% of the dry weight of the tissues. This concentration of prostaglandin methyl esters could not exist in aqueous solution - presumably the prostaglandin esters occur as droplets of oil.

In the course of the next few years it was discovered that substantial amounts of PGE_2 are also present in *P. homomalla* (9, 10). Also, specimens of the coral collected from different locations contained either 15R or 15S prostaglandins, and very occasionally both were found in the same colony. The prostaglandins with the 15S configuration are identical in structure to the prostaglandins which occur in mammalian systems.

Non-cyclooxygenase Prostaglandin Biosynthesis

The first studies on the biosynthesis of prostaglandins in *P. homomalla* were reported by E. J. Corey and colleagues. A key finding was that there was no evidence for cyclooxygenase activity in the coral (11). Also, there was no evidence for specific metabolism of exogenously added prostaglandin endoperoxides (12). Thus, it appeared that another metabolic pathway led to prostaglandin biosynthesis in the coral, the question remained as to the nature of this pathway?

Prostaglandin-Like Analogues in Plants

The first glimpse into a possible non-cyclooxygenase route to the prostaglandins came from studies on the metabolism of lipid hydroperoxides in plant tissues. In 1978, Zimmerman and Feng reported on the conversion of the 13S-hydroperoxide of linolenic acid to a PGA-like analogue in flaxseed (13). This transformation is now known to be a key step in the formation of the plant growth hormone, jasmonic acid (14).

Earlier work by Zimmerman and Vick (15,16) and Gardner (17) had established that linolenic acid 13S-hydroperoxide is avidly metabolized by flaxseed and by several other plants, and the major metabolite was known to be an α-ketol (12-keto,13-hydroxy). Important insight into the mechanism of formation of this α-ketol came from the work of Veldink et al (18), who demonstrated that the 13-hydroxyl is derived from water. Also, Gardner showed that the 13-hydroxyl is predominantly of the 13R configuration - opposite in chirality to the 13S-hydroperoxy substrate. He was the first to implicate a 12,13-epoxy intermediate in the transformation of hydroperoxide to α-ketol (19). Further studies established that the enzymic activity which formed the cyclopentenone (PGA analogue) could not be separated from the activity giving rise to the α-ketol (20). The reason for this is now clear - the enzyme which metabolizes the hydroperoxide forms an allene oxide, and the cyclopentenone and α-ketol are breakdown products of this unstable intermediate.

It was also shown that the enzyme would utilize the hydroperoxy derivatives of both C18 and C20 polyunsaturated fatty acids (21). As described later in this chapter, in our very recent work on the non-cyclooxygenase pathway to the prostaglandins, we have utilized the metabolism of the 15S-hydroperoxides of arachidonic and eicosapentaenoic acids by the hydroperoxide dehydrase (allene oxide synthase) in flaxseed.

Formation of Cyclopentenones in Coral

The first evidence in corals for a pathway analogous to the plant system came with the report by E. J. Corey and colleagues that the Japanese coral *Clavularia viridis* converted radiolabeled arachidonic acid to a cyclopentenone product they named "Pre-clavulone A" (22). By analogy to the plant system, Corey et al. made the prediction that this PGA analogue was derived via an 8-lipoxygenase pathway of arachidonic acid metabolism. The discovery of an 8-lipoxygenase in coral was reported by Bundy et al. (23), and it was shown that the enzyme is an 8R-lipoxygenase. We detected 8R-lipoxygenase metabolism in *P. homomalla*, and from analysis of hydrolysis and cyclization products we could implicate conversion of the 8R-hydroperoxide to an allene oxide (24).

Detection of Allene Oxides as Metabolites of Lipid Hydroperoxides

In the period of 1987-1988, there were reports from four groups who had deduced independently that the metabolism of hydroperoxides to α-ketols and cyclopentenones involves the direct conversion of hydroperoxide to an unstable allene epoxide (24-27). Further hydrolysis or cyclization reactions then give rise to the ketols and PGA analogues, respectively.

The most direct evidence for the formation of an allene oxide was provided by the elegant work of Mats Hamberg of the Karolinska Institute (26). Hamberg conducted short (30 second) incubations of the 13S-hydroperoxide of linoleic acid with a preparation of corn which contained the allene oxide synthase (hydroperoxide dehydrase). Reaction was terminated by addition of cold acetonitrile, immediately followed by rapid filtration and examination of the filtrate by uv spectroscopy. By this approach, the existence of the allene oxide could be demonstrated by the fleeting appearance of a characteristic conjugated diene chromophore. The chromophore disappeared over the course of a few minutes as the allene oxide degraded. By methanol trapping, the half-life of the allene oxide at 0°C and pH 7.4 was estimated as approximately 30 seconds (26).

Isolation and Characterization of Natural Allene Oxides

In the course of our studies on the mechanism of biosynthesis of the PGA related cyclopentenones, we obtained direct evidence that the immediate precursor of these derivatives is indeed the allene epoxide. The experimental approach we adopted involved isolation and purification of the allene oxide, followed by investigation of the products formed from the purified compound (28). After experimenting with and attempting to optimize many different parameters, we succeeded in recovering the allene oxides of linoleic and linolenic acids, as their methyl esters, using the following procedure:

(i) 5 second incubation of a high concentration of the 13S-hydroperoxy substrate with flaxseed proteins at 0°C.
(ii) Addition of ice cold hexane and rapid extraction.
(iii) Evaporation of the hexane down to a smaller volume, then addition of 1% ethanol plus diazomethane in ether.
(iv) After 10 seconds on ice, evaporation to dryness, immediately followed by reconstitution of the sample in cold hexane and storage at temperatures below -20°C.

Further analytical developments allowed the chromatographic purification of the allene oxide methyl esters by SP-HPLC. We found that the use of short columns, fast flow rates, and in particular, subzero temperatures would allow satisfactory chromatography of the allene oxide methyl esters, both on an analytical and semi-preparative (≈0.5mg) scale.

Use of these procedures permitted the isolation of the allene epoxides from incubations with flaxseed. The compounds were stable in hexane at -20°C or below, and they could be characterized by UV (Figure 1), ^{1}H-NMR and other spectroscopic methods (28).

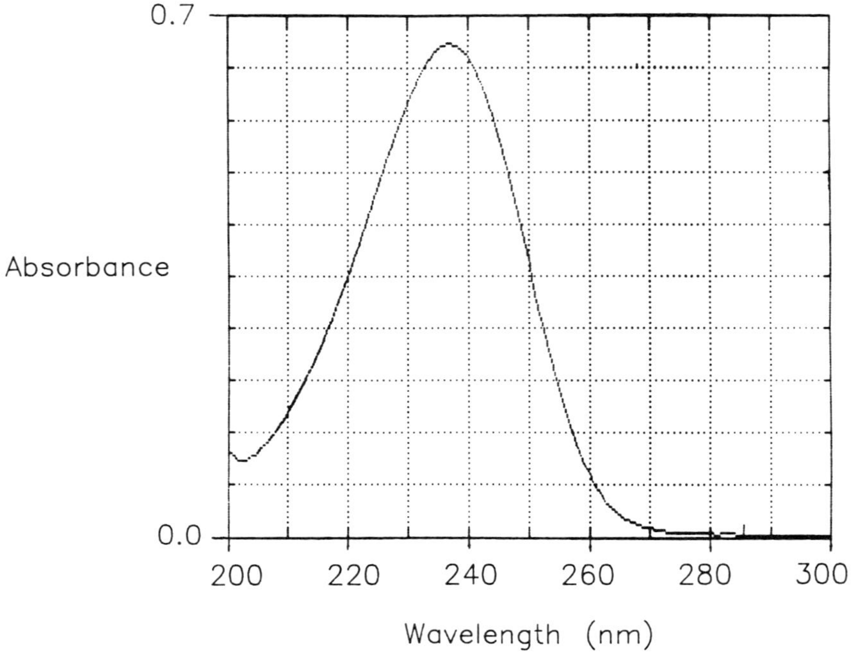

Figure 1. *UV Spectrum of Purified C18.3 Allene Oxide (Methyl Ester).* The spectrum was recorded in cold hexane. This allene oxide is closely related in structure to the product shown at the bottom of Scheme 1. The conjugated diene chromophore of the allene epoxide (lambda max at 237nm) has a characteristic smooth and rounded profile, lacking the angular shoulders typically seen in HETEs and HPETEs.

Development of a Biphasic System for Allene Oxide Biosynthesis

The above methods worked well when there was very fast metabolism of the hydroperoxide substrate. But application of these incubation and extraction procedures to the metabolism of 8R-HPETE in preparations of the coral *P. homomalla* did not give a significant yield of the desired allene oxide product. It was reasoned that a longer incubation time was required, and yet the allene oxide had to be protected from the rapid hydrolysis. As a potential solution to the problem, we experimented with continuous extraction with organic solvent *during* the incubation of hydroperoxide substrate with enzyme. The substrate hydroperoxide is substantially more polar than the allene oxide product, and consequently would have a significant presence in the aqueous phase. Furthermore, the enzyme preparations were acetone-washed protein, and therefore should survive contact with a nonpolar solvent such as hexane. Thus, the enzyme could metabolize the hydroperoxide, while virtually at the same instant, the allene oxide was extracted with hexane.

It was found that the optimal yield of allene oxide was obtained using around pH6, which is well below the pH optimum of the allene oxide synthase in the coral. At higher pH values the allene oxide is not extracted so completely, and hence it is exposed to the aqueous phase during the preparation. More important than the pH optimum is to allow a steady rate of metabolism together with a very efficient extraction of the product into hexane. The conditions finally adopted involved an incubation at 0°C and pH6, for 2 minutes, with continuous hexane extraction, followed by rapid centrifugation and recovery of the organic layer. In this way, isolation of an allene oxide from *P. homomalla* was achieved (29).

Enzymic Metabolism of Allene Oxides

Allene oxides are subject to nonenzymatic hydrolysis and cyclization reactions which can proceed with partial or complete loss of chirality of the parent molecule. The natural cyclopentenone products such as jasmonic acid and the prostaglandin derivatives in coral are chiral molecules (8, 30, 31). As we pointed out in describing the conversion of an allene oxide to a racemic cyclopentenone in acetone powders of flaxseed (32), for allene oxide involvement in these biosynthetic pathways, the cyclization must be directed to a single enantiomer, or, further metabolic steps must select out the correct enantiomer. The resolution of this issue in the plant systems has been clarified by the detection of an enzyme which will selectively cyclize the allene oxide formed from the 13S-hydroperoxide of linolenic acid (33). The cyclopentenone product is formed with retention of chirality of the parent allene oxide. It is not yet established whether an analogous enzyme can account for the formation of the single enantiomers of the clavulones and the prostaglandins in coral. Although the *nonenzymic* cyclization of certain substituted allene oxides to chiral products has been reported (5), the existence of novel enzymes for allene oxide metabolism seems distinctly likely.

Metabolism of More Complex Hydroperoxide Substrates

Potentially there are many pathways for conversion of arachidonic acid to prostaglandins via an allene oxide intermediate. One of the most straightforward possibilities involves the route shown on the left-hand side of Scheme 2. The 8R-HPETE has been converted to a 15-hydroxy derivative (alternatively a 15-hydroperoxide) *prior* to conversion to the allene oxide. Cyclization of the allene oxide might then lead directly to PGA_2. We made several attempts to detect this type of metabolism in *P. homomalla*, but it appeared that the 15-hydro(pero)xy derivatives of 8R-HPETE are not substrates for the allene oxide synthase in the coral.

Scheme 2

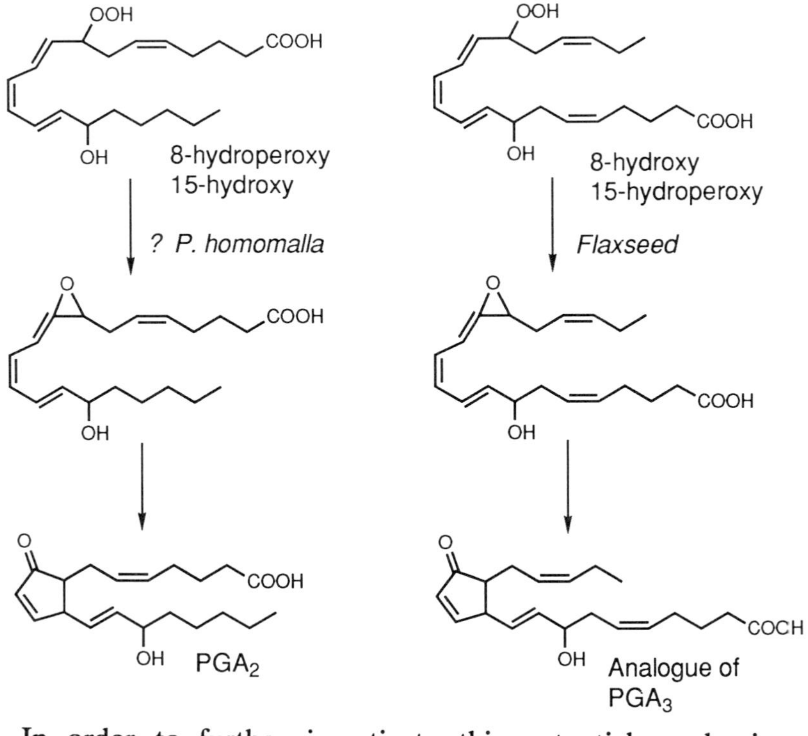

In order to further investigate this potential mechanism of prostaglandin biosynthesis, we turned to an analogous reaction in flaxseed. In flaxseed, the allene oxide synthase will metabolize 15S-hydroperoxides, and the additional hydroxy group is at C-8 (Scheme 2, right-hand side).

We found that the 8-hydroxy,15S-hydroperoxy derivatives of arachidonic and eicosapentaenoic acids are substrates for the allene oxide synthase in flaxseed. A complex mixture of products is formed (34). Interestingly, the configuration of the hydroxyl group at C-8 exerted a marked influence on the rate of reaction *and* on the pattern of products formed. The 8S-hydroxy diastereomers were substantially more reactive. The main products were α-ketols (hydrolysis products of the allene oxide) and a mixture of prostaglandin A analogues. The more slowly reacting 8R-hydroxy substrates were converted to several isomeric 13-hydroxy,14,15-epoxy derivatives in addition to the α-ketols and PGA analogues. Figure 2 summarizes the main pathways for conversion of the 8-hydroxy, 15S-hydroperoxy substrates to analogues of PGA_3. The evidence indicates that under the conditions of our experiments, two allene oxides (11**Z** and 11E) are formed from a single substrate, and that these give rise to PGA_3 analogues with *cis* and *trans* side chain configurations, respectively. The original report includes a more detailed discussion of these findings (34).

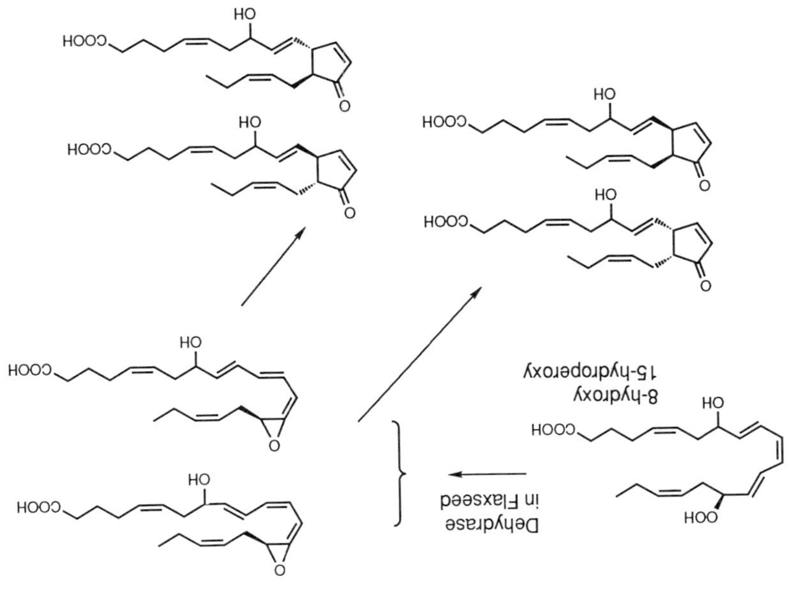

Figure 2. *Formation of PGA_3 Analogues from an 8-hydroxy, 15S-hydroperoxy Substrate in Flaxseed:* There is evidence for formation of two allene oxides. Nonenzymatic cyclization of the 11**Z** allene oxide gives PGA_3 analogues with *cis* side chain configuration, whereas the *trans* PGA_3 derivatives are formed from the 11E allene epoxide.

Implications for the Biosynthesis of Prostaglandins in Coral

An important point is established by the above reactions in flaxseed, namely that the hydroxylated allene oxide can cyclize to cyclopentenone derivatives which are close analogues of the A series prostaglandins. In the coral, there are two potential routes to the analogous hydroxylated allene oxide. We have been unable to detect its formation via the sequence analogous to the reactions in flaxseed (Scheme 2, left-hand side). Nevertheless, an alternative route remains as a feasible alternative. This could involve the conversion of 8R-HPETE to allene oxide, *followed by* formation of the 15-hydroxy allene oxide, and then by its cyclization to PGA_2.

Acknowledgments

The excellent technical assistance of William E. Boeglin was greatly appreciated. Our work on allene oxides is supported by NIH grant DK-35275 and center grant ES-00267.

References

1. Chan, T. H., and Ong, B. S. (1980) *Tetrahedron* **36,** 2369-2289
2. Camp, R. L., and Greene, F. D. (1968) *J. Am. Chem. Soc.* **90,** 7349
3. Roumestant, M. L., Malacria, M., Gore, J., Grimaldi, J., and Bertrand, M. (1976) *Synthesis* 755-757
4. Malacria, M., and Roumestant, M. L., (1977) *Tetrahedron* **33,** 2813-2817
5. Bertrand, M., Dulcere, J. P., and Gil, G. (1977) *Tetrahedron Lett.* **50,** 4403-4406
6. Doutheau A., Gore, J., and Diab, J. (1985) *Tetrahedron* **41,** 329-338
7. Kim, S. J., and Cha, J. K. (1988) *Tetrahedron Lett.* **29,** 5613-5616
8. Weinheimer, A. J., and Spraggins, R. L. (1969) *Tetrahedron Lett.* **59,** 5185-5188
9. Light, R. J., and Samuelsson, B. (1972) *Eur. J. Biochem.* **28,** 232-240
10. Schneider, W. P., Hamilton, R. D., Rhuland, L. E. (1972) *J. Am. Chem. Soc.* **94,** 2122-2123
11. Corey, E. J., Washburn, W. N., and Chen, J. C. (1973) *J. Am. Chem. Soc.* **95,** 2054-2055
12. Corey, E. J., Ensley, H. E., Hamberg, M., and Samuelsson, B. (1975) *J. Chem. Soc., Chem. Commun.* 277-278
13. Zimmerman, D. C., and Feng, P. (1978) *Lipids* **13,** 313-316
14. Vick, B. A., and Zimmerman, D. C. (1984) *Plant Physiol.* **75,** 458-461
15. Zimmerman, D. C. (1966) *Biochem. Biophys. Res. Commun.* **23,** 389-402
16. Zimmerman, D. C., and Vick B. A. (1970) *Plant Physiol.* **46,** 445-453

17. Gardner, H. W. (1970) *J. Lipid Res.* **11,** 311-32
18. Veldink, G. A., Vliegenthart, J. F. G., and Boldingh, J. (1970) *FEBS Lett.* **7**, 188-190
19. Gardner, H. W. (1979) *Lipids* **14,** 208-211
20. Vick, B. A., and Zimmerman, D. C. (1981) *Plant Physiol.* **67,** 92-97
21. Vick, B. A., and Zimmerman, D. C. (1979) *Plant Physiol.* **63,** 490-494
22. Corey, E. J., Lansbury, P. T. Jr, and Yamada, Y. (1985) *Tetrahedron Lett.* **26**, 4171-4174
23. Bundy, G. L., Nidy, E. G., Epps, D. E., Mizsak, S. A., and Wnuk, R. J. *J. Biol. Chem.* **261,** 747-751
24. Brash, A. R., Baertschi, S. W., Ingram, C. D., Harris, T. M. (1987) *J. Biol. Chem.* **262,** 15829-15839.
25. Corey, E. J., d'Alarcao, M., Matsuda, S. P. T., Lansbury, P. T. Jr. (1987) *J. Am. Chem. Soc.* **109,** 289-290
26. Hamberg, M. (1987) *Biochim. Biophys. Acta* **920,** 76-84
27. Crombie, L., and Morgan, D. O. (1988) *J. Chem. Soc., Chem. Commun.* 558-560
28. Brash, A. R., Baertschi, S. W., Ingram, C. D., and Harris, T. M. (1988) *Proc. Natl. Acad. Sci. U.S.A.* **85,** 3382-3386
29. Brash, A. R. (1989) *J. Am. Chem. Soc.* **111,** 1891-1892
30. Demole, E., Lederer, E., and Mercier, D. (1962) *Helv. Chim. Acta* **45,** 675-685
31. Hill, R. K., and Edwards, A. G. (1965) *Tetrahedron* **21,** 1501-1507
32. Baertschi, S. W., Ingram, C. D., Harris, T. M., and Brash, A. R. (1988) *Biochemistry* **27,** 18-24
33. Hamberg, M. (1989) *Biochem. Biophys. Res. Commun.* **156,** 543-550
34. Brash, A. R., Baertschi, S. W., and Harris, T. M. (1990) *J. Biol. Chem.*, in press

Non-Cyclooxygenase Oxidative Formation of a Series of Novel Prostaglandins: Ramifications for the Measurement of Eicosanoids in Biological Fluids

Jason D. Morrow, M.D. and L. Jackson Roberts, II, M.D.
Division of Clinical Pharmacology, Departments of Pharmacology and Medicine
Vanderbilt University, Nashville, TN 37232 USA

Abstract

When fresh plasma from normal volunteers is analyzed by gas chromatography negative ion chemical ionization mass spectrometry, multiple PGF_2 compounds can be detected with levels ranging from 5-40 pg/ml. Interestingly, storage of plasma at -20°C for several months markedly increases the levels of these compounds to 1000-4000 pg/ml. Further studies attempting to understand this observation disclosed that alkaline hydrolysis of plasma lipids also yielded quantities of these compounds in the range of 1000-4000 pg/ml. Utilizing a number of techniques including deuteriated derivatives, hydrogenation, immunoreactivity with a $9\alpha,11\beta$-PGF_2 antibody, and electron ionization mass spectral analysis, convincing evidence was obtained that these compounds in both stored and base-treated plasma were, in fact, PGF_2 compounds. Formation of these compounds was found to occur by a non-enzymatic oxidative process in that the antioxidant, butylated hydroxytoluene, and the reducing agent, triphenylphosphine, markedly suppressed their formation. A mechanism for formation of these compounds is proposed involving endoperoxide intermediates which are directly reduced by naturally occurring biological substances to PGF_2 compounds. Formation of these compounds occurs very readily in biological fluids. This finding has important analytical ramifications for analysis of enzymatically derived PGF_2 compounds and probably other eicosanoids measured by immunoassay methods and physical methods of analysis such as gas chromatography mass spectrometry.

Introduction

Prostaglandin (PG) D_2 is metabolized in humans *in vivo* to $9\alpha,11\beta$-PGF_2 (1). In the course of developing a gas chromatography

Biological Oxidation Systems, Volume 2

negative ion chemical ionization mass spectrometric (GC NICI/MS) assay for this compound, analysis of plasma from normal human volunteers for 9α,11β-PGF_2 revealed the presence of what were presumed to be multiple PGF_2 compounds. Interestingly, storage of plasma at -20°C for several months was found to increase levels of these compounds in plasma approximately 50-fold compared to amounts detected in fresh plasma not subjected to prior storage.

The purpose of the studies reported herein, therefore, was to determine the processes involved leading to the detection of markedly elevated levels of these putative PGF_2 compounds in plasma during storage. We report that these prostaglandins are formed by a non-enzymatic oxidative process. A mechanism is proposed involving initial free radical attack of arachidonic acid with subsequent incorporation of oxygen and endoperoxide formation. These PGG_2-like compounds are then reduced by naturally occurring biological substances directly to PGF-ring compounds.

The observation that these compounds can be formed in biological fluids by non-cyclooxygenase mechanisms has important analytical ramifications for the measurement of eicosanoids.

Experimental Methods

Analysis of Plasma PGF_2 Compounds: One to 3 ml of plasma or plasma lipid extract was analyzed by GC NICI/MS for PGF_2 compounds as a pentafluorobenzyl ester, trimethylsilyl ether derivative using [$^2H_7$] 9α,11β-PGF_2 as an internal standard as previously described (2). Cyclic butylboronate derivatives were formed and analyzed as described (3).

Extraction, Purification, and Base Treatment of Plasma Lipids: Lipids from 1 to 5 ml of plasma from normal healthy volunteers were extracted according to the method of Bligh and Dyer (4). After evaporation of the chloroform phase, the residue was resuspended in 2 ml of methanol to which 2 ml of 15% KOH was added. The mixture was incubated at 40° C for one hour. Plasma lipid subclasses were treated in an identical manner after thin layer chromatography (TLC) separation and purification as described (5). The solutions were then acidified to pH3 with 1 N HCl and extracted, purified, and assayed for PGF2 compounds as described (1). In some cases, 0.002% butylated hydroxytoluene (BHT) or 5 mg triphenylphosphine was added to the plasma prior to lipid extraction.

Hydrogenation of Plasma: PGF_2 compounds from 5 ml of stored plasma or from the base-treated lipid extract of 1 ml of fresh normal plasma were resuspended in 200μl of ethanol to which 10 mg of platinum oxide was added and the solution bubbled with hydrogen for 2 minutes. Ethyl acetate was then added, centrifuged briefly, and the supernatant dried under N_2.

Radioimmunoassay (RIA) Analysis of Plasma: PGF_2 compounds from 1 ml aliquots of fresh plasma, stored plasma or base-treated plasma lipid extracts were purified by TLC and subjected to analysis using a commercial RIA kit with an anti-9α,11β-PGF_2 antibody (Amersham International, Amersham, U.K.)

Results

SIM Analysis of Plasma for PGF_2: Figure 1A shows the selected ion current chromatograms obtained from analysis of PGF_2 compounds in fresh plasma assayed without storage. The m/z 576 chromatogram represents the deuteriated 9α,11β-PGF_2 internal standard and the m/z 569 chromatogram demonstrates a series of peaks presumed to be PGF_2 compounds. Levels of compounds in the m/z 569 chromatogram range from 5-40 pg/ml. For this and subsequent chromatograms, levels of PGF_2 compounds will be based on the major peak eluting just to the left (4 seconds earlier) of the internal standard.

Figure 1B shows the SIM chromatograms of plasma obtained from a normal individual which had been stored for several months at -20°C. As is evident, the profile of m/z 569 peaks is virtually identical as that obtained in fresh plasma. However, levels of compounds in stored plasma were 1500 pg/ml, approximately 50-fold higher compared to levels in fresh plasma. This finding of markedly increased levels of presumed PGF_2 was confirmed in additional normal plasma samples which had been stored for several months.

Evidence that the m/z 569 Peaks Represent PGF_2 Compounds: Even though the patterns of m/z 569 peaks obtained from analysis of fresh plasma and stored plasma are similar, suggesting that these peaks represented the same compounds, and presumably PGF_2 compounds, selected ion monitoring mass spectrometry is not definitive proof of compound identification. Further, the relative abundance of different m/z 569 peaks has been observed to vary somewhat between individuals and in the same individual from day to day. Thus, additional proof was sought that the m/z 569 peaks in fresh and stored plasma actually represented PGF_2 compounds.

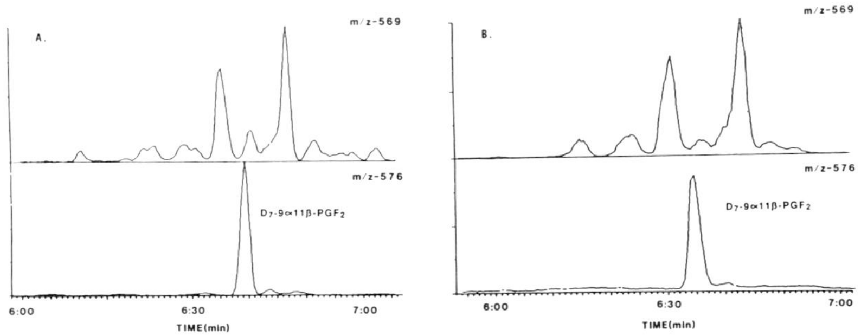

Figure 1. *Analysis of Fresh Plasma (A) and Plasma Which Had Been Stored at -20°C for Several Months (B) by GC NICI/MS.* The m/z 569 selected ion current chromatograms represent PGF_2 compounds and the m/z 576 chromatograms represent the $[^2H_7]$ $9\alpha,11\beta$-PGF_2 internal standard. Levels of PGF_2 compounds in fresh plasma (A) range from approximately 5 to 40 pg/ml and the levels in stored plasma (B) range from 200 to 2000 pg/ml.

First, no peaks were present when m/z 568 was monitored arguing that m/z 569 peaks were not natural isotope peaks of compounds with an ion of lesser weight. Second, analysis of the compounds in fresh and stored plasma as a $[^2H_9]$ trimethylsilyl ether derivative shifted the m/z 569 peaks upward 27 Da to m/z 596, indicating that the compounds had three hydroxyl groups. Third, the compounds were analyzed following hydrogenation to determine the number of double bonds present. Results of this experiment showed that virtually all m/z 569 compounds appeared 4 Da higher after hydrogenation, indicating that two double bonds were present.

Fresh plasma and stored plasma were then analyzed by RIA using an anti-$9\alpha,11\beta$-PGF_2 antibody. Although multiple isomeric forms of PGF_2 may be present in plasma, we believed there should be some cross reactivity with an anti-$9\alpha,11\beta$-PGF_2 antibody since the PGF ring is likely an important determinant of immune reactivity. Even though the degree of cross reactivity was unknown, we expected a substantial increase in immunoreactive levels of PGF_2 in stored plasma compared to fresh plasma. RIA levels in fresh plasma were 16 pg/ml. In stored plasma, the levels were substantially increased at 350 pg/ml providing further evidence that storage of plasma is associated with elevated levels of what appear to be PGF_2 compounds.

Although the above results provided very suggestive data that the compounds in fresh and stored plasma were PGF_2 compounds, additional evidence was considered important. Such evidence was obtained by analysis of stored plasma (50 ml) by electron ionization

MS (EI/MS). After purification, the compounds were analyzed as the methyl ester, trimethylsilyl ether derivative. Multiple scans of compounds were obtained over a 25 second period eluting from the capillary GC column which had the characteristic electron ionization mass spectrum of the methyl ester, trimethylsilyl ether derivative of $PGF_{2\alpha}$ and 9α, 11β-PGF_2, which have essentially identical mass spectra. Figure 2 shows one of these scans. Of note, relative ion abundances differed from scan to scan and some scans were obtained, which in addition to having ions characteristic of $PGF_{2\alpha}$, also had other prominent ions (see Figure 6). The importance of this finding is discussed later.

Nonetheless, the above data provide extremely strong evidence that the compounds in fresh and stored plasma are PGF_2 compounds.

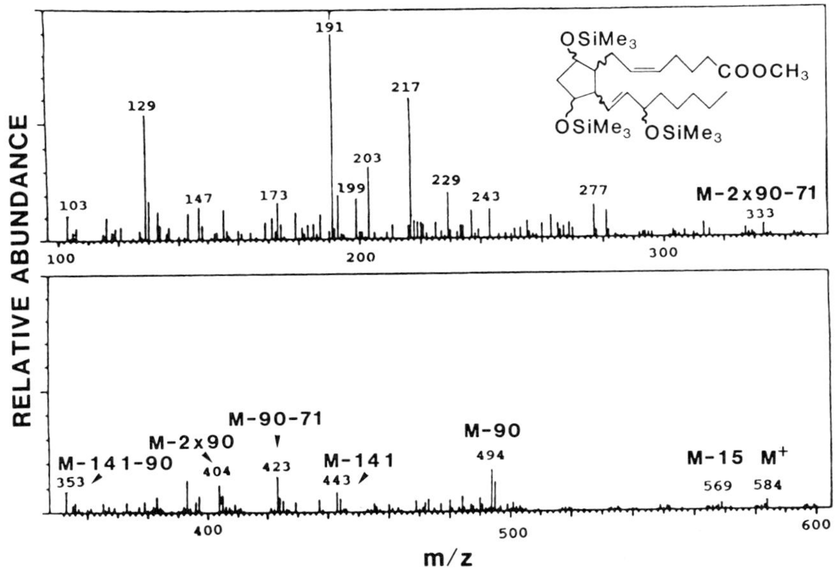

Figure 2. *Electron Ionization Mass Spectrum of a PGF_2 Compound as a Methyl Ester, Trimethylsilyl Ether Derivative from Analysis of Stored Plasma*

Studies Regarding the Mechanism of Formation of PGF_2 Compounds.

Once we had substantial evidence that the m/z 569 peaks in fresh and stored plasma represented PGF_2 compounds, we undertood studies to explain the formation of these compounds and how they could increase with storage of plasma.

Failure to increase levels in fresh plasma following incubation of plasma with proteases, precipitation of proteins with ethanol, and addition of detergents suggested that these compounds were not simply noncovalently bound to plasma proteins.

The possibility that these compounds could be present in the form of a labile conjugate was then considered. In support of this possibility it was found that treatment of fresh plasma with methanolic KOH increased levels of these presumed PGF_2 compounds approximately 50-fold with a virtually identical pattern of m/z 569 peaks.

Further analysis disclosed that, as with fresh and stored plasma, all m/z 569 peaks in base treated plasma had 3 hydroxyls and 2 double bonds. Further, EI/MS analysis of base treated plasma showed multiple PGF_2 compounds eluting from the capillary GC column over a 25 second period.

We then addressed the possibility that these compounds may be acylated into phospholipids and found that base treatment of a Folch lipid extract of plasma consistently yielded markedly increased levels of the compounds (2300 ± 640 pg/ml, mean ± S.D.) (n=11). Further supporting a connection between these compounds and phospholipids were the facts that 1) base treatment of commercially prepared arachidonyl phosphatidylcholine yielded 400pg of presumed PGF_2 compound per mg of phospholipid and 2) treatment of plasma lipid subclasses with KOH after TLC separation yielded nanogram levels of these compounds with the majority derived from the major arachidonic acid containing phospholipid, phosphatidylcholine.

These data thus collectively supported the hypothesis that PGF_2 compounds may be acylated into phospholipids. However, incubation of radiolabelled $PGF_{2\alpha}$ and $9\alpha,11\beta$-PGF_2 with HTC cells, a liver epithelial tumor cell line, did not result in uptake of radiolabel into cells. Although this does not exclude the possibility that PGF_2 compounds are acylated into phospholipids in plasma, other possibilities were considered.

That the PGF_2 compounds observed in plasma may be formed by nonenzymatic oxidation of phospholipid arachidonic acid was then considered, especially in view of the fact that others had previously shown that autoxidation of linolenic acid or hydroperoxylinolenate yields PGG-like endoperoxides (6-9). Therefore, we examined the effect of the addition of BHT or triphenylphosphine on the formation of PGF_2 compounds in plasma. Addition of BHT to the Folch lipid extract procedure before base hydrolysis reduced the yield of PGF_2 compounds after base hydrolysis by 86 ± 3% (n=4) ($p < 0.001$). Likewise triphenylphosphine decreased the yield by 91 ± 3% (n=4) ($p<0.001$).

Previous studies have shown that endoperoxides formed from autooxidation isomerize to PGD and PGE compounds but PGF compounds did not form unless an exogenous reducing substance was added (6,10-12). In this regard, Nugteren had shown that the reduction of endoperoxides to PGF_2 compounds can occur if the naturally occurring substances glutathione and hematin are added to a solution containing PGG_2 (12). Since plasma may contain potential reducing substances, we reasoned that endoperoxides formed may be directly reduced to PGF_2 compounds in human plasma. To address this possibility, we incubated a mixture of 10 μg of PGG_2 and PGH_2 with plasma at 20° for 4 hours and found a 1-11% conversion to $PGF_{2\alpha}$ in 4 experiments. Interestingly, consistent with Nugteren's previous findings, addition of glutathione (100 μM) and hematin (1 mM) to plasma containing PGG_2/PGH_2 increased the conversion to PGF_2 to 28%.

The above results thus suggested that the formation of PGF_2 compounds in plasma occurs by a non-enzymatic oxidative process involving formation of endoperoxide intermediates which are reduced directly to PGF_2 compounds. A mechanism of formation is proposed in Figure 3 which leads to the formation of 4 regioisomers of PGF_2. Each regioisomer consists of a mixture of 8 racemic diastereomers and thus, at least theoretically, a total of 64 compounds can be formed.

According to the above proposed mechanism in which PGG-like compounds are directly reduced to PGF_2 compounds, only PGF_2 in which the cyclopentane ring hydroxyls are oriented *cis* are formed. In contrast, if the PGG-like compounds isomerize to PGD and PGE-like compounds which are then reduced to PGF_2 compounds, a mixture of epimers of the C-9 and C-11 hydroxyls of the PGF_2 compounds would be present. We therefore examined the ability of PGF_2 compounds in fresh, stored and base treated plasma to form a cyclic boronate derivative when treated with butylboronic acid. As Figure 4 shows, a cyclic butylboronate derivative will form bridging the C9 and C11 oxygens of PGF_2 compounds only if the 2 hydroxyls are oriented *cis*. The major ion generated by the pentafluorobenzyl ester, butylboronate, trimethylsilyl ether derivative when analyzed by GC NICI/MS is m/z 491 which represents the loss of the pentafluoro-benzyl group from the derivatized PGF_2 molecule (2). Figure 5 shows the result of butylboronic acid treatment of stored plasma. In Figure 5A the results obtained from analysis of one half of the sample that was not treated with butylboronic acid are shown. Thus the typical pattern of PGF_2 compounds is present at m/z 569 and the deuteriated $9\alpha,11\beta$-PGF_2 compound is represented by the peak in the m/z 576 chromatogram. As is evident no compounds are present in the butylboronate chromatogram at m/z 491. In Figure 5B are shown results

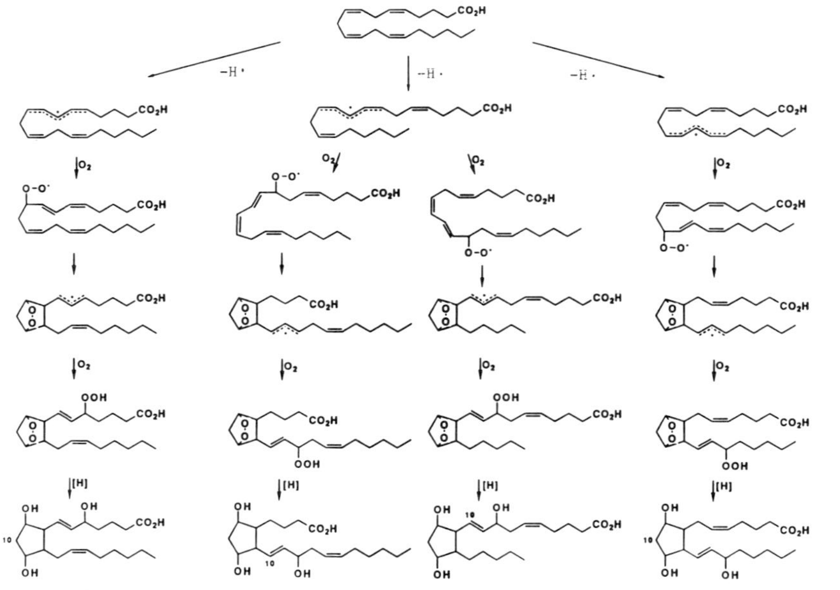

I II III IV

Figure 3. *Proposed Mechanism for Non-Cyclooxygenase Oxidative Formation of PGF_2 Compounds.* This pathway leads to the formation of four regioisomers of PGF_2 (I-IV). For simplicity, stereochemical orientation is not indicated in the figure. Each regioisomer theoretically is composed of a mixture of eight racemic diastereomers.

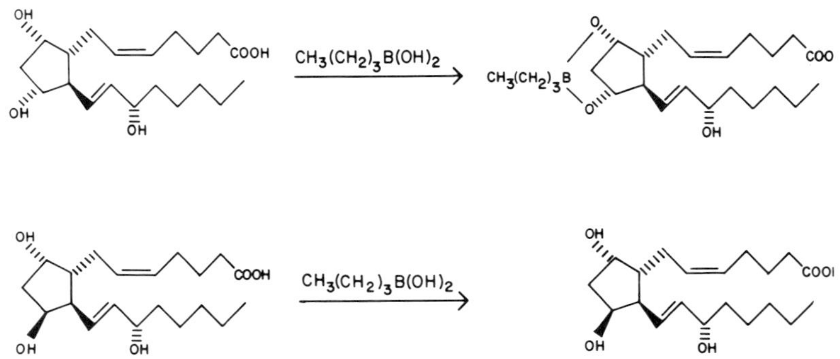

Figure 4. *Butylboronic Acid Treatment of PGF Ring Compounds Containing Cyclopentane Ring Hydroxyls Which Are Oriented cis (top) and trans (bottom).* Only compounds with *cis* ring hydroxyls are capable of forming a cyclic boronate derivative.

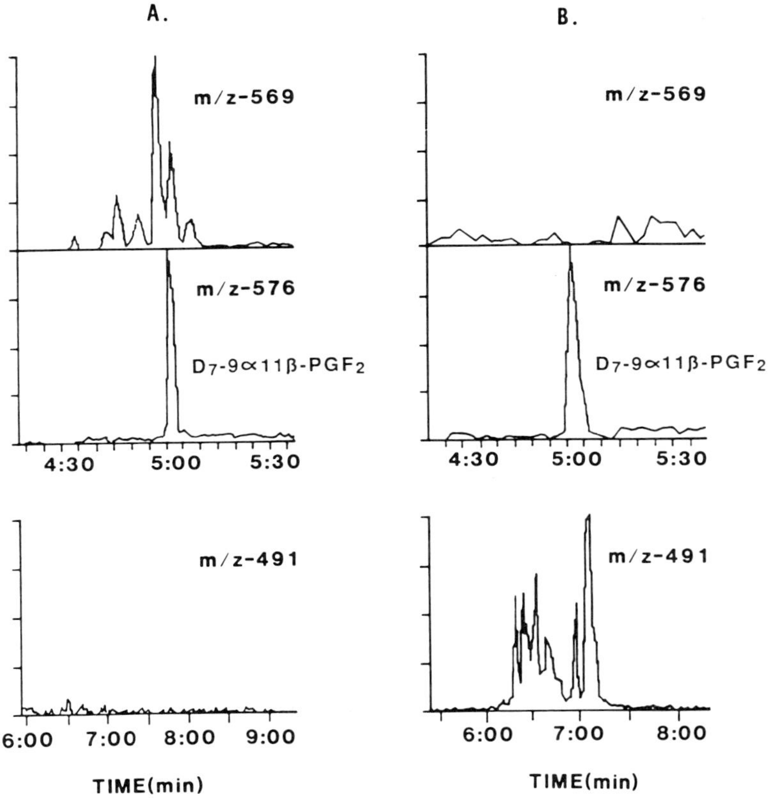

Figure 5. *Formation of Cyclic Boronate Derivatives of PGF$_2$ Compounds Present in Stored Plasma Following Treatment with n-Butylboronic Acid.* The m/z 569 and m/z 576 selected ion current chromatograms represent the pentafluorobenzyl ester, trimethylsilyl ether derivative of PGF$_2$ compounds and the [$^2H_7$] 9α,11β-PGF$_2$ internal standard, respectively. The m/z 491 chromatograms represent the pentafluorobenzyl ester, butylboronate, trimethylsilyl ether derivative of PGF$_2$ compounds. (A) Analysis of plasma which was not treated with butylboronic acid and (B) analysis of plasma which had been treated with butylboronic acid. The pattern of m/z 569 peaks in (A) differs somewhat from the pattern of m/z 491 peaks in (B). This is likely attributed to differences in GC characteristics of the cyclic boronate derivative of PGF$_2$ compounds which have not been converted to a boronate derivative.

obtained after treatment of the other one half of the sample with butylboronic acid. This resulted in the appearance of new intense peaks in the m/z 491 chromatogram with concomitant loss of peaks in the m/z 569 chromatogram. These results indicated essentially all

PGF_2 compounds present formed a cyclic boronate derivative and thus support the above proposed mechanism.

Additional support for the proposed mechanism was obtained from the results of the analysis of the PGF_2 compounds by EI/MS which was described previously. As mentioned, many of the mass spectra obtained contained ions characteristic of $PGF_{2\alpha}$ but, in addition, also contained ions not present in the mass spectra of $PGF_{2\alpha}$. An example of this is shown in Figure 6 in which there is a very prominent ion at m/z 243. Since this analysis was performed on a biological sample that was not extensively purified, it is possible these ions were generated from other coeluting impurities. However, the ion at m/z 243 would be expected in the mass spectrum of PGF_2 regioisomer III shown in Figure 3, since fragmentation is favored alpha to carbon atoms linked to a trimethylsilyl ether group (13). Two additional prominent ions in Figure 6 that are not present in the mass spectrum of $PGF_{2\alpha}$ are m/z 203 and m/z 213. The m/z 203 ion would be expected in the mass spectrum of regioisomer I resulting from fragmentation between C5 and C6 and m/z 213 would be expected in the mass spectrum of regioisomer II as a result of fragmentation between C11 and C12. These same ions, m/z 203, 213, and 243, were present in varying abundance in multiple scans over the 25-30 second period in which the other characteristic ions of $PGF_{2\alpha}$ were obtained. The presence of these ions is consistent with the existence of the 4 predicted regioisomers of $PGF_{2\alpha}$ shown in Figure 3 and suggest co-elution from the GC of a mixture of PGF_2 regioisomers.

Discussion

These studies show that PGF_2 compounds can be formed readily by a noncyclooxygenase oxidative mechanism involving plasma lipids. A mechanism for their formation is proposed involving endoperoxide intermediates which are directly reduced to PGF_2 compounds. Four regioisomers of PGF_2 are thus produced, each of which can be composed of 8 racemic diastereomers.

Formation of mono and bicyclic endoperoxides has been shown to occur *in vitro* from autoxidation of unsaturated fatty acids (6-9,11). In these studies, PGF-ring compounds did not form in significant amounts unless an exogenous reducing substance was added. However, we have presented evidence that naturally occurring biological substances in plasma reduce endoperoxides to PGF_2 compounds. This is not an unexpected finding in that plasma contains several reducing agents. For example, Nugteren has shown glutathione and hematin reduce PGG_2 to $PGF_{2\alpha}$ and Marnett *et al.* found that lipoic acid catalyzed endoperoxide reduction (12,14).

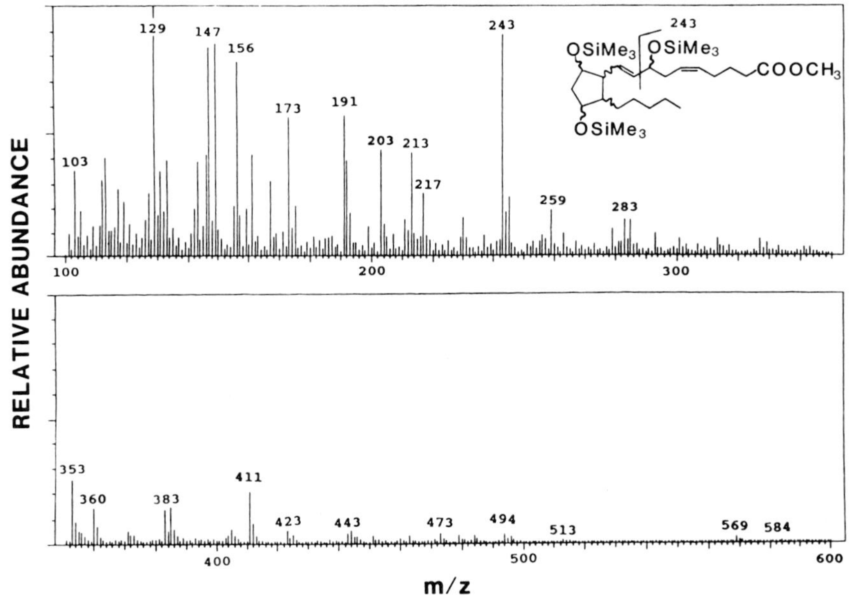

Figure 6. *Electron Ionization Mass Spectrum Obtained of a PGF_2 Compound in Base Treated Plasma as a Methyl Ester, Trimethylsilyl Ether Derivative*

Although the proposed mechanism of formation yields 4 regioisomers and a total of 64 compounds, it seems likely that some of the compounds are preferentially formed compared to others which may be formed in relatively minor quantities for kinetic or thermodynamic reasons. In this regard, it has been shown that bicyclic endoperoxides formed from autoxidation have side chain oriented *cis* almost exclusively (15).

This nonenzymatic oxidative formation of PGF_2 compounds appears to occur readily in biological fluids. Possibly relevant to this is the recent report that phospholipids are easily oxidized and that oxidation occurs more readily than unesterified arachidonic acid (16). Thus, in biological samples, the initial formation of the peroxy radical may occur more readily on a phospholipid, although this remains speculative.

Perhaps most importantly, analytic ramifications arise from these findings. Formation of PGF_2 compounds in plasma, and presumably other biological fluids, occurs very readily. Substantial quantities of these compounds form during storage. Therefore precautions need to be taken when analyzing any biological sample containing lipids for these compounds. Such precautions may include analyzing the sample immediately or addition of antioxidants. Although such precautions may minimize non-enzymatic formation of these compounds, it is important to recognize that they may not totally prevent their formation.

Although these studies have focused on the nonenzymatic formation of PGF_2 compounds, the analytical ramification of these findings probably also extends to the quantification of other eicosanoids. Although these PGF_2 compounds are formed by direct reduction of bicyclic endoperoxide intermediates, it seems not unlikely that the endoperoxides also isomerize to some extent to PGD_2 and PGE_2 compounds. The mechanism involved in formation of PGF_2 compounds is a multistep process involving peroxidation of arachidonic acid, cyclization, insertion of a second molecule of O_2, and reduction of the peroxy groups. Since prostaglandins are formed readily in biological fluids by this process, it would seem that one needs to be particularly concerned about nonenzymatic formation of hydroperoxy and hydroxyeicosatetraenoic acid products, which are also formed by lipoxygenase enzymes, since the formation of these compounds only requires the initial step of peroxidation of arachidonic acid (17,18). Clearly, for similar reasons, one must also be extremely cautious regarding nonenzymatic formation of epoxide P-450 metabolites of arachidonic acid.

The potential analytical problems that can arise from nonenzymatic formation of eicosanoids encompasses both physical methods of analysis such as GC/MS as well as immunoassay methods. For example, we found that an anti-$9\alpha,11\beta$-PGF_2 antibody which has only 0.0006% cross-reactivity with $PGF_{2\alpha}$ possessed considerable cross-reactivity with the nonenzymatically formed PGF_2 compounds, even though these compounds have *cis* cyclopentane ring hydroxyls as does $PGF_{2\alpha}$. Chromatographic purification of compounds prior to GC/MS analysis may separate out some of the PGF_2 compounds formed nonenzymatically, depending on the extent of purification. This may in part explain why RIA methods frequently measure higher levels than GC/MS assays and why levels measured by RIA following HPLC purification are frequently lower compared to RIA that does not involve HPLC purification prior to analysis. In addition, it is possible that nonenzymatic formation of prostaglandins may occur during the RIA assay procedure since samples are frequently incubated for several hours to permit antibody binding.

In summary, these studies describe noncyclooxygenase oxidative formation of PGF_2 compounds which is proposed to involve the formation of endoperoxide intermediates and then direct reduction by naturally occurring substances present in biological fluids. Furthermore, this appears to be a facile process which occurs readily in biological fluids. In fact, we have obtained recent evidence suggesting that noncyclooxygenase formation of these compounds is also occurring *in vivo* and that the levels reported here in fresh plasma appear to be formed endogenously (19). Clearly, these findings have important analytical ramifications regarding analyses of cyclooxygenase-derived $PGF_{2\alpha}$, PGF_2 compounds derived from 11-ketoreductase metabolism of PGD_2, and potentially other prostaglandins and eicosanoids.

Acknowledgements

This work was supported by Grant GM 42056 from the National Institutes of Health. JM was supported by Training Grant GM 07569 for the National Institutes of Health. The skillful technical assistance of William Zackert was greatly appreciated as was the help of Amanda Simpson in preparing this manuscript.

References

1. Liston, T.E., and Roberts, L.J. (1985) *J. Biol. Chem.* **260**, 13172-13180
2. Wendelborn, D.F., Seibert, K., and Roberts, L.J. (1988) *Proc. Natl. Acad. Sci. USA* **85**, 304-308
3. Pace-Asciak, C., and Wolfe, L.S. (1971) *J. Chromatogr.* **56**, 129-135
4. Bligh, E.G., and Dyer, W.J. (1959) *Canad. J. Biochem. Physiol.* **37**, 911-917
5. Moore, L., Chen, T., Knapp, H.R., and Landon, E.J. (1975) *J. Biol. Chem.* **250** 4562-4568
6. Pryor, W.A., Stanley, J.P., and Blair, E. (1976) *Lipids* **11**, 370-379
7. Pryor, W.A., and Stanley, J.P. (1975) *J. Org. Chem.* **40**, 3615-3617
8. Porter, N.A. Funk, M.O., Gilmore, D.W., Issac, S.R., Menzel, D.B., Nixon, J.R., and Roycroft, J.H. (1977) *Biochemical Aspects of Prostaglandins and Thromboxanes.* (Kharasch, N., and Fried, J., Eds) pp. 39-53, Academic Press, New York
9. O'Connor, D.E., Mihelich, E.D., and Coleman, M.C. (1984) *J. Amer. Chem. Soc.* **106**, 3577-3584
10. Hamberg, M., Svensson, J., Wakabayashi, T., and Samuelsson, B. (1974) *Proc. Natl. Acad. Sci. USA* **71**, 345-349
11. Nugteren, D.H., Vonkeman, H., Van Dorp, D.A. (1967) *Recl. Trav. Chim. Pays-Bas.* **86**, 1237-1245
12. Nugteren, D.H., and Hazelhof, E. (1973) *Biochim. Biophys. Acta* **326**, 448-461

13. Watson, J.T. (1985) *Introduction to Mass Spectrometry*, 2nd ed., pp. 172-173, Raven Press, New York
14. Marnett, L.J., and Bienkowski, M.J. (1977) *Biochemistry* **16**, 4303-4307
15. O'Connor, D.E., Mihelich, E.D., and Coleman, M.C. (1981) *J. Amer. Chem. Soc.* **103**, 224-226.
16. Frei, B., Stocker, R., and Ames, B.N. (1988) *Proc. Natl. Acad. Sci. USA* **85**, 9748-9752
17. Porter, N.A., Lehman, L.S., Weber, B.A., and Smith, K.J. (1981) *J. Amer. Chem. Soc.* **103**, 6447-6455
18. Porter, N.A. (1986) *Acc. Chem. Res.* **19**, 262-268
19. Morrow, J.D., and Roberts, L.J., manuscript in preparation.

Lipoxins: Bioactive Eicosanoids Generated by Lipoxygenase Interactions

Charles N. Serhan
Hematology Division, Department of Medicine, Brigham and Women's Hospital
75 Francis Street, Boston, MA 02115 USA

Abstract

Of the major lipoxygenases (5-, 12-, and 15-LO) observed in human tissues, the 5-LO is a key enzyme involved in the generation of leukotrienes, which are potent arachidonate-derived mediators (1). In view of the potential importance of LO products, studies were initiated to examine the consequences of initial lipoxygenation at carbon-15 of arachidonic acid as well as to investigate interactions between the major LO pathways (2,3). These studies led to the identification of a new series of biologically active eicosanoids termed lipoxins (Fig. 1). This chapter summarizes their isolation and biosynthesis as well as reviews recent results concerning their formation and biological activities.

Isolation and Structural Elucidation of the Lipoxins

While studying the transformation of [1-^{14}C]arachidonate (C20:4) in suspensions of mixed human leukocytes (*i.e.* neutrophils, eosinophils, basophils, etc.) with Samuelsson and Hamberg at the Karolinska Institute, we observed that labeled C20:4 was transformed and associated with polar compounds that had not been previously described. Since human leukocytes rapidly oxygenate exogenous C20:4 (without addition of other stimuli, *i.e.* A23187) at the C-15 position (4,5), we investigated the products associated with 15-lipoxygenation. In addition, results from several laboratories suggested that initial lipoxygenation at carbon-15 could lead to the formation of eicosanoids of biological interest (6-8). 15S-hydroxy-5,8,11-cis-13-trans-eicosatetraenoic acid (15-HETE), for example, had been identified as a major product of C20:4 in both normal and asthmatic human lung tissue (6). 15-HETE is also found in microgram quantities in bronchoalveolar lavage fluids obtained from patients with chronic stable asthma following antigenic challenge (7). Although 15-lipoxygenation of C20:4 is a major route of its

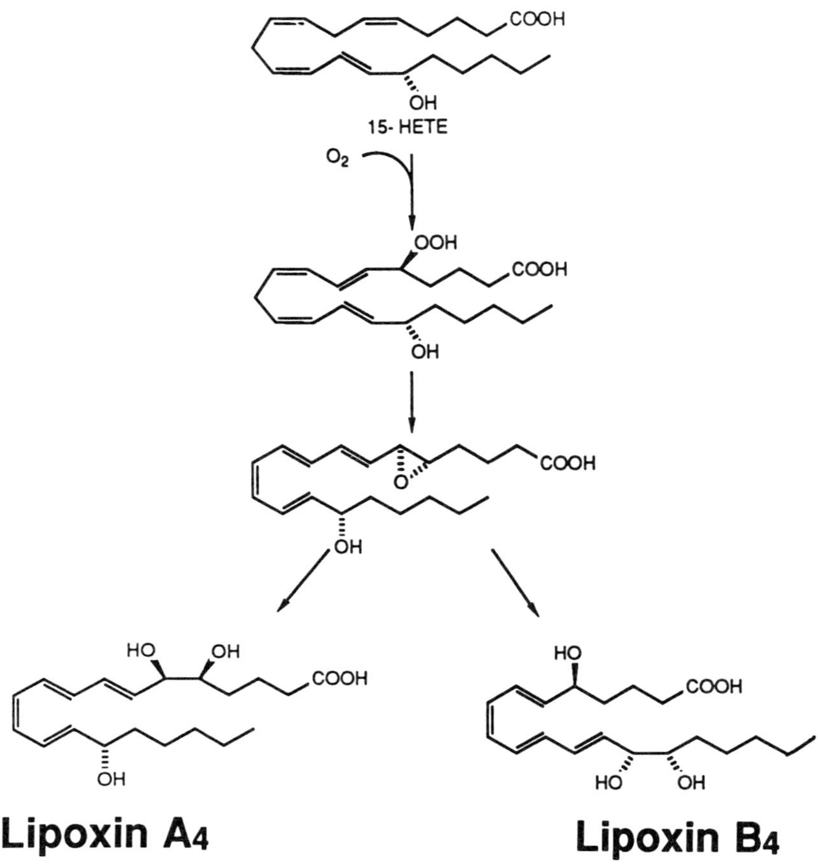

Figure 1. *Formation of Lipoxins by Activated Human Leukocytes* (transformation of 15-HETE; see text for details)

transformation in a wide range of mammalian tissues (reviewed in ref. 8), receptor-mediated activation of this enzyme system and general physiological role of its products has remained an area of interest.

To mimic cellular events and reaction sequence(s) that could have given rise to these novel polar compounds, as well as to prepare large amounts of the materials for further studies, 15-HPETE and 15-HETE were prepared and the products formed upon their exposure to human leukocytes were studied (2,3,9-12). This approach revealed a new series of oxygenated products of C20:4 that contain a unique conjugated tetraene. Since these compounds arose via interaction(s) between LO pathways (Figure 1) and displayed novel biological properties, the trivial name lipoxins (lipoxygenase interactions products) was proposed (3,10).

Activation of human leukocytes with an ionophore of divalent cations (A23187) led to increased formation of these tetraene-containing eicosanoids, which facilitated their isolation (2,9). A fraction containing several of these compounds was obtained from incubations with leukocytes following silicic acid chromatography and thin layer chromatography of material, which was treated with diazomethane. The basic structures of the first two members of this series were determined by physical methods that included HPLC, UV spectroscopy, gas chromatography-mass spectrometry (GC-MS) and oxidative ozonolysis. One compound was identified as 5,6,15L-trihydroxy-7,9,11,13-eicosatetraenoic acid (2), and the other proved to be a positional isomer 5D,14,15-trihydroxy-6,8,10,12-eicosatetraenoic acid (3). When these biologically derived products were isolated following HPLC and added to either human neutrophils, natural killer (NK) cells or guinea pig lung strips, they elicited bioactions different from those previously observed with other eicosanoids. Therefore these new bioactive compounds were termed lipoxin A_4 (LXA_4) and lipoxin B_4 (LXB_4), respectively (3,11).

Biosynthesis and Stereochemistry of LXA_4 and LXB_4

Next, we determined the complete stereochemistry of the biologically active LXA_4 and LXB_4, as well as identified the structures and actions of their naturally occurring isomers. Results from isotopic oxygen studies guided these studies, since they revealed that each of the tetraene-containing compounds generated by intact leukocytes carried an ^{18}O atom at C-5 and that the oxygen atoms at either C-6 of LXA_4 or C-14 of LXB_4, as well as their isomers, were not exclusively derived from molecular oxygen (11). These findings and results of alcohol trapping studies suggested that the formation of lipoxins in human leukocytes proceeds via an epoxide-containing intermediate (9-13).

At this point, a synthetic approach was undertaken to establish the stereochemistry of LXA_4 and LXB_4 (*i.e.* geometry of the double bonds and the chirality of the carbon-6 and carbon-14 position alcohol groups; cf. ref. 2 and 3). Several criteria and matching studies were employed with materials prepared by total synthesis. Synthetic and biologically-derived materials were subject to analysis by UV spectroscopy, HPLC (co-chromatography in several systems), GC-MS of several derivatives, and, most importantly, they were assessed for biological activities in parallel studies. LXB_4 proved to be

5S,14R,15S-trihydroxy-6,10,12-trans-8-cis-eicosatetraenoic acid (12), and LXA_4 is 5S,6R,15S-trihydroxy-7,9,13-trans-11-cis-eicosatetraenoic acid (9). A 6S isomer of LXA_4 (6S-LXA_4) and two all-trans epimers (6S-11-trans-LXA_4 and 11-trans-LXA_4) were also identified. However, these compounds have proven to be substantially less active than native LXA_4 in several bioassay systems (see Tables 1-3). Synthetic approaches and matching analysis were also attempted by others (14,15).

Although it is clear that multiple biosynthetic routes as well as cell-cell interactions can lead to formation of tetraene-containing eicosanoids, the finding that 15-HETE is rapidly transformed by activated leukocytes provided a model for studying one pathway in their formation (9,12). In this route, schematically summarized in Fig. 1, 15-HETE is converted to 5(S)-hydroperoxy-15(S)-hydroxy-6,13-trans-8,11-cis-eicosatetraenoic acid by activated human leukocytes and further transformed to a 5(6)-epoxytetraene. This epoxide or its equivalent could be enzymatically transformed to either LXA_4 (by the action of an epoxide hydrolase to yield a 5S,6R diol) or LXB_4 (by attack of the C-14 position with the generation of an 8-cis double bond). Further evidence for a 5(6)-epoxytetraene intermediate in the biosynthesis was obtained with 15(S)-hydroxy-5(6)-oxido-7,9,13-trans-11-cis-eicosatetraenoic acid prepared by total chemical synthesis. This epoxide was rapidly and quantitatively converted to LXA_4 (*in vitro*) by a purified cytosolic epoxide hydrolase (13). It remains to be determined whether a similar enzyme is solely responsible for the formation of LXA_4 by human leukocytes. Others have also postulated the role of epoxytetraenes in the formation of LXA_4, LXB_4 and their isomers (14-17) in model systems. In particular, Kühn *et al.* (17), Fitzsimmons *et al.* (14) and Adams *et al.* (15) considered several theoretical schemes to generate lipoxins, one involving an epoxide intermediate and another involving three consecutive oxygenations catalyzed by LO's. It is clear from these results as well as results obtained from our biosynthetic studies with human leukocytes that the generation of a 5(6)-epoxytetraene from 15-HETE is an enzymatic event (9,11,12). Fitzsimmons *et al.* (14) suggested that, once formed, the 5(6)-epoxytetraene could undergo nonenzymatic hydrolysis to yield a mixture of products including 14(S)-8-trans-LXB_4, 8-trans-LXB_4, and a racemic mixture of both LXA_4 and its 6(S) isomer (6(S)-LXA_4), but not LXB_4. This suggestion was based upon the results of experiments in which synthetic 5(6)-epoxytetraene was added to leukocytes in suspension (14). However, it is not clear from these results whether the 5(6)-epoxytetraene entered the cells or if it underwent hydrolysis in the extracellular medium before

Table 1. *Biological Activities of Lipoxins: in vivo Models*

Bioassay System	Compounds	Response	Reference
Microcirculation: Hamster cheek pouch	LXA_4	Arteriolar dilation without vascular permeability or leukocyte adherence	(41)
Systemic and renal hemodynamics in rats	LXA_4	Selective fall in afferent arteriolar resistance, glomerular hyperperfusion, hypertension and hyperfiltration (arteriolar dilation)	(42)
Renal hemodynamics	LXA_4	Antagonizes leukotriene D_4-induced fall in glomerular filtration rate but not renal plasma flow (counterregulatory actions)	(43)
Cerebral arterioles of newborn pigs	LXA_4 and LXB_4	Stimulate dilation of cerebral arterioles not protanoid mediated	(44)
Mesenteric circulation in conscious rats	LXA_4 and LXB_4	I.V. administration constricts mesenteric vessels	(45)
Hamster cheek pouch	Lipoxin A_4	Inhibits leukotriene B_4-induced inflammation LXA_4 blocks plasma leakage	(46)
Mouse hematopoietic in stem cells	Lipoxin B_4, but not lipoxin A_4	Radioprotective actions without diarrhea or ataxia	(47)

Table 2. *Biological Activities of Lipoxins: Isolated Organ Baths*

Bioassay System	Compounds	Response	Reference
Guinea pig lung strips	Lipoxin A_4	Contracts with slow onset	(9)
Guinea pig lung strips	Lipoxin A_4	Cyclooxygenase-independent contraction	(41)
Guinea pig lung strips	LXA_4 geometric isomers	Structure-function relationship with smooth muscle contraction	(30,48)
Guinea pig lung strips	Lipoxin A_4 but not LXB_4	Contracts parenchymal strips, and not tracheal spirals; LXA_4 contractions were not blocked by a 5-LO inhibitor	(49)
Guinea pig lung parenchyma and ileum smooth muscle	LXA_4	Contracts lung strips but not ileum smooth muscles in cascade superfusion	(50)
Rat tail artery	LXA_4	Contracts	(40)
Guinea pig, rabbits, and rat lung strips	LXA_4 and LXB_4	Bronchoconstrictors and exert endothelium-dependent vasorelaxation	(51)
Guinea pig lung	LXA_4	Stimulates thromboxane production	(52)

Table 3. *Biological Activities of Lipoxins: in vitro Systems*

Bioassay System	Compounds	Response	Reference
Isolated cell systems:			
Human neutrophils	$LXA_4 > LXB_4$	Chemotaxis without aggregation	(3,53)
Human neutrophils	LXA_4	Inhibits leukotriene and FMLP-induced chemotactic responses, Ca^{2+} mobilization and IP_3 generation	(54,55)
Human neutrophils	LXA_4 and LXB_4	Do not inhibit FMLP- or A23187-induced superoxide anion generation	(57)
Human neutrophils	LXA_4 and LXB_4	Stimulate lipid remodeling and release C20:4 without LO activation	(56)
Human neutrophils	5(6)-epoxytetraene	Stimulates Ca^{2+} mobilization	(58)
Human endothelial cells	LXA_4 and LXB_4	Stimulates prostacyclin generation	(59)
Rat mesangial cells	LXA_4	Inhibits LTD_4-induced IP_3 generation and LTD_4 binding	(43)
Human mononuclear cells	LXB_4	Stimulates colony formation in diffusion chambers	(60)
Human natural killer cells	LXA_4 and LXB_4	Blocks cytoxicity without altering target-cell binding	(61)
Isolated enzymes:			
Protein kinase C (PKC) human placental-derived	LXA_4	Activates PKC in the presence of Ca^{2+} and selects substrate specificity	(62)
PKC γ-subspecies bovine cerebellum	LXA_4	Selective activator	(63)

gaining access to cellular compartments. Moreover, the results of time course studies are not consistent with this proposal, namely, equal amounts of LXA_4 and 6S-LXA_4 would be expected, rather than the quantitatively higher amounts of LXA_4 that are observed with cells that have not undergone lysis (9,18).

Enzymes Which Generate Lipoxins and Related Compounds: In Vitro Models

The proposal that tetraene-containing eicosanoids, in particular LXA_4 and LXB_4, could be generated by several distinct biosynthetic routes (2,3) has now been explored in several model systems using both purified and partially purified enzymes. Studies by Kühn *et al.* (17) indicate that a purified LO obtained from reticulocytes can convert 5S,15S-DiHETE to trihydroxytetraenes. Since these transformations were blocked by inhibitors, heat denaturing, and by anaerobic conditions, it was suggested that the products were enzymatically generated by LO reactions. Two independent mechanisms were proposed (17). One involves a 5(6)-epoxide (*vide supra*), while the other implicates the formation of an intermediate that is generated by the action of a LO on either 5,15-DHPETE or its reduced form 5,15-DHETE. This second route can be considered a sequential triple lipoxygenation where a product of dioxygenation becomes a substrate for the third attack by a LO. Recently, these investigators (19,20) demonstrated that the reticulocyte enzyme can convert 5,15-DHETE to LXB_4, illustrating this triple LO route.

Ueda *et al.* (21) have shown that a 5-LO purified from porcine leukocytes can enzymatically generate a 5(6)-epoxytetraene with either 15-HPETE or 15-HETE. Studies by the same group, using a purified 12-LO from porcine leukocytes, indicate that LXB_4 can be generated by both the 14R-oxygenase activity as well as the LTA_4 synthase activity of this enzyme with 5,15-DiHPETE (22). Thus, this enzyme can generate LXB_4 by both triple oxygenation and by the formation of a putative 14(15)-epoxytetraene. It should be noted, however, that, unlike the porcine leukocyte, human leukocytes do not possess a 12-LO, and when 5,15-DiHPETE was added to intact human platelets, which contain 12-LO, LXB_4 was not generated (12). Nevertheless, results obtained with purified 12-LO from porcine leukocytes have given a potential enzymatic route to LXB_4 with intriguing mechanistic implications.

5-LO of potato tubers can also generate LXB_4, as well as a novel 6,7-dihydro-lipoxin B_4, when incubated with 15-HPETE (23,24). In addition to utilizing hydroxy- and hydroperoxy-containing compounds as substrates, recent results indicate that LO's can also transform epoxide-containing compounds to lipoxin precursors. For example, the soybean 15-LO rapidly transforms LTA_4 to a 5(6)-epoxytetraene (25), suggesting that 15-lipoxygenation of LTA_4 could give rise to lipoxins in various cells.

A potential role for initial oxygenation of C20:4 by 5-LO, followed by its interaction with either the 12-LO or 15-LO, has been explored in studies with intact human platelets (25,26), rat mesangial cells (27) and neutrophils (28). Recent results with human platelet-derived 100,000 g supernatants, which display 12-LO activity, show that the platelet 12-LO can transform LTA_4 to lipoxins (25,28). This action of the human 12-LO with LTA_4 exemplifies its ω-6 oxygenase activity and provides a role for this enzyme in lipoxin formation. Once generated, the 5(6)-epoxytetraene could be substrate for lipoxin production by several distinct mechanisms (9,12). It is clear, at least *in vitro*, that synthetic epoxytetraenes (*e.g.* 5(6)-epoxytetraene and 14(15)-epoxytetraene) can undergo nonenzymatic hydrolysis to generate both LXA_4 and LXB_4, respectively, as well as some but not all of their isomers (14,29,30). However, further transformations of the epoxytetraenes by various cell types have not been examined in detail. Together, results obtained with isolated enzymes have illustrated the potential routes for lipoxin production, some of which may be utilized by specific cell types.

Cellular Origins of Lipoxins

Cell types from several species have now been shown to generate lipoxins from either endogenous and/or exogenous sources of C20:4. The first reports of lipoxin formation from endogenous sources of either C20:4 or eicosapentaenoic acid were presented by Wong *et al.* (31). Lam *et al.* (32), utilizing an isolated PLA_2 isoenzyme added to porcine leukocytes, demonstrated the formation of LXB_4 and its isomers, while eosinophil-rich granulocyte suspensions from humans generate LXA_4, which correlated with the number of eosinophils present (33). The biosynthetic route for LXA_4 formation from endogenous sources could not be concluded from these results; that is, was LXA_4 produced exclusively from C20:4 released and processed within a single cell? Upon activation of eosinophil-enriched cell suspensions, suitable substrates could be mobilized by one cell and transformed to LXA_4 by another (33). The 5(6)-epoxytetraene, LTA_4 or 15-HETE are each potential candidates and hence may shuttle between individual cells for further processing. In addition, recent findings indicate that human eosinophils can also transform 15-

HETE to peptido-lipoxins, which appear to be generated via conjugation of a 5(6)-epoxytetraene intermediate (34). The biological significance of these new products is currently under investigation.

Canine mastocytoma cells in culture have been shown to generate lipoxins from C20:4 (35). Recently, evidence has been presented for triple lipoxygenation to generate lipoxins with bovine leukocytes (36). Unlike human neutrophils, which possess both 5- and 15-LO activities, the bovine leukocyte also displays 12-LO activity. As a result, bovine leukocytes can generate lipoxins from either C20:4, 15-HETE, 5,15-DHETE, 5-HPETE or 5-HETE (36). Since human neutrophils do not possess this 12-LO activity, which can be provided to neutrophils by contact with other cells (*i.e.* human platelets), lipoxin biosynthesis in neutrophils alone appears to proceed by a different route (18).

Interactions between platelets and neutrophils can lead to the transcellular metabolism of eicosanoids (37). Recently, Edenius *et al.* (26) found that mixed platelet-granulocyte suspensions stimulated with the ionophore A23187 generate LXA_4, as well as four other tetraene-containing compounds, from endogenous sources. They also showed that LTA_4 is converted to LXA_4 by platelets and suggested that the reported 15-LO activity of platelets may be involved in this enzymatic conversion (26). Garrick *et al.* (27) found that mesangial cells from rat kidney can also utilize LTA_4 to form lipoxins.

Although ionophore-stimulated cells can generate lipoxins without addition of exogenous C20:4 (18,26,33), it remained to be determined if receptor activation could also lead to their formation. Indeed, simultaneous activation of human neutrophils and platelets by receptor-mediated agonists (FMLP and thrombin) leads to formation of both LXA_4 and LXB_4 (28). In these studies, LTA_4 was transformed to LXA_4 and the recently identified 7-cis-11-trans-LXA_4 (30) by cell-free 100,000 g supernatants of human platelets, which displayed 12-LO activity (28). Together these results suggest that lipoxins can also be formed by initial oxygenation by the 5-LO and transcellular metabolism of LTA_4 (26-28). This new route of lipoxin formation is of particular interest, since recent results indicate that, when neutrophils utilize 15-HETE to generate lipoxins, an inverse relationship is observed between leukotriene and lipoxin generation (18).

Biological Actions of Lipoxins

The activities of lipoxins have been evaluated in both *in vivo* and *in vitro* systems (Tables 1-3). In each system, the responses evoked by either LXA_4 or LXB_4 have proven to be stereospecific and distinct from those of other eicosanoids. In several *in vivo* systems, LXA_4 and LXB_4 stimulate changes in microcirculation (41-46). For example, LXA_4 induces rapid arteriolar dilation in both the hamster cheek pouch and renal hemodynamics of the rat (41,42). LXA_4 can also antagonize LTD_4-induced vasoconstriction and block the binding of LTD_4 to mesangial cells (in culture), which are responsible, in part, for regulating hemodynamics. These findings suggest that LXA_4 may regulate the actions of vasoconstrictor leukotrienes *in vivo* (43). Along these lines, LXA_4 has also been shown to block LTB_4-induced inflammation in the hamster cheek pouch (46). In this model, LXA_4 blocked both LTB_4-induced plasma leakage and leukocyte migration.

The actions of lipoxins have been studied by several groups utilizing isolated organs (Table 2). Unlike leukotrienes, LXA_4 contracts guinea pig lung strips but not ilium smooth muscle preparations, which suggests a degree of organ selectivity (9,41,48-50). The contraction induced by LXA_4 is slow in onset and not blocked by either cyclooxygenase or LO inhibitors (41), although LXA_4 can stimulate thromboxane production (52). In addition to bronchoconstrictor activities, both LXA_4 and LXB_4 stimulate endothelium-dependent vasorelaxation (51), which may be mediated, in part, by the ability of lipoxins to stimulate prostacyclin release with cultured human endothelial cells (59).

Lipoxins also display unique actions on specific cell types *in vitro* (Table 3). For example, LXA_4 elicits selective actions on human neutrophils (3,53-58). Recent results indicate that, in addition to inducing chemotaxis (53), LXA_4 can also block both LTB_4- and FMLP-induced responses (*i.e.* IP_3 generation and Ca^{2+} mobilization) with these cells (54,55). However, neither LXA_4 nor LXB_4 can inhibit FMLP-stimulated superoxide anion generation with neutrophils (57). Both LXA_4 and LXB_4 selectively stimulate lipid remodeling with neutrophils and release C20:4 without stimulating LO activity or adherence (56). Also, their precursor, the 5(6)-epoxytetraene (in the micromolar range) can mobilize Ca^{2+} (58). LXA_4 and LXB_4 also display selective actions with NK cells and block cytotoxicity (61). Recent results indicate that LXB_4 stimulates colony formation with human mononuclear cells (60) and that it possesses a selective radioprotective effect with mouse hematopoietic stem cells (47).

In addition to its actions on various tissues and potential extracellular roles, LXA_4 may also have an intracellular role within its cell type of origin before its release to the extracellular milieu. In this regard, LXA_4 activates isolated PKC and proved to be more potent than either diacylglyceride or native arachidonate (62). The actions of LXA_4 appear to be more specific for the γ-subspecies of PKC, which is found in brain and spinal cord (63).

In summary, lipoxins can be generated by several biosynthetic routes, which can include lipoxygenase as well as cell-cell interactions. LXA_4 and LXB_4 each carry selective biological activities in both *in vivo* and *in vitro* systems, which suggests that they may be involved in regulating cellular responses of interest in inflammation and host defense.

Acknowledgments

The author thanks Mary Halm Small for skillful preparation of this manuscript and my collaborators in this area (see references). Studies conducted in Boston were supported in part by NIH grants #AI26714 and GM38765. CNS is a recipient of the J.V. Satterfield Arthritis Investigator Award from the National Arthritis Foundation and is a Pew Scholar in the Biomedical Sciences.

References

1. Samuelsson, B. (1983) *Science* **220,** 568-575
2. Serhan, C.N., Hamberg, M., and Samuelsson, B. (1984) *Biochem. Biophys. Res. Commun.* **118,** 943-949
3. Serhan, C.N., Hamberg, M., and Samuelsson, B. (1984) *Proc. Natl. Acad. Sci. USA* **81,** 5335-5339
4. Borgeat, P., Fruteau de Laclos, and Maclouf, J. (1983) *Biochem. Pharm.* **32,** 381-387
5. Serhan, C.N., Lundberg, U., Weissmann, G., and Samuelsson, B. (1984) *Prostaglandins* **27,** 563-581
6. Hamberg, M., Hedqvist, P., and Radegen, K. (1980) *Acta Physiol. Scand.* **110,** 219-223
7. Murray, J.J., Tonnel, A.B., Brash, A.R., Roberts, L.J., II, Gosset, P., Workman, R., Capron, A., and Oates, J.A. (1986) *N. Engl. J. Med.* **315,** 800-805
8. Needleman, P., Turk, J., Jakschik, B.A., Morrison, A.R., and Lefkowith, J.B. (1986) *Ann. Rev. Biochem.* **55,** 69-102
9. Serhan, C.N., Nicolaou, K.C., Webber, S.E., Veale, C.A., Dahlén, S.-E., Puustinen,T.J., and Samuelsson,B. (1986) *J. Biol. Chem.* **261,** 16340-16345

10. Serhan, C.N., Wong, P.Y.-K., and Samuelsson, B. (1987) *Prostaglandins* **34,** 201-204
11. Serhan, C.N., Fahlstadius, P., Dahlén, S.-E., Hamberg, M., and Samuelsson, B. (1985) in *Advances in Prostaglandin, Thromboxane, and Leukotriene Research* (Hayaishi, O., and Yamamoto, S., eds) Vol 15, pp. 163-166. Raven Press, New York
12. Serhan, C.N., Hamberg, M., Samuelsson, B., Morris, J., and Wishka, D.G. (1986) *Proc. Natl. Acad. Sci. USA* **83,** 1983-1987
13. Puustinen, T., Webber, S.E., Nicolaou, K.C., Haeggström, J., Serhan, C.N., and Samuelsson, B. (1986) *FEBS Lett.* **207,** 127-132
14. Fitzsimmons, B.J., Adams, J., Evans, J.F., Leblanc, Y., and Rokach, J. (1985) *J. Biol. Chem.* **260,** 13008-13012
15. Adams, J., Fitzsimmons, B.J., Girard, Y., Leblanc, Y., Evans, J.F., and Rokach, J. (1985) *J. Am. Chem. Soc.* **107,** 464-469
16. Corey, E.J., and Mehrotra, M.M. (1986) *Tetrahedron Lett.* **43,** 5173-5176
17. Kühn, H., Wiesner, R., and Stender, H. (1984) *FEBS Lett.* **177,** 255-259
18. Serhan, C.N. (1989) *Biochim. Biophys. Acta* **1004,** 158-168
19. Kühn, H., Wiesner, R., Alder, L., Schewe, T., and Stender, H. (1986) *FEBS Lett.* **208,** 248-252
20. Kühn, H., Wiesner, R., Alder, L., Fitzsimmons, B.J., Rokach, J., and Brash, A.R. (1987) *Eur. J. Biochem.* **169,** 593-601
21. Ueda, N., Yamamoto, S., Fitzsimmons, B.J., and Rokach, J. (1987) *Biochem. Biophys. Res. Commun.* **144,** 996-1002
22. Ueda, N., Yokoyama, C., Yammoto, S., Fitzsimmons, B.J., Rokach, J., Oates, J.A., and Brash, A.R. (1987) *Biochem. Biophys. Res. Commun.* **149,** 1063-1069
23. Ho, H.-Y., and Wong, P.Y.-K. (1989) in *Advances in Prostaglandin, Thromboxane, and Leukotriene Research* (Samuelsson, B., Wong, P.Y.-K., and Sun, F.F., eds) Vol. 19, pp. 137-140. Raven Press, New York
24. Reddy, C.C., Whelan, J., Rao, M.K., and Reddanna, P. (1989) in *Advances in Prostaglandin, Thromboxane, and Leukotriene Research* (Samuelsson, B., Wong, P.Y.-K., and Sun, F.F., eds) Vol. 19, pp. 132-136. Raven Press, New York
25. Serhan, C.N., Sheppard, K.-A., and Fiore, S. (1990) in *Advances in Prostaglandin, Thromboxane, and Leukotriene Research* Vol. 20. Raven Press, New York, in press
26. Edenius, C., Haeggström, J., and Lindgren, J.Å. (1988) *Biochem. Biophys. Res. Commun.* **157,** 801-807
27. Garrick, R., Shen, S.-Y., Ogunc, S., and Wong, P.Y.-K. (1989) *Biochem. Biophys. Res. Commun.* **162,** 626-633
28. Serhan, C.N., and Sheppard, K.-A. (1990) *J. Clin. Invest.,* in press
29. Leblanc, Y., Fitzsimmons, B.J., Rokach, J., Ueda, N., and Yamamoto, S. (1987) *Tetrahedron Lett.* **28,** 3449-3452

30. Nicolaou, K.C., Marron, B., Veale, C., Webber, S.E., Dahlén, S.E., Samuelsson,B., and Serhan,C.N. (1989) *Biochim.Biophys.Acta* **1003,** 44-53
31. Wong, P.Y.-K., Spur, B., Hirai, A., Yoshida, Tamura, Y., and Lam, B.K. (1986) *Fed. Proc.* **45,** 927A
32. Lam, B.K., Serhan, C.N., Samuelsson, B., and Wong, P.Y.-K. (1987) *Biochem. Biophys. Res. Commun.* **144,** 123-131
33. Serhan, C.N., Hirsch, U., Palmblad, J., and Samuelsson, B. (1987) *FEBS Lett.* **217,** 242-246
34. Steinhilber, D., and Roth, H.J. (1989) *FEBS Lett.*, **255,** 143-148
35. Lazarus, S.C., and Zocca, E. (1988) *FASEB J.* **2,** A409
36. Walstra, P., Verhagen, J., Vermeer, M.A., Klerks, J.P.M., Veldink, G.A., and Vliegenthart, J.F.G. (1988) *FEBS Lett.* **228,** 167-171
37. Marcus, A.J., Brockman, M.J., Safier, L.B., Ullman, H.L., Islam, N., Serhan, C.N., Rutherford, L.E., Korchak, H.M., and Weissmann, G. (1982) *Biochem. Biophys. Res. Commun.* **109,** 130-135
38. Kim, S.J. (1988) *Biochem. Biophys. Res. Commun.* **150,** 870-876
39. Lee, T.H. (1989) *Allergologie, Jahrgang 12, Kongressausgabe* 89-90
40. Ng, C.F., Lam, B.K., Pritchard, K.A., Jr., Stemerman, M.B., Hejny, P., and Wong, P.Y.-K. (1989) *Biochim. Biophys. Acta,* in press
41. Dahlén, S.-E., Raud, J., Serhan, C.N., Björk, J., and Samuelsson, B. (1987) *Acta Physiol. Scand.* **130,** 643-647
42. Badr, K.F., Serhan, C.N., Nicolaou, K.C., and Samuelsson, B. (1987) *Biochem. Biophys. Res. Commun.* **145,** 408-414
43. Badr, K.F., DeBoer, D.K., Schwartzberg, M., and Serhan, C.N. (1989) *Proc. Natl. Acad. Sci. USA* **86,** 3438-3442
44. Busija, D.W., Armstead, W., Leffler, C.W., and Mirro, R. (1989) 468-471
45. Feuerstein, G., and Siren. A.-L. (1988) *FEBS Lett.* **232,** 51-55
46. Hedqvist, P., Raud, J., Palmertz, U., Haeggström, J., Nicolaou, K.C., and Dahlén, S.E. (1989) *Acta Physiol. Scand.*, **137,** 571-572
47. Walden, T.L., Jr. (1988) *J. Radiat. Res.* **29,** 255-260
48. Dahlén, S.-E., Veale, C.A., Webber, S.E., Marron, B.E., Nicolaou, K.C., and Serhan, C.N. (1989) *Agents and Actions* **26,** 93-95
49. Jacques, C.A.J., Spur, B.W., Crea, A.E.G., and Lee, T.H. (1988) *Br. J. Pharmacol.* **95,** 562-568
50. Cristol, J.-P., and Sirois, P. (1988) *Res. Commun. Chem. Pathol. Pharmacol.* **59,** 423-426
51. Lefer, A.M., Stahl, G.L., Lefer, D.J., Brezinski, M.E., Nicolaou, K.C., Veale, C.A., Abe, Y., and Smith, J.B. (1988) *Proc. Natl. Acad. Sci. USA* **85,** 8340-8344
52. Wikström, E., Westlund, P., Nicolaou, K.C., and Dahlén, S.-E. (1989) *Agents and Actions* **26,** 90-92
53. Palmblad, J., Gyllenhammar, H., Ringertz, B., Serhan, C.N., Samuelsson, B., and Nicolaou,K.C. (1987) *Biochem.Biophys.Res.Commun.* **145,**168-175

54. Lee, T.H., Horton, C.E., Kyan-Aung, U., Haskard, D., Crea, A.E.G., and Spur, B.W. (1989) *Clin. Sci.* **77,** 195-203
55. Lee, T.H., Grandordy, B., Horton, C.E., Crea, A.E.G., and Spur, B.W. (1989) *FASEB J.* **3,** A1277
56. Nigam, S., Fiore, S., Lucinskas, F.W. and Serhan, C.N. (1990) *J. Cell. Physiol.*, in press
57. Serhan, C.N., and Reardon, E. (1989) *Free Rad. Res. Comms.*, (1990) **Vol 7,** 341-345
58. Luscinskas, F.W., Nicolaou, K.C., Webber, S.E., Veale, C.A., Gimbrone, M.A., Jr., and Serhan, C.N. (1990) *Biochem. Pharm.*, **39,** 355-365
59. Brezinski, M.E., Gimbrone, M.A., Jr., Nicolaou, K.C., and Serhan, C.N. (1989) *FEBS Lett.* **245,** 167-172
60. Popov, G.K. Nekrasov, A.S., Khshivo, A.L., Pochinskii, A.G., Lankin, V.Z., and Vikhert, A.M. *Biull. Eksp. Biol. Med.* (1989) **107,** 93-95 (From Russian)
61. Ramstedt, U., Serhan, C.N., Nicolaou, K.C., Webber, S.E., Wigzell, H., and Samuelsson, B. (1987) *J. Immunol.* **138,** 266-270
62. Hansson, A., Serhan, C.N., Haeggström, J., Ingelman-Sundberg, M., and Samuelsson, B. (1986) *Biochem. Biophys. Res. Commun.* **134,** 1215-1222
63. Shearman, M.S., Naor, Z., Sekiguchi, K., Kishimoto, A., and Nishizuka, Y. (1989) *FEBS Lett.* **243,** 177-182

Novel Hepoxilin Glutathione Conjugates with Biological Activity

C.R. Pace-Asciak, O. Laneuville, N. Gurevich, P. Wu,
P.L. Carlen, W-G. Su, E.J. Corey, M. Chang, and C.C. Reddy
Division of Neurosciences, Research Institute, Hospital for Sick Children,
555 University Avenue, Toronto, Ontario, Canada M5G 1X8

Abstract

The mode of synthesis of the hepoxilins and the involvement of ferriheme proteins has been investigated. The metabolism of hepoxilins into novel glutathione conjugates is described, occurring in rat liver and brain; this metabolism is analogous to that occurring with purified glutathione S-transferase. Novel biological activity data on the hepoxilins and their metabolites indicates that they modulate synaptic activity and neuronal excitability in mammalian CA1 neurons.

Hepoxilins are hydroxy-epoxide derivatives of arachidonic acid formed via the 12-lipoxygenase pathway (1). There are two hepoxilins, A_3 and B_3, depending on whether a hydroxyl group is located at the 8-carbon position or the 10-carbon position respectively (1). The term 'hepoxilin' was coined by Pace-Asciak and Martin to relate the structure of these products (Hydroxy-EPOXy) with the first biological effect demonstrated (InsuLIN secretion) (2). Hepoxilins have been isolated from a variety of tissues; hepoxilin A_3 was first isolated from rat lung preparations (1, 3, 4), although subsequent studies showed both hepoxilin A_3 and B_3 to be present in pancreatic islets (5) and brain (6). Both hepoxilins have also been isolated from *Aplysia* sensory neurons (7). Hepoxilin B_3 was first isolated from platelets and evidence for the isolation of the trihydroxy derivatives of hepoxilin A_3 was also derived from this tissue (8-13) . The latter products were formed nonenzymatically, through facile hydrolysis of hepoxilin A_3 during an acidic workup. Hepoxilin B_3 is more resistant to hydrolysis during acidification (1). A specific enzyme, hepoxilin epoxide hydrolase, has been isolated which hydrolyzes hepoxilin A_3 specifically over other epoxides as leukotriene A_4 and styrene oxide, into trihydroxy derivatives termed trioxilins (14). Thus hepoxilin A_3 gives rise to trioxilin A_3 and hepoxilin B_3 gives rise to trioxilin B_3.

Biological Oxidation Systems, Volume 2

Mechanistically, ^{18}oxygen studies have shown that hepoxilins A_3 and B_3 are formed from 12-HPETE through an intramolecular rearrangement in which complete retention of ^{18}oxygen was observed in the hydroxyl group at position 8 (in hepoxilin A_3) or position 10 (in hepoxilin B_3) and the epoxide moiety at positions 11,12 (in both hepoxilins)(15, 16). It was also shown that hemin or hemoglobin catalyze this transformation, suggesting that Fe^{+3} was an essential component for this reaction (15, 16).

The biological activities of the hepoxilins appear to be mediated through their effects on calcium. Thus, hepoxilin A_3 increases the glucose-dependent secretion of insulin from pancreatic islet cells (2), an effect which is related to the glucose concentration, being maximal at glucose concentrations around 10 mM (17). Hepoxilin B_3 is also active in this assay. Hepoxilin A_3 facilitates the transport of calcium across the guinea pig yolk sac membrane, an effect which is not dependent on the side (maternal or fetal) of the membrane where hepoxilin is administered (18). Hepoxilin A_3 is capable of significant inhibitory and excitatory actions in mammalian hippocampal CA1 neurons (19) and in *Aplysia* sensory neurons (7). In hippocampal CA1 neurons, it causes hyperpolarization and sometimes a later depolarization of the membrane potential, and increases the amplitude and duration of the post-spike train afterhyperpolarization and the inhibitory post-synaptic potential although it does not affect the excitatory post-synaptic potential (19).

We have recently shown that hepoxilin A_3 is metabolized by a purified preparation of glutathione S-transferase into a glutathione-conjugated product which we refer to as hepoxilin A_3-C (20), in keeping with the leukotriene nomenclature for the glutathione-containing leukotriene C_4. Thus hepoxilin A_3 is metabolized via two independent pathways, an epoxide hydrolase pathway through a specific enzyme which we termed, hepoxilin epoxide hydrolase (HEH) to form the trihydroxy product trioxilin A_3, and a glutathione conjugating pathway via glutathione S-transferase (GST) to form hepoxilin A_3-C. These pathways are analogous to those described for leukotriene A_4 metabolism although different enzymes appear to be involved.

The present report describes the effects of a variety of metal protoporphyrins and ferriheme proteins on the formation of hepoxilin A_3 and its metabolism into hepoxilin A_3-C by homogenates of rat liver and brain. We also present data showing that hepoxilin A_3-C displays potent effects on hippocampal CA1 neurons.

Materials and Methods

Metal protoporphyrins (Mn, Sn, Zn, and Co) were purchased from Porphyrin Products Inc., Logan, Utah, USA; all hemoproteins were purchased from Sigma; [1-^{14}C]-12-HPETE was prepared from [1-^{14}C]-arachidonic acid (Amersham, UK) as previously described using rat platelets as source of 12-lipoxygenase (16); [1-^{14}C]-hepoxilin A_3 was prepared from [1-^{14}C]-12-HPETE as previously described using hemin as catalyst (16).

Incubations

Formation of [1-^{14}C]-Hepoxilin A_3: [1-^{14}C]-12-HPETE (10,000 cpm) was added in 20 μl of ethanol to 1 ml of 0.05 M KH_2PO_4-NaOH buffer, pH 7.4, containing the various individual protoporphyrins (4 μg each) or the various ferriheme proteins (50 μg each) and incubated at 37 °C for 20 minutes. The incubation was terminated by cooling on ice and by dilution with an equal volume of ice-cold methanol and 4 volumes of freshly distilled diethyl ether. The mixture was acidified to pH 3 with 1N HCl to convert hepoxilin A_3 into trioxilin A_3, mixed and centrifuged at 2000 rpm in a laboratory centrifuge for 30 seconds. The ether layer was transferred to another tube, washed with 1 ml of water, centrifuged, and transferred to a flask and evaporated *in vacuo*. The residue was dissolved in 200 μl of ether/methanol (1/1 v/v) and analyzed by TLC using ethyl acetate/acetic acid (99/1 v/v) as developing solvent. After 60 minutes development, the plate was removed, dried free of solvent, and scanned for radioactive products with a Berthold thin layer chromatogram scanner. Appropriate sections of the plate were marked, the silica gel was scraped into scintillation vials with a blade, 1 ml of methanol/water was added to elute the radioactive products from the silica followed by 10 ml of PCS scintillation cocktail (Amersham, UK). Vials were counted for radioactivity with a Beckman scintillation spectrophotometer.

Metabolism of [1-^{14}C]-Hepoxilin A_3: When tissue homogenates were used, tissues (liver or brain) were obtained from rats (250-300 g) which were sacrificed by decapitation. The tissues were freed from blood by careful washing with ice-cold artificial cerebrospinal fluid (ACSF)(brain) or 0.1 M Tris.HCl buffer, pH 8.0 (liver). Brain was dissected into 400 micron thick slices using a vibratome and the hippocampus was further dissected and placed in ACSF buffer at 30 °C for 60 minutes. Both the liver and the hippocampus were homogenized in 1 ml of the respective buffer and incubated with

[1-^{14}C]-hepoxilin A_3 (50,000 cpm) for 30 minutes at 37 °C. The incubation was terminated by cooling the mixture on ice, and addition of 50 μl of 4 N formic acid. The sample was centrifuged for 30 seconds at 3000 rpm, and the acidic supernatant was passed through a C18 SepPak (Waters) preactivated by washing with 20 ml of methanol, then 20 ml of water. After the sample had been loaded, the SepPak cartridge was eluted with 10 ml of water, and then 10 ml of methanol. The methanol eluate which contained better than 95% of the radioactivity loaded on the SepPak, was taken to near dryness in vacuo and the residue was resuspended in 1 ml of methanol. Aliquots of this solution were analyzed by TLC (as above) and by reversed phase HPLC on a C18-microBondapak column (Waters) using as mobile phase a solution of methanol/water/acetic acid (60/40/0.01 v/v/v). Eluate was detected by mixing with PCS scintillation cocktail and passage through an on-line Berthold radioactivity detector.

Electrophysiological Experiments

Intracellular electrophysiological recordings were carried out in an interface-type chamber at 34 °C using standard techniques as previously described (19) with glass micropipettes filled with 3M KCl or 3M K acetate (resistances 60 - 150 Meg Ohms). Orthodromic stimulation for the excitatory postsynaptic potentials (EPSPs) and inhibitory postsynaptic potentials (IPSPs) were done by mono- or bipolar tungsten electrodes. Post-spike train afterhyperpolarizations (AHPs) were generated by 100 ms constant current depolarizing pulse. Following drug application by perfusion, the pulse was adjusted if necessary to obtain the same number of spikes as required to generate the pre-drug AHP. All post-drug measurements of the AHP, postsynaptic potentials and the input resistance (R_{in}) were done at the same membrane potentials as the control pre-drug membrane potential, which sometimes required constant current injections. Measurements included the maximum amplitude of the EPSP, IPSP, and AHP from the pre-stimulus baseline. The duration of the AHP was measured following the offset of the 100 ms maximum stimulation pulse until the AHP reached the level of the pre-stimulus baseline. Stock solution of hepoxilin A_3-C was dissolved in DMSO (1 μg/μl) and diluted to the appropriate concentrations (3 - 30 nM) with ACSF.

Results and Discussion

Formation of Hepoxilin A_3

The catalytic effect of ferric protoporphyrin on the conversion of 12-HPETE into the hepoxilins appears quite specific. Figure 1 demonstrates that hemin (Fe-PP) is the best catalyst for hepoxilin formation, with Co-PP and Mn-PP being much less active. Zn-PP and Sn-PP are inactive. We previously showed that hemoglobin, an Fe^{+3}-containing protein was as effective as hemin in converting 12-HPETE into the hepoxilins (21). We therefore investigated whether this transformation was also possible when other ferriheme proteins were used. Indeed, although there is some specificity, most ferriheme proteins are capable of catalyzing this transformation. Of the ferriheme proteins used (Fig. 2), catalase and lactoperoxidase appear to form hepoxilin best.

The diversity of products possessing the Fe^{+3}-protoporphyrin group that can transform 12-HPETE into the hepoxilins further confirms our reported observations with ^{18}oxygenated 12-HPETE and hemin. We suggested that a cage-like complex is formed between 12-HPETE and the Fe^{+3}-protoporphyrin group (16), with the latter facilitating the intramolecular transfer of the terminal hydroxyl of the hydroperoxy group of 12-HPETE to the 8-carbon position of hepoxilin A_3 or to the 10-carbon position of hepoxilin B_3 leading to the formation of an 11,12-epoxide functionality (in both hepoxilins A_3 and B_3).

Metabolism of Hepoxilin A_3

Incubation of [1-^{14}C]-hepoxilin A_3 with rat liver homogenates revealed the presence by TLC of a metabolite more polar than the expected epoxide hydrolase metabolite, trioxilin A_3. The polar metabolite was analyzed further by reversed phase HPLC and shown to migrate earlier than trioxilin A_3 (Fig. 3). Comparison with authentic standard of a glutathione conjugate of hepoxilin A_3 (hepoxilin A_3-C) indicated identity of the two products. Furthermore this new metabolite was also formed when a purified preparation of GST was used (20). Additional studies with brain hippocampal homogenates revealed that hepoxilin A_3-C was also formed by this tissue, albeit in small amounts. The formation of hepoxilin A_3-C was dependent on the addition of glutathione, and was eliminated when heat-inactivated enzyme was used. These observations indicate a dual pathway of metabolism for hepoxilin A_3, involving hepoxilin epoxide hydrolase to form trioxilin A_3 and glutathione conjugation to form the new metabolite, hepoxilin A_3-C.

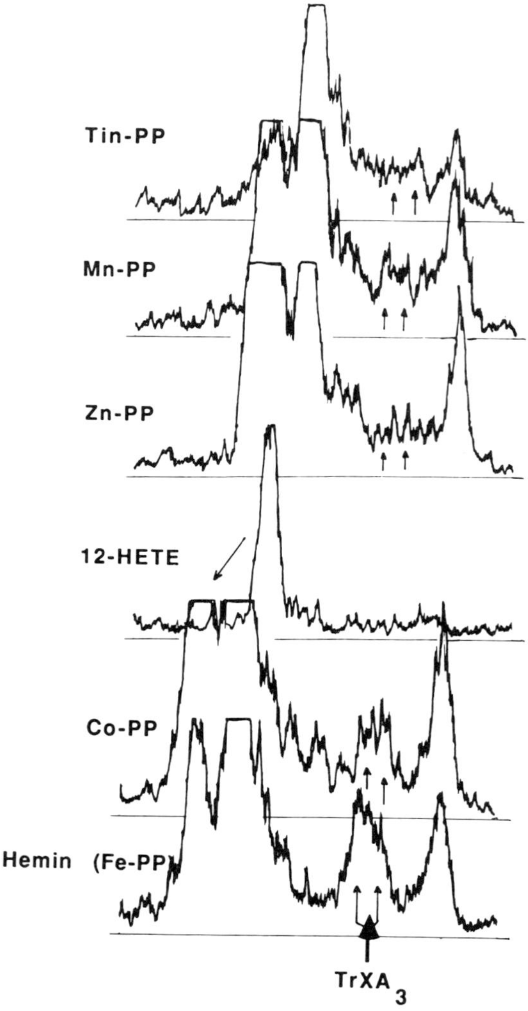

Figure 1. *Radio TLC of Extracts of Incubation of [1-^{14}C]-12-HPETE with Various Metal Protoporphyrins Compared with Fe^{+3}-Protoporphyrin (hemin).* Results show that hemin is the best catalyst for the formation of hepoxilin A_3 shown here as its trioxilin A_3 derivative formed through acidification of the reaction mixture.

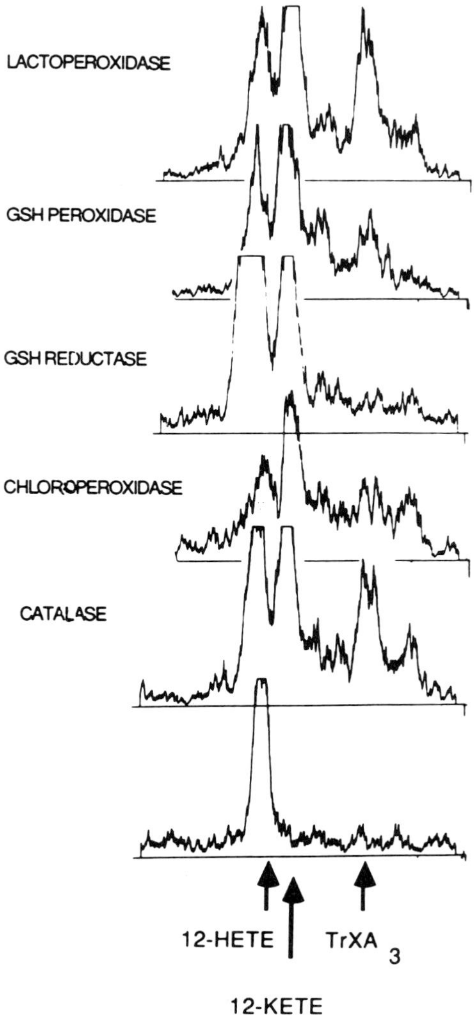

Figure 2. *Radio TLC of Extracts of Incubation of [1-^{14}C]-12-HPETE with Various Ferriheme Proteins.* Results show that lactoperoxidase and catalase are also active in conversion of 12-HPETE into hepoxilin A_3 shown here as its stable trioxilin derivative formed through acidification of the reaction mixture.

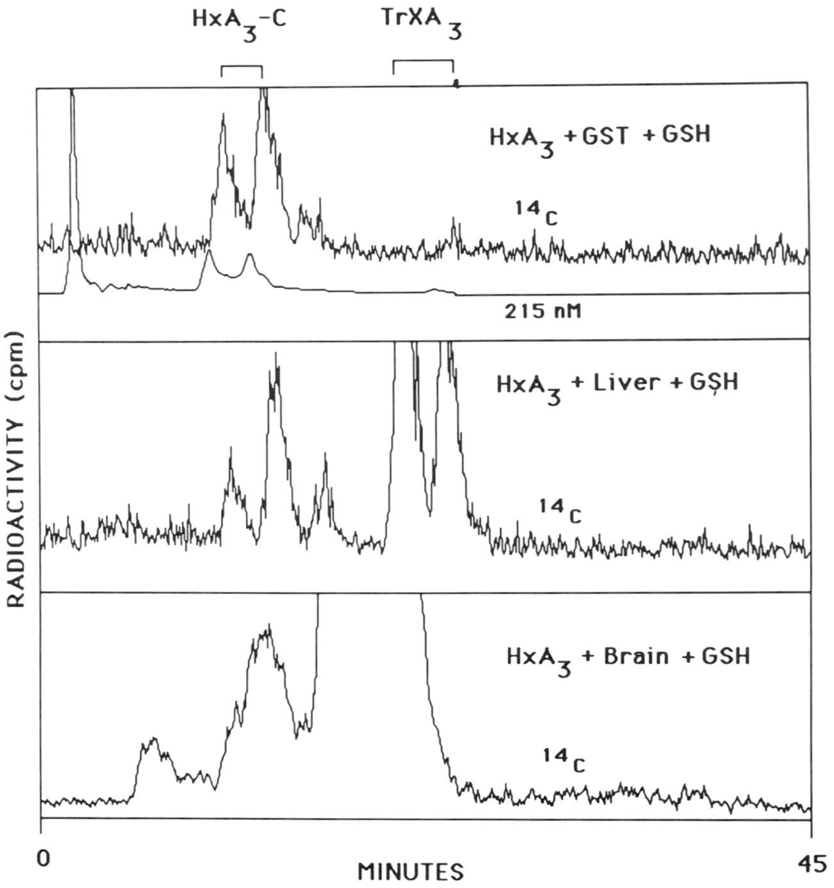

Figure 3. *Radio HPLC (reversed phase) of Extracts of Incubation of [1-^{14}C]-Hepoxilin A_3 with A) Purified Glutathione S-Transferase and **B**) Rat Liver Homogenate and C) Brain Hippocampal Homogenate.* Results show that formation of hepoxilin A_3-C by the tissues corresponds to the product formed by purified GST. The lack of alignment of products in panel C with those of panels A and B is because analysis of the samples was performed on different days and slightly different column conditions.

Electrophysiological Effects of Hepoxilin A_3-C

The effects of hepoxilin A_3-C (30 nM) on the membrane potential, post-spike train afterhyperpolarization, and the inhibitory postsynaptic potential of hippocampal CA1 neurons are shown in Figure 4. The threshold concentration for the observed effects of hepoxilin A_3-C was 3 nM when applied by perfusion. The effects of this drug were somewhat slow in onset (5-10 minutes after start of perfusion) and persistent. The drug effects were not reversed after 60 minutes of washout. Hepoxilin A_3-C caused hyperpolarization (4 mV) of the membrane potential (Fig. 4A), increased the amplitude and duration of the post-spike afterhyperpolarization (Fig. 4B) and increased the amplitude of the inhibitory postsynaptic potential without affecting the excitatory postsynaptic potential (Fig. 4C). To compare the AHPs and IPSPs at the resting membrane potential, repolarizing DC current was injected (Fig 4B and C) which resulted in much spontaneous spiking. This form of increased excitability was also noted with hepoxilin A_3 (19).

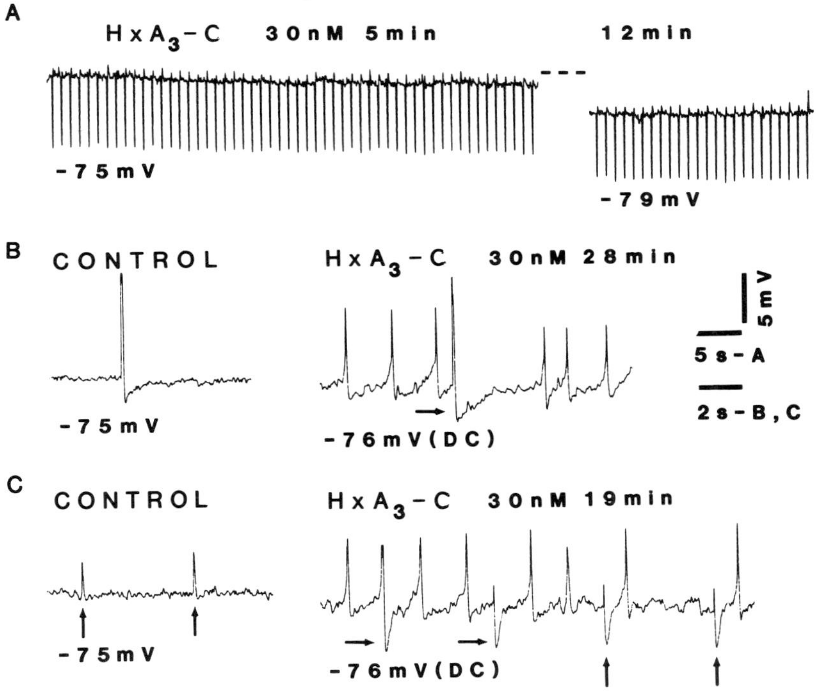

Figure 4. *Electrophysiological Effects of Perfusion of Hepoxilin A_3-C (30 nM) on the A) Membrane Potential, B) Post-Spike Train Afterhyperpolarization (AHP), and C) Excitatory/Inhibitory Post-Synaptic Potentials (EPSP/IPSP).* Results show that hepoxilin A_3-C hyperpolarizes the membrane potential by 4 mV, increases the amplitude and duration of the AHP, and increases the amplitude of the IPSP without much effect on the EPSP. Note that with repolarization of the membrane potential towards the pre-drug (control) resting potential with injected DC current, many spontaneous spikes were seen (B,C)

These findings demonstrate that hepoxilin A_3 is metabolized into a biologically relevant product, hepoxilin A_3-C, by brain and liver and possibly other tissues which possess a glutathione conjugating pathway. Hepoxilin A_3-C is capable of expressing synaptic neuromodulatory effects on hippocampal CA1 neurons. We have recently shown that hepoxilin A_3 is capable of mobilizing bound calcium in human neutrophils to increase the cytosolic calcium concentrations in the cell (22). Preliminary studies have shown that hepoxilin A_3-C is also capable of mobilizing calcium in these cells but its effect is less potent than that of hepoxilin A_3 although the effect is slower in onset and more long-lasting.

These observations suggest that products in the hepoxilin pathway may represent new lipidic modulators of calcium homeostasis in the cell, effects being initiated by hepoxilin A_3 and likely prolonged by its conversion within the cell cytosol into the chemically more stable glutathione-conjugated metabolite, hepoxilin A_3-C (Fig. 5).

These studies were supported by grants to CRP-A (MRC), OL (FCAR) PLC (MRC, OMH), SG (MRC), EJC (NIH) and CCR (NIH).

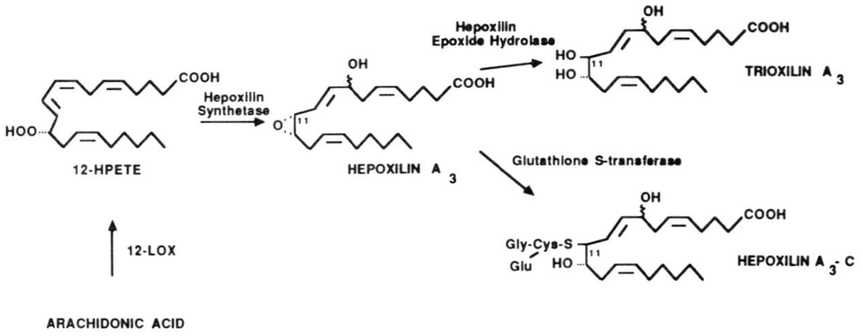

Figure 5. *Scheme Showing the Two Pathways of Metabolism of Hepoxilin A_3 Via the Hepoxilin Epoxide Hydrolase Pathway and the Glutathione Conjugating Pathway to form Trioxilin A_3 and Hepoxilin A_3-C Respectively*

References

1. Pace-Asciak, C.R., Granstrom, E. and Samuelsson, B. (1983) *J. Biol. Chem* **258**, 6835-6840
2. Pace-Asciak, C.R. and Martin, J.M. (1984) *Prostagl. Leukotriene and Med.* **16**, 173-180
3. Pace-Asciak, C.R., Mizuno, K. and Yamamoto, S. (1983) *Prostaglandins* **25**, 79-84

4. Pace-Asciak, C.R., Mizuno, K. and Yamamoto, S. (1982) *Biochim. Biophys. Acta* **712**, 142-145
5. Pace-Asciak, C.R., Martin, J.M., Corey, E.J. and Su, W.-G. (1985) *Biochem. Biophys. Res. Commun.* **128**, 942-946
6. Pace-Asciak, C.R. (1988) *Biochem. Biophys. Res. Commun.* **151**, 493-498
7. Piomelli, D., Shapiro, E., Zipkin, R., Schwartz, J.H. and Feinmark, S.J. (1989) *Proc. Natl. Acad. Sci. (USA).* **86**, 1721-1725
8. Jones, R.L., Kerry, P.J., Poyser, N.L., Walker, I.C. and Wilson, N.H. (1978) *Prostaglandins* **16**, 583-590
9. Bryant, R.W. and Bailey, J.M. (1979) *Prostaglandins* **17**, 9-18
10. Bryant, R.W. and Bailey, J.M. (1978) *Fed. Proc.* **37**, 1317
11. Bryant, R.W. and Bailey, J.M. (1980) *Biochem. Biophys. Res. Commun.* **92**, 268-276
12. Bryant, R.W. and Bailey, J.M. (1980) *Adv. Prostagl. Thromb. Res.* **6**, 95-100
13. Bryant, R.W., Simon, T.C. and Bailey, J.M. (1982) *J. Biol. Chem.* **257**, 14937-14943
14. Pace-Asciak, C.R. and Lee, W.S. (1989) *J. Biol. Chem.* **264**, 9310-9313
15. Pace-Asciak, C.R. (1984) *Biochim. Biophys. Acta* **793**, 485-488
16. Pace-Asciak, C.R. (1984) *J. Biol. Chem.* **259**, 8332-8337
17. Nathan, M.H. and Pek, S.B. (1989) *Diabetes* **Suppl 2**, Abstract #784
18. Derewlany, L., Pace-Asciak, C.R. and Radde, I.C. (1984) *Can. J. Physiol. Pharmacol.* **62**, 1466-1469
19. Carlen, P.L., Gurevich, N., Wu, P.H., Su, W.-G., Corey, E.J. and Pace-Asciak, C.R. (1989) *Brain. Res.* **497**, 171-176
20. Pace-Asciak, C.R., Laneuville, O., Chang, M., Reddy, C.C., Su, W.-G. and Corey, E.J. (1989) *Biochem. Biophys. Res. Commun.* **163**, 1230-1234
21. Pace-Asciak, C.R. (1984) *Biochim. Biophys. Acta* **793**, 485-488
22. Dho, S., Grinstein, S., Corey, E.J., Su, W.-G. and Pace-Asciak, C.R. (1990) *Biochem. J.* **266**, 63-68

Optically Pure Allylic Hydroperoxides: Mechanisms for Hydroperoxide Rearrangements

Ned A. Porter, Jere K. Kaplan, Patrick H. Dussault and Robert A. Breyer
Department of Chemistry, Paul M. Gross Chemical Laboratory
Duke University, Durham, NC 27706 USA

Abstract

We have discovered what appears to be a general method for preparing optically pure hydroperoxides: the resolution of unsaturated hydroperoxide enantiomers by liquid chromatography of diastereomeric derivatives. Preliminary screens of several hydroperoxide derivatives of the structure R-OO-CR'XOR" and perketals with R'=X=alkyl show that they: are readily prepared from hydroperoxides; are stable to normal- or reverse-phase chromatography, and can be deprotected under very mild acid conditions. In addition, when R" is derived from 2-phenylcyclohexanol, diastereomers of the perketals can be separated by normal- or reverse-phase HPLC. This approach has facilitated preparation of optically pure hydroperoxides derived from linoleic, arachidonic acid, and oleic acid. Previous investigations showed that dienyl peroxyls rearrange *via* a β-scission pathway, while we show here that simple allyl peroxyls rearrange by a concerted, highly enantioselective 3,2-pathway. Hence, we conclude that the mechanistic pathways of dienyl peroxyls differ from those for simple allyl peroxyls.

Autooxidation

The autooxidative process consists of three steps: (1) chain initiation, (2) propagation, and (3) termination steps (1-3) as shown in Eq. (**1**)-(**4**):

Initiation: $$\text{In}\bullet + \text{RH} \rightarrow \text{R}\bullet + \text{InH} \quad (1)$$

Propagation: $$\text{R}\bullet + \text{O}_2 \rightarrow \text{ROO}\bullet \quad (2)$$

$$\text{ROO}\bullet + \text{RH} \xrightarrow{k_p} \text{R}\bullet + \text{ROOH} \quad (3)$$

Termination: $2ROO\cdot \rightarrow [ROOOOR]$

$$[ROOOOR] \rightarrow O_2 + \text{non radical products} \qquad (4)$$

The key event in initiation is the formation of R• [Eq. (**1**)]. There are many sources of radical species which may serve to abstract a hydrogen atom from RH. Nitric oxide (NO), nitrogen dioxide (NO_2), and ozone are all common environmental pollutants and have been shown to initiate autooxidation *via* hydrogen atom abstraction. Another method of generating R• is *via* thermal or photochemical induced homolytic scission.

In the initial propagation step of autooxidation, molecular oxygen adds to R• [Eq. (**2**)]. At partial oxygen pressures above 100 mm Hg, this addition approaches the diffusion controlled limit (*ca.* 10^9 1/ mol^{-s}) (4). Under such conditions, the major radical in solution is peroxy radical ROO• and not R•, so it is unlikely that any termination reactions involving R• will take place.

In the second propagation reaction [Eq.(**3**)], ROO• abstracts a hydrogen atom from RH at rate k_p to generate more R•. This is the rate-determining step in the propagation sequence. For each hydroperoxide product formed, another radical R• is generated. This process proceeds *ad infinitum* until interrupted by the termination step.

Eq. (**4**) shows one potential termination step, a peroxy radical coupling to give oxygen and non-radical organic products, the Russell mechanism (5).

The rate constant k_p [Eq. (**2**)] for hydrogen atom abstraction depends primarily on the strength of the C-H bond being broken. ROO• is strongly resonance stabilized and is a comparatively unreactive radical, and so prefers to act on the most weakly bound hydrogen atom. Thus, products of hydrocarbon autooxidation can be understood based on the stability of intermediate radicals (6). Scheme 1 shows products of several hydrocarbon autooxidations and the intermediate carbon radicals involved in the autooxidation chain propagation. In each example, the resulting products can be understood on the basis of the most stable intermediate radicals formed. Each intermediate radical has an adjacent π-system that stabilizes the radical. This radical stability is subsequently reflected in the selectivity of the autooxidation process.

Autooxidation takes a different course with substrates capable of aromatization by the loss of H_2. For compounds such as 1,4-cyclohexadiene and 9,10-dihydroanthracene, autooxidation gives benzene and anthracene, along with hydrogen peroxide.

Scheme 1

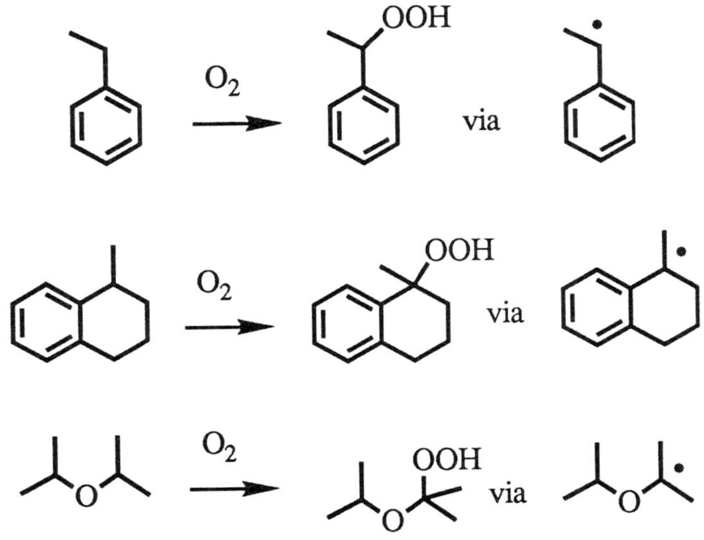

Lipid Autooxidation

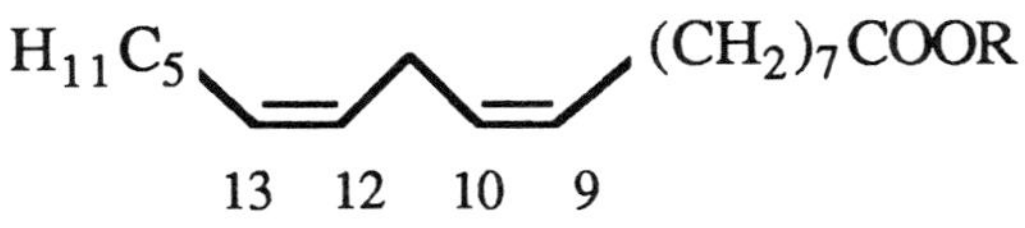

Linoleate

Linoleates autoxidize readily at room temperature. The major products formed under atmospheric oxygen are hydroperoxides formed by oxygen addition at the 9- and 13 positions of the carbon chain (7-8). Two products of autooxidation have oxygen substitution at the 9 position of the chain (**1** and **2**), while two products have substitution at the 13 position of the chain (**3** and **4**). Compounds **3** and **4** each have a conjugated diene unit with different stereochemistry: **3** has *trans-cis* diene stereochemistry, while **4** has *trans-trans* diene geometry. (Compounds **1** and **2** also have different geometries at the conjugated diene.)

While the major products **1-4** result from abstraction of a C-11 H atom *vide infra*, the minor products, such as non-conjugated hydroperoxides (but in very small amounts), result from abstraction of H-atoms at C-8 and C-14. At higher temperatures or with low oxygen concentrations present, a more complex product mixture results from linoleate.

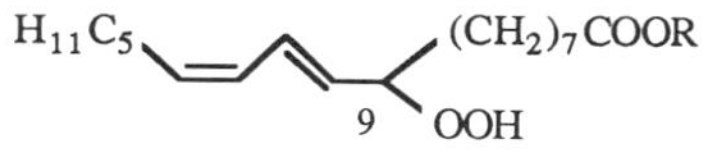

1

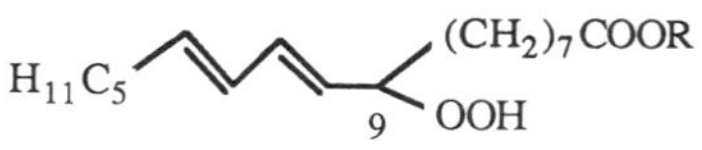

2

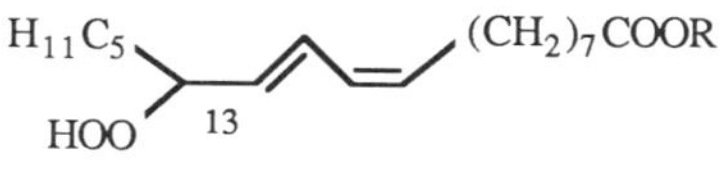

3

4

The mechanism for the formation of the four major linoleate products has been discussed in detail elsewhere and will be described here in outline form (2). Briefly stated, the mechanism proceeds by H-atom removal from C-11 to generate a pentadienyl radical, **5**, with stereochemistry, as shown in Scheme 2. The geometry of the pentadienyl radical is written as *cis-trans-trans-cis*, since the two terminal partial double bonds of the radical were formerly *cis* double bonds of linoleate. The other two bonds at Δ 10-11 and Δ 11-12 have partial double bond character and are formed in the lower energy *trans* geometry.

Scheme 2

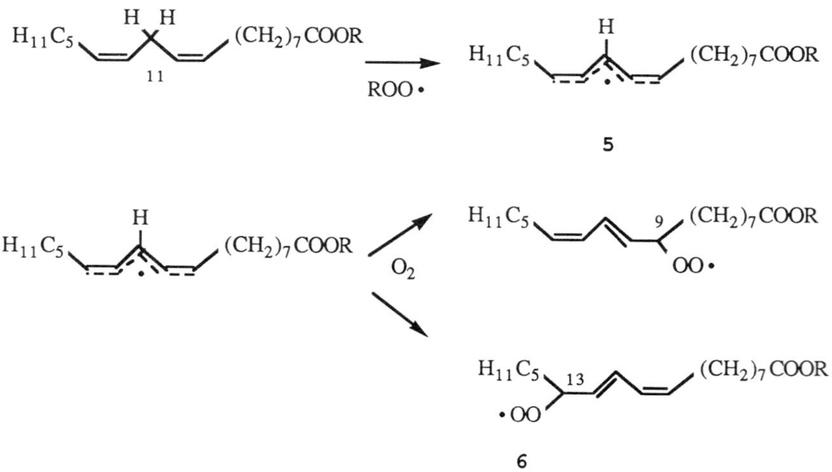

6

The second step in autooxidation of linoleate is addition of oxygen to the intermediate radical **5**. Oxygen addition is observed only at positions 1 and 13 of the delocalized radical and the products derived from this radical are observed only with oxygen substitution at these carbons. The fact that no oxygen addition occurs at position 11 of the pentadienyl radical is surprising because this site has the highest odd electron density in the pentadienyl radical. While it seems likely that oxygen addition does occur at C-11, the peroxyl radical product would probably be extremely reactive with respect to 3,2-rearrangement (*vide infra*) or β-fragmentation, as described in the next section. In other pentadienyl systems, products are observed from reaction of oxygen at the center carbon of the pentadienyl system, but in those systems the terminally substituted hydroperoxides analogous to **1-4** are sterically crowded.

To simplify the discussion, consider only oxygen coupling at C-13 (recognizing that similar chemistry occurs at C-9), which gives the peroxyl radical **6**, a radical with *trans-cis* conjugated diene geometry. Trapping **6** before β-fragmentation affords the kinetic product **3**, the *trans-cis* compound. If good H-atom donors are not present (R-H or antioxidants, such as tocopherol), then β-fragmentation leads to radicals **5** and **7**. Oxygen coupling at C-9 of the carbon radical **7** then leads to the thermodynamically more stable product, the 9 *trans-trans* product **2**. The competition between H-atom transfer to give kinetic product and β-fragmentation to give thermodynamic product thus accounts for the four major products from linoleate autooxidation.

The mechanism outlined in Scheme 3 is consistent with several experimental observations (1). Higher concentrations of linoleate lead to more *trans-cis* products because at higher [R-H] the reaction is driven toward kinetic control, since linoleate is an H-atom donor (2). Addition of other good H-atom donors, such as 1,4-cyclohexadiene, causes the product mixture to consist of primarily *trans-cis* products (kinetic control) (3). Product distributions are independent of oxygen pressure between 10 and 1000 mm Hg. Oxygen addition to carbon radicals is not rate limiting at these pressures and β-fragmentation is independent of oxygen concentration.

In summary, autooxidation propagation mechanisms of diene lipids consist of three steps: (1) H-atom transfer from substrate to peroxy radical; (2) oxygen coupling to intermediate carbon radicals; and (3) β-fragmentation of intermediate peroxyl radical. The kinetics and product distributions of autooxidation can be understood by this fundamental mechanism.

Scheme 3

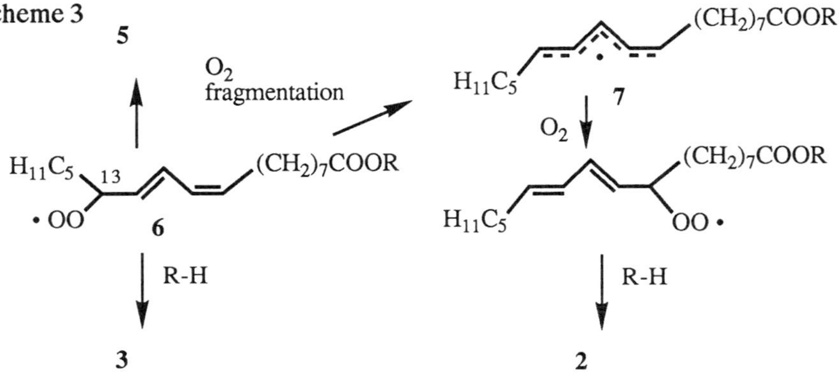

A mechanistic alternative to the β-fragmentation sequence outlined in Scheme 3 is a 3,2-concerted rearrangement interconnecting *trans-cis* and *trans-trans* peroxyl radicals (Scheme 4). This mechanism is consistent with the experimental data described above for the autooxidation of linoleate because a β-fragmentation oxygen addition is essentially equivalent to the 3,2-concerted rearrangement. The fragmentation sequence is favored for linoleate hydroperoxides, and the evidence for this conclusion is that linoleates undergo rearrangement by incorporating atmospheric oxygen into the hydroperoxide. While this indicates that fragmentation does occur in these systems, it does not rule out the possibility of 3,2-rearrangements under some conditions. Arachidonic acid auto-oxidation has been extensively investigated (9-11), since this compound is an important precursor to a variety of important natural products containing peroxide functionality.

The product mixtures obtained in arachidonate autooxidation are extremely complex. Six simple hydroperoxide products (**8-13**) have been isolated, all with *trans-cis* diene geometry. The alcohols formed from **8-13** by reduction can be separated by HPLC, allowing for extensive investigation of the autooxidation mechanism. Smaller amounts of *trans-trans* products can be isolated (particularly the 15-*trans-trans* compound).

For arachidonate, there are three centers (C-7, C-10, and C-13) that have carbons flanked by two double bonds. These carbons are analogous to the C-11 carbon of linoleate and are readily attacked by chain-carrying peroxyl radicals. Thus, three pentadienyl radicals are formed as intermediates in arachidonate autooxidation. These three pentadienyl radicals lead to products **8-13**. In Figure 1, the HPLC chromatogram shows the separation of the hydroperoxides from methyl arachidonate.

Scheme 4

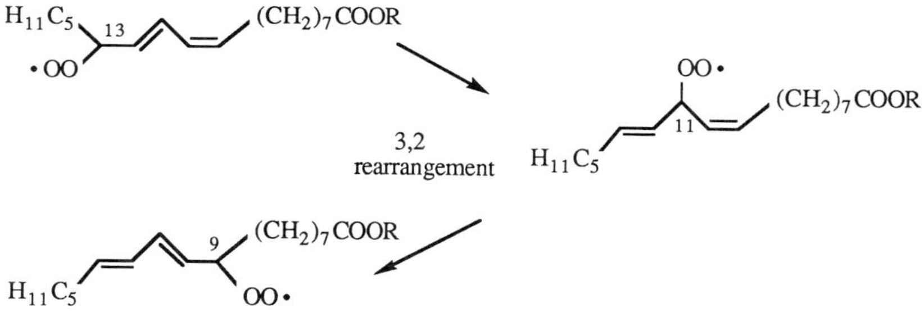

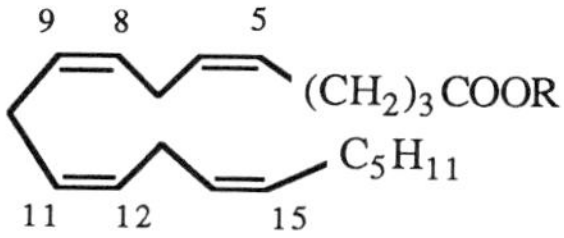

Arachidonate

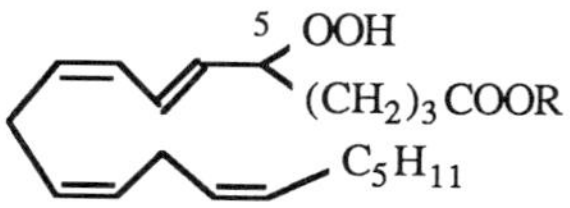

8

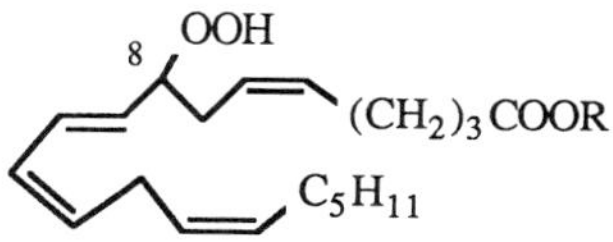

9

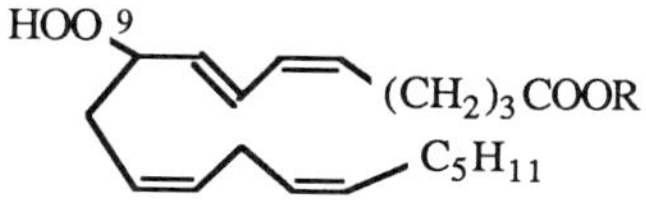

10

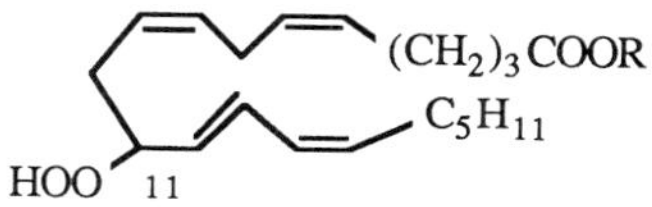

11

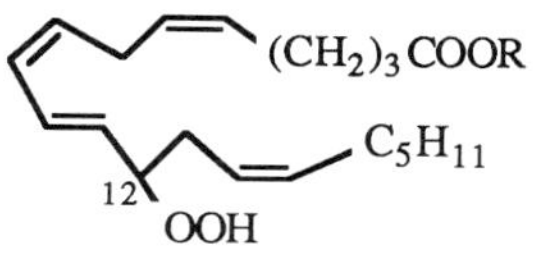

12

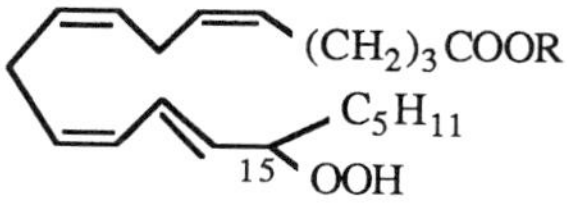

13

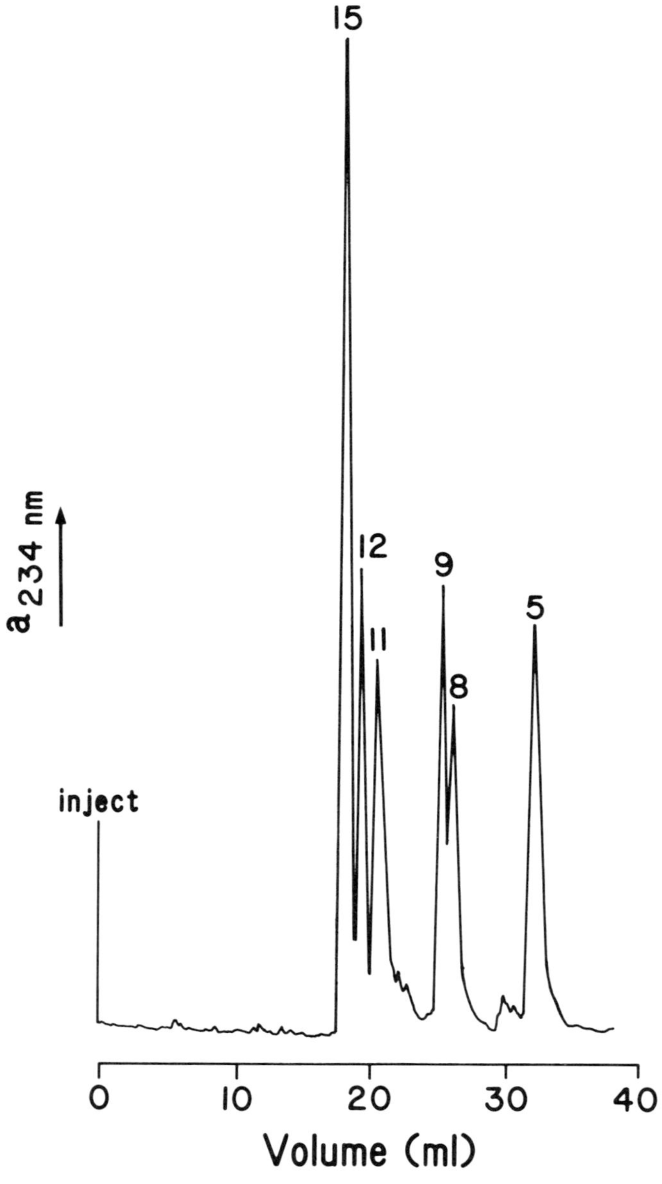

Figure 1. *HPLC for Separation of Methyl Arachidonate Hydroperoxides*. Normal phase separation on silica with hexane:isopropanol, 99:1. Numbers indicate the position of OOH substitution. Detection with 234 nm light.

Hydroperoxide Resolution

We have discovered what appears to be a general method for the preparation of optically pure hydroperoxides: the resolution of unsaturated hydroperoxide enantiomers by liquid chromatography of diastereomeric derivatives. This method allows, for the first time, the nonenzymatic preparation of optically pure allylic or dienylic hydroperoxide natural products.

Preliminary screens (12) of several hydroperoxide derivatives of the structure R-OO-CR'XOR", **14**, were carried out to achieve three specific goals: (1) to easily convert the hydroperoxide to **14**; (2) to easily remove the protecting group without racemization or destruction of the unsaturated hydroperoxide substructure, and (3) to determine acceptable chromatographic characteristics of the derivative including resolution of diastereomers and stability to the conditions of chromatography. Four hydroperoxide derivatives were likely candidates: peracetals (**14** with R'=alkyl or H, X=H), which are readily formed and stable to chromatography, but require harsh conditions for deprotection; perorthoesters (R'=H, aryl or alkyl and X=OR"); peraminals (R'=H, aryl or alkyl and X=NR_2"), which are unstable to chromatography; and perketals (**14** with R'=X=alkyl), which are readily prepared from hydroperoxides, are stable to normal or reverse-phase chromatography, and can be deprotected under very mild acid conditions. Hence, perketals were chosen for further experimentation.

Scheme 5

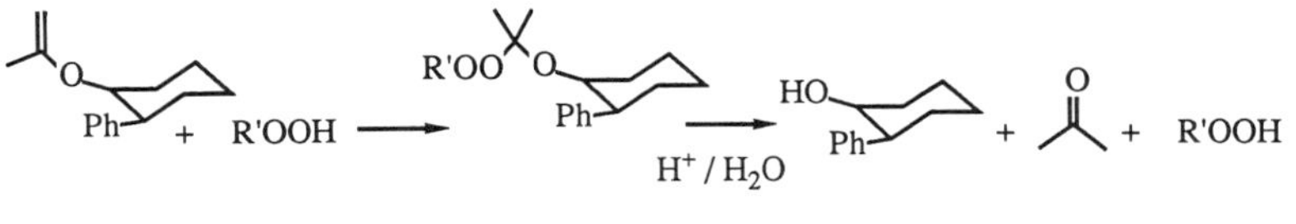

15 14

Several hydroperoxides (R_1=H, R_2=Me) were converted to the perketals **14**. In each case, the perketals were readily formed, were stable to chromatography under appropriate conditions, and the perketal protecting group was removed with acetic acid/THF/water. (Although perketals are completely stable to normal-phase chromatography, some decomposition occurs in reverse-phase solvents, such as acetonitrile/water or methanol/water unless 0.01% Et_3N is added.) Diastereomers could be isolated with isomeric purities of 99% or better. After recovery of hydroperoxide from the perketal by hydrolysis with acetic acid/THF/water, the optical purity of the hydroperoxide was assayed by rederivatization with **15** and analytical chromatography, or by comparison of the specific rotation of the resolved enantiomers with a known natural product (*i.e.*, 15-HPETE from soybean or 5-HPETE from potato lipoxygenase).

Among the hydroperoxides investigated were arachidonate hydroperoxides including 15-HPETE, 5-HPETE, 8-HPETE, and 9-HPETE. These hydroperoxides were first separated as their methyl esters (as shown in Figure 1) and then derivatized and separated as the perketals, and finally reconverted to the optically pure HPETE. In every case studied, the R perketal was the first to elute on reverse-phase HPLC. A mixture of all of the arachidonate HPETE's was also derivatized by the vinyl ether **15**. Figure 2 shows the normal-phase chromatogram of this mixture of perketals. While the separation of all 12 perketals (from the 6-HPETE's) is not complete, we do see complete separation for the 11-, 9-, 8-, and 5-HPETE's.

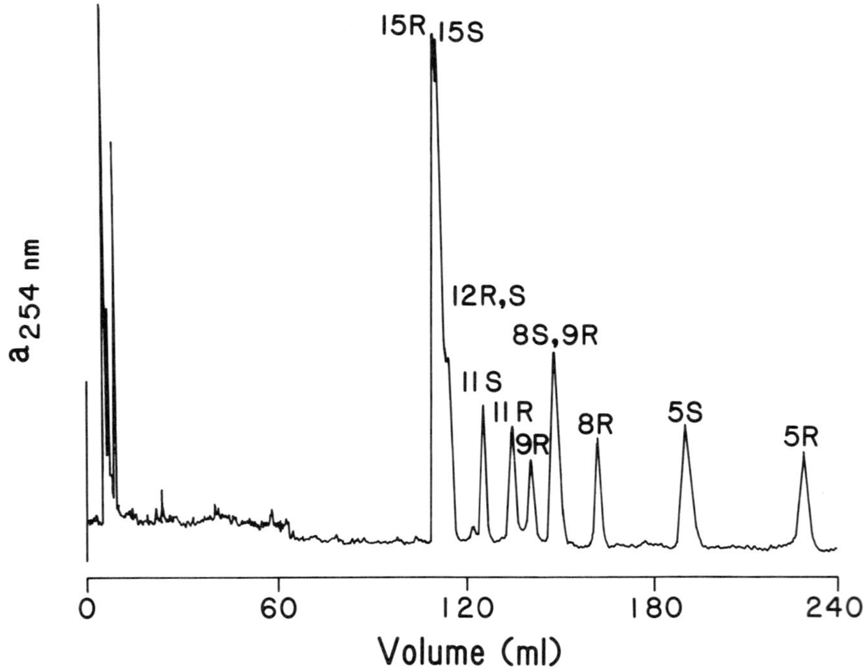

Figure 2. *HPLC for Separation of All Perketals Analogous to 15 Derived from Methyl Arachidonate Hydroperoxides.* Normal-phase separation on silica with hexane:ethyl acetate, 999:1. Numbers indicate the position of OOH substitution and absolute configuration of the stereocenter. Detection with 254 nm light.

Allylic Hydroperoxide Rearrangements

Allylic hydroperoxides undergo structural rearrangement. This rearrangement has been known since 1957, when Schenck reported that the tertiary C-5 α-hydroperoxide of cholesterol rearranges to its C-7 α-allylic isomer (13-16). The allylic rearrangement of hydroperoxides is catalyzed by UV light and free radical initiators and is inhibited by 2,6-di(*tert*-butyl)-4-methylphenol. One step in chain propagation presumably involves generation of an allylic peroxyl radical from the corresponding hydroperoxide by H-atom transfer. The H-atom abstracting agent may be the initiator radical or another peroxyl radical present in solution. The ROO-H bond is known to be weak and transfer of a hydroperoxyl H to alkoxyl radicals or other peroxyls is a facile process. Once formed, the allylic peroxyl radicals rearrange by an allylic shift. The mechanism of this rearrangement has been the subject of much debate.

At least three mechanisms for the allylic hydroperoxide rearrangement have been proposed. These three mechanisms involve: (1) formation of a cyclic 5-membered ring peroxide with a free radical at position 4 of the ring. (2) Formation of a cyclic 5-membered ring transition state that links the two allylic hydroperoxyl radicals. This mechanism is a concerted mechanism in which there is not an authentic reaction intermediate. (3) β-Fragmentation of an allylic peroxyl radical to form molecular oxygen and an allyl carbon radical, which can recombine with oxygen at either end of the radical to give the starting and rearranged peroxyl radicals. Each of these mechanisms involves intermediate peroxyls, and is consistent with this is the fact that free radical initiators facilitate the reaction and phenolic inhibitors stop the rearrangement.

We report here a study of the optically pure hydroperoxides prepared by the perketal resolution method described above. The hydroperoxides (substituted at positions 9 and 10), formed from the reaction of singlet oxygen and oleic acid were separated by normal-phase HPLC and the optically pure hydroperoxides were rearranged in hexane with di-*tert*-butyl hyponitrite initiator. After 5 h at 40°C, the starting material and rearrangement products were present in approximately equal amounts. For the 9 hydroperoxide starting material, the product is the 11 hydroperoxide, while the 10 hydroperoxide rearranges to the 8 hydroperoxide product. If the rearrangement is followed over time and the 9- and 11 hydroperoxides are isolated and assayed for isomeric purity over the course of the rearrangement, it is found that the rearrangement occurs with high stereoselectivity. Thus, if the starting 9 substituted hydroperoxide is

>99% R configuration, the 11 hydroperoxide formed in the rearrangement is enriched in the S enantiomer. In the early stages of rearrangement the enantiomeric excess of the 11-hydroperoxide formed approaches 96% ee at 40° and the enantiomeric purity of the 11 hydroperoxide and the starting 9 substituted compound slowly decrease at longer reaction times. A plot of enantiomeric purity of starting hydroperoxide and rearrangement hydroperoxide as a function of extent of rearrangement is shown in Figure 3. The results are dramatic. The rearrangement occurs with significant transfer of chirality in these acyclic systems and at lower temperatures for rearrangement, the selectivity of the rearrangement is even greater.

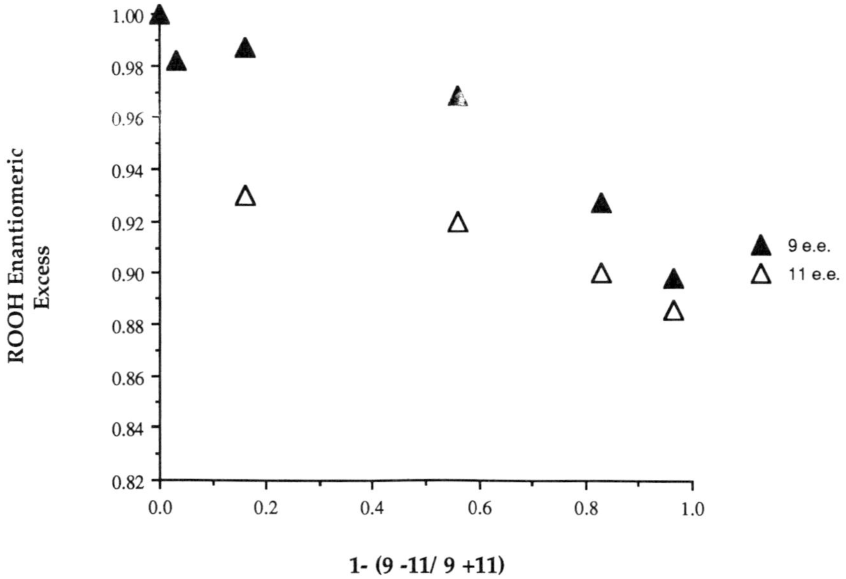

Figure 3. *Enantiomeric Excess of Starting Hydroperoxide (9-OOH) and Rearranged Hydoperoxide (11-OOH) vs. Extent of Rearrangement in Hexane at 40°*

Scheme 6

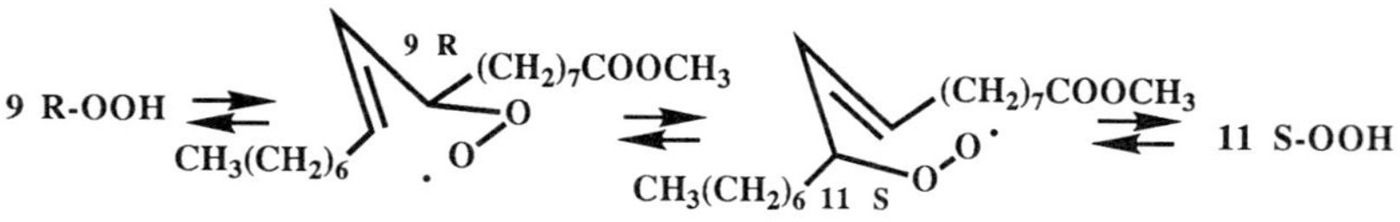

The results of our rearrangement studies support the concerted radical rearrangement pathway *via* a 5-membered ring transition state, as shown in Scheme 6. The rearrangement apparently occurs *via* the transition state, as shown with the two alkyl substituents occupying pseudoequatorial positions on the envelope conformation.

We conclude that the dienyl peroxyl rearrangements previously investigated differ mechanistically from the rearrangement of simple allyl peroxyls. The dienyl peroxyls rearrange by a β-scission pathway, while the simple allyl peroxyls rearrange by a concerted 3,2-pathway.

References

1. Ingold, K.U. (1969) *Acc. Chem. Res.* **2**, 1-9
2. (a) Porter, N.A. (1980) In *Free Radicals in Biology* Vol. IV, (ed. Pryor, W.A.) New York: Academic Press, 261-291; (b) Porter, N.A. (1986) *Acc. Chem. Res.* **19**, 262-268
3. McBrien, D., and Slater, T. F (1982) *Free Radicals, Lipid Peroxidation, and Cancer* Academic Press, Eds.,
4. Maillard, B., Ingold, K.U., and Scaiano J.C. (1983) *J. Am. Chem. Soc.* **105**, 5095-5099
5. Russell, G.A. (1957) *J. Am. Chem. Soc.* **79**, 3871-3877
6. Howard, J.A. (1972) In *Adv. in Free Radical Chem.* New York: Academic Press, Volume 4, 49-173
7. Porter, N.A., Lehman, L.S., Weber, B.A., and Smith, K.J. (1981) *J. Am. Chem. Soc.* **103**, 6447-6455
8. Porter, N.A., Weber, B.A., Weenen, H., and Khan, J.A. (1980) *J. Am. Chem. Soc.* **102**, 5597-5601
9. Porter, N.A., Wolf, R.A., Yarboro, E.M., and Weenen, H. (1979) *Biochem. Biophys. Res. Comm.* **89**, 1058-1064
10. Khan, J.A., and Porter, N.A. (1982) *Angew. Chem., Int. Ed. Engl.* **21**, 513-522
11. Porter, N.A., Logan, J., and Kontoyiannidou, V. (1979) *J. Org. Chem.* **44**, 3177-3181
12. Dussault, P.H., and Porter, N.A. (1988) *J. Am. Chem. Soc.* **110**, 6276-6277
13. Schenck, G.D., Neumuller, O.A., and Eisfeld, W. (1958) *Angew. Chem.* **70**, 595
14. Brill, W.F. (1984) *J. Chem. Soc., Perkin Trans II* 621-627
15. Beckwith, A.L.J., Davies, A.G., Davison, I.G.E., Maccoll, A., and Mruzek, M.H. (1988) *J. Chem. Soc., Chem. Comm.* 475-476
16. Porter, N.A., and Sullivan Wujek, J. (1987) *J. Org. Chem.* **52**, 5085-5089

Arachidonate 12-Lipoxygenases

S. Yamamoto, T. Yoshimoto, N. Ueda, A. Hiroshima, Y. Takahashi, H. Suzuki, K. Natsui, T. Maruyama, F. Shinjo, K. Ii, N. Komatsu, K. Watanabe, K. Gerozissis, F. Dray, C. Yokoyama, T. Takai, and T. Tanabe

Department of Biochemistry, Tokushima University School of Medicine Tokushima 770 Japan

Abstract

Two types of arachidonate 12-lipoxygenase (platelet-type and leukocyte-type), which were immunologically distinguishable from each other, were immunoaffinity-purified. The platelet-type enzyme (from bovine platelets) was selectively active with eicosapolyenoic acids whereas the leukocyte-type enzyme (from bovine and porcine leukocytes) had a broad substrate specificity reacting with polyenoic acids of C_{18}, C_{20} and C_{22}. As an example of the leukocyte-type 12-lipoxygenase, the purified enzyme of porcine leukocytes was thoroughly investigated. Catalytically, the porcine leukocyte 12-lipoxygenase is a multi-functional enzyme with the oxygenase activities of various regiospecificities and the 14,15-leukotriene A synthase activity. By a peroxidase-linked immunoassay utilizing monoclonal antibodies against the porcine leukocyte enzyme, considerable amounts of 12-lipoxygenase were detected not only in porcine leukocytes but also in small intestine, thymus, lymph node and pituitary gland. In these porcine organs except for pituitary, their parenchymal cells were not significantly immuno stained. Only the pituitary parenchymal cells with granules were positively stained. The DNA complementary to the mRNA of porcine leukocyte 12-lipoxygenase was cloned, and subjected to the nucleotide sequence analysis, from which the amino acid sequence of the enzyme was deduced. The 12-lipoxygenase cDNA recognized a 3.4-kilobase mRNA in the largest amount in leukocytes, followed by pituitary, lung, jejunum and spleen.

Introduction

Arachidonate 12-lipoxygenase is a dioxygenase which incorporates one molecule of oxygen into arachidonic acid regiospecifically and stereospecifically, and produces 12S-hydroperoxy-5,8,10,14-eicosatetraenoic acid (Fig. 1). This compound (referred to as 12-

Biological Oxidation Systems, Volume 2

HPETE in this chapter) is converted by reduction to a corresponding hydroxy acid (12-HETE).

Since the enzyme was found in platelets as the first mammalian lipoxygenase (1,2), the 12-lipoxygenase has been found in a variety of tissues of a number of animal species (3). In addition to 12-lipoxygenase several more lipoxygenases have been found in mammalian tissues. It is well known that the leukotriene synthesis is initiated by 5-lipoxygenase and the biosynthesis of prostaglandin and thromboxane by cyclooxygenase. These compounds have specific biological activities which regulate the functions of various animal tissues, and their biological functions have been generalized for a variety of animal species.

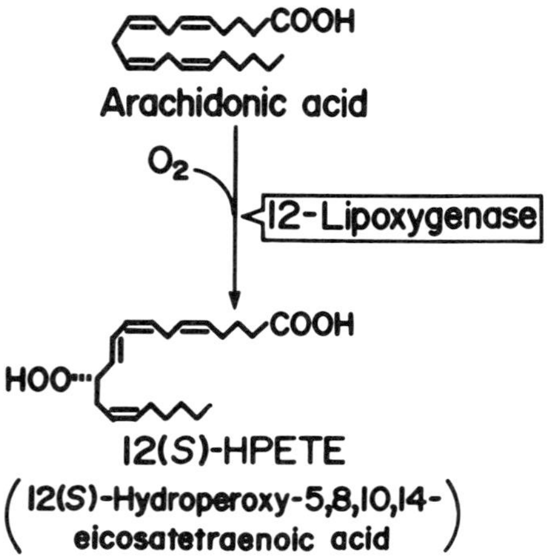

Figure 1. *Arachidonate 12-Lipoxygenase Reaction*

Goetzl and others reported a chemotactic activity of various HETEs, but considerably high concentrations were required to stimulate leukocyte chemotaxis (4). Later Murota and co-workers demonstrated a potent activity of 12-HETE to stimulate the migration of cultured smooth muscle cells of rat aorta. The activity was shown with 10 fM 12-HETE, and was specific for 12-HETE rather than 5-HETE and 15-HETE (5). Such a potent activity of the 12-HETE has not been further reported with different cells of other animal species, and the mechanism of the chemotactic activity has not been elucidated. Secondly, a role of the 12-lipoxygenase metabolites in neurotransmission was discovered with *Aplysia* ganglion by Feinmark and Schwartz and their colleagues (6). Bioactive metabolites were identified as 12-keto-5,8,10,14-eicosatetraenoic acid (7) and 8-

hydroxy-11,12-epoxy-5,9,14-eicosatrienoic acid (8). The latter compound referred to as hepoxilin A_3 was shown to be a neurotransmitter in rat hippocampal neuron (9). Hepoxilin A_3 was earlier shown to stimulate insulin secretion in rat (10). However, the enzymatic origin of these 12-HPETE metabolites remains to be further investigated. Thirdly, according to Honn and co-workers, the cancer cells have a glycoprotein receptor like platelet GpIIb/IIIa, which is involved in the cancer cell adhesion to vascular endothelial cells. 12-HETE mediates either the expression or the activation of the GpIIb/IIIa-like receptor of the cancer cell (11).

Unlike the cyclooxygenase and 5-lipoxygenase pathways, however, no bioactive compound with such a general function has so far been found in the arachidonate metabolism initiated by 12-lipoxygenase (12), and the 12-lipoxygenase has been ignored by most investigators.

When we looked for the source of 5-lipoxygenase, which was physiologically more important and apparently more interesting, we found a high activity of 12-lipoxygenase in porcine leukocytes (13). As shown in Fig. 2, the 12-lipoxygenase activity was not detected in porcine platelets in contrast to other mammalian platelets (13). When we tried to purify the 12-lipoxygenase from porcine leukocytes, we found that the enzyme was readily inactivated by conventional purification methods. Therefore, we changed our strategy, and a crude enzyme preparation was given to a mouse as antigen to prepare a monoclonal anti-12-lipoxygenase antibody by the hybridoma technique (14). An antibody thus prepared was utilized for immunoaffinity purification of the 12-lipoxygenase. A crude enzyme preparation containing 12-lipoxygenase was applied to an immunoaffinity column containing agarose to which a monoclonal anti-12-lipoxygenase antibody was immobilized. A bulk of protein without 12-lipoxygenase activity passed through the column, and the enzyme was eluted from the column by increasing the concentration of non-ionic detergent and raising the pH of the buffer to 10.4. By this step the enzyme could be purified to near homogeneity, and gave a specific activity of about 3 μmol/min/mg protein at 30° C (15).

When we investigated the catalytic properties of the purified enzyme, we found that the enzyme was inactive with 12-HPETE, which was produced from arachidonic acid. This was in sharp contrast to 5-lipoxygenase which transformed 5-HPETE to leukotriene A_4 (16). Instead, the 12-lipoxygenase reacted with 15-HPETE as substrate nearly as fast as arachidonic acid. The enzymatic products from 15-HPETE were identified in collaboration with A.R. Brash, who isolated and identified various dihydroxy acids produced from

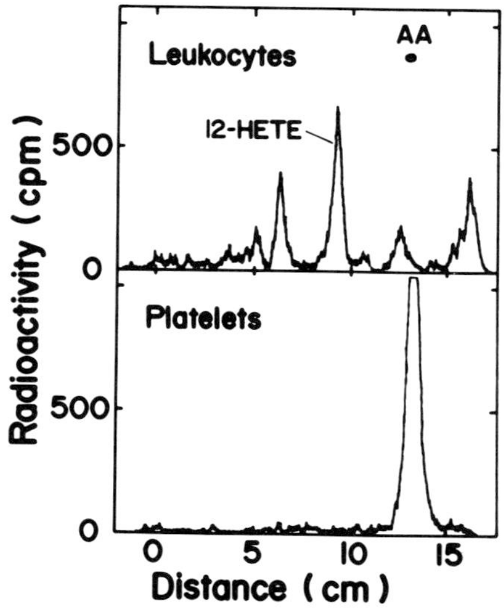

Figure 2. *Arachidonate Metabolites of Porcine Leukocytes (upper) and Platelets (lower).* ^{14}C-Labeled arachidonic acid was incubated with a cell suspension, and the radioactive products were separated by thin layer silica gel chromatography.

arachidonic acid by whole cells of leukocytes and proposed the participation of 12-lipoxygenase in the formation of these dihydroxy acids (17). The use of our purified enzyme allowed the isolation of hydroperoxy acids, and our new findings were confirmatory and supplementary to the reaction mechanism previously proposed by Brash and associate (17). One of the predominant products was 14R,15S-dihydroperoxyeicosatetraenoic acid (diHPETE) with a conjugated triene, which was accompanied by a smaller amount of 8S,15S-diHPETE. Thus, 12-lipoxygenase reacted with 15-HPETE as 14R- and 8S-dioxygenases. 8S,15S- and 8R,15S-Dihydroxy-eicosatetraenoic acids (diHETE) with all-trans triene were also produced in comparable amounts, and these diHETEs appeared even in the absence of air. These findings showed that the purified 12-lipoxygenase was a multifunctional enzyme. In addition to the oxygenation at position 12 of arachidonic acid, the enzyme catalyzed oxygenation of 15-HPETE at positions 8 and 14. On the other hand, the enzyme produced a 14,15-epoxy compound of leukotriene nature by an overall dehydration from 15-HPETE. The epoxide was unstable, and transformed non-enzymatically to a mixture of 8S,15S- and 8R,15S-diHETEs. Fig. 3 summarizes the reactions catalyzed by 12-lipoxygenase with 15-HPETE as substrate.

The substrate specificity studies demonstrated another interesting catalytic property of 12-lipoxygenase. Our group (15) and Nugteren's group (18) reported that the 12-lipoxygenase of porcine leukocytes had a broad substrate specificity in terms of the carbon chain length. As illustrated in Fig. 4, the porcine leukocyte 12-lipoxygenase was active not only with eicosapolyenoic acids but also with octadeca- and docosa-polyenoic acids. This broad substrate specificity was in contrast to the narrow specificity of platelet 12-lipoxygenase as earlier described by Nugteren (2). A paper from Vliegenthart's group reported that the bovine platelet 12-lipoxygenase was inactive with linoleic acid while the enzyme of bovine leukocytes was active with linoleic acid as well as arachidonic acid (19).

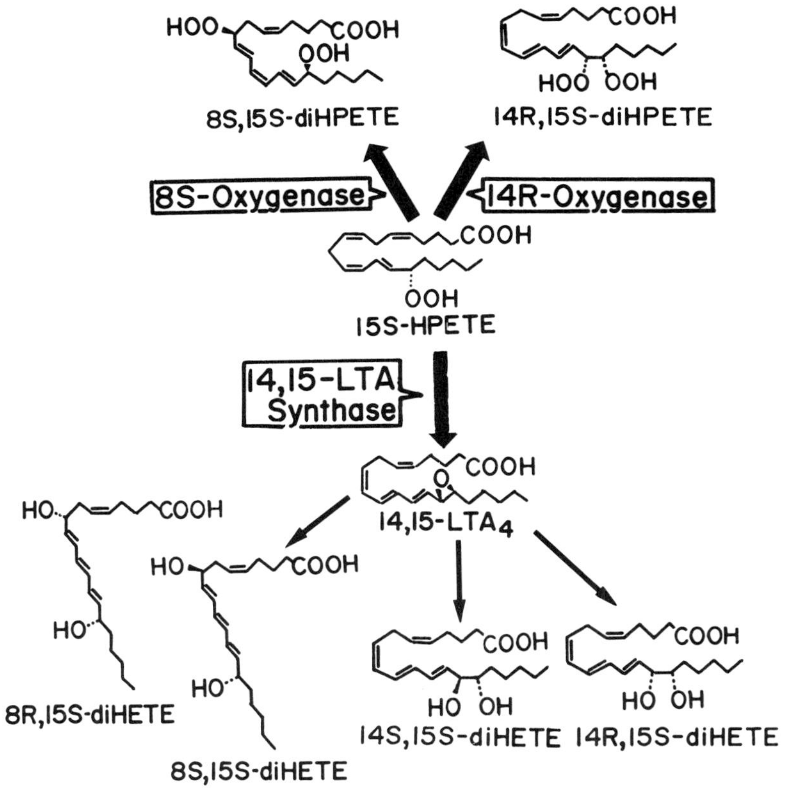

Figure 3. *Reactions of Porcine Leukocyte 12-Lipoxygenase with 15-Hydroperoxy-5,8,11,13-Eicosatetraenoic Acid*

We had two useful tools to precisely analyze this interesting subject of substrate specificity of 12-lipoxygenase; namely, a monoclonal antibody against *porcine leukocyte* enzyme and a monoclonal antibody against *human platelet* enzyme. These antibodies cross-reacted with the enzymes of *bovine leukocytes* and *platelets*, respectively, and were utilized for immunoaffinity purification of bovine enzymes (20). Fig. 5 shows the substrate specificities of the two distinct bovine 12-lipoxygenases thus purified by immunoaffinity chromatography. The platelet enzyme was specific for eicosapolyenoic acids while the leukocyte enzyme had a broad substrate specificity reacting with docosa-, eicosa- and octadeca-polyenoic acids (20). Thus, the two 12-lipoxygenases of the same animal species were distinguishable each other not only immunologically but also catalytically. Both of the 12-lipoxygenase were also multifunctional enzymes as tested with 15-HPETE (20).

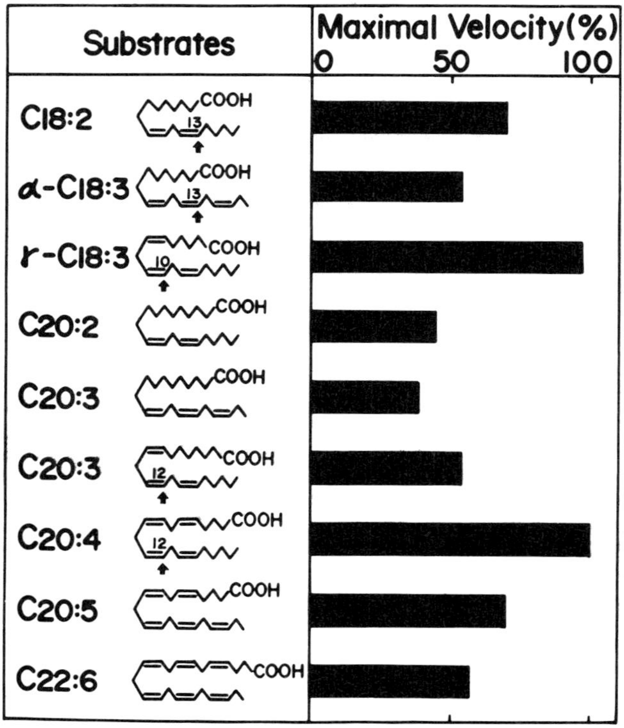

Figure 4. *Reactivities of Immunoaffinity-Purified 12-Lipoxygenase of Porcine Leukocytes as Tested with Various Unsaturated Fatty Acids.* Maximum velocities determined at saturating substrate concentrations are presented.

As summarized in Table 1, two types of 12-lipoxygenase are found in mammalian tissues. The leukocyte-type enzyme has a broad substrate specificity in terms of the carbon chain length of unsaturated fatty acid, while the platelet-type enzyme is selectively active with eicosapolyenoic acid. The leukocyte-type enzyme has been found also in porcine pituitary (to be discussed later) and bovine tracheal epithelial cells (21) in addition to bovine and porcine leukocytes. The platelet-type enzyme has been so far found only in platelets. Bovine tissues have both types of 12-lipoxygenase. However, 12-lipoxygenase is present in porcine leukocytes (and pituitary) but undetectable in porcine pletelets, while the enzyme is present in human platelets but undetectable so far in human leukocytes.

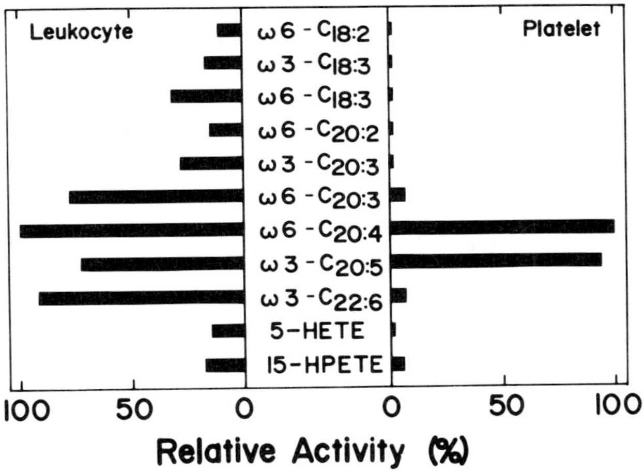

Figure 5. *Reactivities of Immunoaffinity-Purified 12-Lipoxygenase of Bovine Leukocytes and Platelets as Tested with Various Unsaturated Fatty Acids.* Relative values of maximum velocity (arachidonic acid, 100) are presented.

Table 1. *Two Types of 12-Lipoxygenase*

Type	Leukocyte-Type	Platelet-Type
Substrate Specificity	C_{18}, C_{20}, C_{22}	C_{20}
Distribution		
Bovine	Leukocyte, Trachea	Platelet
Porcine	Leukocyte, Pituitary	---
Human	---	Platelet

Another application of our monoclonal anti-12-lipoxygenase antibodies was a peroxidase-linked immunoassay of sandwich-type, which allowed a quantitative determination of 12-lipoxygenase in various porcine tissues (14). This method determines the quantity of the 12-lipoxygenase protein rather than the enzyme activity. Fig. 6 (right panel) presents the results of the peroxidase-linked immunoassay of 12-lipoxygenase in various porcine tissues. By far the highest enzyme content was found in leukocytes, followed by small intestine, stomach, thymus and lymph node. No significant amount of 12-lipoxygenase was detected in platelets by the enzyme immunoassay. At that time the sensitivity of the assay was too low to detect 12-lipoxygenase in porcine brain tissues.

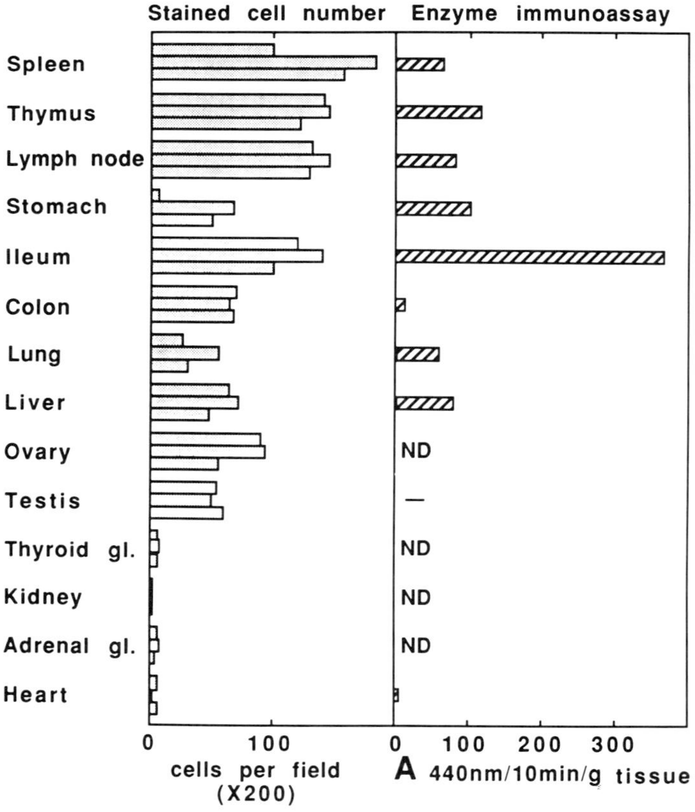

Figure 6. *The 12-Lipoxygenase Contents Correlated with the Numbers of Immunohistochemically Stained Infiltrating or Resident Leukocytes in Various Porcine Tissues*

Recently we improved the sensitivity of the assay by introducing the solid phase method and the avidin-biotin method (22). The new method was about 10 times more sensitive than the previous method, and applied to various parts of porcine brain as shown in Fig. 7. Anterior pituitary showed the highest content of 12-lipoxygenase. This value was about 5% of the enzyme content of leukocyte which was the richest source of 12-lipoxygenase. Posterior pituitary had about one-tenth of the enzyme in anterior pituitary. When the substrate specificity was examined, the 12-lipoxygenase of porcine pituitary was found to be of the leukocyte-type reacting with linoleic, α-linolenic and γ-linolenic acids as fast as arachidonic acid (22).

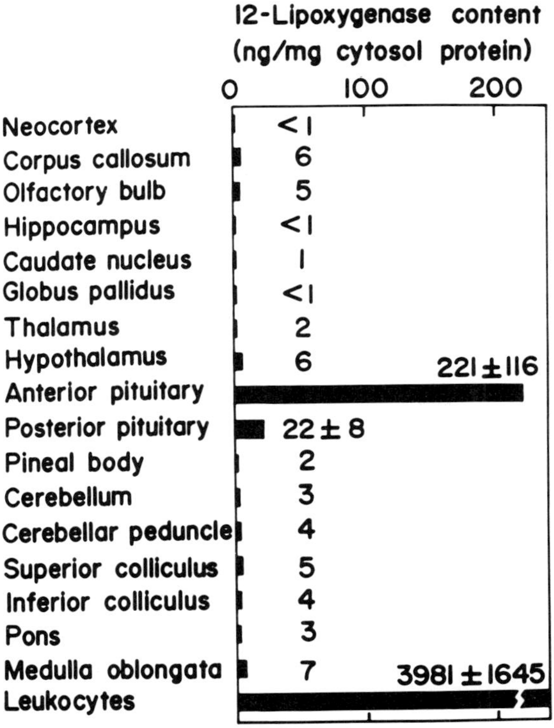

Figure 7. *The 12-Lipoxygenase Contents in Porcine Brain Tissues Estimated by the Peroxidase-Linked Immunoassay*

As mentioned above (Fig. 6), significant amounts of 12-lipoxygenase were detected by the enzyme immunoassay in several tissues in addition to leukocytes. For immunohistochemical studies we prepared a polyclonal antibody against the 12-lipoxygenase which was immunoaffinity-purified by the use of a monoclonal antibody against the enzyme. When we examined these tissues, their parenchymal cells were not stained significantly, and, as shown in Fig. 6 (left panel), the positively stained cells were resident mast cells or infiltrating leukocytes (23). Therefore, we subjected porcine pituitary to histological examination to clarify whether or not the positive values of the enzyme immunoassay was attributed to the infiltrating or contaminating leukocytes. We found that certain parenchymal cells rather than infiltrating leukocytes were positively stained. Electron microscopic observation showed positively stained parenchymal cells with granules, and the enzyme was located in the cytosol of these cells (22). The positively stained cells have not yet been identified.

Another method to screen 12-lipoxygenase is the measurement of mRNA for the enzyme. For cloning the cDNA for 12-lipoxygenase, as illustrated in Fig. 8, we constructed the cDNA libraries from poly A containing RNA from porcine leukocytes. The cDNA mixture was inserted into pBluescript. For screening the 12-lipoxygenase cDNA, two oligonucleotide probes were prepared on the basis of the amino acid sequence of one of the proteolytic fragments of the purified 12-lipoxygenase of porcine leukocytes. Transformed *E. coli* cells were screened by the use of ^{32}P-labeled oligonucleotide probes A and B (Fig. 8), and a clone pLOX89 was obtained. Additionally, we constructed the primer-extention libraries employing pUC19 and two primers with the nucleotide sequence found in the cDNA insert of pLOX89, and obtained two clones pLOX96 and pLOX134. The nucleotide sequences of the cDNA inserts of these clones were determined by the method of Sanger, and the amino acid sequence of the enzyme was deduced from the nucleotide sequence.

The deduced amino acid sequence included the amino terminal region of the enzyme and the sequences of the proteolytic fragments which were determined by automated Edman degradation. The cDNA had an open reading frame encoding 662 amino acid residues with a molecular weight of 74,911. In terms of the primary structure, the porcine leukocyte 12-lipoxygenase exhibited 86% identity with human reticulocyte 15-lipoxygenase (24) and 41% identity with human leukocyte 5-lipoxygenase (25). The cDNA clone thus prepared was utilized for screening 12-lipoxygenase mRNA by Northern blotting. A 3.4-kb mRNA was detected not only in leukocytes but also in pituitary, lung, jejunum, spleen and so on. The results were consistent with those observed by the enzyme immunoassay.

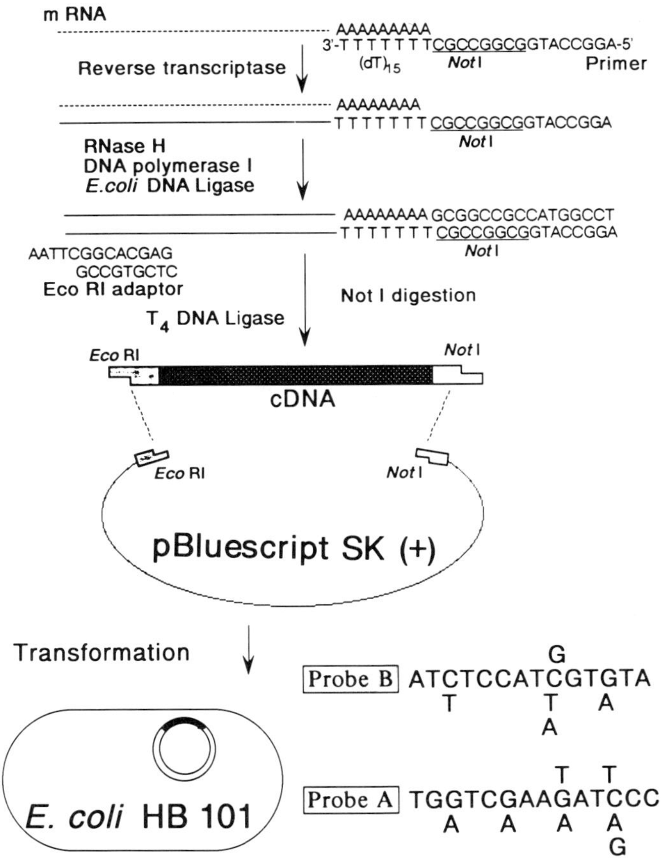

Figure 8. *Cloning of the cDNA for 12-Lipoxygenase of Porcine Leukocytes*

As discussed at the beginning of this chapter, we still do not know a general theory for the physiological and pathological function of 12-lipoxygenase. A screening of 12-lipoxygenase by the enzyme immunoassay in various tissues revealed a relatively high content of the enzyme in pituitary. This finding was supported by the measurement of the 12-lipoxygenase mRNA by the use of the cDNA for the enzyme. Several previous papers described the transformation of exogenous arachidonic acid to 12-HETE by rat pituitary (26-30). Our work now confirmed these earlier observations ruling out a possibility that the 12-lipoxygenase was derived from infiltrating or contaminating leukocytes. The occurrence of 12-lipoxygenase in porcine pituitary suggests an endocrine role of a certain 12-lipoxygenase metabolite of arachidonic acid. Our finding may be relevant to the previous findings by Dray and co-workers that LH-RH excretion from rat median eminence (31) and melatonin excretion from rat pineal gland (32) are stimulated by 12-lipoxygenase metabolites.

References

1. Hamberg, M. and Samuelsson, B. (1974) *Proc. Natl. Acad. Sci. USA* **71**, 3400-3404
2. Nugteren, D.H. (1975) *Biochim. Biophys. Acta* **380**, 299-307
3. Yamamoto, S. (1983) in *Prostaglandins and Related Substances* (Pace-Asciak, C. and Granström, E., eds.), pp. 171-202, Elsevier, Amsterdam
4. Goetzl, E.J., Woods, J.M. and Gorman, R.R. (1977) *J. Clin. Invest.* **59**, 179-183
5. Nakao, J., Ooyama, T., Ito, H., Chang, W.C. and Murota, S. (1982) *Atherosclerosis* **44**, 339-342.
6. Piomelli, D., Volterra, A., Dale, N., Siegelbaum, S.A., Kandel, E.R., Schwartz, J.H. and Belardetti, F. (1987) *Nature* **328**, 38-43
7. Piomelli, D., Feinmark, S.J., Shapiro, E. and Schwartz, J.H. (1988) *J. Biol. Chem.* **263**, 16591-16596
8. Piomelli, D., Shapiro, E., Zipkin, R., Schwartz, J.H. and Feinmark, S.J. (1989) *Proc. Natl. Acad. Sci. USA* **86**, 1721-1725
9. Carlen, P.L., Gurevich, N., Wu, P.H., Su, W., Corey, E.J. and PaceAsciak, C.R. (1989) *Brian Res.* **497**, 171-176
10. Pace-Asciak, C.R., Martin, J.M. and Lee, S. P. (1988) *Biochem. Cell Biol.* **66**, 901-909
11. Grossi, I.M., Fitzgerald, L.A., Umbarger, L.A., Nelson, K.K., Diglio, C.A., Taylor, J.D. and Honn, K.V. (1989) *Cancer Res.* **49**, 1029-1037
12. Brash, A.R. (1985) *Circulation* **72**, 702-707
13. Yoshimoto, T., Miyamoto, Y., Ochi, K. and Yamamoto, S. (1982) *Biochim. Biophys. Acta* **713**, 638-646
14. Shinjo, F., Yoshimoto, T., Yokoyama, C., Yamamoto S., Izumi, S., Komatsu, N. and Watanabe, K. (1986) *J. Biol. Chem.* **261**, 3377-3381
15. Yokoyama, C., Shinjo, F., Yoshimoto, T., Yamamoto, S., Oates, J.A. and Brash, A.R. (1986) *J. Biol. Chem.* **261**, 16714-16721
16. Ueda, N., Kaneko, S., Yoshimoto, T. and Yamamoto, S. (1986) *J. Biol. Chem.* **261**, 7982-7988
17. Maas, R.L., and Brash, A.R. (1983) *Proc. Natl. Acad. Sci. USA* **80**, 2884-2888
18. Claeys, M., Kivits, G.A.A., Christ-Hazelhof, E. and Nugteren, D.H. (1985) *Biochim. Biophys. Acta* **837**, 35-51

19. Walstra, P., Verhagen, J., Vermeer, M.A., Veldink, G.A. and Vliegenthart, J.F.G. (1987) *Biochim. Biophys. Acta* **921**, 312-319
20. Takahashi, Y., Ueda, N. and Yamamoto, S. (1988) *Arch. Biochem. Biophys.* **266**, 613-621
21. Hansbrough, J.R., Takahashi, Y., Ueda, N., Yamamoto, S. and Holtzman, M.J. (1990) *J. Biol. Chem.* **265**, 1771-1776
22. Ueda, N., Hiroshima, A., Natsui, K., Shinjo, F., Yoshimoto, T., Yamamoto, S., Ii, K., Gerozissis, K. and Dray, F. (1990) *J. Biol. Chem.* 265, 2311-2316
23. Maruyama, T., Ueda, N., Yoshimoto, T., Yamamoto, S., Komatsu, N. and Watanabe, K. (1989) *J. Histochem. Cytochem.* **37**, 1125-1131
24. Sigal, E., Craik, C.S., Highland, E., Grunberger, D., Costello, L.L., Dixon, R.A.F. and Nadel, J.A. (1988) *Biochem. Biophys, Res. Commun.* **157**, 457-464
25. Dixon, R.A.F., Jones, R.E., Diehl, R.E., Bennett, C.D., Kargman, S. and Rouzer, C.A. (1988) *Proc. Natl. Acad. Sci. USA* **85**, 416-420
26. Pilote, S., Vallerand, P. and Borgeat, P. (1982) *Biochem. Biophys. Res. Commun.* **104**, 867-873
27. Vanderhoek, J.Y., Kiesel, L., Naor, Z., Bailey, J.M. and Catt, K.J. (1984) *Prostaglandin and Leukotriene Medicine* **15**, 375-385
28. Capdevila, J., Snijder, G.D. and Falck, J.R. (1984) *FEBS Lett.* **178**, 319-322
29. Rabier, M., Chavis, C., Crastes de Paulet, A. and Damon, M. (1987) *Prostaglandin Leukotriene and Medicine* **27**, 27-42
30. Yoshimoto, T., Kusaka, M., Shinjo, F., Yamamoto, S. and Dray, F. (1984) *Prostaglandins* **28**, 279-285
31. Gerozissis, K., Vulliez, B., Saavedra, J.M., Murphy, R.C. and Dray, F. (1985) *Neuroendocinology* **40**, 272-276
32. Sakai, K., Fafeur, V., Vulliez-le Normand, B. and Dray, F. (1988) *Prostaglandins* **35**, 969-976

The Unique Characteristics of the Purified 5-Lipoxygenase from Potato Tubers and the Proposed Mechanism of Formation of Leukotrienes and Lipoxins

J. Whelan, P. Reddanna, V. Nikolaev, G.R. Hildenbrandt
and *C.C. Reddy*
Department of Veterinary Science and Environmental Resources Research Institute
The Pennsylvania State University, University Park, PA 16802 USA

Summary

Lipoxygenases are a group of closely related enzymes responsible for the dioxygenation of various fatty acids containing all-*cis*-methylene-interrupted double bonds. In the study reported here, we provide the evidence that an electrophoretically pure lipoxygenase from potato tubers exhibits nonregiospecificity towards different polyenoic fatty acids, such as linolenic acid, arachidonic acid, eicosapentaenoic acid and docosahexaenoic acid, to produce all possible fatty acid hydroperoxides. For example, it catalyzes the oxidation of arachidonic acid to produce all six hydroperoxyeicosatetraenoic acids (HPETEs), viz, 5-, 8-, 9-, 11-, 12- and 15-HPETE. In addition, all possible secondary oxidation products, in this case diH(P)ETEs, each exhibiting a distinct U.V. absorption spectrum, are isolated from the reaction. The diH(P)ETEs include those directly generated by double dioxygenation reactions as well as those derived from non-enzymatic hydrolysis of LTA_4s. The nonregiospecificity of the potato lipoxygenase is responsible for the synthesis of all possible LTA_4s, 8,9-, 11,12- and 14,15-LTA_4, in addition to the previously reported 5,6-LTA_4. Further confirmation of the LTA_4 synthase activity is obtained by demonstrating the formation of lipoxins A and B and their isomers from such precursors as 5(OH),5(OOH)- and 5(OOH),15(OH)-AA. Stereochemical analysis by HPLC on a chiral column indicates that 5-, 8- and 11-HPETE have an S-configuration, while 9-, 12- and 15-HPETE are mixtures of both S- and R-isomers. Interestingly, the proportions of the S- and R-isomers are not equal. Potato enzyme exhibits similar nonregiospecificity towards dihomo-γ-linolenic acid, eicosapentaenoic acid and docosahexaenoic acid as substrates. When EPA is incubated with the enzyme, the pattern of oxidation products is remarkably similar to that of arachidonic acid;

Biological Oxidation Systems, Volume 2

however, there are two additional hydroperoxy products formed from EPA as a result of dioxygenation at C-14 and C-18.

Introduction

Lipoxygenases (EC1.13.11.12) catalyze the dioxygenation of polyenoic fatty acids containing all-*cis*-methylene-interrupted double bonds (1-4). The most common substrate for lipoxygenases in animals is arachidonic acid which is oxygenated to hydroperoxyeicosatetraenoic acids (HPETEs). The nomenclature for these enzymes is based on the site of insertion of molecular oxygen onto arachidonic acid molecule. Accordingly, they have been classified as 5-, 8-, 9-, 11-, 12-, and 15-lipoxygenases. Until recently it was generally assumed that lipoxygenases exhibit stringent positional specificity with respect to the removal of hydrogen atom from a double allylic carbon (1,2). Also, it was thought for a long time that most lipoxygenases introduce molecular oxygen stereospecifically forming only one of the possible optical isomers (2). Nonregiospecificity, however, has been demonstrated recently with a number of lipoxygenase preparations from both plant and animal sources (1,2). Furthermore, the steric analysis of the reaction products revealed the presence of a mixture of optical isomers (1,2). Interestingly, the occurrence of different optical isomers seems to coincide with the nonregiospecificity of lipoxygenases.

Lipoxygenases are widely distributed in both plants and animals. One of the most physiologically important lipoxygenases and the subject of greater interest in recent years is 5-lipoxygenase. This enzyme catalyzes the first committed step in a series of enzymatic transformations of arachidonic acid leading to leukotrienes (LTs), lipoxins and other classes of biologically active compounds (5-7). However, this is not the rate-limiting step *in vivo* because phospholipase A_2-catalyzed release of arachidonic acid must precede dioxygenation reaction. 5-Lipoxygenase has been purified from a number of mammalian cells, including human (8) and porcine (9) leukocytes, rat basophilic leukemia cells (10), and from plants (11-13). The purified enzyme from mammalian sources requires Ca^{++} and ATP for its maximal activity (8,14). Recently some information on the primary structure of the mammalian 5-lipoxygenase has been derived from cDNA clones of mRNAs isolated from various sources (15-17). Nevertheless, information on the mechanistic details of this enzyme-catalyzed reaction is still very limited. This is mainly attributable to limitations in obtaining sufficient quantity of the purified enzyme in stable form from mammalian sources. In this regard, potato tubers are frequently employed as the source of large amounts of 5-lipoxygenase. Most recently we have demonstrated that highly purified lipoxygenase from red skin potato tubers catalyzes the dioxygenation of arachidonic acid at all six possible positions (18-20). Furthermore, we have shown that nonregiospecificity of this enzyme is responsible for the synthesis of the four possible LTA_4s as well as lipoxins (18,20). This chapter is intended to give an overview of the structural and unique catalytic properties of potato lipoxygenase.

Isolation and Molecular Properties of Potato Lipoxygenase

Fresh red potato tubers are the preferred source of 5-lipoxygenase. In general, white potato tubers have much lower activity than the red tubers. Compared to the fresh potato tubers, those stored at 0°C for more than a week have significantly less enzyme activity. There is a significant amount of α-linolenic acid in potato tubers. This endogenous fatty acid can serve as a substrate in cell-free extracts, thus inactivating the enzyme by a self-catalyzed mechanism. Therefore, it is essential to include EDTA, sodium metabisulfite and ascorbic acid in the homogenizing buffer to prevent the inactivation of enzyme during isolation. Lipoxygenase is readily purified from potato tubers by a simple and convenient purification procedure involving three main steps: 1) $(NH_4)_2 SO_4$ fractionation, 2) ion-exchange chromatography on a DEAE-cellulose column, and 3) hydrophobic interaction HPLC. The details of the purification procedure, which is a modification of the one reported by Sekiya *et al.* (21), are published elsewhere (12). HPLC purification of the lipoxygenase preparation obtained after anion-exchange chromatographic step resulted in two enzyme activity peaks. They are designated as L_1 and L_2, in the order of their elution from HPLC column (12). The L_1 peak has a specific activity of ~14 μmoles/min/mg protein and accounts for greater than 80% of the total lipoxygenase activity obtained from this procedure. Whereas the L_2 peak has a lower specific activity (6 μmoles/min/mg protein) and accounts for less than 20% of the total activity. It is determined that L_2 is a degradation product of L_1 (12). This is evident from the fact that, upon storage at 0°C for a week, the L_1 enzyme form is transformed into L_2 form. Both the L_1 and L_2 lipoxygenase preparations obtained by this procedure are greater than 99% pure as judged by analytical and SDS-gel electrophoresis (12).

The molecular weights of L_1 and L_2 determined by SDS-gel electrophoresis are estimated to be 85,000 and 38,000 respectively. The native molecular weights as determined by gel filtration are the same as those estimated by SDS-gel electrophoresis, indicating that they are monomeric proteins. Both L_1 and L_2 are very acidic, as indicated by their isoelectric points of 4.5 and 4.25, respectively (12). The purified enzyme, both L_1 and L_2, are inactivated readily at 0°C within a week. The loss in enzyme activity cannot be prevented even if it is stored at -20°C or -80°C. However, when the enzyme is stored in 40 mM potassium phosphate buffer, pH 6.3, containing 2M $(NH_4)_2SO_4$, it is stable for months at -20°C. Even the degradation of L_1 to L_2 is minimized. Although, both L_1 and L_2 forms yield similar product profiles with arachidonic acid (12), it is advisable to employ L_1 form for all the kinetic and mechanistic studies. Immunologically, L_1 and L_2 appear to be different as determined by western blot analysis. The antiserum raised against L_1 protein cross-reacts only with L_1 form but not L_2 form. Whereas the antiserum raised against a mixture of L_1 and L_2 forms cross-reacts with both L_1 and L_2 forms. This suggests that the antigenic determinants may be different for L_1 and L_2, even though the catalytic properties appear to be similar. Interestingly, the antiserum raised against a mixture of L_1 and L_2 forms cross-reacts with the crude preparation of 5-lipoxygenase from bovine polymorphonuclear leukocytes.

Catalytic Properties

In 1984 Shimizu and his associates reported that an enzyme purified from potato tubers possess both the 5- and 8-lipoxygenase activities, and this duel lipoxygenase activity of the enzyme has been implicated in the synthesis of 5,6-LTA_4 from arachidonic acid (13). As illustrated in Figure 1, *pro*-S hydrogen atom at C-7 is stereospecifically removed by 5-lipoxygenase activity generating a carbon centered radical that couples to O_2 at C-5 to form 5-HPETE (13,22), while the 8-lipoxygenase activity associated with the same enzyme removes *pro*-R hydrogen atom from C-10 (13). Because of the conjugated diene structure ($\Delta^{6,8}$) of 5-HPETE, a radical at C-6 might react with the hydroperoxy oxygen, resulting in O-O scission to yield 5,6-LTA_4, instead of insertion of molecular O_2 at C-8 (13). A similar stereospecific elimination of the *pro*-R hydrogen atom at C-10 has been demonstrated for the synthesis of 5,6-LTA_4 in neutrophils (23,24). The identity of the oxidizing agent on the protein that removes the hydrogen is uncertain, although it has recently been proposed that it is a pyrroloquinoline quinone in soybean lipoxygenase (25). However, it is very well established that all lipoxygenases contain non-heme iron in the active site. It exists as high spin Fe^{III} in the active form of the enzyme (26,27).

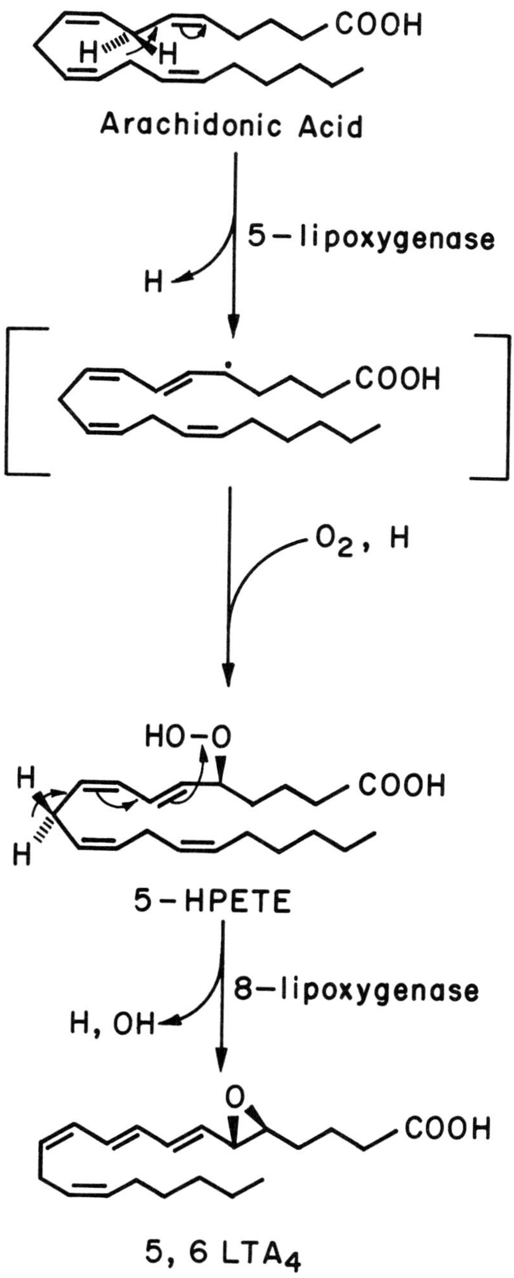

Figure 1. *Mechanism of Synthesis of 5,6-LTA$_4$ from Arachidonic Acid by Sequential Action of 5- and 8-Lipoxygenases*

As shown in Figure 2, an electrophoretically pure lipoxygenase from potato tubers exhibits not only the 5- and 8-lipoxygenase activities reported by Shimizu *et al.* (13) but also the 9-, 11-, 12-, and 15-lipoxygenase activities, thus generating all six possible HPETES from arachidonic acid (12, 18-20, 28). The individual peaks are characterized by GC-MS analysis of the TMS ether derivatives of the respective hydroxy fatty acid methy esters. They are identified as 12-, 15-, 11-, 9-, 8-, and 5-HPETES, in the increasing order of elution time on HPLC. 5-HPETE is the most abundant dioxygenation product, representing approximately 60% of the total HPETES formed. The relative concentrations of other HPETES are determined to be approximately 8- (12%), 9- (10%), 11- (7%), 15- (5%), and 12-HPETE (5%). Interestingly, the product profile generated by L_2 form is very similar to that of the L_1 protein, although they appear to be immunologically different. Stereochemical analysis by HPLC on a chiral column indicates that the 5-, 8-, and 11-HPETES have the S-configuration whereas the 9-, 12-, and 15-HPETES are mixtures of both S and R; however, they are not racemic. Based on these results, it appears that when a (-2) rearrangement of the fatty acid radical takes place the enzyme exerts strict control over the insertion of O_2 from only one direction and the products have S stereochemistry. However, when the enzyme catalyzes a (+2) rearrangement of the fatty acid radical the stereospecificity of O_2 insertion is lost and the products are a mixture of optical isomers. A reverse is true with the soybean lipoxygenase catalyzed oxidation of polyeroic fatty acids (Vladimir and Reddy, unpublished results).

The L_1 form has an apparent K_m of 33 μM and an apparent V_{max} of 14 μmoles/min/mg protein with arachidonic acid as substrate. The maximum O_2 uptake with the L_1 form occurs between the pH range of 5.0-6.5, with the optimum pH using arachidonic acid being 6.3. Enzyme activity is markedly decreased at neutral and basic pH values (pH>7.0); however, such dramatic effects are not observed at acidic pHs. The product profiles are also altered significantly by pH. For example, at pH 9.0, the formation of 8-HPETE exceeds that of 5-HPETE, while the other HPETEs drop to undetectable levels. Surprisingly, the pH at which maximizes HPETE formation is pH 7.3, even though the activity is only 30% of that at pH 6.3.

When the reaction time of arachidonic acid and the L_1 form of the potato LOX is extended the 235 nm/268 nm ratio progressively decreases. This occurs as the pool of HPETEs increase in the reaction mixture and begin to compete with arachidonic acid as substrates for the enzyme. At least six diHPETEs can be formed as a result of double lipoxygenase reactions: 5,12-, 8,15-, 9,15-, 5,11-, 9,11-

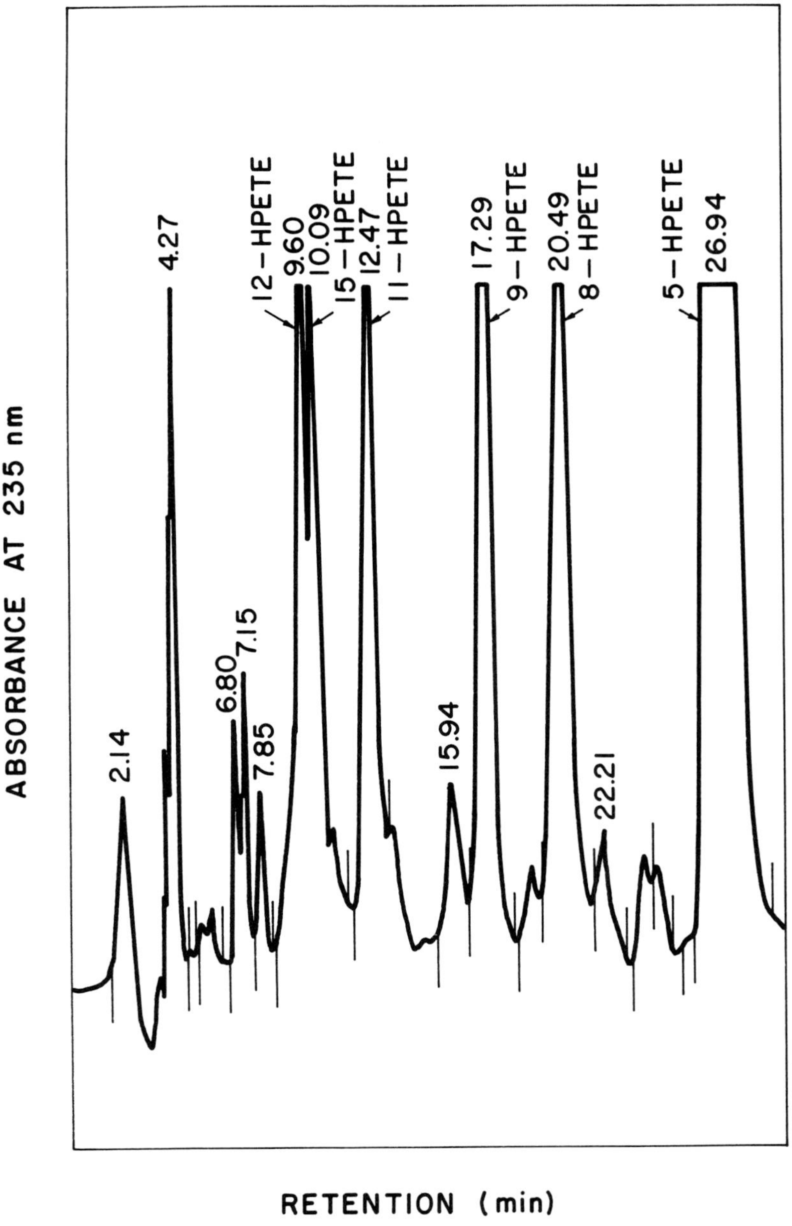

Figure 2. *SP-HPLC Analysis of Monohydroperoxy Products of Arachidonic Acid Oxidation Catalyzed by the Purified Lipoxygenase from Potato Tubers*

Figure 3.

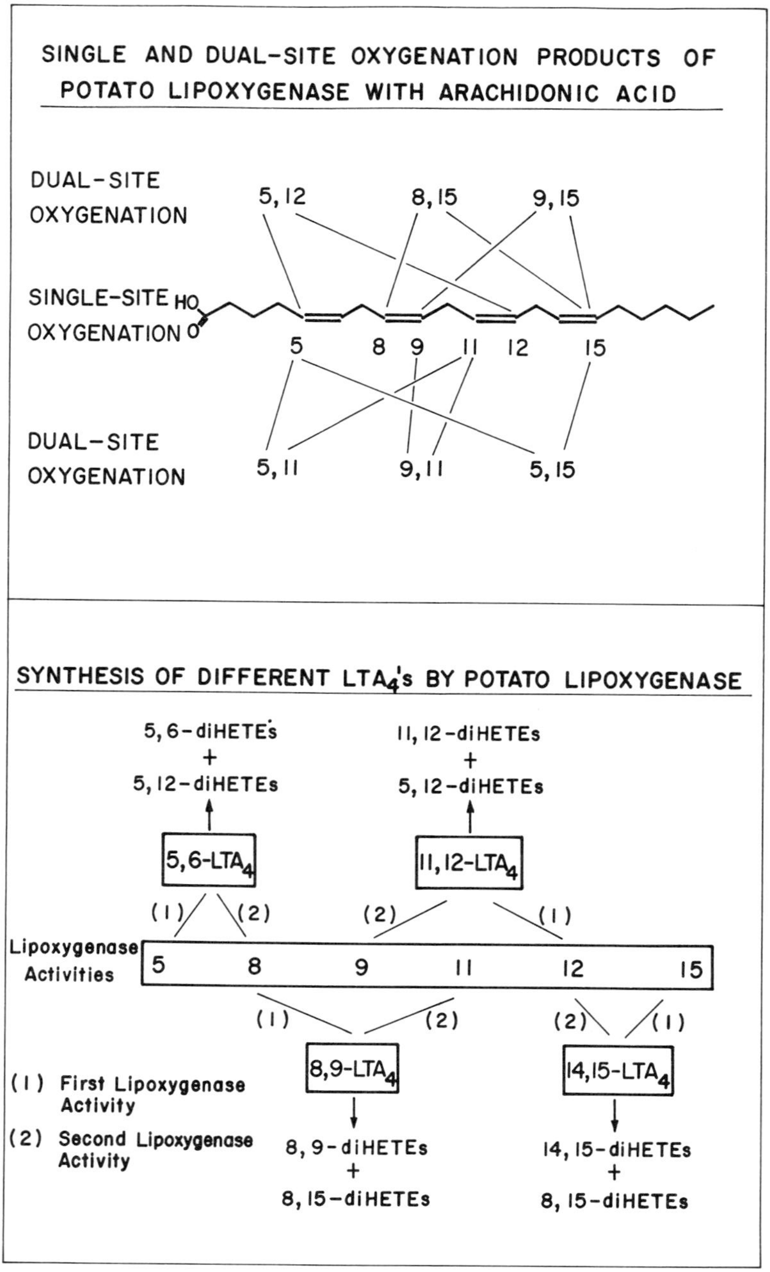
SINGLE AND DUAL-SITE OXYGENATION PRODUCTS OF POTATO LIPOXYGENASE WITH ARACHIDONIC ACID
DUAL-SITE OXYGENATION
5,12
8,15
9,15
SINGLE-SITE OXYGENATION
HO
O
5
8
9
11
12
15
DUAL-SITE OXYGENATION
5,11
9,11
5,15
SYNTHESIS OF DIFFERENT LTA4's BY POTATO LIPOXYGENASE
5,6-diHETEs + 5,12-diHETEs
11,12-diHETEs + 5,12-diHETEs
5,6-LTA4
11,12-LTA4
(1)
(2)
Lipoxygenase Activities
5 8 9 11 12 15
8,9-LTA4
14,15-LTA4
(1) First Lipoxygenase Activity
(2) Second Lipoxygenase Activity
8,9-diHETEs + 8,15-diHETEs
14,15-diHETEs + 8,15-diHETEs

and 5,15-diHPETEs (Fig. 3a). When the lipoxygenase activities act in a specific order, this can result in the formation of all possible LTA_4s (Fig. 3b). For example, the sequential action of 5- and 8-lipoxygenases results in 5,6-LTA_4 formation as detected by its nonenzymatic hydrolysis products, primarily 6-trans LTB_4 and its C-12 epimer. Similarly, 12-lipoxygenase followed by 9-lipoxygenase yields 11,12-LTA_4, 8-lipoxygenase followed by 11-lipoxygenase forms 8,9-LTA_4, and the sequential action of 15- and 12-lipoxygenase results in the formation of 14,15-LTA_4. Since these allylic epoxides are unstable under aqueous conditions, their nonenzymatically-derived diHETE products aid in the identification of LT formation. Each compound isolated from the incubation of arachidonic acid with the potato 5-lipoxygenase has a distinct U.V. absorption spectra (Fig. 4). Those compounds containing a conjugated triene system in their structure, i.e., the 5,12- and 8,15-diH(P)ETEs formed from dual lipoxygenase activity and those derived nonenzymatically from LTA_4 hydrolysis, exhibit characteristic chromophores with λ_{max} at 268.5 and shoulders at 259 nm and 279 nm. Four diHPETEs, formed as a result of dual lipoxygenase activity, contain two conjugated diene systems. The two conjugated dienes associated with 5,15- and 9,11-diHPETEs are arranged somewhat symmetrically around the parent arachidonic acid molecule, while the conjugated dienes of 9,15- and 5,11-diHPETEs have a staggered orientation. The result of these differences is evident in the U.V. absorption spectra. The 5,15- and 9,11-diHPETEs have almost identical U.V. spectral patterns with λ_{max} of approximately 242 nm and a hypochronic shoulder around 226 nm. In a similar manner, the U.V. spectral patterns for 9,15- and 5,11-diHPETEs are equivalent and distinctly different from 5,15- and 9,11-diHPETEs, with λ_{max} 236nm and a shoulder of almost equal intensity at 229 nm.

The origin of oxygen atoms associated with the H(P)ETEs and diH(P)ETEs can be determined when arachidonic acid and the potato lipoxygenase are incubated under anaerobic conditions. The characteristic conjugated diene chromophore, indicative of HPETE formation, is generated when the reaction proceeds under an atmosphere of air. However, under an atmosphere of argon, conjugated diene formation is negligible, indicating molecular oxygen is incorporated during the lipoxygenase reaction. If, under anaerobic conditions, 5-HPETE is the substrate, the 235 nm/268 nm ratio drops dramatically, indicating a loss of the conjugated diene system in favor of the conjugated triene system associated with diHETEs. This suggests that LTA_4 synthesis from 5-HPETE is not dependent upon molecular oxygen and that the oxygen originates from water.

Figure 4.

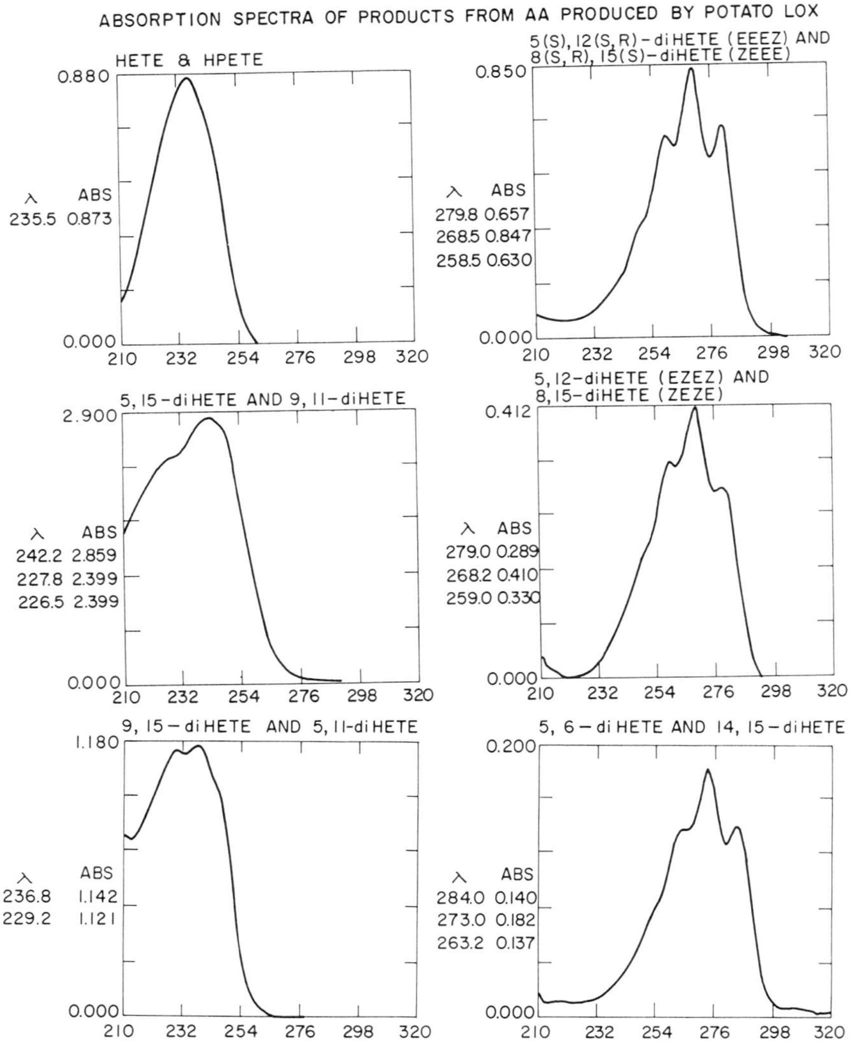

Previous $^{18}O_2$ studies also demonstrated that the oxygen at C-12 of the 6-trans LTB_4 and its C-12 epimer are derived from water; however, 30% was derived from molecular oxygen (13). since these reactions were run under aerobic conditions, it is possible that the 12-lipoxygenase activity associated with the potato lipoxygenase could have contributed to those results.

As shown in Figure 5, with 5(OH), 15(OOH)-ETE and 5(OOH), 15(OH)-ETE as substrates with the potato lipoxygenase clearly show

Figure 5.

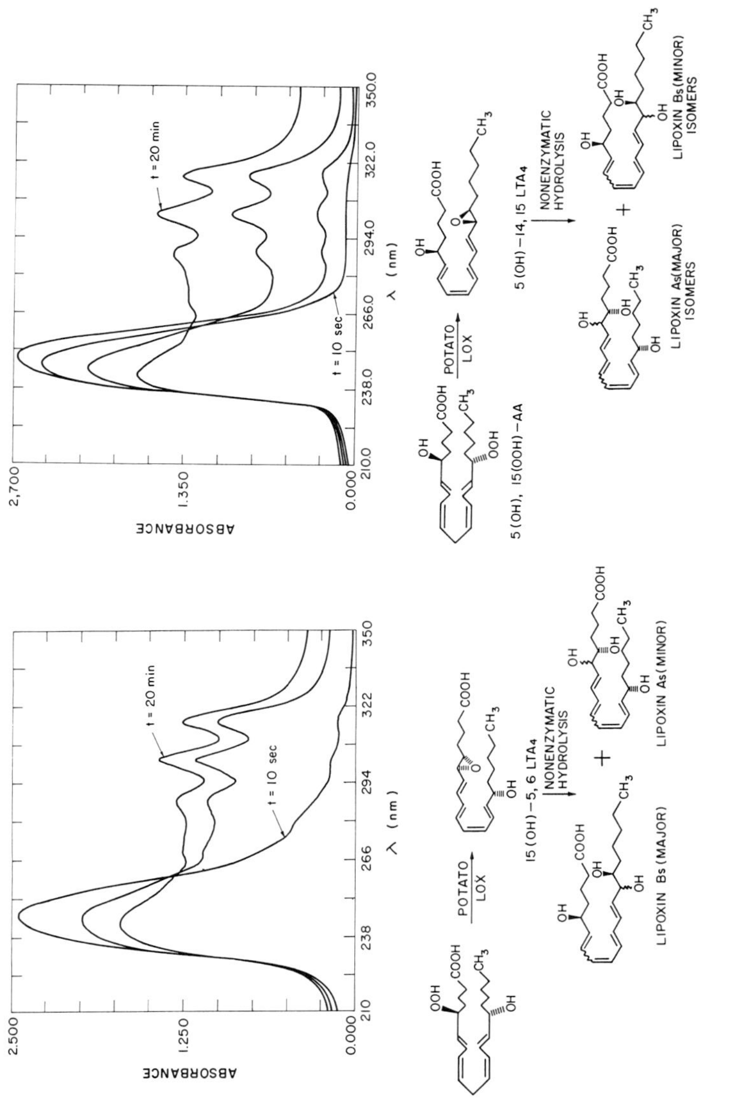
ABSORBANCE
2.500
1.250
0.000
210
238
266
294
322
350
λ (nm)
t = 20 min
t = 10 sec
POTATO
LOX
15(OH)−5, 6 LTA4
NONENZYMATIC
HYDROLYSIS
LIPOXIN Bs (MAJOR)
LIPOXIN As (MINOR)
2.700
1.350
210.0
238.0
266.0
294.0
322.0
350.0
5(OH), 15(OOH)−AA
5(OH)−14, 15 LTA4
LIPOXIN As(MAJOR)
ISOMERS
LIPOXIN Bs(MINOR)
ISOMERS

the formation of characteristic chromophores indicative of conjugated tetraenes, identified as the all-trans-isomers of lipoxin A_4 (LXA_4) and LXB_4 (19). It is also reasonable to expect that the potato lipoxygenase can catalyze the generation of LXA_4 and LXB_4. For example, when 5(S),15(S)-diHETE (EZZE) is acted upon by the 8- and 12-LOX activities associated with the potato enzyme, followed by reduction, it should yield 5(S), 6(R), 15(S)-trihydroxy-7,9,13-trans-11-cis-eicosatetraenoic acid (LXA_4) and 5(S),14(R),15(S)-tri-hydroxy-6,10,12-trans-8-cis-eicosatetraenoic acid (LXB_4), respectively (29).

When dihomo-γ-linolenic acid (DHGLA) is incubated with the potato enzyme, 8(S)-hydroperoxyeicosatrienoic acid (8(S)-HPETrE) is the most abundant dioxygenation product, representing 70% of the total HPETrEs formed, followed by 11-, 15-, and 12-HPETrEs in decreasing order of abundance (30). Additionally, four major secondary oxidation products of DHGLA are isolated from the reaction. These products include two double dioxygenation products, 8(S), 15(S)- and 8(S), 15(R)-DiHPETrES (EZE), and two nonenzymatic hydrolysis products of 8,9-LTA_3, 8(S), 15(S)- and 8(S), 15(R)-DiHETrEs (EEE) (30). Evidence for the formation of 8,9-LTA_3 is obtained by incubating 8(S)-HPETrE with potato lipoxygenase under anerobic conditions. This yielded mainly the two isomers of 8,15-DiHETrEs (EEE). These results suggest that 8,9-LTA_3 is synthesized from DHGLA by the sequential action of 8- and 11-lipoxygenase activities associated with the potato lipoxygenase (30).

When eicosapentaenoic acid (EPA) is incubated with the potato enzyme under similar conditions, the product profiles for the hydroperoxy and hydroxy (following reduction) derivatives of EPA are remarkably similar to the product profiles derived from arachidonic acid, with the C-5 oxygenated derivative is the major product formed. When docosahexaenoic acid (DHA) is used as the substrate, the product profiles also demonstrate multiple regiospecificity; however, the major oxygenation product at C-4 is noticeably reduced.

In summary, we have presented a detailed description of the unique characteristics of the purified potato 5-lipoxygenase system. We have demonstrated the nonspecific nature of the purified enzyme. Although dominated by 5-lipoxygenase activity, thus the name, it is responsible for multiple site oxygenation of arachidonic acid, DHGLA, EPA and DHA to form all possible mono-, di and trioxygenated derivatives. It is conceivable that the mechanism of biosynthesis of LTs and LXs catalyzed by mammalian lipoxygenases may be analogous to the potato enzyme.

References

1. Schewe, T., Rapaport, S.M., and Kühn, H. (1986) *Adv. Enzymol.* **58**, 191-272
2. Kühn, H., Schewe, T., and Rapaport, S.M. (1986) *Adv. Enzymol.* **58**, 273-311
3. Vliegenthart, J.F.G., and Veldink, G.A. (1982) in *Free Radical in Biology*, Pryor, W. A., Ed., Academic Press, Inc., Vol. V, pp. 29-66
4. Hanson, G., Malonsten, C., and Rådmark, O. (1983) in *Prostaglandins and Related Substances*, Pace-Asciak, C. R., and Granstrom, E., eds., Elsevier, Amsterdam, 127-169
5. Borgeat, P., Hamberg, M., and Samuelsson, B. (1976) *J. Biol. Chem.* **251**, 7816-7820; Correction (1977) *IBID*, **252**, 8772
6. Samuelsson, B. (1983) *Science* **220**, 568-575
7. Samuelsson, B., Dahlen, S.-E., Lindgren, J.A., Rouzer, C.A., and Serhan, C.N. (1987) *Science* **237**, 1171-1176
8. Rouzer, C.A., Matsumoto, T., and Samuelsson, B. (1986) *Proc. Natl. Acad. Sci.* USA **83**, 857-861
9. Ueda, N., Kaneko, S., Yoshimoto, T., and Yamamoto, S. (1986) *J. Biol. Chem.* **261**, 7982-7988
10. Goetze, A.M., Fayer, L., Bousaka, J., Bornemeier, D., and Carter, G.W. (1985) *Prostaglandins* **29**, 689-701
11. Reddanna, P., Whelan, J., Reddy, P.S., and Reddy, C.C. (1988) *Biochem. Biophys. Res. Commun.* **157**, 1348-1351
12. Reddanna, P., Whelan, J., Maddipati, K.R., and Reddy, C.C. (1990) *Methods Enzymol.* **187**, 268-277
13. Shimizu, T., Radmark, O., and Samuelsson, B. (1984) *Proc. Natl. Acad. Sci.* USA **81**, 689-693
14. Rouzer, C.A., Thornberry, N.A., and Bull, H.G. (1988) *Ann. N.Y. Acad. Sci.* **524**, 1-11
15. Matsumoto, T., Funk, C.-D., Radmark, O., Hoog, J.-O., Jornvall, H., and Samuelsson, B. (1988) *Proc. Natl. Acad. Sci.* USA **85**, 26-30; Correction (1988) *IBID*, **85**, 3406
16. Dixon, R.A.F., Jones, R.E., Diehl, R.E., Bennet, C.D., Kargman, S., and Rouzer, C.A. (1988) *Proc. Natl. Acad. Sci.* USA **85**, 416-420
17. Balcarek, J.M., Theisen, T.W., Cook, M.N., Varrichio, A., Hwang, S.-M., Strohsacker, M.W., and Crooke, S.T. (1988) *J. Biol. Chem.* **263**, 13937-13941
18. Whelan, J., Reddanna, P., Prasad, G., Rao, M.K., and Reddy, C.C. (1988) *Ann. N.Y. Acad. Sci.* **524**, 391-392
19. Whelan, J., Reddanna, P., Prasad, G., and Reddy, C.C. (1987) in *Proc. of the Short Course on Polyunsaturated Fatty Acids and Eicosanoids*, Lands, W.E.M., Ed., AOCS Press, Champaign, IL, pp. 468-471
20. Reddy, C.C., Whelan, J., Rao, M.K., and Reddanna, P. (1988) in *Adv. Prostaglandins, Thromboxanes and Leukotriene Research*, Samuelsson, B., Wong, P.Y.-K., and Sun, F.F., eds., Raven Press, New York, pp. 132-136

21. Sekiya, J., Aoshima, H., Kajiwara, T., Togo, T., and Hatanaka, A. (1977) *Agric. Biol. Chem.* **41**, 827-832
22. Borgeat, P., and Samuelsson, B. (1979) *Proc. Natl. Acad. Sci.* **76**, 3213-3217
23. Panossian, A., Hamberg, M., and Samuelsson, B. (1982) *FEBS Lett.* **150**, 511-513
24. Mass, R.L., Ingram, C.D., Taker, D.F., Oates, J.A., and Brash, A.R. (1982) *J. Biol. Chem.* **257**, 13515-13519
25. van der Meer, R.A., and Duine, J.A. (1988) *FEBS Lett.* **235**, 194-200
26. Cheesbrough, T.M., and Axelrod, B. (1983) *Biochemistry* **22**, 3837-3840
27. Pistorius, E.K., Axelrod, B., and Palmer, G. (1976) *J. Biol. Chem.* **251**, 7144-7148
28. Reddanna, P., Rao, M.K., and Reddy, C.C. (1985) in *Proc. Intl. Symp. on Leukotrienes and Prostaglandins in Health and Disease*, Tel-Aviv, Israel, October 20-25
29. Serhan, C.N., and Samuelsson, B. (1988) *Adv. Exptl. Med. and Biol.* **229**, 1-14
30. Reddy, C.C., Bertler, C., and Hammarstrom, S. (1990) *Arch. Biochem. Biophys.* **229**, 211-217

Arachidonic Acid Oxidation and Platelet Function

Gundu H. R. Rao, Vhundi G. Mahadevappa, Timothy D. Hill and James G. White
Department of Laboratory Medicine and Pathology, University of Minnesota
Box 198 Mayo Memorial Bldg., 420 Delaware Street S.E., Minneapolis, MN 55455 USA

Abstract

Arachidonic acid metabolites seem to play a critical role in human platelet activation leading to shape change, development of stickiness, secretion of granule contents and irreversible aggregation. We have evaluated various inhibitors that selectively block different steps of arachidonic acid metabolism in order to determine the importance of the eicosanoid pathway in platelet activation. Results of our studies suggest that heme or a heme protein associated with cyclooxygenase interacts with arachidonic acid and facilitates its oxidation. Although transient formation of an arachidonic acid free radical can occur, little evidence exists for any role of oxygen radicals in arachidonic acid metabolism. Inhibitors of peroxidase, cyclooxygenase and thromboxane synthetase blocked the second wave response of platelet aggregation. However, none of the inhibitors caused disaggregation of preformed platelet aggregates. Agents that depleted glutathione levels and inhibited glutathione peroxidase enhanced metabolism of arachidonic acid by cyclooxygenases. When threshold concentration of physiological agonists are used, products of arachidonate metabolism play a critical role in platelet activation. However, when a high concentration of a potent agonist like thrombin is used or the newly discovered intrinsic mechanism of membrane modulation is stimulated in refractory platelets by treatment with epinephrine or norepinephrine, activation and irreversible aggregation can be secured independent of eicosanoid synthesis (1, 2).

Introduction

Platelets are activated by a variety of stimuli. Irrespective of the source of this stimulus, these cells react with characteristic responses, including shape change, development of stickiness, spreading, secretion of granule contents and irreversible aggregation (3). All physiological agonists exert their effect by interacting with specific receptors or special glycoprotein-rich domains on the plasma membrane. Biochemical events that occur following platelet activation are similar whether the stimulation occurs via receptor interaction or by as yet unknown mechanisms. Recent studies suggest that agonist-induced transmembrane signal occurs via GTP-binding proteins (G-proteins). Human platelet activation seems to be mediated by at least two pertussis toxin sensitive G-proteins (4). When platelets are stimulated, phospholipase C associated with the surface membrane is activated via a GTP binding protein. This enzyme causes the hydrolysis of phosphatidyl inositol 4,5-bisphosphate and liberates two messenger molecules (Figure 1), a hydrophobic 1,2 diacylglycerol (DG) which stays in the membrane and a hydrophilic inositol 1,4,5-trisphosphate (IP_3) (5, 6). Diglycerides like putative tumor promoters (phorbol esters), directly activate a multifunctional enzyme, protein kinase C. Whereas IP_3 has been demonstrated to be a potent mobilizer of cytosolic free calcium from internal membrane stores, activation of protein kinase C facilitates the formation of binding sites/receptors for adhesive proteins on the surface membranes of platelets by an as yet unknown mechanism (7).

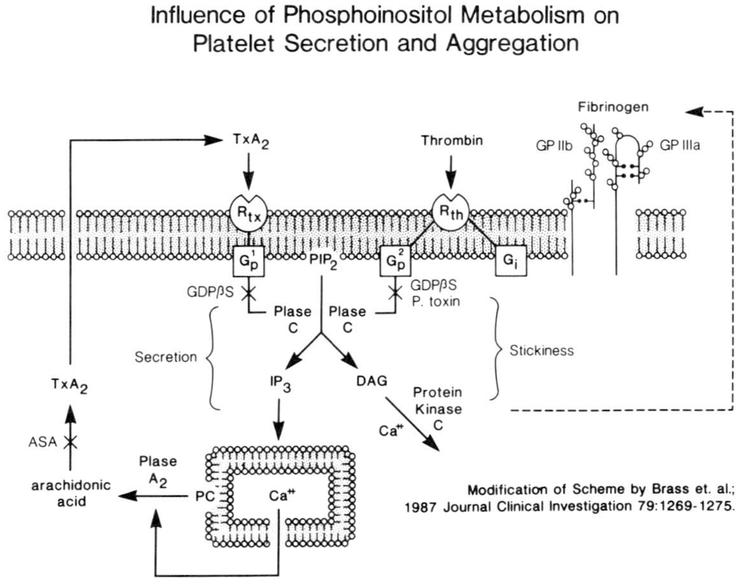

Figure 1. *In Human Platelets Agonist-Induced Receptor Mediated Transmembrane Signalling Seems to Occur Via Two Different Pertussis Toxin Sensitive GTP-Binding Proteins (G Proteins).* Resulting activation of phospholipase C leads to the hydrolysis of phosphatidyl inositol and formation of two second messengers, 1,2-diacyl glycerol (DG) and inositol 1,4,5-trisphosphate (IP_3). DG directly activates protein kinase C, a multifunctional enzyme. IP_3 facilitates mobilization of cytosolic free calcium from the internal membrane stores.

Increase in cytosolic free calcium induced by IP_3 activates phospholipase A_2 which liberates fatty acids from the second position of various phospholipid species. The bulk of the liberated fatty acid in platelets is arachidonic acid, a preferred substrate for cyclooxygenase and lipoxygenase enzymes.

Arachidonic acid is metabolized by cyclooxygenase to prostaglandin (PG) endoperoxides (PGG_2/PGH_2) and these metabolites are further transformed into thromboxane A_2 (8). The major metabolite formed in activated platelets via this pathway is thromboxane, a potent vasoconstrictor and a powerful platelet stimulator. Weak agonists, such as epinephrine, adenosine diphosphate and platelet activating factor, are dependent on the mediation of thromboxane for mobilization of cytosolic calcium and induction of secretion of granule contents. In this review we will discuss some aspects of arachidonic acid oxidation and its influence on platelet function. For additional information on this subject, readers are referred to special reviews (9-11).

Arachidonic Acid Liberation

There are four major phospholipid species associated with platelet membranes. They are phosphatidyl inositol, phosphatidyl choline, phosphatidyl ethanolamine and phosphotidyl serine. Several types of phospholipases can hydrolyze phospholipids from cell membranes. Phospholipase A_1 specifically removes the fatty acids from the carbon 1 position, whereas phospholipase A_2 exerts its effect on fatty acids attached to the second carbon of the phospholipids. Phospholipase C liberates phosphorylated bases from the diglyceride portion of the phospholipids. Phospholipase D releases the bases such as inositol, choline, ethanolamine and serine from the phosphatidic acid moiety (11).

In human platelets various stimuli that induce activation stimulate phospholipase C, which is closely coupled to specific receptors or sites of action on the plasma membrane. Platelet membranes contain large amounts of phosphatidyl inositol (PI) and relatively smaller amounts of phosphatidyl inositol-4 phosphate (PIP) and phosphatidyl inositol 4,5-bisphosphate (PIP_2). All three species can serve as substrates for the action of phospholipase C. However, physiological agonists seem to cause hydrolysis of PIP_2, which liberates two putative messengers, 1,2-diacyl glycerol and inositol 1,4,5-trisphosphate (11). If all three species are hydrolyzed, then three molecules of diglyceride are formed for every one molecule of inositol 1,4,5-trisphosphate generated (Figure 2). Diglyceride produced could be further metabolized to form phosphatidic acid via the action of specific kinases or acted on by a diglyceride lipase and a monoglyceride lipase to release free fatty acids. Diglyceride may also play a critical role in activation of the multifunctional enzyme protein kinase C.

Phosphoinositide Metabolism and Formation of Intracellular Messengers

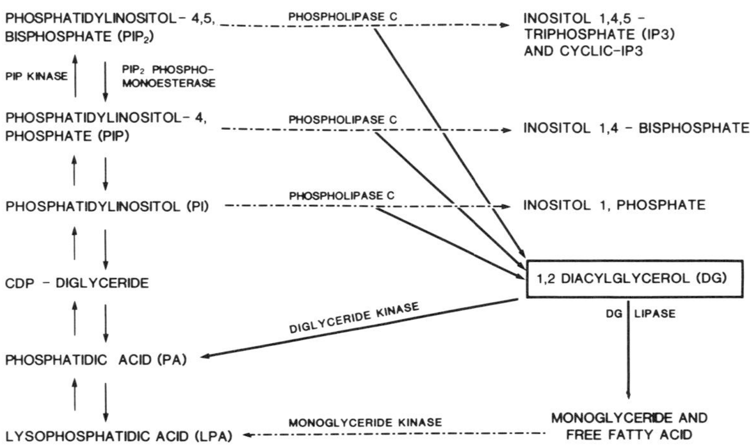

Modification from Gerrard,J.M. 1985, Prostaglandins and Leukotrienes.

Figure 2. *There are Three Species of Phosphatidyl Inositols in Platelet Membranes.* The bulk of this phospholipid consists of phosphatidyl inositol (P). Relatively smaller amounts of phosphatidyl inositol 4-phosphate (PIP) and phosphatidyl inositol 4,5-bisphosphate coexist along with PI. If all three species are hydrolyzed by the action of phospholipase C, then three molecules of diglyceride are formed for every one molecule of inositol 1,4,5-trisphosphate generated. Diglyceride produced could be further transformed to phosphatidic acid or acted upon by a diglyceride lipase and a monoglyceride lipase to release free fatty acids.

In addition to phosphatidyl inositols, phosphatidyl ethanolamine and phosphatidyl choline may serve as a source for arachidonic acid. Phospholipase A_2 is a membrane associated enzyme responsible for the liberation of arachidonic acid from these phospholipids. Although it is capable of releasing any polyunsaturated fatty acid from the second carbon of these phospholipids, by and large, agonist induced stimulation of platelets seems to preferentially yield arachidonic acid. The mechanisms by which phospholipase C and phospholipase A_2 exert their specific effect on various phospholipids associated with the platelet membranes is not clear. Free arachidonic acid liberated during cell activation is the major substrate for the synthesis of eicosanoids (9).

Arachidonic Acid Oxidation: Role for Free Radical Involvement

In vitro studies from several laboratories demonstrated a facilitating effect of hemoglobin or metalloporphyrin for the activation of fatty acid cyclooxygenase or prostaglandin synthetase (12). Based on these observations the enzymes were localized in the internal membrane systems of human platelets by using iron/heme complexing agents(13). Since iron or heme can participate in the formation of oxygen radicals, many researchers suggested a role for the free radicals in prostaglandin synthesis and platelet activation (12).

In a study designed to evaluate the role of oxygen radicals, Rao *et al.*, showed that a combination of nitroblue tetrazolium and vitamin E inhibited platelet arachidonic acid oxidation and platelet function (14). Utilizing nitroblue tetrazolium reduction to monitor ferrous iron mediated oxidation of arachidonic acid, Peterson *et al.* demonstrated that the substrate interaction with the oxidant occurred at a mole ratio of 1:1 suggesting that one fatty acid molecule reacted with one iron molecule (15). Since this reaction was not inhibitable by scavengers of oxygen radicals, such as superoxide dismutase, catalase, B-carotene, 1,3-diphenylbenzofuran, tetraphenyl cyclopenta-dienone, n-butanol, methional, dimethyl sulfoxide, mannitol, and benzoates, the authors hypothesized formation of an arachidonic acid free radical during this rapid reaction. Though no direct proof was provided for the existence of such a species, studies by Mason *et al.* showed formation of a carbon centered radical intermediate in the prostaglandin synthetase oxidation of arachidonic acid by ESR spectroscopy (16). Available data suggests the transient formation of an arachidonic acid radical before conversion to eicosanoids.

Arachidonic Acid Oxidation by Prostaglandin Synthetase

To elucidate the mechanism by which cyclooxygenase oxidizes arachidonic acid, a cell free assay system was developed in which ferrous iron induced oxygenation of arachidonic acid could be followed in the presence of nitroblue tetrazolium (14, 17). Several known cyclooxygenase inhibitors were used in this system to evaluate affinity of these drugs for ferrous iron. Results demonstrated that a well known cyclooxygenase inhibitor like indomethacin can form a complex with ferrous iron and inhibit arachidonic acid oxidation and, thereby, the reduction of nitroblue tetrazolium (15).

In *in vitro* assays, ferric heme is used for stimulating prostaglandin synthetase enzymes. Studies by Rao *et al.* showed that ferrous iron facilitated conversion or oxidation of polyenoic acid (18). It is, therefore, reasonable to speculate that there are mechanisms capable of converting ferric heme to ferrous heme. Further studies showed that fatty acid peroxides as well as physiological compounds such as epinephrine, tryptophan and ascorbic acid, could reduce ferric heme to the ferrous state and facilitate the oxidation of fatty acids by cyclooxygenases (19).

The Role of Glutathione/Glutathione Peroxidase in Arachidonic Acid Metabolism

Several studies using agents capable of interfering or depleting glutathione have suggested a critical role for glutathione in the normal function of cells (20). Platelets stored under various conditions rapidly lose their glutathione content and become less responsive to agonists upon *in vitro* aging. However, no definite role for glutathione/glutathione peroxidase has been demonstrated in modulating human platelet function. Arachidonic acid released during platelet activation is metabolized by cyclooxygenases as well as lipoxygenases. The platelet lipoxygenases convert the fatty acid to 12-hydroperoxyeicosatetraenoic acid (12- HPETE). The hydroperoxy moiety of this metabolite is reduced to a hydroxy acid. The glutathione-glutathione peroxidase (GSH peroxidase) system seems to facilitate this conversion. It has been reported that rats fed a selenium deficient diet develop low GSH-peroxidase activity and as a result have impaired arachidonic acid metabolism via this pathway (21).

Selective removal of intracellular levels of glutathione and inhibition of GSH-peroxidase could be achieved by using 1-chloro-2,4-dinitroenzene (CDNB). In human platelets exposed to CDNB, although a decrease in arachidonic acid conversion occurred, significant reduction of hydroperoxide could be demonstrated. The finding suggested that reduction of hydroperoxide does not require GSH/GSH-peroxidase activity (22, 23). Upon replenishment of glutathione, normal conversion of arachidonic acid via lipoxygenase could be demonstrated in spite of reduced GSH-peroxidase activity. Results of studies with CDNB showed that glutathione and GSH peroxidase facilitate conversion of arachidonic acid via lipoxygenase, but do not play a significant role in the reduction of hydroperoxy acid to hydroxy acids.

Agents such as N-ethylmaleimide (NEM) and diamide have been used to deplete intracellular levels of glutathione. Based on the results obtained by use of such agents, it has been concluded that depletion of cellular glutathione levels leads to platelet dysfunction. However, Hill *et al.*, using CDNB to deplete glutathione levels, have shown that the platelets respond to the action of agonists in a normal manner in spite of reduced glutathione levels (24). It has been shown that a normal response to arachidonate stimulation, conversion of the substrate to metabolites, aggregation and secretion of granule contents occurs in stored platelets which have significantly low intracellular glutathione levels. Also, platelets exposed to drugs that interfere with glutathione levels become hypersensitive to the action of arachidonic acid. This may be related to the inhibition of GSH-peroxidase and may not be due to the depletion in the levels of glutathione. Such a treatment may favor substrate metabolism by cyclooxygenases by altering enzyme kinetics as well as substrate availability. Available data shows a very limited role for GSH/GSH peroxidase in arachidonic acid metabolism and platelet activation (22-24).

Role of Cyclooxygenase Metabolites in Platelet Activation

Free arachidonic acid released during agonist induced activation is rapidly converted to cyclic endoperoxides. These transient intermediates are as potent as thromboxane in causing platelet activation (8). Cyclic endoperoxides are further transformed to a novel metabolite, thromboxane A_2, by thromboxane synthetase (Figure 3). Thromboxane is the major metabolite of this synthetic pathway. The receptors for these agonists are probably located in plasma membranes. Further evidence for this suggestion comes from the observations that receptor antagonists such as 13-azaprostanoic acid (13-APA), SKF-96148 and SKF-95587 inhibit endoperoxide mimic, U46119, induced platelet activation, secretion of granule contents and irreversible aggregation. Furthermore, docosahexaenoic acid, a 22-carbon polyenoic acid, also seem to antagonize these receptors, in addition to its putative action on cyclooxygenases.

Weak agonists like epinephrine, ADP, platelet activating factor, and collagen depend on arachidonic acid metabolites for causing secretion of granule contents. However, once the activation has reached its maximum, addition of inhibitors of peroxidase, cyclooxygenase, thromboxane synthetase or receptor antagonists such

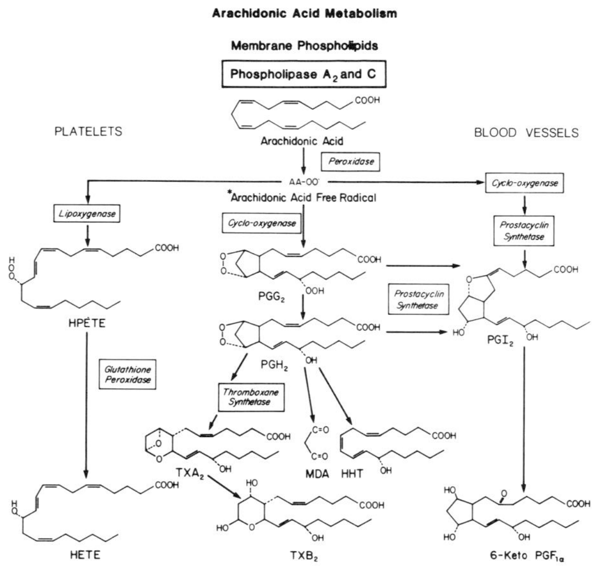

Figure 3. *Arachidonic Acid is the Major Polyenoic Acid Released During Physiological Activation of Human Platelets.* Free arachidonic acid is converted to various biologically active metabolites by three different enzymes: lipoxygenase, cyclooxygenase and thromboxane synthetase. Thromboxane is the major biologically active metabolite generated by human platelets.

as 13-APA or SKF compounds will not induce disaggregation of platelet aggregates, suggesting that continuous formation or sustained activation by these metabolites is not essential for maintaining irreversible aggregation (25). In addition, it has been demonstrated that inhibition of arachidonic acid metabolism will not totally block platelet function (26).

Both adrenalin and noradrenalin can modulate the alpha 2 adrenergic receptor and restore the sensitivity of cyclooxygenase deficient platelets to the action of other agonists. Arachidonic acid metabolites, endoperoxides and thromboxanes play an essential role in inducing formation of binding sites for adhesive proteins such as fibringogen, fibronectin and von Willebrand factor, and in the development of stickiness, secretion of granule contents and irreversible aggregation. However, human platelets exposed to potent agonists like thrombin or in the presence of epinephrine or norepinephrine become responsive to the action of agonists by an as yet unknown mechanism.

The Role of Lipoxygenase Metabolites in Platelet Activation

Arachidonic acid is metabolized by platelet lipoxygenases to 12-hydroperoxyeicosatetraenoic acid and its hydroxy acid. Inhibitors of cyclooxygenase block conversion of arachidonic acid to endoperoxides and thereby prevent formation of thromboxanes. However, these inhibitors do not prevent release of arachidonic acid from platelet membranes or the formation of lipoxygenase metabolites. Aspirin treated platelets adhere to the exposed subendothelium better than untreated controls. Platelet stickiness has been attributed to the action of hydroxy acids and elevated levels of cyclic GMP. Studies employing a specific lipoxygenase inhibitor 4,7,10,13-eicosatetraenoic acid (U53119) and an inhibitor of both cyclooxygenase and lipoxygenase ETYA (5,8,11,14-eicosatetraenoic acid) showed that lipoxygenase metabolites do not play a critical role in irreversible aggregation of human platelets (27).

Lipoxygenase metabolites are rapidly incorporated by endothelial cells, converted to novel metabolites by neutrophils. Platelets have enzymes capable of converting 5-lipoxygenase metabolites to leukotrienes. Although available data do not suggest a critical role for lipoxygenase metabolites in causing irreversible aggregation of platelets, they may very well play some modulating role in cell-cell communication via formation of novel metabolites or by facilitating the formation of appropriate metabolites.

Polyenoic Acids and Arachidonic Acid Oxidation

Several epidemiologic studies have suggested beneficial effects of fish diet in lowering the risk for cardiovascular diseases (28, 29). Fish lipids are rich in omega 3 polyenoic acids such as eicosapentaenoic acid (EPA 20:5) and docosahexaenoic acid (DHA 22:6).

Earlier studies suggested that the beneficial effects of these polyenoic acids were related to the formation of triene prostanoids such as prostacyclin (PGI_3) and thromboxane A_3 (TXA_3), which differ in their properties from the natural diene prostanoids (PGI_2 and TXA_2), products of arachidonic acid metabolism (30). Although thromboxane A_3 can be synthesized from EPA, it seems to lack any aggregatory effect on human platelets. Studies from our laboratory have shown that both EPA and DHA are poorly metabolized by human platelet cyclooxygenases (31). Some studies have suggested that the inhibitory effect of these polyenoic acids may be due to the hydroperoxy acids formed via the lipoxygenase pathway. Results from our laboratory, as well as others, suggest that polyenoic acid induced inhibition of platelet function may be more related to their ability to compete for the heme site on cyclooxygenase as well as their competition at the endoperoxide/ thromboxane receptors (32). Furthermore, the incorporation of these fatty acids in the platelet membrane may alter the property of the plasma membrane to the action of agonists and various artificial and natural surfaces. *In vivo* studies suggest that ingestion of large amounts of these fatty acids may be essential to obtaining any beneficial effect.

In conclusion, results of our studies and the studies of others, suggest that free arachidonic acid released during platelet activation is transformed to a transient radical by the heme proteins associated with cyclooxygenases and then converted to prostaglandin endoperoxides, PGG_2 and PGH_2. These intermediate metabolites are further converted to thromboxane by thromboxane synthetase. Substantial evidence does not exist for any significant role for oxygen radicals in this fatty acid oxidation and biosynthesis of eicosanoids. Arachidonic acid metabolites play an active role in the platelet activation caused by weak agonists such as epinephrine, adenosine diphosphate and platelet activating factor. The specific mechanism by which they induce formation of binding sites for adhesive proteins, promote development of stickiness, mobilize granules and initiate release of their contents is not clear. Potent agonists like thrombin and intrinsic mechanisms such as membrane modulation can secure irreversible aggregation of human platelets independent of any mediation by eicosanoids.

Acknowledgments

The authors wish to thank Ms. Susan Schwarze for her help in the preparation of the manuscript. This work was supported by USPHS grants HL-27355, HL-11880, CA-21737, GM-22167, and HL-30217, a March of Dimes grant (MOD 1-886), and the Minnesota Medical Foundation (HRF-57-88).

References

1. Rao, G.H.R., Reddy, K.R., and White, J.G. (1981) *Prost. and Med.* **6**, 75-90
2. Rao, G.H.R., Gerrard, J.M., Witkop, C.J., and White, J.G. (1981) *Prost. and Med.* **6**, 459-472
3. Zucker, M.B., and Nachmias, V.T. (1985) *Arteriosclerosis* **5**, 1-18
4. Brass, L.F., Woolkalis, M.J., and Manning, D.R. (1988) *J. Biol. Chem.* **263**, 5348-5355

5. Berridge, M.J., and Irvine, R.F. (1984) *Nature* **312**, 315-321
6. Nishizuka, Y. (1984) *Nature* **308**, 693-698
7. Plow, E.F., Srouji, A.H., Meyer, D., Marguerrie, G., and Ginsberg, M.H. (1984) *J. Biol. Chem.* **259**, 5388-5391
8. Hamberg, M., Svensson, J., and Samuelsson, B. (1975) *Proc. Natl. Acad. Sci. USA* **72**, 2994-2998
9. Marcus, A.J. (1978) *J. Lipid Res.* **19**, 793-826
10. Elattar, T.M.A. (1978) *J. Oral Pathol.* **7**, 175-207
11. Siess, W. (1989) *Physiol. Rev.* **69**, 58-141
12. Rao, G.H.R., and White, J.G. (1985) in *Prostaglandins Leukotrienes and Lipoxins* (Bailey, J.M., ed), pp. 357-370. Plenum Press, New York
13. Gerrard, J.M., White, J.G., Rao, G.H.R., and Townsend, D. (1976) *Am. J. Pathol.* **83**, 283-298
14. Rao, G.H.R., Gerrard, J.M., Eaton, J.W., and White, J.G. (1978) *Photochem. and Photobiol.* **28**, 845-850
15. Peterson, D.A., Gerrard, J.M., Rao, G.H.R., and White, J.G. 91980) *Adv. Prost. Thromb. Res.* **6**, 157-161
16. Mason, R., Kalyanaraman, B., Tainer, B.E., and Eling, T.E. (1980) *J. Biol. Chem.* **255**, 5019-5022
17. Peterson, D.A., Gerrard, J.M., Rao, G.H.R., and White, J.G. (1979) *Prost. Med.* **2**, 97-108
18. Rao, G.H.R., Gerrard, J.M., Eaton, J.W., and White, J.G. (1978) *Prost. Med.* **1**, 55-70
19. Peterson, D.A., Gerrard, J.M., Rao, G.H.R., and White, J.G. (1980) *Prost. Med.* **4**, 73-79
20. Meister (1981) *Trends Biochem. Res.* **6**, 231-234
21. Bryant, R.W., Simon, T.C., and Bailey, J.M. (1983) *Biochem. Biophys. Res. Comm.* **117**, 183-189
22. Hill, T.D., White, J.G., and Rao, G.H.R. (1989) *Thromb. Res.* **53**, 447-457
23. Hill, T.D., White, J.G., and Rao, G.H.R. (1989) *Thromb. Res.* **53**, 457-467
24. Hill, T.D., White, J.G., and Rao, G.H.R. (1989) *Prostaglandins* (in press)
25. Rao, G.H.R., and White, J.G. (1985) *Am. J. Pathol.* **19**, 339-347
26. Rao, G.H.R., Johnson, G.J., and White, J.G. (1980) *Prost. Med.* **5**, 45-58
27. Rao, G.H.R., Radha, E., and White, J.G. (1985) *Biochem. Biophys. Res. Comm.* **131**, 50-57
28. Bang, H.O., Dyerberg, J.D., and Nielsen, B.A. (1971) *Lancet* **1**, 1143-1146
29. Dyerberg, J.D., and Bang, H.O. (1979) *Lancet* **2**, 433-435
30. Needleman, P., Minkes, M.S., and Raz, A. (1976) *Science* **193**, 163-165
31. Rao, G.H.R., Kishore, N.P., Peller, J.D., and White, J.G. (1987) in *Cardiovascular Disease* (Gallo, L.L. ed) pp. 495-505, Plenum Press, New York
32. Rao, G.H.R., Radha, E., and White, J.G. (1983) *Biochem. Biophys. Res. Comm.* **117**, 549-556

Role of Eicosanoids of the Lipoxygenase Pathway in Vertebrate Reproduction

P. Reddanna, G. Panduranga Reddy, G. Ramesh Reddy and M. Prasad
Department of Zoology, School of Biological and Earth Sciences
S.V. University, Tirupati - 517 502, India

Introduction

Eicosanoid is the term given to describe the various oxidative derivatives of eicosapolyenoic fatty acids, especially arachidonic acid. After an appropriate physiologic transmembrane signal or after treatment with a calcium ionophore, cellular phospholipases and lipases of distinct pathways cleave arachidonic acid from membrane phospholipids making it available as a substrate for eicosanoid production (1,2). Besides normal β-oxidation, arachidonic acid is enzymatically oxygenated by two important pathways: the cyclooxygenase pathway leading to the formation of prostaglandins, thromboxanes and prostacyclins and the lipoxygenase pathway generating the hydroperoxy- and hydroxy-eicosatetraenoic acids (HPETEs and HETEs), the leukotrienes (LTs), and the lipoxins (LXs). Modulation of cellular functions by these eicosanoids occurs in a variety of ways. These compounds act as local "hormones" or autocoids. The vast literature available on the metabolism, biochemistry and pathophysiological effects of cyclooxygenase products reveals that they affect almost all body functions at least in pharmacological doses.

The lipoxygenase pathway, on the other hand, is less well characterized, and most of its bioactive products have only recently become available to scientists. The biological activity of one major class of lipoxygenase products, the peptido leukotrienes, was originally described as slow reacting substances of anaphylaxis (SRS-A) and has been known for over 50 years (3,4). Following the elucidation of the chemical structures of SRS-As and the enzymatic pathways involved in peptido leukotriene biosynthesis (5-9), eicosanoids of the lipoxygenase pathway received much attention which lead to numerous discoveries about their biochemistry and pathophysio-logical effects. Several studies have revealed that different cells and tissues transform arachidonic acid into a variety of products via the lipoxygenase pathway (10-20). Besides their

Biological Oxidation Systems, Volume 2

ubiquitous distribution, lipoxygenase eicosanoids have been shown to be extremely potent biologically active compounds with a bewildering variety of actions. Several of the monohydroxy compounds like 5-HETE, 8-HETE, 11-HETE and 12-HETE (21-23) and LTB_4 (24, 25) are the most potent chemotactic agents for several immune cells. These chemotactic agents are also shown to be secretagogues for lysosomal enzyme secretion (26, 27) superoxide anion radical generation (28-30), smooth muscle contraction (31-33), vascular permeability (34) and other physiological processes.

This review will consider only the pathophysiological effects of lipoxygenase products in vertebrate reproduction. The distribution and effects of eicosanoids in invertebrates was recently reviewed by Stanley-Samuelsson (35). The role of lipoxygenase products in vertebrate reproduction is an actively evolving area of research. This review is intended as a guide for further research on lipoxygenase products in verbrate reproduction.

Release/Secretion of Gonadotrophins :

The secretion of gonadotrophin in the pituitary is under the control of hypothalamic releasing hormones. The action of gonadotrophin releasing hormone (Gn-RH) on the secretion of pituitary gonadotrophins involves calcium in a receptor-mediated mechanism. The binding of the hypothalamic peptide to its receptor at the gonadotroph cell membrane initiates events that result in the secretion of gonadotrophins. Various potential mediators such as cyclic nucleotides (36) and prostaglandins (37) have been implicated in the receptor-mediated secretion of gonadotrophins. Several studies have demonstrated that the activation of phospholipids (38) and the cleavage of arachidonic acid followed by its oxidation via the lipoxygenase pathway (39) or epoxygenase pathway (40) might be involved in the Gn-RH-induced release of leuteinizing hormone (LH) and follicle stimulating hormone (FSH). Among the lipoxygenase products tested, 5-HETE, 15-HETE and LTC_4 stimulated the release of FSH and LH in rat pituitary cells (38). Leukotriene C_4 action on the release of LH was biphasic with low concentrations (around 10^{-10} M) being 10 fold more potent than that of the physiological stimulus of Gn-RH and high concentrations (10^{-9} M) showing no effect (38). Pretreatment of perfused rat anterior pituitary cells with NDGA, an inhibitor of lipoxygenase pathway, decreased the total LH response to Gn-RH (41,42), also suggesting the involvement of lipoxygenase products in the mediation of gonadotrophin secretion induced by releasing hormones. The studies on purified rat Leydig cells revealed the blocking of LH-RH and calcium ionophore-stimulated steroidogenesis by calmodulin inhibitors (43). Further quinacrine, a phospholipase A_2 inhibitor, was shown to reduce both basal and growth hormone releasing factor (GRF)-stimulated free arachidonic acid levels (44). These results suggest that mobilization and metabolism of arachidonic acid are important in the rapid initial phase of the LH response to Gn-RH and that activiation of a protein kinase C-dependent mechanism participate in the maintenance of gonadotrophin secretion.

Similarly, the inhibition of lipoxygenase pathway with inhibitors like NDGA was shown to inhibit both basal and neurotensin stimulated prolactin release (44,45). The lipoxygenase products like 5-HETE and 15-HETE stimulated the release of prolactin in superfused rat pituitary cells in a dose dependent manner (46). These studies also revealed that LTA_4, LTB_4, LTC_4 and LTE_4, but not LTD_4, provoked a marked biphasic and dose dependent secretion of prolactin from superfused cells, with maximum effects at concentrations of 3×10^{-11} to 3×10^{-10} M. Arachidonic acid metabolites were shown to be more potent than arachidonic acid itself.

From the above studies it is clear that arachidonic acid and its lipoxygenase metabolites play an important role in the secretory mechanism of several hormones in the pituitary gonadotrophs. However, most of the studies on pituitary were on *in vitro* effects of metabolites of the lipoxygenase pathway. Further studies to directly demonstrate the lipoxygenase pathway in pituitary cells are needed.

Male Reproduction

Testicular tissue has an unusually high concentration of polyunsaturated fatty acids (PUFAs) such as arachidonic acid (47, 48). Hence, it is natural that these PUFAs might be playing an important role in testicular metabolism. The presence of lipoxygenase like enzyme was described in the microsomes of rat testis (49-51). Later studies from the same laboratory revealed the formation of 15-HPETE and 15-HETE, through the 15-lipoxygenase pathway in the rat testis (52). ETYA, a general inhibitor of AA metabolism, and NDGA, an inhibitor of the lipoxygenase pathway, were shown to inhibit hCG-induced testosterone secretion in rat Leydig cells with no effect of indomethacin (40,53). However the effects of these inhibitors on the testicular steroidogenic enzyme, 3β-Hydroxysteroid dehydrogenase (3β-HSD) showed a different pattern (55). Indomethacin as well as NDGA both increased the activity level of

3β-HSD in rat testicular slices. The effects were more pronounced when the testicular slices were incubated with indomethacin and NDGA. Also vitamin E, a potent inhibitor of the 5-lipoxygenase pathway (17), showed similar trend on 3β-HSD activity level (55). Surprisingly the incubation of testicular slices with 5-HPETE, 15-HPETE, 5-HETE and 15-HETE also increased the activity levels of 3β-HSD at low concentrations and decreased or no effect at higher concentrations, warranting more detailed studies.

These findings suggest a regulatory role for lipoxygenase products in the biosynthesis of testosterone. 5-HPETE, the product of 5-lipoxygenase pathway, enhanced testosterone production stimulated by LH in a dose dependent manner up to 1 μM (56). Their studies revealed that 5-HPETE may be enhancing steroidogenesis at a point subsequent to cyclic AMP production. LTA_4 and LTB_4 also were shown to mediate LH-stimulated testosterone production in rat testis Leydig cells (56). Their studies with ^{14}C arachidonic acid in Leydig tumor cells and rat testis Leydig cells in the presence of LH indicated incorporation of ^{14}C AA into LTB_4, 5-HETE and 12-HETE. Leydig cells from rat testis treated *in vitro* with hCG, A231187 and LH were also shown to produce increased amounts of LTB_4 (56,57). Cooke and Sullivan (57) studied the production of LTB_4 in different degrees of purified Leydig cells and confirmed the production of LTB_4 in Leydig cells. They also pointed out that the interstitial cells are not the only cells to produce LTB_4. Phospholipase C, the enzyme involved in arachidonic acid release, or arachidonic acid directly stimulated hCG-induced testosterone secretion from Leydig cells by about 2 fold (58). This effect of phospholipase C and AA was biphasic, in that, at low concentrations both phospholipase C and AA augmented hCG-induced testosterone secretion, while at higher concentrations they inhibited steroid production. Arachidonic acid metabolites were also shown to be involved in the action of LH-RH in testosterone secretion (59,43).

From the few studies available on lipoxygenase products and Leydig cell function, it is clear that these arachidonic acid metabolites may be involved in the mediation of LH or hCG-induced testosterone secretion. It is well established that steroidogenesis in Leydig cells is regulated by LH. Following binding of LH to its specific receptors on Leydig cells, adenyl cyclase is activated to produce cyclic AMP. Cyclic AMP then activates specific protein Kinase (s), which inturn phosphoylate the proteins to trigger steroidogenesis by as yet unknown mechanism(s) (60-62). Though this cyclic AMP dependent pathway of LH action is well accepted, it seems that cyclic AMP-independent pathway also exists in Leydig cells. This was demonstrated in experiments where inhibition of adenylate cyclase did not block LH-induced pregnenolone secretion, though cyclic AMP levels were drastically decreased (63). The cyclic AMP independent or alternate pathway proposed was the phosphatidyl inositol (PI) cycle. PI cycle produces its second messengers, diacylglycerol and inositol triphosphate through the action of phospholipase C. Diacylglycerol can subsequently be hydrolyzed to release free AA which can be further metabolized to give various eicosanoids which ultimately trigger the steroidogenesis. Phospholipase C, in fact, stimulated testosterone secretion, suggesting that the PI cycle may be involved in AA release (63). The other mechanism of AA release *i.e.*, through the action of phospholipase A_2, does not seem to operate as it had no effect on testosterone secretion (56).

Besides the role of lipoxygenase products in the steroidogenesis of Leydig cells, they also have been implicated in hCG-induced inflammatory type of reaction in the testis. Since LTB_4 is secreted in Leydig cells (56, 57) and is also a potent chemotactic agent to various immune cells, it is proposed that LTB_4 may mediate hCG-induced inflammatory type of-reaction.

Gossypol, a polyphenolic compound, is a known inhibitor of spermatogenesis and sperm motility (NO REFERENCE - Changamma & Reddanna, 1987). It inhibits both basal and LH stimulated release of testosterone from Leydig cells *in vitro* at concentrations between 21 and 170 μM (64). Gossypol, was shown to be a potent inhibitor of purified 5-lipoxygenase (65) suggesting that part of the action of Gossypol might be mediated through inhibition of lipoxygenase pathway. However further work is required on the production of eicosanoids in testis during the course of Gossypol treatment.

There is not much information available on the role of lipoxygenase products in male accessory glands. The occurrence of 15- and 11-lipoxygenase enzyme activities in sheep vesicular glands was reported as early as in 1967 (14). The occurrence of specific LTC_4 receptors on the ductus deferens of Syrian hamster (66) was reported. These LTC_4 receptors did not compete with the naturally occurring sulfidopeptide leukotrienes, LTD_4, and LTE_4 suggesting the presence of individual receptor sites for LTC_4, LTD_4 and LTE_4.

Female Reproduction

Ovary

It is reported earlier that the products of cyclooxygenase pathway, especially prostaglandins are obligatory mediators in several events of female reproduction (67-69). The information on the involvement of lipoxygenase metabolites on the other hand, is limited and is of recent origin only. Ovarian follicles from proestrous rats were shown to transfrom arachidonic acid into various hydroxyeicosatetraenoic acids (HETEs) via the lipoxygenase pathway (70). Arachidonic acid metabolism via the lipoxygenase pathway was reported in human granulosa cells (71). Their later studies have shown that NDGA, the inhibitor of lipoxygenase pathway blocked ovulation and inhibited hCG-induced ovarian collagenolysis (72), suggesting the possible involvement of lipoxygenase products in the process of ovulation. Recent studies of Laloraya *et al.* (73) have, in fact, indicated the involvement of superoxide anion radicals, probably generated by leukotrienes, in LH induced ovulation in mammalian ovary. Oxytocin, known to induce ovulation in rat ovaries, augmented the basal synthesis and release of $PGF_{2\alpha}$ and of 5-HETE from 14C-arachidonic acid (74). From these results it was suggested that ovarian oxytocin may play a role in the ovulatory process, via the generation of $PGF_{2\alpha}$ to enhance contractions of ovarian smooth muscle and of 5-HETE to promote follicular collagenolysis. Carvalho *et al.* (75) followed the changes in the concentrations of LTB_4 within the wall of the periovulatory ovine follicles before, during and after ovulation and reported a significant rise in LTB_4 only after follicular rupture had occurred. This rise in LTB_4 after ovulation, was attributed towards its role in the formation of corpus luteum (luteinization). Thus lipoxygenase products have been shown to be involved in ovulation and luteinization.

Arachidonic acid and linoleic acid were shown to significantly stimulate oxytocin secretion from the sheep corpus luteum, which was inhibited by NDGA, the lipoxygenase inhibitor but not by indomethacin, the inhibitor of cyclooxygenase (76). This observation has indicated the participation of lipoxygenase metabolites in stimulation of oxytocin secretion from the sheep corpus luteum. Intrauterine administration of NDGA was shown to extend luteal life in cattle (77). Also, the lipoxygenase products 15-HPETE and 5-HETE were shown to reduce the biosynthesis of progesterone in bovine luteal cells (77). However, later studies have revealed that NDGA, the lipoxygenase inhibitor, inhibits progesterone secretion by bovine luteal cells *in vitro* (78) and caused luteolysis in sheep (79).

These reports do not agree with the earlier reports of Milvae *et al.*, (77) in cattle, suggesting species specific effects.

Leukotrienes were shown to exert pronounced effects on *in vitro* motility of hen oviduct (80). LTC_4 and LTD_4 stimulated the motility of all oviducal parts *in vitro*. LTD_4 in the conscious animal induced a short-lasting increase in uterine myoelic activity (80).

Uterus

Besides the metabolism of arachdonic acid via the cyclooxygenase pathway , oxidations via the lipoxygenase pathway were also reported in uterus. Bovine uterus was shown to convert arachidonic acid into 12-lipoxygenase indicating the presence of 12-HETE (81). Intrauterine tissues from women at term of pregnancy metabolized arachidonic acid by way of lipoxygenase pathway into 5-, 12- and 15-HPETES (82). The operation of a lipoxygenase pathway in uterine tissues suggests the probable involvement of its products in uterine functions. Towards this direction several attempts have been made, among which the report of Pesty *et al.* (83) is of prime importance. They incubated rat endometrial cells obtained from the uterine horn between day 3 and 9 of gestation, in 2 ml of Dulbeccos medium for 1 hr at 37° C and the concentrations of various leukotrienes were measured in the medium. The levels of LTB_4 and LTC_4, rose in endometrial cells between day 3 and 5 with a peak at day 4, decreased by day 6 and undetectable by day 9. These changes in leukotriene concentrations closely coincided with acid phosphatase activity levels. Based on these observations, Pesty *et al.*, (83) have concluded that leukotrienes might influence acid phosphatase activity by increased lysosomal enzyme secretion/release. The reported roles of leukotrienes in chemotaxis, lysosomal enzyme secretion and other immunological reactions (24,29,84) support such a possibility of leukotriene induced secretion of acid phosphatase activity in endometrial cells. This type of proinflammatory reaction accompanied by an increased uterine stromal capillary permeability at the site of blastocyst attachment is considered to be one of the events for implantation. As per the reports of Pesty *et al.* (83) leukotrienes in uterine endometrium might be involved in proinflammatory type of reactions. Leukotrienes were also reported to interact with prostaglandins during decidualization process (85). Recent studies of Kennedy (86) have revealed that LTB_4 along with PGF_2 may be involved in endometrial vascular permeability. The superoxide anion radicals, probably generated by eicosanoids, were shown to be increased on the early morning of Day 5 of pregnancy, facilitating increased vascular permeability at the initiation of implantion of process (87). From all these observations it can be concluded that leukotrienes especially LTB_4 may be involved in the process of implantation.

Leukotrienes were also shown to be involved in uterine contractions. LTC_4 and LTD_4 elicited concentration dependent contractions of the guinea pig uterus (88). LTE_4 was shown to be 1000 fold weaker than LTC_4 and LTD_4. The potencies of LTC_4 and LTD_4 were similar to that of acetyl choline and $PGF_{2\alpha}$ but weaker than that of PGE_2 (88). This uterine contractile activity of peptido leukotrienes can be comparable to their general smooth muscle contractile activity like vasoconstriction and bronchoconstriction (33,89-91). The impact of these peptido leukotrienes on uterus may be mediated through their specific receptors. In fact, several studies have demonstrated distinct functional stereoselective, reversible high affinity LTC_4, and LTD_4 receptors in uterine tissue (92-94). Recent studies of Dr. Gardner Russo of Villanova University, have revealed that estradiol induced uterine growth (mediated by epidermal growth factor) can be inhibited only partially by indomethacin or NDGA, but totally by the combination of NDGA and indomethacin (unpublished data). These studies also indicate the involvement of lipoxygenase products in the mediation of effects of estradiol and progesterone on uterus.

Besides uterus, various intrauterine tissues have been shown to metabolize arachidonic acid by way of the lipoxygenase pathway. Maternal and fetal cotyledon, myometrium and fetal membrane of sheep in the third term of gestation exhibited evidence of arachidonic acid oxidation by 5-, 12- and 15-lipoxygenase activities (95). The major product of fetal membrane and fetal cotyledon was LTB_4, whereas maternal cotyledon and myometrium produced mainly 12-HETE (95). The major lipoxygenase product in amnion obtained from women, before labor was LTB_4 and after labor was 12-HETE (82). *Chorion laeve* and *Decidua vera* were shown to synthesize 15-HETE at all times, while placenta produced mainly 12-HETE (82). From these studies, it was suggested that the products of arachidonic acid play a significant role in the control of fetal and uteroplacental hemodynamics and in the mechanism of parturition. Cytosolic fraction of rabbit placenta was shown to convert arachidonic acid into 12-HETE and 15-HETE (96). Formation of 12-HETE and 15-HETE at 21 days was shown to be as effective as at 31 days of gestation.

Incubation of trophoblast cells with eicosatetrayenoic acid ETYA (the general inhibitor of cyclooxygenase and lipoxygenase pathways) but not with indomethacin (the specific inhibitor of cyclooxygenase pathway) resulted in significant inhibition in the secretion of hormones, suggesting the regulatory role of lipoxygenase products in the secretion of hormones from trophoblast cells (97). However, further studies are required to find out the specific lipoxygenase product(s) involved.

In general, the lipoxygenase products in female reproduction seem to be involved either in the release/secretion of ovarian hormones or in the mediation of the effects of ovarian hormones on uterine growth and contractions.

Concluding Comments

The eicosanoids of lipoxygenase pathway, in general, appear to be involved in the mediation of a variety of reproductive processes. They seem to be generated in response to the specific stimuli, elicit the response and are then quickly metabolized. Hence, these compounds are rightly termed 'local hormones' or autocoids.

The lipoxygenase products along with other eicosanoids have been implicated in the release/secretion of a number of hormones from pituitary, testis and ovaries mediated by the releasing hormones and gonadotrophins. Arachidonic acid lipoxygenase products in most of the cases have been shown to enhance the basal as well as agonist mediated secretion of various hormones. It appears that the transmembrane signals generated by the agonists release free arachidonic acid, which then is metabolized depending on the enzymes available in the respective tissues. These arachidonic acid products once released elicit specific responses, for example, release of various hormones. Though a number of reports are available on the regulation of release of various hormones by the eicosanoids, it is not clear whether these are nonspecific effects leading to leakage of cell contents (as shown in case of oxytocin release from sheep corpus luteum) or stimulation of specific biosynthetic processes. Moreover, most of the studies were carried out by studying the *in vitro* effects of purified lipoxygenase products, without much knowledge of the type of eicosanoid produced in the cellular system. Studies on the eicosanoid metabolic pathways and the specific arachidonic acid product(s) systhesized in particular cellular systems would be most helpful in understanding these regulatory mechanism(s).

The other effects of the lipoxygenase products in vertebrate reproduction reported are mostly with regard to smooth muscle contraction, inflammatory type of reactions, capillary permeability etc. These are the general effects exhibited by some of the lipoxygenase products where ever they are produced. For example bronchoconstriction, vasoconstriction and other smooth muscle contraction are the properties exhibited by LTC_4 ,LTD_4 and LTE_4 . Chemotaxis, enzyme secretion, superoxide anion radical generation, vascular permeability and other proinflammation effects are generally mediated by LTB_4 and other mono and dihydroxy acids. In fact the involvement of superoxide anion radical species probably generated by eicosanoids in the ovulation and blastocyst implantation was already reported.

Thus, most of the studies on lipoxygenase products in reproduction, are confined to leukotrienes and other monohydroxy derivatives of arachidonic acid . There is practically no information on the distribution and effects of dihydroxy compounds (diH(P)ETEs) and lipoxins in reproductive tissues which were also shown to be biologically potent molecules. This could be partly due to the non-availabity of the compounds.

Acknowledgements

The financial support of the Department of Biotechnology (Grant No. BT/TF/13/05/89, dated 27-3-1989) and the Department of Science and Technology (Grant No. SP/SO/D-33/88, dated 19-12-1988) to Dr. P. Reddanna is gratefully acknowledged. We also acknowledge the help of Mr. N. Sudhakar Raju for his assistance in the preparation of the manuscript.

References

1. Bell, R.L., Kennerly, D.A., Stanford, N., and Majerus, P.W. (1979) *Proc. Acad. Sci.* USA, **76**, 3238-3241
2. McKean, M. L., Smith, J. B., and Silver, M.J. (1981) *J. Biol. Chem.* **256**, 1522-1524
3. Feldberg, W., and Kellaway, C.H. (1938) *J. Physiol. (Lond.)* **94**, 187-226
4. Kellaway, C.H., and Trethewie, E.R. (1940) *J. Exp. Physiol.* **30**, 121-145
5. Hammarström, S., Murphy, R. C., Samuelsson, B., Clark, D. A., Mioskowski, C., and Corey, E.J. (1979) *Biochem. Biophys. Res. Commun.* **91**, 1266-1272
6. Hammarström, S., Samuelsson, B., Clark, D. A., Goto, G., Marfat, A., Mioskowski, C., and Corey, E.J. (1980) *Biochem. Biophys. Res. Commun.* **92**, 946-953
7. Corey, E. J., Branton, A. E., and Clark, D.A. (1980) *J. Amer. Chem. Soc.* **102**, 6607-6608
8. Chang, M., Rao, M. K., Reddanna, P., Li, C. H., Tu, C.-P. D., Corey, E. J., and Reddy, C.C. (1987) *Arch. Biochem. Biophys.* **259**, 536-547
9. Reddanna, P., Whelan, J., and Reddy, C. C. (1988) *Ann. N.Y. Acad. Sci.* **524**, 391

10. Bach, M. K., Brashler, I. R., Hammarstr−m, S., and Samuelsson, B. (1980) *J. Immunol.* **125**, 115-117
11. Borgeat, P., Hamberg, M., and Samuelsson, B. (1976) *J. Biol. Chem.* **251**, 7816-7820 (See correction (1987) *J. Biol. Chem.*, **252**, 8772)
12. Goetzl, E.J., and Sun, F.F. (1979) *J. Exp. Med.* **150**, 406-411
13. Hamberg, M., and Samuelsson, B. (1967) *J. Biol. Chem.* **247**, 5329-5335
14. Hamberg, M., and Samuelsson, B. (1967) *J. Biol. Chem.* **242**, 5344-5354
15. Nugteren, D.H. (1985) *Biochem. Biophys. Acta* **431**, 651-654
16. Lundberg, U., Rodmark, O., Malmsten, C., and Samuelsson, B. (1981) *FEBS Lett.* **126**, 127-132
17. Reddanna, P., Krishna Rao, M., and Reddy, C.C. (1985) *FEBS Lett.* **93**(1), 39-43
18. Reddanna, P., Whelan, J., and Reddy, C. C. (1986) *Fed. Proc.* **45**(6), 2647
19. Reddanna, P., Whelan, J., Reddy, P. S., and Reddy, C.C. (1988) *Biochem. Biophys. Res. Commun.* **157**, 1348-1351
20. Reddanna, P., and Reddy, C.C. (1988) *Natl. Seminar on Rep. Biol.* held at Mysore
21. Goetzl, E.J. (1980) *Immunology* **40**, 709-719
22. Goetzl, E.J., Brash, A.R., Tabuer, A.I., Oates, J.A., and Hubbard, W.C. (1980) *Immunology* **39**, 491-501
23. Turner, S.R., Tainer, S.A., and Lynn, W.S. (1975) *Nature* **257**, 680-681
24. Carr, S.C., Higgs, G.A., Salmon, J.A., and Spayne, J.A. (1981) *Br. J. Pharmacol.* **73**, 254
25. Bhattacherjee, P., Ekins, K. E., and Hammond, B. (1981) *Br. J. Pharmacol.* **73**, 254
26. Naccache, P.H., Showell, H.J., Backer, E.L., and Sha'afi, R.I. (1977) *J. Cell Biol.* **75**, 635-649
27. Showell, H.J., Freer, R.J., Zigmond, S.H., Schiffman, E., Aswanikumar, S., Carcoran, B., Becker, E.L. (1976) *J. Exp. Med.* **143**, 1154-1169
28. O'Flaherty, J.T., Showell, H.J., Ward, P.A., and Becker, E.L. (1979) *Am. J. Pathol.* **96**, 799-810
29. Palmer, R.M.J., and Yeats, D.A. (1981) *J. Pharmacol.* **73**, 260
30. Palmblad, J., Malusten, C.L., Uden, A.M., Redmark, O., and Samuelsson, B. (1981) *Blood* **58**, 658-661
31. Headquist, P., Dahlen, S.E., Gustafsson, L., Hammarström, S., and Samuelsson, B. (1980) *Acta Physiol. Scand.* **10**, 331-334
32. Peck, M.J., Piper, R.J., and Williams, T.J. (1981) *Prostaglandins* **21**, 315-322
33. Palmer, M.H., Piper, P.J., Samhoun, M.N., and Tippins, J.R. (1981) *Br. J. Pharmacol.* **73**, 213
34. Williams, T.J., and Piper, P.J. (1980) *Prostaglandins* **19**, 779-789
35. Stanley-Samuelson, D.W. (1987) *Biol. Bull.* **173**, 92-109
36. Naor, Z., and Catt, K.J. (1980) *J. Biol. Chem.* **255**, 342-344

37. Naor, Z., Koch, Y., Bauminger, S., and Zor, U. (1975) *Prostaglandins* **9**, 211-219
38. Kiesel, L., Przylipiak, A., Rabe, T., and Runnbeaum, B. (1986) *Human Repro.* **1**, 349-352
39. Naor, Z., Kiesel, L., Vanderhock, J. K., and Catt, K. J. (1985) *J. Steroid Biochem.* **23**, 711-717
40. Snyder, G. D., Capdavila, J., Chacos, N., Menna, S., and Falck, J.R. (1983) *Proc. Natl. Acad. Sci. USA* **80**, 3505-3507
41. Sullivan, M. H. F., Chaudry, L., and Cooke, B.A. (1987) *J. Endocrinology* **112** (suppl.)
42. Chang, J. B., Graeter, J., and Catt, K. J. (1987) *Endocrinology* **120**, 1837-1845
43. Sullivan, M.H.F., and Cooke, B.A. (1985) *Biochem. J.* **232**, 54-55
44. Canonico, P. L., Lug, P., Speciale, C., Sortino, M.A., and Scapagnini, U. (1985) *Am. J. Physiol.* **249**, E257-E263
45. Canonico, P. L., Schettini, G., Valdenegro, C. A., Valdenegro, R. M., and Leod, M. (1983) *Neuroendocrinology* **37**, 212-217
46. Kiesel, L., Przylipiak, A., Rabe, T., Przylipiak, M., and Runnebaum, B. (1987) *Human Reprod.* **2**, 281-285
47. Johnson, A. D. (1970) in *Testicular Lipids*, Johnson, A. D., Gomes, W. R., and Van De Mark, N. L., Eds., Academic Press, New York, Vol. II, pp. 193-258
48. Neil, A.R., and Masters, C.J. (1974) *Int. J. Biochem.* **5**, 137-156
49. Shahin, I., Grossmen, S., and Sredni, B. (1978) *Biochimica et Biophysica Acta* **529**, 300-308
50. Grossman, S., Shahin, I., and Sreini, B. (1979) *Biochemica et Biophysica Acta* **572**, 293-297
51. Grossman, S., Schon, T., Sofer, Y., Magia, N., and Bartoor, B. (1986) *Thromboxane and Leukotriene Research* **16**, 241-252
52. Shahin, I. S., Saffer, Y., Cojacaru, M., and Grossman, S. (1989) *Int. J. Bio.* **21**, 7-13
53. Hamberg (1976) *Biochem. Biophys. Acta* **431**, 651-654
54. Panduranga Reddy, G., and Reddanna, P. (1989) *Int. Symp. on Biol. Oxid. Sys.* held at Bangalore 26th to 29th Oct., 89
55. Sullivan, M.H.F., and Cooke, B.A. (1987) *J. Endocrinology* **115** (suppl.)
56. Abayasekara, D. R. E., Kurlak, K., Cooke, B. A., Teremy, J. Y., Dandona, P., and Sharpe, K.A. (1987) *J. Endocrinology* **115** (suppl.)
57. Cooke, B.A., and Sullivan, M.H.F. (1986) *J. Endocrinology* **III** (suppl.)
58. Didolkar, A.K., and Sundaram, K. (1987) *Life Sciences* **41**, 471-477
59. Lin, T. (1985) *Life Sciences* **36**, 1255-1264
60. Cooke, B. A., Lindh, L. M., and Jonszen, F.H.A. (1976) *Biochem. J.* **160**, 439-446
61. Podesta Derfan, M. L., Solano, A. R., and Catt, K.J. (1978) *J. Biol. Chem.* **253**, 1894-9001

62. Rommerts, F. F. G., Bakker, G. H., and Vander Molen, H.J. (1983) *J. Steroid Biochem.* **19**, 367-373
63. Themman, A. P., Hoogerbrugge, Rommerts, F.F.G., and Vander Brugge, H.J., Molan (1985) *Biochem. Biophys. Res. Commun.* **128**, 1164-1172
64. Pearce, S.B., Sufi, D., and Jeffwate, S.L. (1987) *J. Endocrinology*
65. Reddanna, P., and Reddy, C.C. (1987) *Symp. on Comparative Animal Physiology - Its Application to Human Welfare* held at Osmania University, Hyderabad, 29-31st, Dec.
66. Krilis, S. R. S., Lewis, E. J., Corey, K.F., and Austen (1983) *J. Clin. Invest.* **72**, 1516-1519
67. Goldberg, V.J., and Ramwell, P.W. (1975) *Physiol. Rev.* **55**, 325-351
68. Green, K., Kerterquist, U., Bugdeman, M., Christenssen, N. J., and Bergstrom, S. (1982) *Prostaglandins* **24**, 451-466
69. Roberts, J. S., Barcikowski, B., Wilson, L., Skarnes, R. C., and KcCraken, J.A. (1975) *J. Steroid Biochem.* **6**, 1091-1097
70. Reich, R., Kohen, F., Naor, Z., and Trafriri, A. (1983) *Prostaglandins* **26**, 1011-1020
71. Feldman, E., Haberman, S., Abisogun, A. O., Reich, R., Levran, D., Maschiach, S., Zuekermann, H., Ruda, K. I., and Tsafriri, A. (1986) *Human Repro.* **6**, 353-356
72. Reich, R., Kohen, F., and Trafrin, A. (1985) *Prostaglandins* **30**, 581-590
73. Laloraya, M., Kumar, G. P., and Laloraya, M.M. (1988) *Proc. Natl. Symp. on Gen. Comp. Endocrinol.* 43-45
74. Viggiano, M., Franchi, A. M., Zicari, J. L., Rethri, V., Gimeno, M. A. F., Kozlowsk, G.P., and Gimeno, A.L. (1989) *Prostaglandins* **37**(3), 367-378
75. Carvalho, B. S., Yeik, and Murdoch, W. J. (1989) *Prostaglandins* **37**, 553-557
76. McCann, T.J. (1988) *J. Rep. Ferti.* Series No:1, Ab 19, July, (suppl.)
77. Milvae, R. A., Alila, H. W., and Hansel, W. (1986) *Biol. Reprod.* **35**, 1210-1216
78. Jenner, L. J., Parkinson, T. J., and Lamming, G.E. (1988) *J. Endocr.* **119**, Abstr. B6 (suppl.)
79. Jenner, L. J., Parkinson, T. J., and Lamming, G.E. (1989) *J. Rep. Fert.* Abstract Series No. 4
80. Wechsung, E., and Hourenaghal, A. (1987) *Prostaglandins, Leukotrienes and Med.* **27**, 119-127
81. Reddanna, P., and Reddy, C.C. (1988c) *Proc. Natl. Symp. on Gen. & Comp. Endocrinol.* PF, 46-47
82. Mitchell, M.D., and Grzyboski, C.F. (1987) *Prostaglandins, Leukotrienes and Med.* **28**, 303-312
83. Pesty, A., Grosse, B., Lieberhen, M., and Acker, G.M. (1988) *J. Rep. & Ferti.* Dec., 77
84. Jubiz, W., Rodmark, O., Malmsten, C., Hansson, G., Lindgrm, J. A., Palmblad, J., Uden, A.M., and Samuelsson, B. (1982) *J. Biol. Chem.* **257**, 6106-6110
85. Tawfik, and Dey (1988) *Prostaglandins* **35**, 379

86. Kennedy, T.G. (1989) *J. Rep. Fer.* **3**, 15
87. Laloraya, M., Kumar, G. P., and Laloraya, M. M. (1989) *Int. Symp. on Biol. Oxid. Syst.* held at Bangalore on Oct. 23-26, 1989
88. Barry, M., Weichman, and Tucker, S.S. (1982) *Prostaglandins* **24**, 245-253
89. Peck, M. J., Piper, P. J., and Williams, T.J. (1981) *Prostaglandins* **21**, 315-322
90. Lewis, R. A., Austen, K. F., Drazen, J. M., Clark, D. A., Marfat, A., and Corey, E.J. (1980) *Proc. Natl. Acad. Sci. USA* **77**, 3710-3714
91. Hanna, C. J., Bach, M. K., Rore, P. D., and Schellenbery, R.R. (1981) *Nature* **290**, 343-344
92. Fleisch, J. H., Rinkema, L. E., and Baken, S.R. (1982) *Life Sci.* **31**, 577-581
93. Weichman, B.M., and Tucker, S.S. (1982) *Prostaglandins* **24**, 245-253
94. Chang, J. B., Land, D., Bewtra, A., and Townly, R.G. (1985) *The J. Pharmacol. & Exptl. Therapentin.* **232**, 80-87
95. Mitchell, M. D., Grzyboski, C. F., Dedha, C.M., and Hunter, J.A. (1987) *Prostaglandins, Leukotrienes and Med.* **27**, 197-208
96. Goldin, E., Harel, S., Tomer, A., and Yavin, E. (1987) *J. Neurochem.* **48**(3):695-701
97. Chandana, D. R. (1990) Abstract of the paper presented in *International Conf., on Perspectives in Primate Reproductive Biology*, held at Bangalore, Feb. 5-7, 1990

Endogenous Formation of Leukotriene C_4 *In Vivo* and Characterization of Novel Metabolites Formed *In Vitro* and *In Vivo*

Sven Hammarström, Kerstin Bernström, Andrea Keppler, Lars Örning and Gerald L. Larsen
Department of Cell Biology, Division of Medical and Physiological Chemistry Faculty of Health Sciences, S-581 85 Linköping, Sweden

Introduction

The biological properties of leukotriene $(LT)C_4$, a proposed mediator of anaphylaxis, comprise induction of airway obstruction, hypotension, plasma extravasation, and constriction of coronary arteries (1, 2).

Previous investigations on the metabolism of leukotrine C have revealed that extensive catabolism and rapid clearance from the general circulation occur predominantly by hepatic uptake and biliary/fecal excretion (3). The major biliary leukotriene C metabolites were identified as LTD_4 in the guinea pig (4), LTE_4 in the monkey (5) and N-acetyl LTE_4 in the rat (6). LTE_4 was also identified as a urinary metabolite in man (7). Evidence suggesting enterohepatic circulation of a part of the biliary LTs has been published (8, 9). Studies on the distribution and metabolism of ^3H-LTC_3 in the mouse (10) showed that LTs C_3, D_3 and E_3 could be detected in lung, liver, intestine, kidney and blood for 1 h or more after isotope administration. The biliary LTs were further processed by microorganisms in the gut yielding several unidentified products (10). Moreover, administration of LTC_4 or E_4 to conventional and germfree rats (6) showed that fecal extracts from conventional rats contained lower amounts of N-acetyl LTE_4 and higher amounts of both more and less polar metabolites. This suggested involvement of the intestinal microflora in the metabolism of cysteinyl LTs. The present chapter reviews evidence for endogenous LTD_4 production *in vivo* in anaphylactic guinea pigs. Transformations of LTE_4 and N-acetyl-LTE_4 to ω-oxidized metabolites by subcellular fractions of rat liver are also described and evidence is presented for the conversion of LTC_4 to ω-oxidized metabolites *in vivo*. Finally, evidence is presented for the enzymatic transformation of LTE_4 to a novel,

Biological Oxidation Systems, Volume 2

reactive metabolite by purified cysteine-conjugate β-lyase from *Eubacterium limosum* (11) and by microorganisms present in rat cecum.

Guinea Pig Experiments

Guinea pigs were sensitized 14-16 days before antigen challenge by i.p. injection of 10 mg/kg of ovalbumin or by i.p. injection of 3 μg/kg of ovalbumin plus 300 mg/kg of $Al(OH)_3$. The developing antibodies (12) were IgG with the first procedure and IgE plus IgG in the second. To minimize contribution of trauma-induced LT production (13) bile was collected as follows: a) the bile duct was incised and an anastomosis was inserted between the hepatic part of the bile duct and the duodenum two days before immunological challenge. Upon closure of the abdominal wall, a loop of the anastomosis tubing was placed subcutaneously to allow subsequent access. b) The bile duct of diazepam-treated, pentobarbital-anesthetized animals was cannulated. The animals were kept anesthetized under a heating lamp for 3 hours.

Animals received either an LTD_4/E_4 receptor antagonist (LY 171883 sodium salt, (14)) 30 min before challenge (method a) or an H-2 antagonist (pyrilamine maleate) at the time of antigen administration (method b). Bile was collected in methanol/water (9:1, v/v) and aliquots of the supernatants, were evaporated and subjected to reverse phase-high performance liquic chromatography (RP-HPLC) after addition of $[^3H_8]LTD_4$ (315 pCi, 6.6 fmol) as a standard. RP-HPLC fractions were neutralized with 1M K_2CO_3 and stored at -20°C under argon for up to 5 days prior to radioimmunological analysis.

The LTC_4-antibody used cross-reacted with LTs and analogs as follows: LTC_4, 100%; LTD_4, 44%; LTE_4, 6%; LTC_4 sulphone, 11%; LTD_4 sulphone, 8%; 5-HETE, 0.04%; and LTB_4, 0.003% (relative percent, molar basis at 50% binding). The lower detection limit of the radioimmunoassay was 50 fmol for LTC_4 and 100 fmol for LTD_4. LTC_4 and LTD_4 standard curves, were constructed using standards freshly purified by RP-HPLC. Neutralized RP-HPLC fractions (0.6-1 ml), which coeluted with the $[^3H]LTD_4$ standard, were analyzed.

$[^3H_8]LTC_4$ was injected i.v. on bile fistulated guinea pigs. RP-HPLC analysis of bile extracts indicated extensive conversion to predominantly LTD_4. Polar metabolites eluting close to the solvent front, LTE_4, and LTC_4 were also detected. Calculations of the cumulative excreted radioactivity for each component in bile *vs.* time showed that 15% of the injected dose was LTD_4, after 15 min, 18% after 30 min and 20% after 3 h.

Analyses of biliary cysteine-containing LTs during anaphylactic shock were performed by RP-HPLC/radioimmunoassay in several experiments. Anaphylaxis was elicited by i.v. administration of a low dose of antigen. In most animals, this led to the development of clear signs of shock as evidenced by breathing difficulties, cyanosis, and micturion. Bile sampling began at least 30 min before antigen challenge. The prechallenge bile samples provided control levels of LTs for each animal. Analysis of prechallenge bile by RP-HPLC/radioimmunoassay (Figure 1, *Bottom panel*) showed only low and evenly distributed immunoreactivity. Corresponding analyses of two bile samples (0-15 and 15-45 min after challenge; Figure 1, *Middle* and *Top panels*, respectively) demonstrated a major immunoreactive component of the same retention time as the $[^3H_8]LTD_4$ standard.

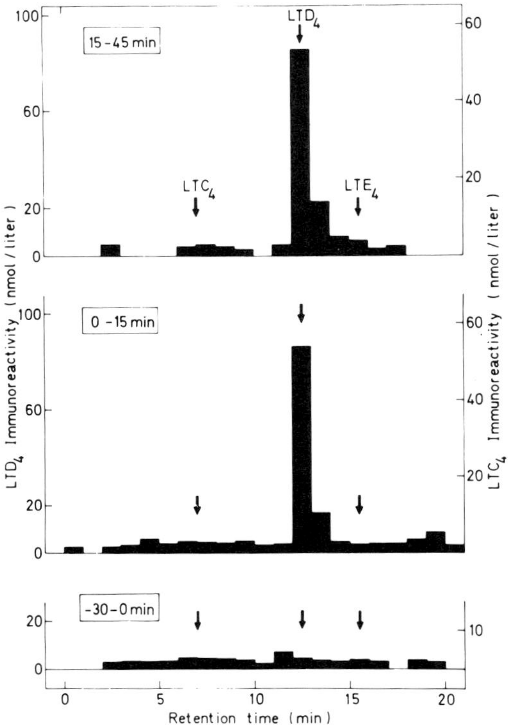

Figure 1. *Radioimmunochromatograms of Biliary Cysteine-Containing LTs in Sensitized Guinea Pigs Before and After Antigen Challenge.* Deproteinized samples, corresponding to 0.1 ml of bile, collected 0-30 min before (Bottom) and 0-15 min (Middle) or 15-45 min after immunological challenge (Top) were analyzed by RP-HPLC/radioimmunoassay (4).

Integration of these chromatograms revealed biliary LTD_4 concentrations of 108 nM and 116 nM, respectively. Radioimmunochromatographic analyses of bile collected 0-15 min after challenge gave similar results. In five separate experiments, a basal LTD_4 concentration in prechallenge bile of 10 ± 1 nM (mean ± SEM, n = 5) rose to 86 ± 10 nM LTD_4 postchallenge (mean ± SEM, P <0.001, n = 5, Figure 2B). After triggering the immunological response the difference was statistically significant. The bile flow decreased to about half, from 83 ± 9 before to 38 ± 4 ml x min^{-1} x kg^{-1} after antigen challenge (mean ± SEM, n = 5, P <0.01, Figure 2A), and the LTD_4 excretion rate rose 3.6-fold from 0.99 ± 0.16 before to 3.18 ± 0.39 pmol x min^{-1} x kg^{-1} after antigen challenge (mean ± SEM, n = 5, P <0.002, Figure 2C).

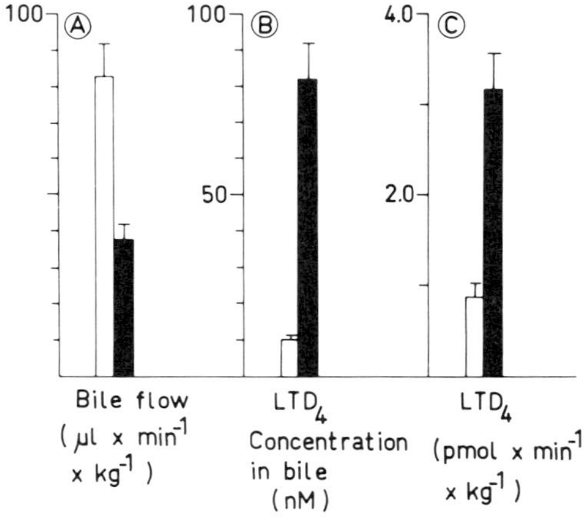

Figure 2. *Effects of Antigen Challenge on Bile Flow (A), biliary LTD_4 concentration (B), and biliary LTD_4 excretion rate (C).* Open bars, prechallenge values; closed bars, postchallenge values. The bars represent mean values ± SEM from five experiments (4).

Part of the pooled LTD_4-containing fractions from RP-HPLC fractionated bile 0-30 min after the challenge was treated with an enzyme preparation containing γ-glutamyltransferase and dipeptidase. Glutathione was also added to force the former reaction in the reverse direction (15). Analysis of the reaction products by RP-HPLC and radioimmunoassay revealed two immunoreactive components of the same retention times as LTE_4 and LTC_4. These results provided further evidence that the endogenously excreted LT was LTD_4.

Rat Experiments

Metabolism of LTE_4 and N-Acetyl LTE_4 by Subcellular Rat Liver Fractions

[$^3H_8$]LTE_4 (1-9 μM; 0-1 μCi), N-acetyl-[$^3H_8$]LTE_4 (1-9 μM;0-0.5 μCi), LTB_4 (4 μM), LTC_4 (5 μM), or LTD_4 (5 μM) were dissolved in ethanol/water 1:3 (v/v) and added to rat liver microsomes. The reaction was started by adding NADPH (2 mM) or an NADPH-regenerating system. Incubations were performed at 37°C for 1-60 min under air, 93% O_2 plus 7% CO_2, or argon. In some experiments, NAD^+ (1 mM) and one volume of rat liver cytosol were added after 30-60 min and the incubation was continued for 15 min.

Figure 3A shows a RP-HPLC chromatogram of labeled products obtained upon incubation of rat liver microsomes, fortified with NADPH, and [$^3H_8$]LTE_4. Compounds I and II had retention times of 0.16 and 0.20 relative to LTE_4. The amount of compound I usually exceeded that of compound II, when regular microsomes were used. When washed microsomes were used, compound II was the major one. Compounds I and II were found to have the structures 5(S)-

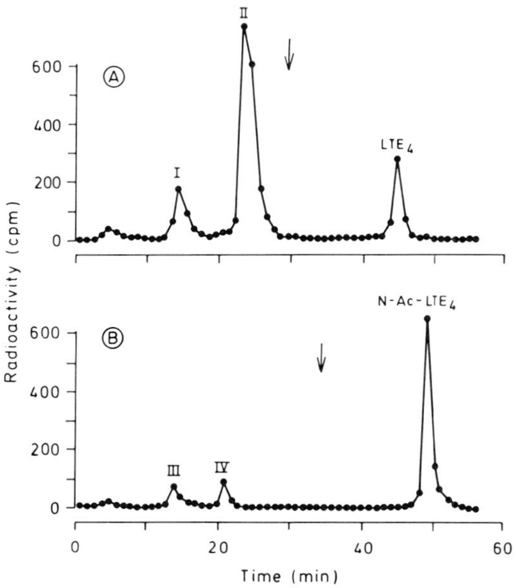

Figure 3. *RP-HPLC Chromatograms of Products Obtained by Incubating Rat Liver Microsomes with (A) [$^3H_8$]LTC_4 and (B) N-acetyl [$^3H_8$]LTE_4.* Incubation conditions: (A), 18 nmol, 0.9 mCi (B), 8 nmol, 0.4 mCi, 37°C, 30 min. Arrows indicate change of solvent. RP-HPLC: see *Materials and Methods* (17).

hydroxy-6(R)-S-cysteinyl-7, 9-*trans*-11, 14-*cis*-eicosatetraen-1, 20-dioic acid (ω-carboxy-LTE_4) and 5(S), 20-dihydroxy-6(R)-S-cysteinyl-7, 9-*trans*-11, 14-*cis*-eicosatetraenoic acid (ω-hydroxy-LTE_4), respectively.

N-acetyl-[$^3H_8$]LTE_4 was converted to two polar products when incubated with rat liver microsomes in the presence of NADPH (Figure 3B). The products, compounds III and IV, were eluted at retention times of 0.16 and 0.20 relative to that of N-acetyl-LTE_4 upon RP-HPLC. The amount of compound III usually exceeded that of compound IV when regular microsomes were used, while compound IV predominated when washed microsomes were used. Compound III was tentatively assigned the structure 5(S)-hydroxy-6(R)-S-N-acetyl-cysteinyl-7, 9-*trans*-11, 14-*cis*-eicosatetraen-1, 20-dioic acid (N-acetyl-ω-carboxy-LTE_4) and the structure of compound IV was established as 5(S), 20-dihydroxy-6(R)-S-N-acetyl-cysteinyl-7, 9-*trans*-11, 14-*cis*-eicosatetraenoic acid (N-acetyl-ω-hydroxy-LTE_4).

Enzymes Catalyzing the Formation of Compounds I-IV

The structures of compounds II and IV indicated that they had been formed by an ω-hydroxylase. The nature of this enzyme (referred to as LTE ω-hydroxylase) was investigated further.

The conversion of LTE_4 or N-acetyl-LTE_4 into ω-hydroxy and ω-carboxy metabolites was dependent on the presence of NADPH and molecular oxygen (Figure 4A). The reaction rate was high for about 10 min and then decreased considerably. Re-addition of NADPH had no significant effect, whereas a second addition of microsomes increased product formation by about 50%.

Substrate specificity studies indicated that LTE_4 was the most efficient substrate: the specific activity with N-acetyl-LTE_4 was 16% of that with LTE_4 and no activity was exhibited towards LTC_4 and LTD_4.

Figure 4B shows the time course for the formation of ω-hydroxy and ω-carboxy metabolites from LTE_4 by microsomes. It depended on the concentrations of microsomes and LTE_4 (Figure 4C). Apparent K_m and V_{max} values were estimated to be 2-3 mM and *ca* 1.2 nmol x min^{-1} x (g of tissue)$^{-1}$, respectively, by non-linear regression analysis.

In order to compare LTE ω-hydroxylase with LTB ω-hydroxylase [$^3H_8$]LTE_4 was incubated with liver microsomes in the presence or the absence of unlabeled LTB_4 (or LTE_4; Figure 4D). In the presence of a 10-fold excess of LTB_4 somewhat enhanced ω-oxidation of [$^3H_8$]LTE_4 was observed. By contrast, the same excess of unlabeled LTE_4 decreased the formation of ω-oxidized metabolites by 73%. The unlabeled LTB_4 was converted into ω-hydroxy-LTB_4 (>90%) + ω-carboxy-LTB_4 (<10%) with a specific activity of 3.4 nmol x min^{-1} x (g liver)$^{-1}$ and the mixture of labeled and unlabeled LTE_4 was converted at a rate of 0.9 nmol x min^{-1} x (g liver)$^{-1}$ (inset, Figure 4D). These results strongly suggest that ω-hydroxylations of LTB_4 and LTE_4 are catalyzed by separate enzymes.

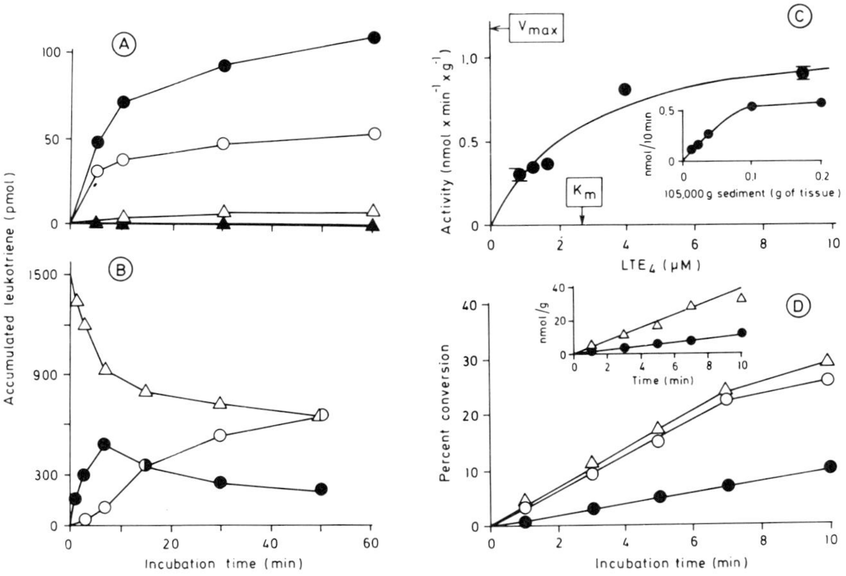

Figure 4. *Characteristics of ω-Oxidation Reaction.* A: Time course of ω-oxidation of N-acetyl-LTE_4 by liver microsomes in the absence (▲), or presence of NADPH, under argon (△), in the presence of NADPH, under normal atmosphere (○), and in the presence of NADPH, under 93% O_2 plus 7% CO_2 (●). B: Time course of ω-oxidation of LTE_4 by liver microsomes supplemented with a NADPH-regeneration system. LTE_4 (△), ω-hydroxy-LTE_4 (●), and ω-carboxy-LTE_4 (○). C: Effect of concentrations of LTE_4 and liver microsomes on the formation of ω-hydroxy-LTE_4 + ω-carboxy-LTE_4. The inset shows the effect of microsome concentration on activity at fixed concentrations of LTE_4 (1 mM) and NADPH (2 mM). D: Effects of LTB_4 and LTE_4 on ω-oxidation of [$^3H_8$]LTE_4. [$^3H_8$]LTE_4 (65 nCi, 414 pmol) was incubated with liver microsomes (corresponding to 80 mg tissue/ml) at 37°C for 1-10 min, in the absence (○) or in the presence of unlabeled LTB_4 (4.15 nmol; △) or unlabeled LTE_4(4.15 nmol; ●). The inset shows formation of ω-hydroxy-LTB_4 + ω-carboxy-LTB_4 (△) and ω-hydroxy-LTE_4 + ω-carboxy-LTE_4 (●) in the incubation mixtures (17).

Liver cytosol, fortified with NAD^+, quantitatively oxidized ω-hydroxy-LTE_4 and N-acetyl-ω-hydroxy-LTE_4 to the corresponding ω-carboxy compounds.

When washed microsomes, supplemented with acetyl-coenzyme A (50 μM), were incubated with ω-hydroxy-LTE_4 or ω-carboxy-LTE_4 (1 nmol) at 37°C for 15 min (*cf.* 16) these compounds were converted to products which co-chromatographed with N-acetyl-ω-hydroxy-LTE_4, and N-acetyl-ω-carboxy-LTE_4 respectively.

In Vivo Excretion of ω-Oxidized Metabolites of LTC_4 in Bile and Feces

Following intraveneous injection of 40 nmol $[^3H_8]LTC_4$ to conventional rats having a bile duct cannula, 28% of the given radioactivity appeared in bile after 30 min, and 68% after 6 hours. RP-HPLC analyses of bile, collected from 0 to 30 min after isotope administration showed 5 radioactive components (Compounds 1-5, Figure 5).

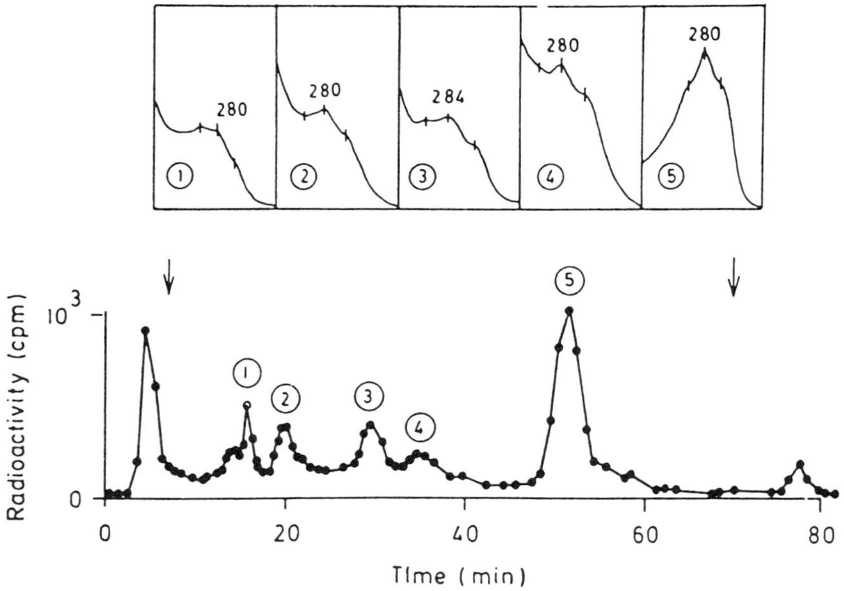

Figure 5. *RP-HPLC Analysis of LTC_4 Metabolites from Rat Bile.* Insets, UV spectra of separated metabolites. Arrows indicate changes of mobile phase, from solvent 1 to solvent 3 (first arrow) and from solvent 3 to methanol (second arrow) (56).

When re-analyzed by RP-HPLC, compound 1 had a retention time of 0.15 relative to that of N-acetyl-LTE_4. The UV spectrum suggested that the conjugated triene ($\Delta^{7\ 9\ 11}$) and the allylic thioether substituent at C-6 had been retained. Compound 1 was identified as N-acetyl-ω-carboxy-LTE_4 by cochromatography with biosynthetic reference compound using regular, ion-pair, and Radial Pak RP-HPLC.

Compound 2 was resolved into two components (2A and 2B) upon re-analysis by RP-HPLC. Compound 2B was identified as N-acetyl-ω-hydroxy-LTE_4 by cochromatography with biosynthetic reference compound using regular RP-HPLC, ion-pair and Radial Pak RP-HPLC.

Compounds 3 and 4 consisted of two components each. Approximately 2/3 of compound 3 was due to LTC_4 sulfoxide whereas the remaining 1/3 was eluted at a retention time of 0.92 relative to LTC_4 sulfoxide. Compound 4 consisted to about 70% of LTC_4 and 30% of N-acetyl LTE_4 sulfoxide.

Compound 5 consisted of 80% N-acetyl-LTE_4 and 20% LTD_4. Sixteen percent of the administered dose was recovered as N-acetyl-LTE_4 plus LTD_4 in 0-3 hour bile.

Previously, N-acetyl-LTE_4 and N-acetyl-11-*trans*-LTE_4 were identified as major fecal LTC_4 metabolites in germfree and conventional rats (6). Several other metabolites, including those designated as compound I, IV and V were also observed (6). The UV spectrum of compound I had a λ_{max} at 280 nm and shoulders at 270 and 292 nm. Reductive desulfurization and analysis of the methylester, trimethylsilyl ether derivative by gas-liquid radio-chromatography showed an equivalent chain length of 25.1, *i.e.* the same as that of the methylester, trimethylsilyloxy derivative of 5-hydroxy-1, 20-eicosadioic acid (17). Compound I coeluted with N-acetyl-ω-carboxy-LTE_4 on regular and ion-pair RP-HPLC. Based on these results, compound I was identified as N-acetyl-ω-carboxy-LTE_4. After 72 hours, 3.5% of the administered radioactivity had been eliminated as N-acetyl-ω-carboxy-LTE_4. Compounds IV and V (6) eluting at retention times of 0.65-0.75 relative to that of N-acetyl-LTE_4 were identified as N-acetyl-LTE_4 sulfoxide and N-acetyl-11-*trans*-LTE_4 sulfoxide by UV spectroscopy (λ_{max}; at 284 and 282 nm, respectively) and coelution with synthetic reference compounds.

Metabolism of LTE_4 by Bacterial Cysteine- Conjugate β-Lyase

When $[^3H_8]LTE_4$ was incubated with β-lyase, isolated from *E. limosum* (18), a rapid disappearence of LTE_4 from the incubation mixture was observed by RP-HPLC analyses of aliquots. No distinct metabolites of LTE_4 were detected under these conditions. When the reaction was stopped by the addition of iodoacetic acid (19), 80% of the added tritium was recovered: 23% as two new metabolites and 18% in the methanol eluate from RP-HPLC. Figure 6 shows a chromatogram of the products (compounds Ia and Ib: retention times relative to that of LTE_4: 1.40 ± 0.03 (mean ± SEM, n=10) and 1.46 ± 0.03 (n=7), respectively). When the enzymatic reaction was carried out under argon in the presence of dithiothreitol and stopped with iodoacetic acid the amount of compound Ia increased and that of compound Ib decreased. If LTE_4 was incubated with boiled (10 min) enzyme compounds Ia and Ib were not detected but around 35% of LTE_4 was converted into 11-*trans*-LTE_4.

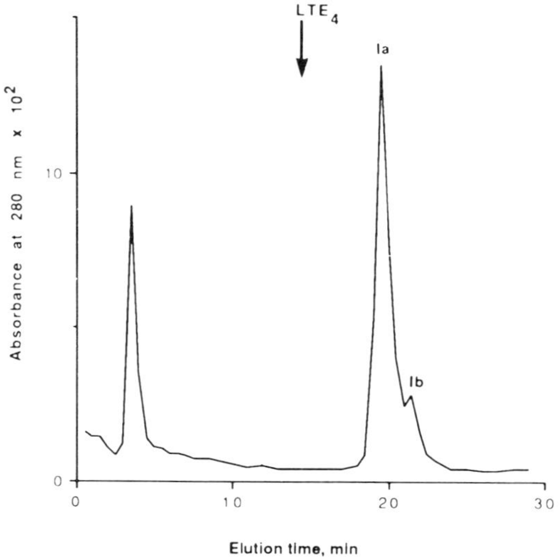

Figure 6. *RP-HPLC of Products Obtained After Incubating LTE_4 with β-Lyase from* E. limosum *(37°C, 30 min), and Treating the Reaction Mixture with 4 μmol Iodoacetic Acid in 0.1 M $NaHCO_3$ for 30 min, Room Temperature.* Conditions for HPLC: Column, C_{18}-Nucleosil (5 μm particles; 250 x 4.6 mm); solvent, methanol/water/acetic acid/-phosphoric acid (75/25/0.07/0.03) adjusted to pH 5.4 with NH_4OH (1 ml/min) (57).

Compound Ia had an ultraviolet spectrum identical to that of LTE_4. Upon analysis of the methyl ester, trimethylsilyl derivative of compound Ia by gas-liquid radiochromatography, the equivalent chain-length was 27.6. This is 6.1 C higher than the equivalent chain-length of the product obtained from desulfurized and hydrogenated LTE_4 (methyl ester, trimethylsilyl ether derivative; 21.5 C (20)), and 4.1 C higher than that of LTB_4 (methyl ester, trimethylsilyl ether derivative; 23.6 C (21)). The fact that compound Ia could be analyzed by GC indicated that the cysteinyl-part of LTE_4 had been modified by β-lyase. 5-Hydroxy-6-S-carboxymethylthio-7, 9-*trans*-11, 14-*cis*-eicosatetraenoic acid monomethyl ester (reference compound 1 methyl ester) was synthesized from LTA_4, methyl ester and α-mercaptoacetic acid (22). The product was purified by RP-HPLC and its structure was verified by UV spectroscopy and gas-liquid chromatography-mass spectrometry (GC-MS). The electron impact mass spectrum of the methyl ester, trimethylsilyl ether derivative showed ions at m/z 479 $[M-OCH_3]^+$, 452 $[M-58]^+$, 437 $[M-CH_2COOCH_3]^+$, 421 $[M-(58+OCH_3)]^+$, 404 $[M-SHCH_2COOCH_3]^+$, 389 $[M-(OCH_3+HOSi(CH_3)_3)]^+$, 362 $[M-(58-HOSi(CH_3)_3]^+$, 331 $[M-(58+OCH_3+HOSi(CH_3)_3)]^+$, 314 $[M-(SHCH_2COOCH_3+HOSi(CH_3)_3)]^+$, 307 $[M-203]^+$, 262, 203 (base peak), 171 (203-$HOCH_3$), 129 $((CH_3)_3Si+$

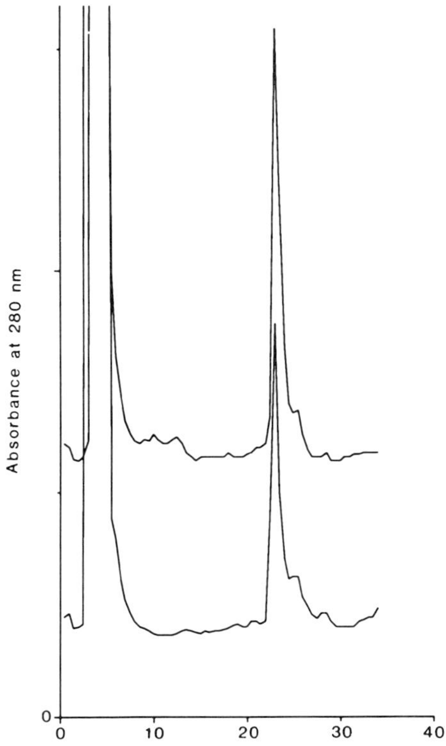

Figure 7. *RP-HPLC of Compound Ia (see Figure 6) and 5-Hydroxy-6-S-carboxymethylthio-7,9-trans-11,14-cis-eicosatetraenoic Acid* (reference compound 1). Conditions for HPLC: the same as described in the legend to Figure 6, but the methanol/water ratio was 73/27. Lower chromatogram: compound Ia alone; upper chromatogram: compound Ia plus reference compound 1 (57).

$OCH_2CH=CH_2$), 113 (203-$HOSi(CH_3)_3$), 105 ($CH_3OOCCH=S$ +H), 99 ,75 ($(CH_3)_2Si+OH$) and 73 ($(CH_3)_3Si)^+$. This verifies the structure 5-hydroxy-6-S-carboxymethylthio-7, 9-*trans*-11, 14-*cis*-eicosatetraenoic acid. Reference compound 1 and compound Ia had identical RP-HPLC retention times (Figure 7) and UV spectra. Based on these results compound Ia was characterized as having the structure 5-hydroxy-6-S-carboxymethylthio-7, 9-*trans*-11, 14-*cis*-eicosatetraenoic acid. Compound Ib also had a triene UV spectrum but the λ_{max} had shifted to 278 nm, suggesting that it was the 11-*trans* isomer of compound 1a (23).

When the β-lyase reaction was stopped with dimethylsulfate (24) rather than iodoacetic acid, two products (compounds IIa and IIb, figure 8) eluting after LTE_4 on RP-HPLC were formed. Relative elution times for compounds IIa and IIb were 2.6 ± 0.2 (n=3) and 2.9 ± 0.2 (n=3) compared to LTE_4 and their UV spectra were similar to those of LTE_4 and 11-*trans*-LTE_4, respectively.

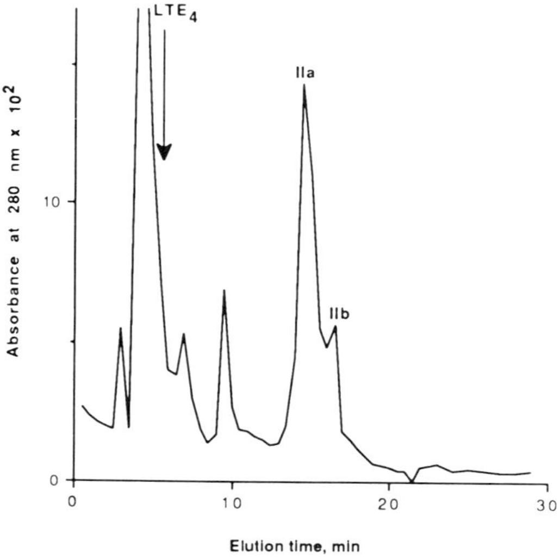

Figure 8. *RP-HPLC of Products Obtained After Incubating LTE_4 with β-lyase from* E. limosum *(37°C, 30 min), and Treating the Reaction Mixture with 2.3 μmol Dimethylsulfate in Dioxane for 30 min at Room Temperature.* Ammonia was added to terminate the latter reaction. Conditions for HPLC: the same as in Figure 6, but the methanol/water ratio was 4/1 (57).

5-Hydroxy-6-S-methylthio-7, 9-*trans*-11, 14-*cis*-eicosatetraenoic acid (methyl ester) (reference compound 2 (methyl ester)) was obtained from LTA_4 (methyl ester of free acid) and methanethiol.

The product was purified by RP-HPLC and its structure was verified by UV spectroscopy, lipoxygenase conversion (20), fast atom bombardment, (FAB), mass spectrometry and reductive desulfurization (25) followed by GC-MS analyses. The electron impact mass spectra of the hydrogenated and desulfurized product (methyl ester, trimethylsilylether derivative, C-value: 21.6) showed ions at m/z 399 $[M-15]^+$, 367 $[M-47]^+$, 313 $[M-101 (CH_2(CH_2)2COOCH_3)]^+$ and 203, the base peak, $[(CH_3)_3SiO{+}{=}CH(CH_2)_3COOCH_3]$, indicating a fatty acid with 20 carbons and a hydroxy group at C-5 (20). The structural assignment was confirmed by negative ion FAB mass spectrometry of reference compound 2. Prominent ions at m/z=365 (base peak), 347, 319 and 301 were interpreted as having the stuctures $[M\text{-}H]^-$, $[M\text{-}H_2O]^-$, $[M\text{-}SCH_3]^-$and $[M\text{-}(SCH_3+H_2O)]^-$, respectively, indicating a molecular weight of 366 and suggesting the structure 5-hydroxy-6-S-methylthio-7, 9-*trans*-11, 14-*cis*-eicosatetraenoic acid. Reference compound 2 had the same retention time on RP-HPLC and the same UV spectrum (Figure 9) as compound IIa. This indicated that the struc-

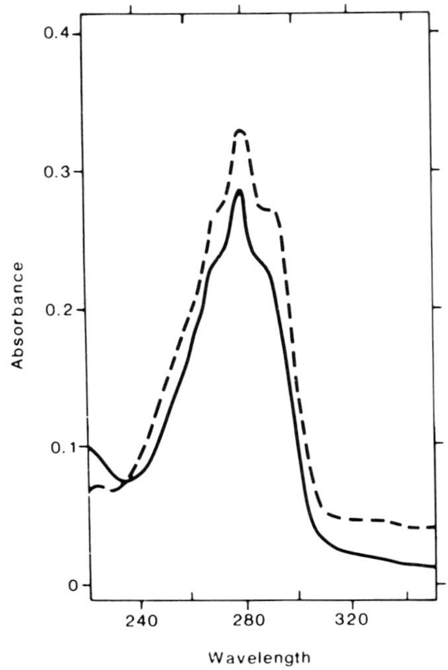

Figure 9. *UV Spectra of Compound IIa (solid line) and Reference Compound 2 (dotted line)* (57)

ture of compound IIa was 5-hydroxy-6-S-methylthio-7, 9-*trans*-11, 14-*cis*-eicosatetraenoic acid. Based on its UV spectrum and retention time on RP-HPLC compound IIb was most likely the 11-*trans* isomer of compound IIa.

The structure of the product formed from LTE_4 by cysteine-conjugate β-lyase was 5-hydroxy-6-mercapto-7, 9, 11, 14-eicosatetraenoic acid, based on the formation of two new adducts when the reaction was stopped with iodoacetic acid or dimethylsulfate. The latter were characterized as having the structures 5-hydroxy-6-S-carboxymethylthio-7, 9-*trans*-11, 14-*cis*-eicosatetraenoic acid and 5-hydroxy-6-S-methylthio-7, 9-*trans*-11, 14-*cis*-eicosatetraenoic acid, respectively.

Cecal contents from rats was incubated with 1.5 μCi $[^3H_8]LTE_4$ and the reactions were terminated with iodoacetic acid. One third of the radioactivity was soluble in methanol. Twenty-seven % of the tritium (0.4 μCi) was found in the 80% ethanol fraction after XAD 8 chromatography and 17% (0.25 μCi) in the methanol fraction after Sep-Pak C_{18} purification. The methanol eluate was analyzed by RP-HPLC. A product constituting 10% of the eluted radioactivity appeared at a retention time of 16 min, or 1.3 times the retention time of LTE_4 (Figure 10, lower chromatogram). This metabolite comigrated with tritium labeled 5-hydroxy-6-S-carboxymethylthio-7, 9-*trans*-11, 14-*cis*-eicosatetraenoic acid, prepared from $[^3H_8]LTE_4$ by incubations with β-lyase as described above on both regular RP-HPLC (Figure 10, upper chromatogram) and ion pair RP-HPLC (not shown). The comigrating metabolite and reference compound together constituted 21% of the eluted radioactivity (compared to 10% in the absence of reference compound, *cf.* above). These results indicated that the untrapped metabolite formed by rat cecal contents was 5-hydroxy-6-mercapto-7, 9, 11, 14-eicosatetraenoic acid.

Discussion

Endogenous Leuotriene D Formation

LTC_4 metabolism in the guinea pig was investigated with the aim of identifying suitable metabolites for analyzing LT formation during anaphylaxis. The results demonstrated that LTD_4 is a major biliary LTC_4 metabolite in the guinea pig. The respiratory, cardiovascular, and microcirculatory effects of LTC_4 and LTD_4 suggest that these compounds can produce the major consequences of acute systemic anaphylaxis (26-29). It has also been documented that inhibitors of LT formation can prevent several anaphylactic responses (30-32). The purpose of this investigation was to develop a suitable animal

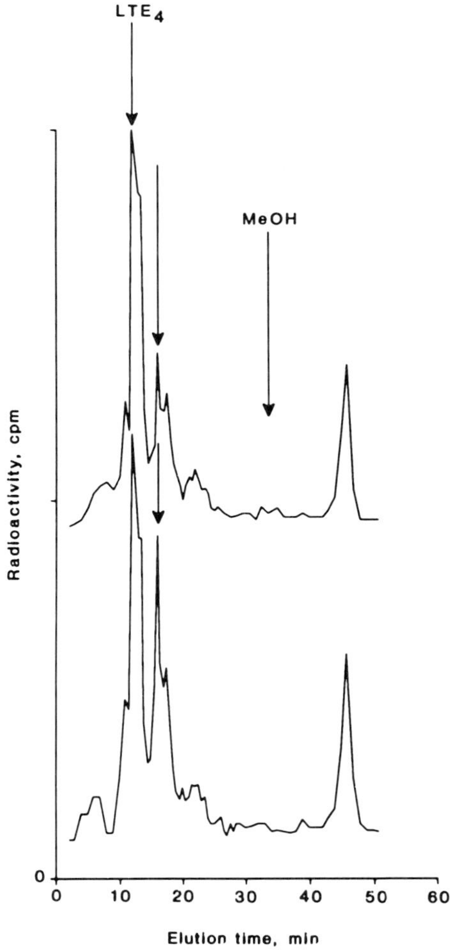

Figure 10. *RP-HPLC of 5-Hydroxy-6-S-carboxymethylthio-7,9-trans-11,14-cis-[^{3}H] eicosatetraenoic Acid Prepared from [$^3H_8$]LTE$_4$ by Treatment with Cysteine-Conjugate β-Lyase and a Metabolite Formed from LTE$_4$ by Rat Cecum Contents.* In both reactions the products were trapped by treatment with iodoacetic acid. Conditions for HPLC: see Figure 6. Lower chromatogram: products from cecum incubation alone (arrow indicates the retention time of reference compound 1); upper chromatogram: products from cecum incubation plus reference compound 2 (arrow indicates retention time of cecum metabolite plus reference compond 1) (57).

model, which would enable the demonstration of *in vivo* lekotriene synthesis during systemic anaphylaxis. Due to the hepatobiliary elimination of LTC_4 in the guinea pig, bile was chosen for these measurements. LTD_4 was identified as the major immunoreactive component in anaphylactic guinea pig bile. The analyses showed an 8.6-fold increase in LTD_4 concentration in bile after challenge compared to bile before the challenge. Antigen provocation was also accompanied by decreased bile flow (to about half) as observed in rats undergoing endotoxin shock (33). After correction for the reduced bile flow, the LTD_4 excretion rate had increased 3.6-fold after antigen challenge of sensitized guinea pigs. The observation that LTD_4 is the predominant endogenous leukotriene metabolite in guinea pig bile was in good agreement with the metabolic tracer studies performed as part of this investigation and in obvious contrast to results obtained in rats and monkeys. In the latter species N-acetyl-LTE_4 (6, 34) and LTE_4 (5), respectively, were the dominating LTC_4 metabolites. It is interesting that LTD_4 predominates in the guinea pig since LTD_4 is more active than LTC_4, N-acetyl-LTE_4, and LTE_4 on several biological parameters (35, 36) and because guinea pigs are especially sensitive to exogenous cysteine-containing LTs (37) as well as to conditions that induce endogenous synthesis of these compounds such as trauma and anaphylaxis.

Excretion of ω-oxidized Metabolites

Incubation of LTE_4 and N-acetyl-LTE_4 with rat liver microsomes and NADPH each led to the formation of two polar metabolites. The results showed that the metabolites were ω-hydroxy-LTE_4, ω-carboxy-LTE_4, N-acetyl-ω-hydroxy-LTE_4, and N-acetyl-ω-carboxy-LTE_4. The reactions demonstrated are analogous to those previously shown for fatty acids (38), prostaglandins (39, 40), 12-hydroxy-eicosatetraenoic acid (41), and 5(S), 12(S)-dihydroxy-eicosatetraenoic acid and LTB_4 (42-44). However, distinct enzymes appear to be involved.

The conversions of LTE_4 and N-acetyl-LTE_4 to ω-hydroxy metabolites required NADPH, molecular oxygen, and microsomes. It was enhanced by starvation of the rats, suggesting that a liver cytochrome P450 might be involved (38). LTE ω-hydroxylase was present primarily in liver microsomes. The enzymatic activity was destroyed within 10-15 min at 37°C during incubations of LTs, but was protected against inactivation by EDTA, EGTA, and dithiotreitol. Time course experiments with $[^3H_8]LTE_4$ and liver microsomes indicated that ω-hydroxy-LTE_4 was an intermediate in the conversion to ω-carboxy-LTE_4.

Substrate specificity studies indicated that LTE_4 and N-acetyl-LTE_4 were substrates for LTE ω-hydroxylase, and suggested that LTE_4 was the preferred substrate. LTC_4 and LTD_4 were not transformed by the enzyme. The apparent Km value of LTE_4 for

LTE ω-hydroxylase was 2-3 mM. This is higher than the Km value of LTB_4 for polymorphonuclear leukocyte LTB ω-hydroxylase (0.2-1.0 mM, (45-47)) but lower than the Km value (40 mM) reported for hepatic LTB ω-hydroxylase (48).

An ω-hydroxy-LTE dehydrogenase requiring NAD^+ as cofactor was detected in rat liver cytosol. The substrate specificity and kinetics of this enzyme are unknown. LTE N-acetyltransferase, present in rat liver microsomes and other organs (16), transformed ω-hydroxy-LTE_4 and ω-carboxy-LTE_4 to N-acetyl derivatives (Figure 11). Recently, other groups have reported ω-oxidation of cysteinyl LTs *in vitro* (49, 50).

Following administration of $[^3H_8]LTC_4$ to bile duct cannulated rats, the recovered biliary radioactive material consisted of at least nine metabolites. Six of these compounds were identified as N-acetyl-LTE_4 (12.8% of the administered dose), LTD_4 (3.2%), LTC_4 sulfoxide (2.7%), LTC_4 (2%), and N-acetyl-ω-hydroxy LTE_4 plus N-acetyl-ω-carboxy-LTE_4 (5% together). No ω-hydroxy-or ω-carboxy-LTE_4 were detected.

A novel metabolite of LTE_4 was isolated after incubations with cysteine-conjugate β-lyase, isolated from the bacterium *E. limosum*. The structure of this metabolite was found to be 5-hydroxy-6-mercapto-7, 9, 11, 14-eicosatetraenoic acid (5, 6-HMETE, Figure 11) based on ultraviolet absorption spectroscopy, behaviour on RP-HPLC, gas-liquid radiochromatography, and comparisons with synthetic reference compounds, characterized by mass spectrometry. 5, 6-HMETE could not be isolated from incubation mixtures without prior derivatization by iodoacetic acid or dimethylsulfate. This was probably due to the reactivity of the free SH-group of the molecule, as has been shown for other products formed by β-lyase (51). The same argument might also explain the low recovery of product and the increased amount of radioactivity in the methanol eluate of the RP-HPLC which probably contained 5, 6-HMETE disulfide. Formation of 5, 6-HMETE was also observed when LTE_4 was incubated with cecal contents from rats. The irreversible binding of 2/3 of the radioactivity to cecal material was probably due to formation of disulfide bonds between the thiol group of 5, 6-HMETE and protein components of the cecal contents. Similar observations have been made concerning thiol-containing metabolites of xenobiotics (51). β-Lyase activity is widely distributed among gastrointestinal bacteria (11). It also occurs in rat liver and kidney (51-53). The mammalian enzyme is specific for S-aryl cysteine-conjugates of xenobiotics whereas bacterial β-lyase is not specific for

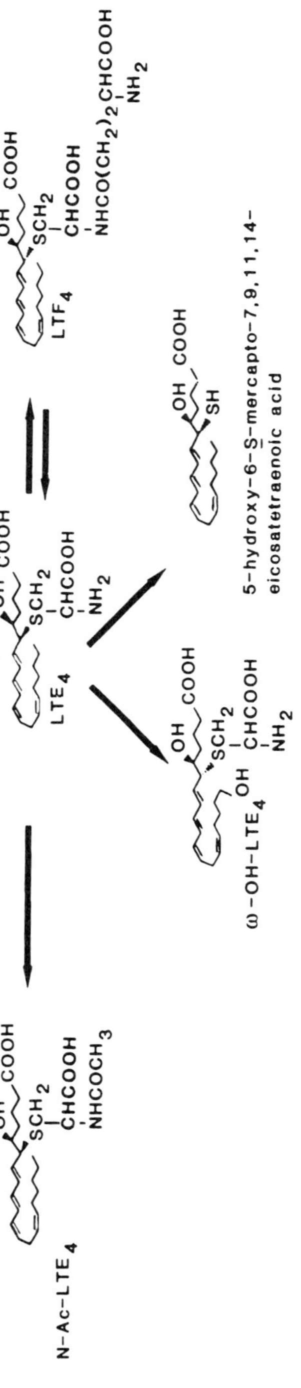
LTC$_4$
LTD$_4$
LTE$_4$
LTF$_4$
N-Ac-LTE$_4$
ω-OH-LTE$_4$
5-hydroxy-6-S-mercapto-7,9,11,14-eicosatetraenoic acid

Figure 11.

aromatic compounds (54). Under the conditions used, rat kidney β-lyase did not convert LTE_4 (Bernström, K. and Ormstad, K., unpublished results). The observation that LTE_4 is a substrate for cysteine-conjugate β-lyase from *E.limosum*, a normal inhabitant of human, rat, poultry and fish bowel (55) suggests that this enzyme participates in the *in vivo* degradation of LTE_4.

Acknowledgements

The work from the authors' laboratories was supported by grants from the Swedish Medical Research Council; project 03X-5914).

References

1. Samuelsson, B. (1983) *Science* **220**, 568-5757
2. Hammarström, S. (1983) *Ann. Rev. Biochem.*, **52**, 355-377
3. Hammarström, S., Örning, L., and Bernström, K. (1985) *Mol. Cell. Biochem.* **69**, 7-16
4. Keppler, A., Örning, L., Bernström, K., and Hammarström, S. (1987) *Proc. Natl. Acad. Sci. USA.* **84**, 5903-5907
5. Denzlinger, C., Guhlman, A., Scheuber, PH., Wilker, D., Hammer, D.K., and Keppler, D. (1986) *J. Biol. Chem.* **261**, 15601-15606
6. Örning, L., Norin, E., Gustafsson, B., and Hammarström, S. (1986) *J. Biol. Chem.* **261**, 766-771
7. Örning, L., Kaijser, L., and Hammarström, S. (1985) *Biochem. Biophys. Res. Commun.* **130**, 214-220
8. Denzlinger, C., Guhlmann, A., Hagmann, W., Scheuber, P.H., Scheyrl, F., Wilker, D., Hammer, D.K. and Keppler, D. (1986) *Prostaglandins, LTs and Medicine* **21**, 321-322
9. Huber, M., Guhlmann, A., Jansen, P.L.M. and Keppler, D. (1987) *Hepatology* **7**, 24-228
10. Appelgren, L.E., and Hammarström, S. (1982) *J. Biol. Chem.* **257**, 531-535
11. Larsen, G.L. (1985) *Xenobiotica* **15**, 199-209
12. Andersson, P. (1982) *Br. J. Pharmacol.* **77**, 301-307
13. Denzlinger, C., Rapp, S., Hagmann, W., and Keppler, D. (1985) *Science* **230**, 330-332
14. Fleisch, J.H., Rinkema, L.E., Haisch, K.D., Swanson-Bean, D., Goodson, T., Ho, P.P.K., and Marshall, W.S. (1985) *J. Pharmacol. Exp. Therap.* **233**, 148-157
15. Hammarström, S. (1981) *J. Biol. Chem.* **256**, 9573-9578
16. Bernström, K. and Hammarström, S. (1986) *Arch. Biochem. Biophys.* **244**, 486-491
17. Örning, L. (1987) *Eur. J. Biochem.* **170**, 77-85

18. Larsen, G.L. and Stevens, J.L. (1986) *Mol. Pharmacol.* **29**, 97-103
19. Reed, D.J., Babson, J.R., Beatty, P.W., Brodie, A.E., Ellis, W.W. and Potter, D.W. (1980) *Anal. Biochemistry* **106**, 55-62
20. Murphy, R., Hammarström, S., and Samuelsson, B. (1979) *Proc. Natl. Acad. Sci. USA.* **76**, 4275-4279
21. Borgeat, P. and Samuelsson, B. (1979) *J. Biol. Chem.* **254**, 7865-7869
22. Corey, E.J., Clark, D.A., Goto, G., Marfat, A., Mioskowski, C., Samuelsson, B. and Hammarström, S. (1980) *J. Am. Chem. Soc.* **102**, 1436-39
23. Crombie, L. and Jacklin, A.G. (1957) *J. Chem. Soc.* **2**, 632-1646
24. Eyem, J., Sjödahl, J. and Sjöquist, J. (1976) *Anal. Biochemistry* **74**, 359-68
25. Balazy, M., and Murphy, R.C. (1986) *Anal. Chem.* **58**, 1098-1101
26. Lewis, R.A., and Austen, K.F. (1984) *J. Clin. Invest.* **73**, 889-897
27. Piper, P.J. (1984) *Physiol. Rev.* **64**, 743-761
28. Feuerstein, G. (1985) *J. Auton. Pharmac.* **5**, 49-168
29. Lefer, A.M. (1986) *Biochem. Pharmacol.* **35**, 123-127
30. Aehringhaus, U., Peskar, B.A., Wittenberg, H.R., and Wölbling, R.H. (1983) *Br. J. Pharmacol.* **80**, 73-80
31. Hedqvist, P., Dahlén, S-E., and Palmertz, U. (1984) *Prostaglandins* **28**, 604-608
32. Dahlén, S-E., Palmertz, U.J., and Hedqvist, P. (1985) *Agents and Actions* **17**, 310-311
33. Hagmann, W., Denzlinger, C., and Keppler, D. (1984) *Circ. Shock.* **14**, 223-235
34. Hammarström, S., Örning, L., Bernström, K., *et al.* (1985) in *Prostaglandins, LTs, and Lipoxins* (Bailey, J.M., ed) pp. 69-77. Plenum Press, New York
35. Lefer, A.M. (1985) *Fed. Proc.* **44**, 275-280
36. Lewis, R.A., Drazen, J.H., Austen, K.F., *et al.* (1981) *Proc. Natl. Acad. Sci. USA.* **78**, 4579-4583
37. Feuerstein, G. (1984) *Prostaglandins* **27**, 781-802
38. Ellin, Å., and Orrenius, S. (1975) *Mol. Cell. Biochem.* **8**, 9-79
39. Oliw, E., Fahlstadius, P., and Hamberg, M. (1986) *J. Biol. Chem.* **261**, 9216-9221
40. Oliw, E., Granström, E., and Ängård, E. (1983) in *Prostaglandins and related substances* (Pace-Asciak, C. and Granström, E., eds) pp. 1-44. Elsevier, Amsterdam
41. Wong, P., Westlund, P., Hamberg, M., Granström, E., Chao, P., and Samuelsson, B. (1984) *J. Biol. Chem.* **259**, 2683-2686
42. Lindgren, J-Å., Hansson, G., and Samuelsson, B. (1981) *FEBS Lett.* **128**, 329-335
43. Hansson, G., Lindgren, J-Å., Dahlén, S-E., Hedqvist, P., and Samuelsson, B. (1981) *FEBS Lett.* **130**, 107-112
44. Harper, T.W., Gauity, M.J., and Murphy, R.C. (1986) *J. Biol. Chem.* **261**, 5414-5418

45. Powell, W.S. (1984) *J. Biol. Chem.* **259**, 3082-3089
46. Soberman, R.J., Harper, T.W., Murphy, R.C., and Austen, K.F. (1985) *Proc. Natl. Acad. Sci. USA.* **82**, 2292-2295
47. Shak, S., and Goldstein, I.M. (1985) *J. Clin. Invest.* **76**, 1218-1228
48. Romano, M.C., Eckardt, R.D., Benders, P.E., Leonard, T.B., Straub, K.M., and Newton, J.F. (1987) *J. Biol. Chem.* **262**, 1590-1595
49. Rokach, J., Foster, A., Fitzsimmons, B., Leblanc, Y., Delorme, D., and Girard, Y. (1989) *Advances in Prostaglandin, Thromboxane and Leukotriene Research* **19**, 102-107
50. Stene, D.O., and Murphy, R.C. (1988) *J. Biol. Chem.* **263** 2773-2778
51. Stevens, J. and Jacoby, B. (1983) *Mol. Pharmacol.* **23**, 761-765
52. Tateishi, M., Suzuki, S. and Shimizu, H. (1978) *J. Biol. Chem.* **253**, 8854-8859
53. Jacoby, W.B., Stevens, J., Duffel, M.W., and Weisiger, R.A. (1984) *Rev. Biochem. Toxicol.* **6**, 7-115
54. Bakke, J. (1986) in *American Chemical Society (ACS) Symposium series* (Paulson, G.D, Caldwell, J., Hutson, D.H., and Henn, J.J., eds.), Vol. 299, pp. 301-321. Van Nostrand, New York
55. Buchanan, R.E., and Gibbons, N.E. (Eds.) (1974) *Bergey's Manual of Determinative Bacteriology*, 8th Ed, p. 644. The Williams and Wilkins Company, Baltimore
56. Örning, L., Keppler, A., Midtvedt, T., and Hammarström, S. (1988) *Prostaglandin* **35**, 493-501
57. Bernström, K., Larsen, G.L., and Hammarström, S. (1989) *Arch. Biochem. Biophys.* **275**, 531-539

CHAPTER VIII

Lipid Peroxidation and other Nonenzymatic Reactions of Oxygen

Vitamin E *in vivo*

K. U. Ingold and G. W. Burton
Division of Chemistry, National Research Council of Canada, Division of Chemistry
Ottawa K1A 0R6 Canada

We have proved that natural vitamin E, *2R,4'R,8'R-α*-tocopherol (Figure 1) is certainly the major and probably the sole lipid-soluble, radical-trapping antioxidant present in human blood (1). That is, vitamin E (α-TOH) inhibits lipid peroxidation *in vivo* by trapping peroxyl radicals generated in biological membranes (eqs. 1 and 2) and it would be fair to say that we do not become rancid until it is our turn to become food thanks to vitamin E.

Figure 1.

Natural (2*R*, 4'*R*, 8'*R*) α-Tocopherol

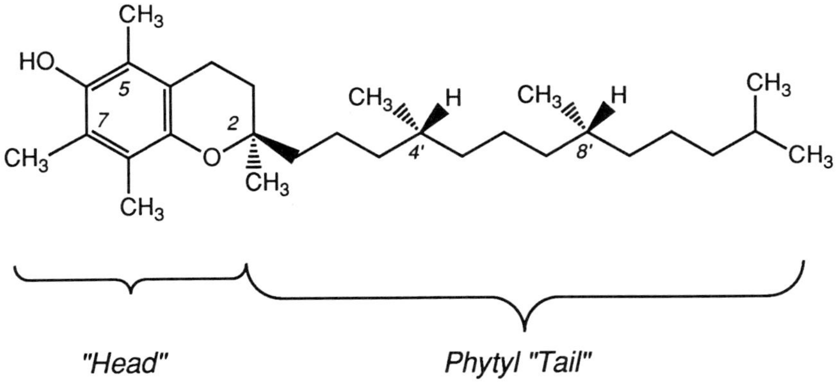

$$\mathrm{ROO\cdot + \alpha\text{-}TOH \rightarrow ROOH + \alpha\text{-}TO\cdot} \quad (1)$$

$$\mathrm{ROO\cdot + \alpha TO\cdot \rightarrow Non\text{-}radical\ products} \quad (2)$$

Our extensive studies on this vitamin (2) began with an *in vitro* demonstration that α-tocopherol was the best peroxyl radical trapping, phenolic antioxidant known at that time. In recent years, our research has moved increasingly to *in vivo* studies in animals and humans.

Which is Best, the Free Phenol or the Acetate?

One of the more intriguing *in vivo* questions we have answered is posed by the title of this section. In its natural form in food, vitamin E is present as α-tocopherol, *i.e.*, as the free phenol. However, commercial vitamin E is generally sold as α-tocopheryl acetate since this is much more airstable than the phenol. The acetate is not a peroxyl radical trap nor is it absorbed from the intestine; instead it is first hydrolyzed to the phenol and it is the phenol which is absorbed.

To the companies that manufacture vitamin E it is naturally a matter of some concern whether the acetate is as effective a source of the vitamin as the free phenol. For 40 years they have dealt with this problem using the classical rat fetal resorption bioassay (3). The protocol for this bioassay involves placing a large number of female rats on a vitamin E-deficient diet for 2-3 months, the rats are then mated with vitamin E-sufficient males, impregnation and implantation proceed normally but, if the female does not receive vitamin E, the fetuses will die and be reabsorbed. The females are divided into six groups, three groups receive α-tocopherol at three different levels as a daily dose in corn oil for four days; the other three groups are similarly dosed with α-tocopheryl acetate at the same three levels. Sacrifice occurs on the fifth day and the "end-point" is determined by counting the number of live fetuses; for a given dose, the more live fetuses the greater the vitamin E activity of the dosed compound. Very surprisingly, this bioassay indicated that the phenol had only some 60% (3a) or 47% (3b) of the activity of the acetate - results which must have brought a sigh of relief to the companies which manufacture vitamin E!

Because "identical" rats are not identical, the fetal resorption bioassay has a statistical nature and goes through rats like the late Ayatollah used to go through ground troops. We decided that in a properly designed, competitive biokinetic experiment, the relative bioactivities of phenol and acetate could be more reliably determined using even a single rat. In the event we actually used four rats, two were vitamin E deficient (as in the bioassay) and two were E sufficient (4). These animals were dosed for four days with an equimolar mixture of d_6-*2R,4'R,8'R*-α-tocopherol (d_6-*RRR*-α-TOH) and d_3-*2R,4'R,8'R*-α-tocopheryl acetate (d_3-*RRR*-α-TOAc) in corn oil, the deuterium labels having been placed in metabolically inactive positions. After sacrifice on the fifth day, the animals were dissected, the lipids from blood and various tissues were extracted, and the ratio d_6-α-tocopherol/d_3-α-tocopherol was measured by GC/MS. The results were the same for all four rats, *i.e.*, the results were not affected by the animals' vitamin E status (4). The mean d_6-/d_3-α-tocopherol ratio for the four rats was 0.49 ± 0.05, (4) in good agreement with the fetal resorption bioassay.

We remained dubious that, under normal conditions, a chemically derivitized form of vitamin E (acetate) could provide more vitamin E than the natural vitamin (phenol). We therefore turned our attention to man. A single capsule containing 50 mg of deuterated phenol and 50 mg of deuterated acetate was swallowed with an evening meal by several volunteers. Blood was drawn on subsequent days and the ratio of α-tocopherol derived from the phenol and from the acetate in the plasma and red blood cells was found to be close to 1.0 (4). This provided the first scientific proof that "man was not a rat". However, certain important ladies rejected this conclusion saying with great vigor: "Nonsense, all men are rats". Clearly, further research was called for.

Rats are night feeders and in the N.R.C.'s animal facility the rats are housed in a basement room with artificial illumination being provided from 6 a.m. to 6 p.m. At *ca.* 10 a.m. the technician doses the animals according to the bioassay protocol with the vitamin E preparation in corn oil. This is not dissimilar to a person swallowing half a pint of corn oil on an empty stomach (ugh!). Five rats were therefore treated rather more like our volunteers. They were fed an equimolar mixture of deuterated phenol and deuterated acetate in an aqueous bolus of laboratory food. Sacrifice occurred 24 hours later and was followed by dissection, lipid extraction and GC/MS analyses. The mean ratio of α-tocopherol derived from phenol to that derived from acetate was 1.06 ± 0.11 (4). Clearly, the ladies have a point; but perhaps more importantly, it is obvious that the long accepted measurement of vitamin E activity by the rat fetal resorption bioassay is irrelevant to the animal under normal dietary conditions and, hence, is quite meaningless insofar as man is concerned.

Is this relevant to you? Yes, if you dose yourself with commercial vitamin E. This can be bought as (natural) *2R,4'R,8'R*-α-tocopheryl acetate or, at a lower price, as (synthetic) all-racemic-α-tocopheryl acetate. The "official" relative vitamin E activities of these two compounds is 1.36:1.00 – but this is based on the rat fetal resorption bioassay! Our own studies (using deuterated natural and synthetic α-tocopherol on human volunteers as well as rats) show that natural α-tocopherol is retained about twice as well as the synthetic, all-racemic material (5).

Does Vitamin C "Spare" Vitamin E in vivo?

We have also recently employed *2R,4'R,8'R-* α-tocopheryl acetate subsituted in a metabolically inactive position with three atoms of deuterium (d_3-*RRR*-α-TOAc) to make the first measurements of the net, longterm uptake of (deuterated) natural vitamin E, d_3-*2R,4'R,8'R*-α-tocopherol (d_3-*RRR*-α-TOH) in the male rat under normal laboratory dietary conditions, using a diet in which the d_3-*RRR*-α-TOAc (36 mg/Kg diet) was the only source of vitamin E (6). We discovered that there were dramatic differences in uptake kinetics between tissues. For example, the equalization time, $t_{1:1}$, which is the time required for the new (deuterium labelled) α-TOH concentration to become equal to that of the old (unlabelled) α-TOH can be estimated to be ca. 7, 23, 40 and 72 days in the lung, heart, brain and spinal cord, respectively.(7)

Since analogous tissue sampling experiments on man are almost inconceivable we decided to examine the biokinetics of natural vitamin E in a different laboratory animal in order to see if the same general pattern of fast uptake tissues and slow-uptake tissues obtained. We chose the guinea pig as our experimental animal and carried out the biokinetics under conditions similar to those employed in the earlier study on rats. However, in order to increase the sensitivity of the measurement of small amounts of deuterated tocopherol in the presence of a large amount of the unlabelled compound we employed *2R,4'R,8'R*-α-tocopheryl acetate labelled in metabolically inactive positions with six atoms of deuterium (d_6-*RRR*-α-TOAc). We found that guinea pigs and rats have a somewhat similar pattern of fast-uptake and slow-uptake tissues. *i.e.*, corresponding tissues could be classified either as "fast" or as "slow". However, the fast-uptake tissues of the guinea pig were somewhat "faster" than those of the rat and, while some of the slow uptake tissues of the guinea pig were "faster" than for the rat, others were dramatically slower, e.g., for the brain, $t_{1:1}$ was 40 and 108 days in the rat and guinea pig, respectively.

The guinea pig was chosen for our second whole animal biokinetic experiment with vitamin E because, unlike a rat but like man, the guinea pig cannot synthesize ascorbic acid (vitamin C, AH_2). The availability of vitamin C depends, therefore, on the dietary level of this compound which can be manipulated over a wide range. This is important because both vitamin E (1, 2, 8) and vitamin C (9-11) are chainbreaking antioxidants and there is a very considerable body of evidence which incidicates that there is a synergistic antioxidant interaction between these vitamins in a wide variety of *in vitro* model system (12-19). There is sound experimental evidence that this synergism is due to the "regeneration" of α-TOH by reduction of its initial oxidation product, the tocopheroxyl radical, α-TO•, by the ascorbate anion, AH^-. That is, the lipid-soluble α-TOH traps lipid peroxyl radicals, ROO•, forming α-TO• and lipid hydroperoxide, ROOH, and the α-TO• is then reduced by the water-soluble ascorbate rather than being irreversibly oxidized by reaction with a second peroxyl radical. The overall synergistic interaction between these two radical-trapping antioxidants *in vitro* can be represented by reactions 1, 3 and 4.

$$\alpha\text{-TO}\cdot + AH^- \rightarrow \alpha\text{-TOH} + A^{\overline{\bullet}} \qquad (3)$$

$$A^{\overline{\bullet}} + A^{\overline{\bullet}} \xrightarrow{H^+} A + AH \qquad (4)$$

There is some much less compelling evidence for an interaction between vitamin E and vitamin C *in vivo* for rats, guinea pigs, and premature infants. Synergistic E/C interactions have generally been reported, a result which lends support to the hypothesis, drawn from *in vitro* experiments, that vitamin C can "regenerate" vitamin E *in vivo,* or at least "spare" vitamin E by some other *in vivo* mechanism. For example, dietary vitamin C has been reported to enhance plasma levels of vitamin E (20-24). However, antagonistic E/C interactions have also been reported with dietary vitamin C lowering plasma levels of vitamin E (25, 26).

Guinea pigs on a diet containing d_6-*RRR*-α-TOAc provide a unique opportunity to search for any *in vivo* protective or destructive effect of vitamin C on vitamin E. An experimental protocol carefully designed to measure vitamin E turnover would be very much more sensitive than any of the earlier whole animal studies in detecting the existence, or otherwise, of an *in vivo* interaction between vitamins E and C. It would therefore provide a far more definitive answer to the important question posed by the title to this section.

We chose essentially the same experimental protocol for the present study of the biokinetics of vitamin E uptake and loss in the guinea pig which we had previously employed with rats (6) with the net rates of uptake of "new" and loss of "old" *RRR*-α-TOH, being measured in the blood and tissues of male guinea pigs over an eight week period by feeding diets containing d_6-*RRR*-α-TOAc. There was an initial two week "lead in" period during which 24 animals (the **HE** group) received diets containing 36 mg of unlabelled (d_0) *RRR*-α-TOAc and 250 mg of ascorbic acid per Kg diet, while another 24 animals (the **LE** group) received diets containing 5 mg (d_0)-*RRR*-α-TOAc and 250 mg ascorbic acid per Kg diet. After two weeks the **HE** group was divided into three equal subgroups which were fed diets containing 36 mg d_6-*RRR*-α-TOAc and 5000 mg (the **HEHC** subgroup), 250 mg (the **HENC** subgroup) and 50 mg (the **HELC** subgroup) ascorbic acid per Kg diet. One animal from each group was sacrificed each week and the blood and tissues were analyzed for d_0- and d_6-*RRR*-α-TOH by gas chromatography-mass spectroscopy. The **LE** group was similarly divided into three equal subgroups with animals receiving diets containing 5 mg d_6-*RRR*-α-TOAc and 5,000 mg (**LEHC**), 250 mg (**LENC**) and 50 mg (**LELC**) ascorbic acid per Kg diet with a similar protocol being followed for sacrifice and analyses. In the **HE** group the total (d_0- + d_6-) *RRR*-α-TOH concentrations in blood and tissues remained essentially constant over the 8 week experiment whereas in the **LE** group the total *RRR*-α-TOH concentrations declined noticeably (except in the brain, an organ with a particularly slow turnover of vitamin E).

The rationale behind this protocol (see Figure 2) is that if vitamin C really does "spare" vitamin E *in vivo* then, at high dietary levels of vitamin C one would expect a slower than normal loss of "old" vitamin E (d_0-*RRR*-α-TOH) from a tissue and a corresponding slower than normal uptake of "new" vitamin E (d_6-*RRR*-α-TOH). Similarly, at low dietary levels of vitamin C one would expect a faster than normal loss of "old" and a faster than normal uptake of "new" vitamin E. There were, however, no significant differences in the concentrations of "old" d_0-*RRR*-α-TOH nor in the concentrations of "new" d_6-*RRR*-α-TOH found in any tissue at a particular time between the **HEHC**, **HENC** and **HELC** subgroups nor between the **LEHC**, **LENC** and **LELC** subgroups. We conclude that the long postulated "sparing" action of vitamin C on vitamin E, which is well documented *in vitro,* is of negligible importance *in vivo* in comparison with the normal metabolic processes which consume vitamin E (e.g., by oxidizing it irreversibly) or eliminate it from the body. This is true both for guinea pigs with an adequate, well maintained vitamin E status and for guinea pigs which are receiving insufficient vitamin E to maintain their body stores.

Stereochemical Discrimination

During the past 2-3 years, we have invested a lot of effort into trying to understand the absorption and transport of vitamin E in rats (6) and in man (27) and in trying to identify the site(s) of chiral discrimination between α-tocopherol stereoisomers in rats (6) and

Figure 2.

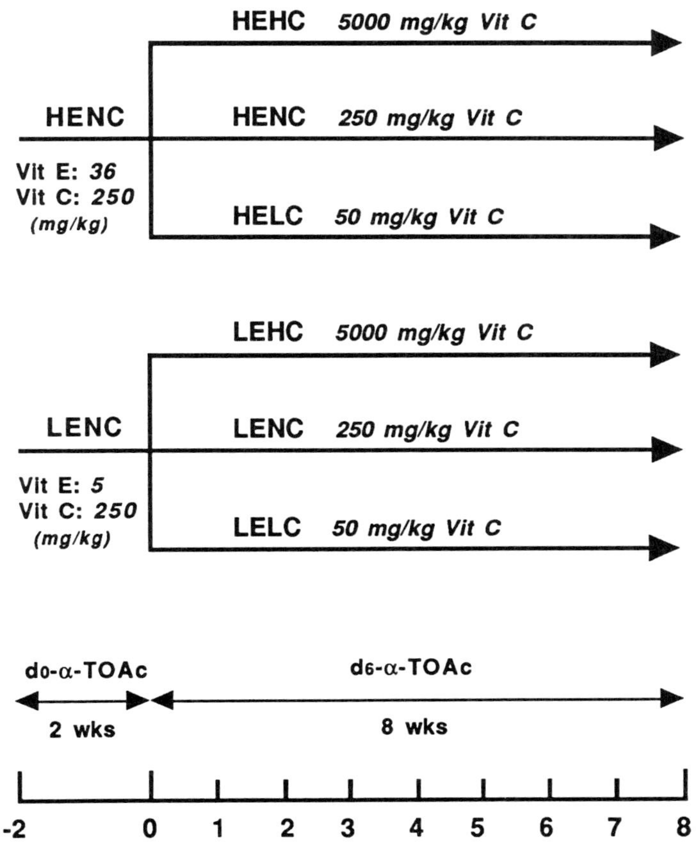

man (28). An early experiment involved rats fed a diet containing equimolar d_6-RRR-α-TOAc and d_3-SRR-α-TOAc (6). Except for the liver, during the first three weeks of this experiment, all tissues favored natural RRR-α-TOAc from day one. For some tissues, this chiral discrimination became quite dramatic, *e.g.*, a factor of 5.3 in the brain after 5 months. The liver is the main organ for chiral discrimination (28, 29) which appears to occur during the manufacture of very low density lipoprotein particles (28). However, there are other sites of discrimination, *e.g.*, red blood cell membranes (30) and the intestine (6).

Not unnaturally our initial research (6) was concerned with the intestine because it is here that the two acetates are hydrolyzed to the corresponding phenols. Hydrolysis is achieved by a pancreatic enzyme, cholesterol esterase, which requires bile salts for activity. In the rats' large intestines we found 1.1 to 1.6 times as much *RRR-α-*TOH and 0.4 to 1.1 times as much *RRR-α-*TOAc as for the corresponding *SRR* compounds, (6) which demonstrates that the acetate of natural *α*-tocopherol is hydrolyzed more rapidly than the acetate of the unnatural stereoisomer.

An attempt to model the hydrolysis of the two acetates *in vitro* using bovine cholesterol esterase, sodium cholate (40 mM) as the obligatory bile salt and with the acetates dispersed in dimyristoylphosphatadyl choline "failed" in that the maximum rate of hydrolysis (V_{max}) for the *SRR-α-*TOAc was about 7 times as great as V_{max} for the *RRR* acetate (31). Subsequent work (29) proved that this "failure" was not due to the use of a bovine rather than a rat pancreatic esterase but rather to the choice of bile salt. In the same model system, except for the replacement of the cholate by its glycine and taurine conjugates (Figure 3), the values of V_{max} for the *2R,4'R,8'R-α-*tocopheryl acetate were 8 and 25 times, respectively, as great as V_{max} for the other stereoisomer. These three bile salts can therefore modulate the stereoselectivity of the bovine esterase by a factor of ca. 200, despite the fact that the bond which is broken is separated by six bonds from the chiral center (Figure 4).

Figure 3.

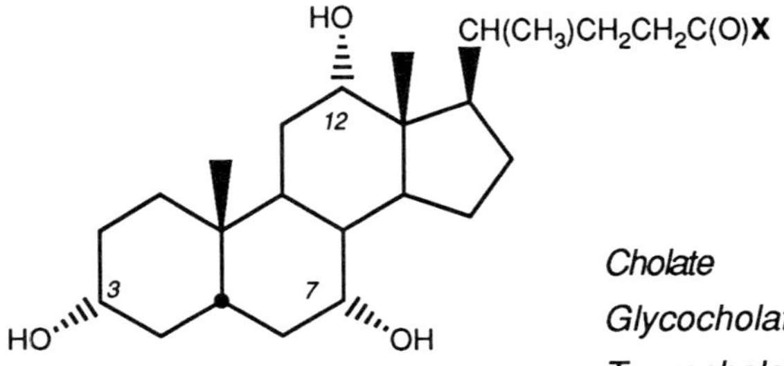

	X
Cholate	O^-
Glycocholate	$NHCH_2CO_2^-$
Taurocholate	$NHCH_2CH_2SO_3^-$

It is clear that each bile salt/substrate combination reacts with the enzyme in such a way that chiral information is conveyed to the active site. Since the *2R,4'R,8'R-* and *2S,4'R,8'R-α-*tocopheryl acetates could form different diastereoisomeric complexes with each of the different bile salts, we suggest that these are selected differently by the enzyme's active site. At 40 mM the bile salts will aggregate and the esterase catalyzed reaction therefore probably involves hydrolysis of the acetates contained in micelles of different structures.

Whatever the precise mechanistic details it is clear that we have stumbled upon a most interesting phenomenon. It would appear that any member of an endogenous family of chiral auxiliaries (the bile salts) can activate the enzyme and determine the stereoselectivity of the enzyme/ (chiral) substrate reaction. *This stereochemical control mechanism would appear to be too valuable and versatile for Nature to use only once.*

Figure 4.

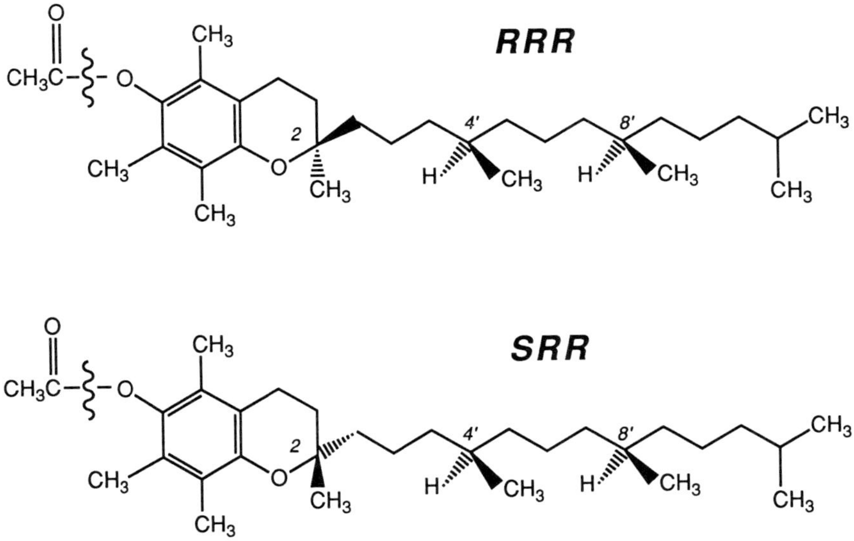

Acknowledgements

This work was made possible through the generosity of the Association for International Cancer Research; the National Foundation for Cancer Research; Eastman Chemical Products, Inc.; Eisai Co. Ltd.; the Henkel Corporation; and the Natural Source Vitamin E Association.

References

1. Burton, G.W., Joyce, A., Ingold, K.U. (1982)*The Lancet* (August) 327.(1983) *Arch. Biochem. Biophys.* **221,** 281-290. Ingold, K.U., Webb, A.C., Witter, D.A., Burton, G.W., Metcalfe, T.A., Muller, D.P.R. (1987) *Ibid.* **259,** 224-225

2. Burton, G.W., Ingold, K.U. (1986) *Accounts Chem. Res.* **19**,194-201
3. (a) Harris, P.L., Ludwig, M.I. (1949) *J. Biol. Chem.* **1980**, 611-614. (b) Wieser, H., Vecchi, M., Schlachter, M. (1986) *Int. J. Vit. Nutr. Res.* **56**, 45-56
4. Burton, G.W., Ingold, K.U., Foster, D.O., Cheng, S.C., Webb, A., Hughes, L., Lusztyk, E. (1988) *Lipids* **23**, 834-840
5. Burton, G.W. Unpublished results
6. Ingold, K.U., Burton, G.W., Foster, D.O., Hughes, L., Lindsay, D.A., Webb, A. (1987) *Lipids* **22**,163-172
7. Ingold, K.U., Burton, G.W., and Siebrand, W. (1989). Unpublished results
8. Burton, G.W., and Ingold, K.U. (1981) *J. Am. Chem. Soc.* **103**, 6472-6477
9. Ingold, K.U., Webb, A.C., Witter, D., Burton, G.W., Metcalf, T.A., and Muller, D.P.R. (1987) *Arch. Biochem. Biophys.* **259**, 224-225
10. Wayner, D.D.M., Burton, G.W., and Ingold, K.U. (1986) *Biochem. Biophys. Acta* **884**,119-123
11. Bendich, A., Machlin, L.J., Scandurra, O., Burton, G.W., and Wayner, D.D.M. (1986) *Adv. Free Radical Biol. Med.* **2**, 419-444
12. Golumbic, C., and Mattill, H.A. (1941) *J. Am. Chem. Soc.* **63**,1279-1280
13. Tappel, A.L., Brown, W.D., Zalkin, H., and Maier, V.P. (1961) *J. Am. Oil Chem. Soc.* **38**, 5-9
14. Packer, J.E., Slater, T.F., and Willson, R.L. (1970) *Nature* **278**, 737-738
15. Barclay, L.R.C., Locke, S.J., and MacNeil, J.M. (1983) *Can. J. Chem.* **61**,1288-1290
16. Scarpa, M., Rigo, A., Maiorino, M., Ursini, F., and Gregolin, C. (1984) *Biochem. Biophys. Acta* **801**, 215-219
17. Doba, T., Burton, G.W., and Ingold, K.U. (1985) *Biochem. Biophys. Acta* **835**, 298-303
18. Niki, E., Kawakami, A., Yamamoto, Y., and Kamiya, Y. (1985) *Bull. Chem. Soc. Jpn.* **58**,1971-1975
19. Barclay, L.R.C., Locke, S.J., and MacNeil, J.M. (1985) *Can. J. Chem.* **63**, 366-374
20. Chen, L.H., Lee, M.S., Hsing, W.F., and Chen. S. (1980) *Internat. J. Vit. Nutr. Res.* **50**,156-162
21. Chen, L.H., and Thacker, R.R. (1985) *FASEB Fed. Proc.* **44**, 934
22. Chen, L.H., and Thacker, R.R. (1987) *Internat. J. Vit.. Nutr. Res.* **57**, 385-390
23. Chen, L.H. and Chang, M.L. (1978) *J. Nutr.* **108**,1616-1620
24. Arad. I.D., Dgani, Y., and Eyal, F.G. (1985) *Internat. J. Vit. Nutr. Res.* **55**, 395-397
25. Chen, L.H. (1981) *Am. J. Clin. Nutr.* **34**,1036-1041
26. Chen, L.H. and Chang, H.M. (1979) *Internat. J. Vit. Nutr. Res.* **49**, 87-91
27. Traber, M.G., Ingold, K.U., Burton, G.W., Kayden, H.J. (1988) *Lipids* **23**, 791-797
28. Traber, M.G. Unpublished results

29. Zahalka, H.A. Unpublished results
30. Cheng, S.C., Burton, G.W., Ingold, K.U., Foster, D.O. (1987) *Lipids* **22**, 469-473
31. Zahalka, H.A., Cheng, S.C., Burton, W.B., Ingold, K.U. (1987) *Biochem. Biophys. Acta.* **921**, 481-485

Interaction of Glutathione and α-Tocopherol in the Inhibition of Lipid Peroxidation in Rat Liver Microsomes

R.W. Scholz, K.S. Graham, M.K. Wynn and C.C. Reddy
Department of Veterinary Science and Environmental Resourses Research Institute
Penn State University, University Park, PA 16802 USA

Abstract

There is increasing evidence that vitamin E, or more specifically α-tocopherol, is the most significant lipid soluble antioxidant in biological membranes. Previous investigations have described a reduced glutathione (GSH)-dependent, heat labile system in rat liver microsomes that inhibited lipid peroxidation. This inhibition was initially considered only dependent on α-tocopherol because microsomes prepared from rats fed a diet deficient in vitamin E were devoid of the protection offered by GSH. However, additional experiments revealed that glutathione disulfide enhanced the protective effects of GSH and that this synergism was also present in microsomes from rats that were deficient in vitamin E. Experiments conducted with an O_2-limiting system revealed changes in O_2 uptake relative to thiobarbituric acid-reactive product formation that argues against a mechanism involving a radical scavenging, GSH-dependent factor in the inhibition of lipid peroxidation.

The data suggest the presence of at least two GSH-dependent, heat labile microsomal factors that inhibit lipid peroxidation. The first involves a role for GSH in the regeneration of α-tocopherol from the α-tocopheroxyl radical. In this case, α-tocopherol would be recycled during oxidative stress to function as a scavenger of lipid peroxy radicals. The second factor does not appear to be dependent on α-tocopherol and may involve the reduction of fatty acid hydroperoxides to their corresponding alcohols. One, or possibly both, of these GSH-dependent factors may be activated by GSSG through thiol/disulfide exchange. The identity of either factor has not been established.

Introduction

Lipid peroxidation occurs by a free radical chain reaction that has been identified with initiation, propagation and termination phases. Reviews on this subject and roles of oxygen and transition metal ions have appeared (1-6). Biological membranes rich in polyunsaturated fatty acids (PUFA) are highly susceptible to free radical-catalyzed oxidation reactions. A general scheme on the peroxidation of PUFA is illustrated in Figure 1 (adapted from references 7-9) and is an oversimplification of the process that would occur in biological membranes.

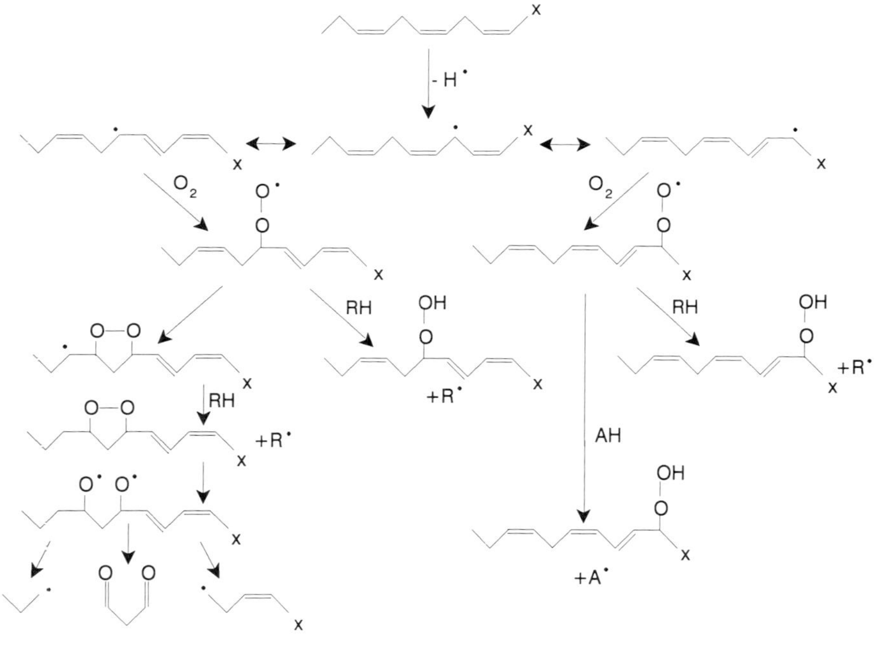

Figure 1. *Peroxidation of Polyunsaturated Fatty Acid* (adapted from refs. 7-9)

Lipid peroxidation is initiated by the abstraction of an allylic hydrogen atom, specifically a divinyl methane hydrogen, by a free radical initiator. Identity of the initiating agent(s) has been subject of extensive study (3,10-12) and has remained elusive primarily because of the presence, even at low levels, of preformed lipid hydroperoxides in biochemical systems that are commonly used to monitor lipid peroxidation (5,13-15). In this regard, iron ions that are generally added to an *in vitro* lipid peroxidation system may degrade existing lipid hydroperoxides and generate additional radical species. Once initiated, lipid peroxidation leads to a number of events, such as oxygen uptake and chemiluminescence, and the formation of various products including conjugated dienes, malonyldialdehyde (MDA), lipofuscin pigments, and volatile hydrocarbons, among others, all of which have been used as indices of lipid peroxidation (6,16-23). During the propagation phase, carbon-centered radicals are resonance stabilized followed by rearrangement of the neighboring double bonds to form conjugated dienes. Molecular oxygen readily reacts at a diffusion-limited rate with these alkyl radicals to form lipid hydroperoxy radicals (ROO•). A chain reaction ensues in which ROO• reacts with another PUFA (RH) producing a lipid hydroperoxide (ROOH) and another alkyl radical (R•). Some of the ROO• may be converted to endoperoxy radicals and proceed to form decomposition products such as MDA. Chain branching occurs when a relatively stable ROOH reacts with iron or superoxide anion to produce additional radical species (4,5) as shown:

$$ROOH + Fe^{3+} \rightarrow Fe^{2+} + H^{+} + ROO\cdot \qquad [1]$$

$$ROOH + Fe^{2+} \rightarrow Fe^{3+} + OH^{-} + RO\cdot \qquad [2]$$

$$ROOH + O_2^{\bar{\cdot}} \rightarrow O_2 + OH^{-} + RO\cdot \qquad [3]$$

The influence of reactions 1-3 on lipid peroxidation is considered significant in that one group of investigators view it as a two-step process of initiation and lipid hydroperoxide-dependent propagation (3,10,11,24). This concept is relevant because many enzyme systems that produce initiators are inactivated during lipid peroxidation, and it is the iron-catalyzed decomposition of lipid hydroperoxides that is considered important for the continuation of the peroxidative reaction (4,6,10,24,25). The importance attributed by Svingen *et al.* (10,24) of these propagation reactions in lipid peroxidation, using reconstituted liposomes, has been questioned by Ernster *et al.* (12), however, where a system containing microsomal membranes was employed. If an antioxidant (AH) such as α-tocopherol is present, propagation of the chain reaction is blocked because AH transfers a hydrogen to the free radical, in this case ROO•, producing ROOH and a relatively nonreactive vitamin E radical. It is important to consider that by promoting the formation of ROOH, vitamin E enhances production of a species capable of generating new radicals via the lipid hydroperoxide-dependent mechanism. Termination of lipid peroxidation may also occur if two radicals react with each other resulting in the production of nonreactive species.

Biological Defense Against Lipid Peroxidation

Cells have evolved systems that either prevent or control lipid peroxidation. These defense systems are partitioned among the cell cytosol and various particulate cellular compartments and may be classified as enzymatic, nonenzymatic or combinations of the two. In general terms, the first category includes enzymes or enzyme systems that control the formation of endogenous initiators of lipid peroxidation. The second category includes chain-breaking antioxidants, or radical scavenging compounds, that sequester or quench radical species that are predominantly involved in propagating reactions of lipid peroxidation.

The enzymes frequently identified in controlling lipid peroxidation are superoxide dismutase, catalase, selenium-dependent glutathione peroxidase (Se-GSHPx) and selenium-independent glutathione peroxidase (nonSe-GSHPx). Both Se-GSHPx and nonSe-GSHPx require the tripeptide glutathione (GSH) for their catalytic activity. Each of the enzymes mentioned play an important role in the control of lipid peroxidation by catalyzing reactions affecting the concentration of superoxide anion, hydrogen peroxide and lipid hydroperoxides. Reviews of the involvement of these enzymes in cellular defense against lipid peroxidation have appeared in the literature (25-30).

Antioxidants, in general, contribute to nonenzymatic cellular defense against lipid peroxidation by donating hydrogen atoms to free radicals, resulting in their inactivation. Biological antioxidants are found both in the aqueous cell cytosol as well as lipid-rich particulate subcellular fractions. Those which are water soluble and exist in the cell cytosol include, among others, ascorbic acid, uric acid, cysteine and GSH (31-36). Ascorbic acid readily quenches singlet oxygen (31) and lipid peroxy and alkoxy radicals (37) in the aqueous phase. Ascorbate can also function as a prooxidant by reducing Fe^{3+} or Cu^{2+}, which may take part in the metal-catalyzed decomposition of lipid hydroperoxides to radical species (37). Uric acid can quench singlet oxygen and the hydroxyl radical in the cell cytosol (34) and, in aqueous dispersions, uric acid, cysteine and GSH have been shown to inhibit lipid peroxidation (35). These results have suggested that the water soluble antioxidants may prevent lipid peroxidation by scavenging radicals in the aqueous phase but are unable to scavenge radicals or prevent lipid peroxidation when it is initiated in the lipid phase (31,34,35). This has significance because various agents thought capable of initiating lipid peroxidation are quite soluble in the organic matrix of cell membranes (9) and to prevent or control lipid peroxidation an effective cellular lipid soluble antioxidant system is required.

Whereas the naturally occurring carotenoids (38,39) and estrogen (40) are considered to have antioxidant properties, vitamin E, or more specifically α-tocopherol, has been described as the only significant lipid soluble, chain-breaking antioxidant present in living systems (41-43). Vitamin E shows a marked capacity to prevent lipid peroxidation *in vitro* and *in vivo* (9,44) and its capacity to do so is positively related to the animal's vitamin E status (45). McCay *et al.* (46) demonstrated that by increasing the concentration of membrane peroxidizable PUFA the amount of α-tocopherol required to prevent lipid peroxidation was also increased. Cohen (47) has reported that the addition of α-tocopherol to liver homogenates decreased lipid peroxidation as measured by TBA-reactive products formation and ethane production. Isolated hepatocytes from vitamin E-deficient rats show enhanced lipid peroxidation compared with control rats supplemented with vitamin E (48). Cadenas *et al.* (49) demonstrated a substantially longer time for the initiation of lipid peroxidation in liver microsomes prepared from vitamin E-supplemented rats as compared with microsomes from rats that were deficient in vitamin E. These investigators also showed that the incorporation of α-tocopherol into liver microsomes from vitamin E-deficient animals delayed the onset of lipid peroxidation. The amount of α-tocopherol incorporated into vitamin E-deficient microsomes was not as effective as the same quantity of α-tocopherol incorporated naturally into the microsomes, however (49). Tissues with higher concentrations of vitamin E are more resistant to lipid peroxidation. For example, Kornburst and Mavis (50) have reported that the relative susceptibility of microsomes derived from rat lung, heart, liver, kidney, brain and testes to lipid peroxidation correlated inversely with the vitamin E content of these tissues.

Whereas few question the role of vitamin E in the prevention of lipid peroxidation, the precise mechanism by which it accomplishes this task has remained elusive. Vitamin E exists in close contact with PUFA's having chain lengths and degree of unsaturation greater than 18:2 (51-54). This association manifests itself by an interaction between the phytyl side chain of the α-tocopherol molecule and the unsaturated portion of PUFA (51,52) and, because of this association, α-tocopherol is in an excellent position to protect PUFA from lipid peroxidation.

Two basic mechanisms have been proposed to explain the role of α-tocopherol in the prevention of lipid peroxidation. The first involves the phytyl side chain of α-tocopherol which, in close contact with the unsaturated portions of PUFA, provides a physical barrier that sterically impedes the attack of free radicals on PUFA. Sunamato *et al.* (55), for example, have shown that esters of α-tocopherol, although lacking radical scavenging ability, can decrease the rate of phosphatidylcholine oxidation in a bilayer system. Other research (56-57), however, does not support this mechanism. α-Tocopherol acetate and phytol were shown to be unable to prevent lipid peroxidation in liposomes (56) and Niki *et al.* (57) demonstrated little effect of the phytyl side chain of α-tocopherol on its ability to prevent lipid peroxidation in liposomes, micelles or in a homogenous solution. The phytyl moiety of α-tocopherol was proposed to effectively maintain the heterocyclic ring in close association with PUFA, rather than to physically hinder the peroxidative process (57).

A second mechanism proposes that α-tocopherol functions as a radical scavenger that quenches free radicals responsible for lipid peroxidation (9,46,56,58-60). Fukuzawa *et al.* (56) demonstrated that α-tocopherol was able to quench lipid peroxy and alkoxy radicals and in the process become oxidized. Earlier McCay *et al.* (46) had shown that α-tocopherol was converted to a compound more polar than α-tocopherol quinone in microsomal membranes undergoing lipid peroxidation. Other investigators (61) have shown that α-tocopherol reacts with peroxy radicals mechanistically in a manner similar to that which occurs in the oxidation of phenols. Whereas the notion of α-tocopherol inhibiting lipid peroxidation via radical scavenging is attractive, an inherent problem in this premise is that its concentration relative to the concentration of peroxidizable PUFA (PUFA > 18:2) is low (50,62). As reported by Evarts and Bieri (62), in rats fed corn oil-based diets, the molar ratio of peroxidizable PUFA to α-tocopherol in tissues varies; with lung and heart this ratio is in the range 300-350:1; in liver and kidney, this ratio is 1200:1 or greater.

Considering the abundance of peroxidizable PUFA relative to α-tocopherol in most tissues, the question then becomes how can vitamin E function so effectively in the inhibition of lipid peroxidation by acting as a radical scavenging antioxidant? In 1968, Tappel (59) proposed what may be important to the answer of this question by suggesting that α-tocopherol could be regenerated from its oxidized form by the more abundant cellular reducing agents, vitamin C and glutathione. Considerable interest has been generated since this proposal in interactions among vitamin E, vitamin C and glutathione in antioxidant defense against free radical-induced lipid peroxidation.

Packer *et al.* (63) observed the quenching of vitamin E radical by vitamin C in an organic solution subjected to pulse radiolysis. Niki *et al.* (64) confirmed this observation using electron spin resonance

(ESR) spectroscopy by demonstrating that vitamin C could interact with vitamin E radical in a homogenous solution and reduce it to α-tocopherol. These investigators also showed that GSH was capable of regenerating α-tocopherol from α-tocopherol radical. The ability of GSH to regenerate α-tocopherol is particularly significant because cytosolic concentrations of GSH make it a more favorable reducing agent *in vivo* than ascorbic acid (65). Nishikimi *et al.* (66) found that in aqueous dispersion ascorbic acid could reduce 8a-hydroxy-α-tocopherone, an oxidation product of α-tocopherol, to α-tocopherol. However, they found that ascorbic acid was unable to reduce α-tocopherol quinone to α-tocopherol.

Niki *et al.* (67) observed that ascorbic acid and α-tocopherol produced an additive effect in the inhibition of peroxidation of methyl linoleate in homogenous solution. Additional studies (68) revealed that when α-tocopherol and vitamin C were present in solution, α-tocopherol was responsible for inhibiting peroxidation and vitamin C extended its ability to do so. These results (68) indicated that vitamin E trapped the peroxy radical responsible for peroxidation and the resulting vitamin E radical was reduced to vitamin E by vitamin C. Barclay *et al.* (69) demonstrated a synergism between vitamin C and vitamin E in inhibiting lipid peroxidation in lipid micelles. Vitamin C alone was ineffective in this system whereas vitamin E alone effectively inhibited the peroxidation of micellar lipid. Studies with liposomes (70,71) have shown that vitamin C and vitamin E act synergistically to inhibit lipid peroxidation. The results reported by Doba *et al.* (71) with multilamellar phospholipid liposomes indicated that ascorbic acid was ineffective in inhibiting peroxidation in the lipid phase. Ascorbic acid was shown to scavenge free radicals in the aqueous phase and, in combination with α-tocopherol, extended inhibition of lipid peroxidation for a period greater than that observed for α-tocopherol alone. Additional experiments utilizing ESR procedures revealed that the α-tocopheroxyl radicals generated by peroxidizing lipids were rapidly quenched by ascorbate (72). Other investigators (73,74) have reported evidence for the radical-scavenging abilities of ascorbic acid and α-tocopherol *in vitro* and have suggested that ascorbic acid regenerates α-tocopherol from the α-tocopherol radical. The results of these *in vitro* studies support the premise that vitamin E in the lipid phase acts to prevent peroxidation and vitamin C in the aqueous phase serves to regenerate α-tocopherol from its oxidized form. Ingold (75), in another section of this symposium, has provided compelling evidence that casts doubt on the ability of vitamin C to spare or regenerate vitamin E *in vivo*.

Glutathione-Mediated Defense Against Lipid Peroxidation

A number of investigators, utilizing both NADPH/Fe-dependent enzymatic and ascorbate/ADP/Fe-dependent nonenzymatic systems to stimulate lipid peroxidation *in vitro*, have observed GSH-dependent cytosolic and membrane-bound factors that inhibit the course of the peroxidative process. McCay *et al.* (76) demonstrated that lipid peroxidation of rat liver microsomes and mitochondria was inhibited by liver cytosol, a process that was also shown to be heat labile. This effect, initially attributed to Se-GSHPx, was not thought to be a result of the reduction of lipid hydroperoxides to alcohols but rather to the prevention of initiation of lipid peroxidation itself. Burk (77) confirmed the heat lability of the cytosolic protective factor and also showed that Se-GSHPx offered no protection against microsomal lipid peroxidation in the *in vitro* system employed. Protection was found, however, in rat liver cytosol fractions containing nonSe-GSHPx activity associated with the glutathione S-transferases (GSTs). McCay *et al.* (78) also demonstrated that partially purified rat liver Se-GSHPx was not responsible for the inhibition of lipid peroxidation and additional studies (79,80) demonstrated that the inhibition of lipid peroxidation present in rat liver cytosol was not associated with the Se-GSHPx, nonSe-GSHPx or GST enzymes. Further, the heat labile, GSH-dependent factor in rat liver cytosol was shown to contain sulfhydryl groups necessary for its activity. These studies (79,80) also provided evidence that whereas the GSH-dependent factor in rat liver cytosol could reduce hydroperoxides of free fatty acids in solution or micellar suspension, it could not react with lipid hydroperoxides bound to membranes. The GSH-dependent inhibition of lipid peroxidation in rat liver cytosol was also demonstrated by Fairhurst *et al.* (81,82). These investigators (82) reported that the protective factor in rat liver cytosol was inhibited by bromosulphophthalein, but did not attribute the GSH-dependent inhibition of lipid peroxidation to GST activity.

The notion that the GSH-dependent inhibitory factor in rat liver cytosol functioned by scavenging radicals responsible for the initiation of lipid peroxidation was studied by Gibson *et al.* (83) using oxygen consumption and conjugated diene formation as indices of lipid peroxidation. If the factor prevented initiation of lipid peroxidation by radical scavenging, no oxygen consumption or formation of conjugated dienes should occur. Some conjugated diene formation and oxygen consumption would occur if the factor functioned to reduce lipid hydroperoxides to their corresponding alcohols. Results obtained by Gibson *et al.* (83) demonstrated suppressed oxygen consumption and no conjugated diene formation in this system. A dependency of the protein on thiol groups was indicated by treatment of the cell cytosol and liver microsomes with iodoacetate. Protection by the cytosolic factor was not exclusively GSH-dependent because 2-mercaptoethanol was as effective as GSH in inhibiting lipid peroxidation. It was concluded that the cytosolic factor associated with inhibition of lipid peroxidation did so by scavenging radicals responsible for the initiation process. The identity of the cytosolic factor has not been established with certainty but may be a protein that enhances radical scavenging properties of GSH (35).

In addition to the cell cytosol, GSH-dependent factors that inhibit lipid peroxidation have been observed in microsomal, mitochondrial and nuclear subcellular fractions. One of the earlier investigations on thiol inhibition of lipid peroxidation was reported by Christopherson (84) in which assays with the ascorbate-induced, nonenzymatic lipid peroxidation system were employed. In these experiments, 1 mM GSH inhibited lipid peroxidation in freshly prepared microsomal and mitochondrial suspensions but not in those that were heat denatured. Later investigations (85-88) confirmed the heat lability of the GSH-dependent factor in rat liver microsomes. In the experiments reported by Reddy *et al.* (85,88), utilizing both enzymatic and nonenzymatic systems, the inhibitory effect of 5 mM GSH on microsomal lipid peroxidation was independent of the selenium status of the rats. This inhibition was, however, dependent upon vitamin E because liver microsomes obtained from rats fed a diet deficient in this vitamin were devoid of the protective effect. These investigators suggested that GSH, through a heat labile factor(s), prevented the irreversible oxidation of α-tocopherol by regenerating it from the tocopheroxyl radical; this concept supports Tappel's (59) original proposal. Homogeneous preparations of Se-GSHPx and nonSe-GSHPx had no effect on inhibiting microsomal lipid peroxidation over that offered by GSH alone (85). Additionally, the inhibitory effect on lipid peroxidation is specific for GSH in that of the other sulfhydryl compounds tested, 2-mercaptoethanol, dithiothreitol and cysteine, none showed dependency on vitamin E (85,86).

In the experiments reported by Burk (89) 0.1 mM GSH inhibited rat liver microsomal lipid peroxidation, an effect that was removed by heat treatment or trypsin digestion of the microsomes. The effect was concentration dependent because increasing GSH concentration from 0.1 to 5 mM resulted in extension in the lag period before the onset of lipid peroxidation. Burk (89) also reported that the O_2 consumed relative to the amount of TBA-reactive products formed by peroxidizing microsomes was the same in the presence or absence of

GSH which indicated that the inhibitory factor functioned to prevent lipid peroxidation through radical scavenging properties. The studies reported by Haenen and Bast (90) also showed a concentration dependency (0.1-1 mM) of GSH on inhibition of lipid peroxidation in rat liver microsomes. These investigators also reported that the GSH-dependent factor was heat labile and that it was inactivated during the lipid peroxidation process. Haenen and Bast (90) indicated further that the GSH-dependent inhibition of lipid peroxidation was not due to radical scavenging properties or GSHPx activity of the microsomal factor but rather was associated with the prevention of radical formation. No direct evidence for vitamin E dependency was claimed, although microsomes from animals deficient in vitamin E were not used in these experiments (90). McCay *et al.* (86,91), however, were able to restore GSH-dependent inhibition of lipid peroxidation by adding α-tocopherol to rat liver microsomes that were depleted of vitamin E. These results are not in support of the suggestion that the factor is inactive in microsomes depleted of vitamin E (90). McCay *et al.* (86,91) also indicated that the GSH-dependent factor was capable of reducing the α-tocopheroxyl radical to α-tocopherol.

Hill and Burk (92) studied the GSH-dependent lag in lipid peroxidation in liver microsomes from rats fed diets deficient in selenium or vitamin E. These investigators found that selenium deficiency did not impair the inhibition of lipid peroxidation offered by GSH, in agreement with previous reports (85,88), but a distinctive lag was shown for microsomes from rats fed their diet deficient in vitamin E. Hill and Burk (92) indicated, therefore, that not all the inhibition by GSH in lipid peroxidation could be attributed to α-tocopherol and that a radical scavenging system independent of α-tocopherol was also associated with the GSH-dependent effect. Wefers and Sies (93) demonstrated protective effects of GSH in inhibiting liver microsomal lipid peroxidation in rats fed diets supplemented or deficient in vitamin E. The lag period in lipid peroxidation was extended by GSH only in microsomes from rats fed the diet supplemented with vitamin E but the time required to reach a given level of chemiluminescence was increased by GSH in both experimental groups. These investigators (93) also reported that ascorbate prolonged the lag phase only in microsomes from vitamin E-supplemented rats; in vitamin E deficiency, ascorbate was shown to have a prooxidant effect that was removed by GSH. Liebler *et al.* (94) demonstrated in their liposomal system that GSH functioned only as a prooxidant and was unable to extend the antioxidant effects of α-tocopherol. This suggests that whereas GSH can regenerate vitamin E in a homogenous solution (64), it may require an enzyme or enzyme system for this process to occur with oxidized α-tocopherol in biological membranes. In experiments reported by Thirot *et al.* (95), GSH was ineffective in inhibiting lipid peroxidation of microsomes that were subjected to γ irradiation. Microsomal vitamin E was markedly reduced by γ irradiation in these studies, a finding that suggests further that GSH is effective in the inhibition of lipid peroxidation only when adequate vitamin E is present in the membranes.

Experiments reported by Robey and Mavis (96), Wefers and Sies (93) and Murphy and Kehrer (97) have shown a direct correlation between microsomal vitamin E content and GSH-dependent protection against lipid peroxidation. Recent investigations have shown that GSH not only extends the lag period prior to the onset of microsomal lipid peroxidation, but it also reduces the loss of α-tocopherol in both NADPH/Fe or ascorbate/Fe stimulated lipid peroxidation systems (86, 96-99). This sparing effect of GSH was also demonstrated in rat liver microsomes that were prelabeled *in vivo* with [^{14}C] α-tocopherol (99).

The research reviewed in this section has dealt primarily with the sparing effect of GSH on lipid peroxidation and α-tocopherol depletion of rat liver microsomes. This protective effect of GSH may be quite specific for rat liver because Murphy and Kehrer (97) have recently presented evidence that it may not be of significance in other tissues or animal species. Nevertheless, identification of the GSH-dependent, heat labile factor or factors responsible for the inhibition of lipid peroxidation in rat liver microsomes has remained an interesting and challenging subject for investigation. Its importance and relevance in other systems have not been established with certainty, and obviously requires further evaluation.

Effects of GSH-Dependent Enzymes on Lipid Peroxidation

Whereas rat liver cytosol contains a high concentration of GST activity, at times comprising approximately 10% of the soluble protein (100), microsomal (101,102), mitochondrial (103) and nuclear (104) subcellular fractions also contain significant amounts of the enzyme. When non Se-GSHPx activity was demonstrated for the GSTs of rat liver microsomes, it was suggested that this activity may protect biological membranes from damage through its ability to reduce lipid hydroperoxides (105). This hypothesis was attractive because of the proximity between potential lipid hydroperoxide substrates and the membrane bound GST enzymes having glutathione peroxidase activity. Peroxidase activity of microsomal GSTs was later confirmed by Morgenstern and DePierre (106), and others have suggested that the GSH-dependent inhibition of lipid peroxidation in rat liver microsomes (107,108), mitochondria (109) and nuclei (104) was associated with GST activity. The microsomal GSTs, unlike their counterparts in the cell cytosol, are activated by limited proteolysis with trypsin (110,111), covalent modification of the single cysteine residue with sulfhydryl compounds such as N-ethylmaleimide (NEM, 112-115), heating (116) and by disulfides such as GSSG and cystine (115,117). Indeed, mixed disulfide formation has been suggested as a means of regulating the microsomal GSTs (112,117) and Aniya and Anders (118) have recently reported that GSSG activates the enzyme when it is membrane bound. The purified enzyme was not activated by GSSG which indicated that the disulfide activation may be catalyzed by microsomal factors that are lost during purification (118). Aniya and Anders (119) have also shown that enzymatically generated reduced oxygen species, such as superoxide anion and hydrogen peroxide, activate microsomal GST by oxidation of the cysteine sulfhydryl moiety.

Inhibitors of microsomal GST activity have been shown to reverse the protective effect of GSH on lipid peroxidation (120,121). This does not necessarily prove a direct involvement of the GSTs in GSH-dependent inhibition of lipid peroxidation, however, because other GSH-dependent factors may also be influenced by these inhibitors. An additional role of the GSTs in microsomal lipid peroxidation may be a protective one through conjugation of toxic products of lipid peroxidation. For example, hydroxyalkenals serve as substrates (121-123) as well as activators (124) of the rat liver microsomal GSTs.

It is generally accepted that the cytosolic GSH-dependent factor that inhibits lipid peroxidation is not associated with the Se-GSHPx, nonSe-GSHPx or GST enzymes (77-82). The Se-GSHPx and GST enzymes of the cell cytosol can inhibit microsomal lipid peroxidation in the presence of phospholipase A_2 (125,126), showing the necessity to release the membrane-bound lipid hydroperoxide substrates for this effect. The specific involvement of the microsomal GSTs in GSH-dependent inhibition of microsomal lipid peroxidation in rat liver has also been questioned. Haenen *et al.* (124) have shown that pretreatment of microsomes with 4-hydroxy-2,3,-*trans*-nonenal reduced the GSH-dependent protection but resulted in the activation of the GSTs by more than two-fold. N-ethylmaleimide, an agent that is known to stimulate microsomal GST activity, also reduced the GSH-dependent protection against lipid peroxidation. These investigators (124) also demonstrated that phospholipase A_2 activity was not required for the protective effect involving the GSH-dependent factor. In related experiments, Nagasaka *et al.* (127) reported inhibition of the GSH-dependent protection against lipid peroxidation by pretreatment of rat liver microsomes with iodoacetate (IA), GSSG, deoxycholate (DOC), bromosulfophthalein (BSP) and NEM. Microsomal nonSe-GSHPx activity was inhibited by IA and GSSG but not by DOC, BSP, or NEM. Microsomal GST activity was inhibited by DOC and BSP, activated by GSSG and NEM, and not affected by IA. The differential effects of these inhibitors argue against either nonSe-GSHPx or GST activities as being protective factors associated with the GSH-dependent inhibition of lipid peroxidation.

The peroxidation inhibiting protein, currently referred to as phospholipid hydroperoxide glutathione peroxidase (PHGPX), which has been isolated and identified by Ursini and colleagues (128,129), has recently been implicated as the protective factor in rat liver microsomes (130). These investigators also demonstrated that the antioxidant effects of vitamin E on lipid peroxidation in microsomes and in Triton-dispersed microsomal lipid micelles occurred only in the presence of PHGPX and GSH (130). This enzyme was originally purified to homogeneity from pig liver (128) and heart (131) cytosol and was also shown to be present in various tissues in a membrane-associated or interfacial nature (129,132,133). Analysis of the homogenous enzyme revealed 1 gram atom Se/mole (129) with a molecular weight of approximately 20 kDal (132). This enzyme reduces lipid hydroperoxides directly at the membrane level and does not require their prior release via the action of phospholipase A_2. Nagasaka *et al.* (127), however, were unable to extract this activity from rat liver microsomes. Further, the protein isolated by Ursini and Bindoli (133) required greater than 1 mM GSH to inhibit lipid peroxidation, and 2-mercaptoethanol could replace GSH for this effect. In the experiments reported by Nagasaka *et al.* (127), however, inhibition was shown at GSH concentration in the range 0.01-0.5 mM and 2-mercaptoethanol could not replace GSH in the protective effect. Additionally, our previous experiments (85,88), and those of Hill and Burk (92), have shown that the GSH-dependent, heat labile protection against lipid peroxidation of rat liver microsomes is independent of the animals' selenium status. It is questionable, therefore, whether the inhibitory effect of GSH on lipid peroxidation in these studies with selenium-deficient animals is associated with the activity of PHGPX. It would be of interest to determine the effectiveness of PHGPX in the GSH-dependent inhibition of lipid peroxidation in rat liver microsomes that are severely depleted of selenium.

GSH Effects on Microsomal Lipid Peroxidation in an Oxygen-Limiting System

Experiments were undertaken in our laboratories to monitor changes in oxygen consumption, TBA-reactive product formation and α-tocopherol concentration during NADPH-dependent lipid peroxidation of rat liver microsomes. Rats were fed vitamin E-supplemented (150 IU/kg) or vitamin E-deficient diets and liver microsomes prepared as described previously (88). Reactions were conducted in an O_2-limiting system that was sealed to the ambient atmosphere. The reaction mixture, 3.0 ml total volume, contained 0.05 M Tris HCl, pH 7.4, 0.25 mM NADPH, 0 or 5 mM GSH, 12 μM $FeSO_4$ and approximately 1.9 mg of microsomal protein. The O_2 initially present in each reaction vial was approximately 670 nmoles. The buffer, NADPH, GSH and microsomes were combined in the reaction vessel and equilibrated at 37° in a circulating water bath for 4 minutes. Following equilibration, reaction vessels were sealed with a Clark-type electrode, and the basal rate of O_2 consumption was monitored. When the O_2 saturation had decreased to 92%, peroxidation was initiated by injecting the $FeSO_4$ solution into the sealed reaction vessel. Companion reaction vessels were used to monitor O_2 consumption continuously in addition to those for the determination of TBA-reactive products and α-tocopherol concentration at specific time points. For the latter, peroxidation was terminated at 90, 85, 75, 50, 25 and 0 percent O_2 saturation and aliquots removed for the determination of TBA-reactive products and α-tocopherol concentration. The rate of O_2 consumption was calculated from the linear region of the oxygraph tracing that occurred after the addition of $FeSO_4$ and after subtracting the basal rate of O_2 consumption that occurred in the absence of added $FeSO_4$.

Rates of O_2 consumption by peroxidizing microsomes are shown in Figure 2. These data show enhanced rates of O_2 consumption by microsomes from rats fed the diet deficient in vitamin E (-E) compared with microsomes from control rats fed the diet supplemented with vitamin E (+E). The enhanced O_2 consumption observed for -E microsomes was accelerated in the presence of 5 mM GSH. The lower rate of O_2 consumption by +E microsomes was not affected by the presence of GSH initially. After approximately 10 minutes, however, the rate of O_2 consumption in the presence of GSH was noticeably decreased.

The time-course of TBA-reactive product formation by rat liver microsomes in these experiments is shown in Figure 3. TBA-reactive product formation, like O_2 consumption, was very rapid for -E

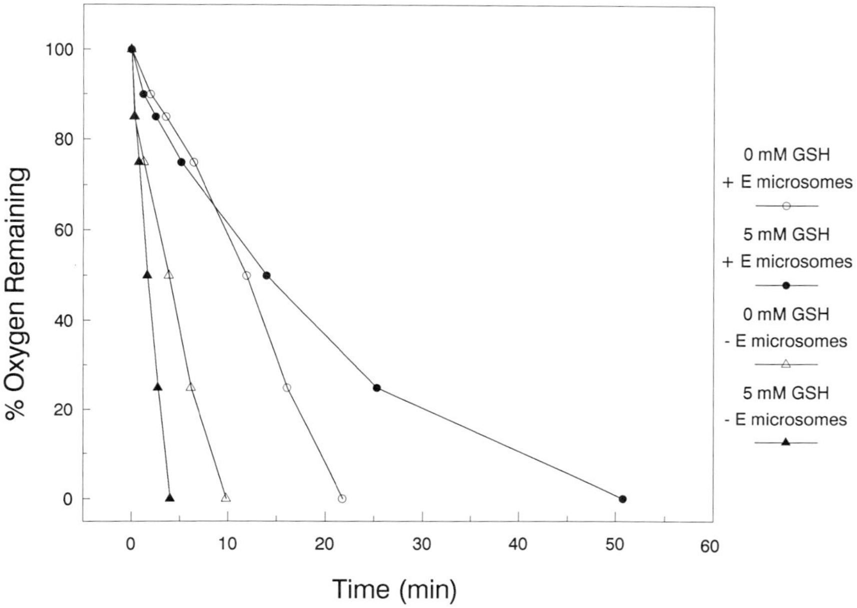

Figure 2. *Effect of GSH on Oxygen Consumption by Liver Microsomes from Rats Fed Diets Supplemented (+E) or Deficient (-E) in Vitamin E.* Each point represents a mean value from 6 rats (reproduced with permission from the New York Academy of Sciences, ref. 120).

microsomes and was increased further by 5 mM GSH. This effect may be explained by the prooxidant ability of GSH in the presence of iron (134). The +E microsomes showed a slow rate of TBA-reactive product formation initially that was independent of GSH. After approximately 5 minutes, however, TBA-reactive product formation increased markedly in reactions that did not contain GSH. In the presence of 5 mM GSH, +E microsomes peroxidized at a relatively low rate during the 50 minute interval required to deplete all the O_2 in this system.

Changes in the concentration of α-tocopherol in +E rat liver microsomes during the time-course of lipid peroxidation are illustrated in Figure 4. In the absence of GSH, approximately 60% of the α-tocopherol was depleted during the 20 minute interval required to consume all O_2 in the system. This contrasts markedly with +E microsomes incubated with 5 mM GSH in which more than 80% of the initial α-tocopherol was still present after the system had been depleted of O_2.

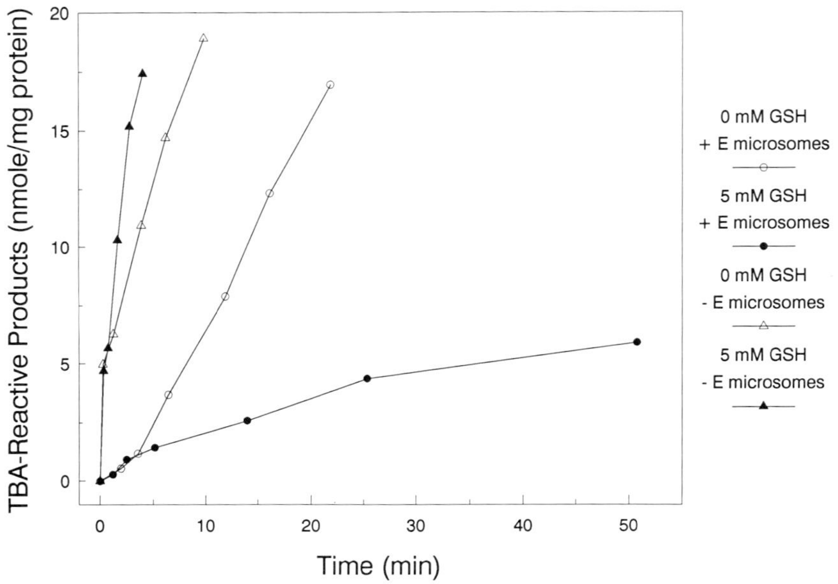

Figure 3. *Effect of GSH on TBA-Reactive Products Formation by Liver Microsomes from Rats Fed Diets Supplemented (+E) or Deficient (-E) in Vitamin E.* Each point represents a mean value from 6 rats (reproduced with permission from the New York Academy of Sciences, ref. 120).

A comparison of TBA-reactive product formation relative to O_2 consumption is shown in Figure 5. The -E microsomes, compared with +E microsomes, showed enhanced rates of TBA-reactive product formation during the early stages of O_2 consumption. This effect occurred whether or not GSH was added to the assay system. Additionally, the presence or deficiency of microsomal α-tocopherol did not limit the maximum amount of TBA-reactive products formed in this sealed system where O_2 was limiting. Rather, it was the presence of GSH that resulted in the marked reduction in TBA-reactive product formation relative to O_2 consumption in the +E rat liver microsomes. It is apparent from these data that the ratio O_2 consumed:TBA-reactive products formed is much higher in the system containing GSH and +E microsomes than in any of the other conditions examined. If this ratio were the same for all conditions (± GSH; ± microsomal α-tocopherol), then inhibition of lipid peroxidation by a mechanism of free radical scavenging could explain the GSH effect (89). Our results do not support this proposal.

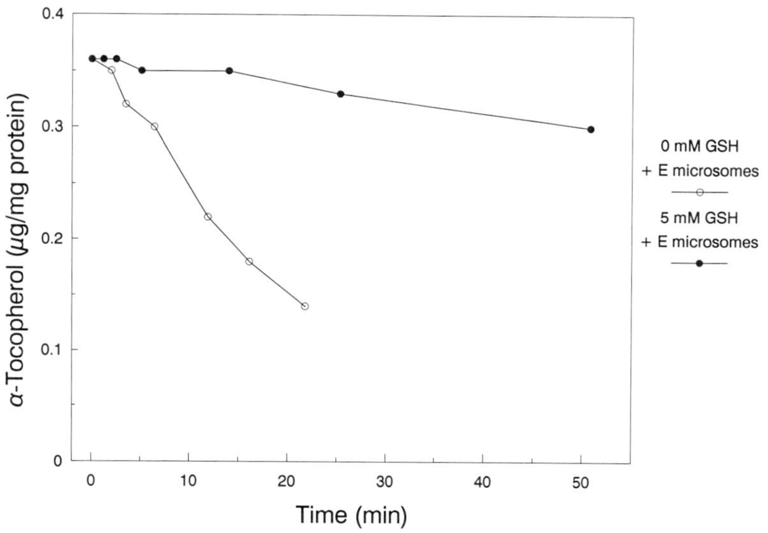

Figure 4. *Effect of GSH on α-Tocopherol Concentration in Peroxidizing Liver Microsomes from Rats Fed Diets Supplemented (+E) with Vitamin E.* Each point represents a mean value from 6 rats (reproduced with permission from the New York Academy of Sciences, ref. 120).

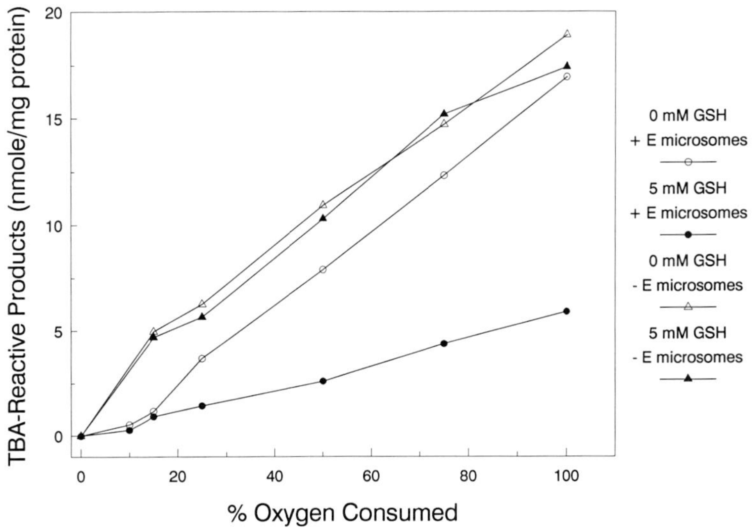

Figure 5. *Effect of GSH on TBA-Reactive Products Formation as a Function of Oxygen Consumption in Liver Microsomes from Rats Fed Diets Supplemented (+E) or Deficient (-E) in Vitamin E.* Each point represents a mean value from 6 rats.

As discussed by others (10-12, 135,136,) initiation of peroxidation of PUFA may involve an activated oxygen-iron complex capable of producing alkyl radicals which then react readily with O_2 with the formation of lipid peroxy radicals. The reactions that follow generate lipid hydroperoxides, radical species that react with more O_2, as well as the iron-catalyzed decomposition of the lipid hydroperoxides to additional reactive radical species and so on. Within this context there are at least three logical ways to inhibit the lipid peroxidation process. The first would be to quench initiators, such as an activated oxygen-iron complex. The second would be to sequester lipid peroxy radicals, thereby inhibiting the propagation phase. The third would be to prevent branching reactions by reducing lipid hydroperoxides to their corresponding unreactive alcohols. The data that we and others have obtained on GSH-dependent inhibition of microsomal lipid peroxidation may conceptually be discussed in a similar manner.

Venekei (137) has presented evidence that α-tocopherol acts to prevent the initiation of lipid peroxidation in rat liver microsomes, rather than by acting as a chain-breaking antioxidant. The effect of GSH was not the subject of this investigation, however. This hypothesis could explain the reduced O_2 consumption observed in our studies. However, if inhibition of lipid peroxidation occurred by quenching of an activated oxygen-iron complex, it would most likely be associated with the GSH-dependent factor independent of α-tocopherol. This is because the concentration of α-tocopherol generally present in membranes is not capable of effective intervention with the oxygen-iron complex (138,139). At this concentration, α-tocopherol reacts primarily with lipid peroxy radicals (139). If microsomal lipid peroxidation were inhibited by removing an initiator species, then the existence of a GSH-dependent, radical scavenging protein (89) would be an attractive mechanism to explain this effect. It is relevant to indicate that Gibson *et al.* (83) have previously reported a GSH-dependent factor in rat liver cytosol capable of inhibiting the initiation of lipid peroxidation in microsomal membranes. However, this mechanism does not explain the dependence on microsomal α-tocopherol for GSH to exert its effect (85,88). Results of our present experiments indicate that inhibition of microsomal lipid prooxidation is not simply accomplished by blocking initiation. These data show a decrease in the rate of O_2 consumption by peroxidizing microsomes in the presence of GSH but not to the extent observed by Gibson *et al.* (83). During the course of our investigations we have observed O_2 consumption initially by peroxidizing microsomes in the relative absence of TBA-reactive product formation. Venekei (137) has explained this type of result by proposing that α-tocopherol interacts with reactive forms of O_2 species capable of initiating lipid peroxidation.

An alternate mechanism for the inhibition of microsomal lipid peroxidation is that α-tocopherol prevents propagation reactions by quenching lipid peroxy radicals and that the GSH-dependent factor acts to regenerate it from the α-tocopheroxyl radical. The mechanism proposed by Burk (89,140) and Hill and Burk (92) on the radical scavenging effect of the GSH-dependent factor could also do so by preventing propagation of lipid prooxidation. This mechanism of radical scavenging does not account for the dependency we have observed on microsomal α-tocopherol, however.

Whereas there is evidence of GSH-dependent regeneration of α-tocopherol in rat liver microsomes, it is not clear whether the radical scavenging abilities of α-tocopherol in association with this factor can account for all the observations on inhibition of lipid peroxidation. The quenching of lipid peroxy radicals by α-tocopherol produces lipid hydroperoxides which in turn can continue to react with transition metal ions and the further production of additional radical species. Some of these species, for example RO•, can react with another PUFA producing R•. The alkyl radical is then able to react with O_2 to form ROO•, thus continuing the chain reation. Therefore α-tocopherol, in association with a GSH-dependent regeneration system, would not prevent the decomposition of lipid hydroperoxides. In this respect vitamin E, as an antioxidant, does not prevent lipid peroxidation but rather minimizes the number of PUFA molecules peroxidized (141). The data presented in Figure 5 show decreased TBA-reactive product formation relative to O_2 uptake by peroxidizing +E microsomes in the presence of 5 mM GSH. This suggests that lipid hydroperoxides are being removed from the branching reactions of lipid peroxidation, possibly by their reduction to alcohols by a GSH-dependent system. We have also recently reported (142) that GSSG, in combination with GSH, was more effective in inhibiting microsomal lipid peroxidation than was GSH alone. The synergism between GSSG and GSH on inhibition of lipid peroxidation was also shown for -E rat liver microsomes. This effect was not observed in heat denatured microsomes using an ascorbate/ADP/Fe nonenzymatic system to study lipid peroxidation. Additionally, no net conversion of GSSG to GSH was found in these experiments, which could otherwise account for the synergistic effects observed. It was proposed that GSSG activated, through thiol/disulfide exchange, an inactive GSH-dependent enzyme capable of 1) regenerating α-tocopherol from the α-tocopheroxyl radical, or 2) reducing lipid hydroperoxides, or both (142).

A system that reduces lipid hydroperoxides to their corresponding alcohols plus one that would regenerate α-tocopherol would be most effective in controlling the magnitude of lipid peroxidation in biological membranes. It would have the ability to impede or prevent chain-branching reactions on the one hand and the ability to block chain-propagating reactions by quenching lipid peroxy radicals on the other. Figure 6 illustrates peroxidation of PUFA in rat liver microsomes and possible sites of intervention of α-tocopherol, GSH, GSSG and the heat labile factors.

The reduction in oxygen consumption observed in native liver microsomes from rats fed the diet supplemented with vitamin E, as compared with native microsomes from rats fed the diet deficient in vitamin E, suggests an interaction at "A". By scavenging radical initiating species before they react with PUFA, vitamin E would decrease the consumption of oxygen. There is, however, little support for a significant effect of vitamin E in inhibiting lipid peroxidation at this site. If the inhibition of peroxidation occurred by quenching of an activated oxygen-iron complex, for example, it would likely be undertaken by the GSH-dependent factor because of the relatively poor ability of α-tocopherol at concentrations found in biological membranes to react with the complex. At physiological concentrations, α-tocopherol reacts effectively as a chain-breaking antioxidant by scavenging lipid peroxy radicals at "B".

As proposed previously (88), we believe that the GSH-dependent factor is involved in inhibition of microsomal lipid peroxidation by regenerating α-tocopherol from the α-tocopheroxyl radical at "C". This is in view of the important function of α-tocopherol at the membrane level in impeding propagating reactions by quenching lipid peroxy radicals. The demonstration by McCay *et al.* (86,91) that vitamin E functions as an electron shuttle for a GSH-dependent "free radical reductase" in microsomal membranes supports this proposal. These investigators have reported that the heat labile factor transfers a hydrogen atom from GSH to the α-tocopherol radical, reducing it to α-tocopherol.

Whereas the radical-scavenging action of α-tocopherol appears responsible for a major portion of the GSH and vitamin E-dependent inhibition of lipid peroxidation in our experiments, it is possible that another GSH-dependent, heat labile factor also exists that is independent of vitamin E. We have observed that lipid peroxidation is inhibited when both GSH and GSSG are added to a peroxidation system containing microsomes from rats that are deficient in vitamin E

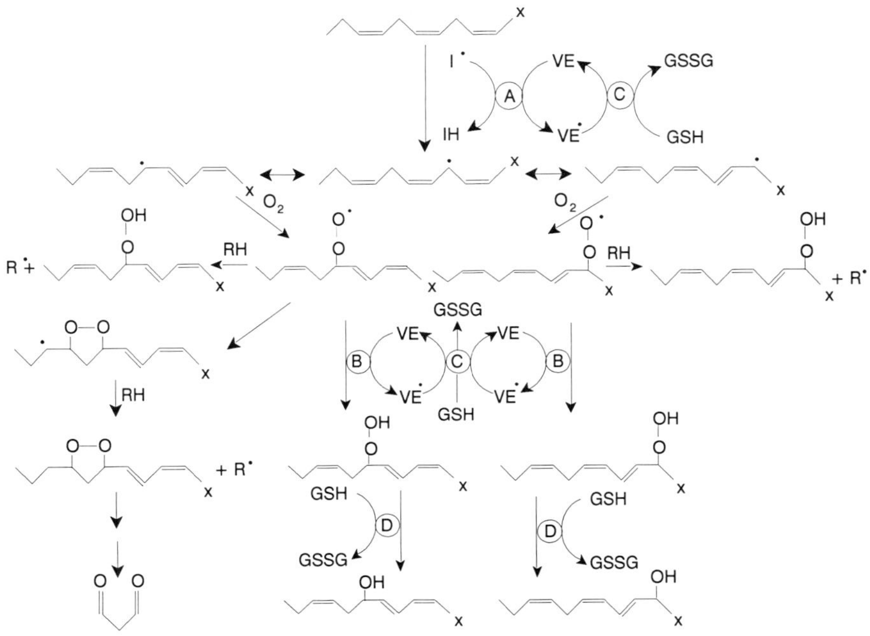

Figure 6. *Possible Sites of Intervention of Vitamin E and GSH in Peroxidation of Polyunsaturated Fatty Acid.* I = free radical initiator involved with hydrogen atom abstraction, VE = vitamin E, VE• = vitamin E radical, RH = polyunsaturated fatty acid, R• = alkyl radical, GSH = reduced glutathione, GSSG = glutathione disulfide.

(142). This factor was absent in heat denatured microsomes in the nonenzymatic lipid peroxidation assay system and therefore precluded a simple chemical interaction between GSH and GSSG. Additionally, the data presented in Figures 2 and 5 for +E microsomes show reduced oxygen consumption and reduced TBA-reactive product formation relative to oxygen consumption in assays in which GSH was included. This second GSH-dependent factor may involve the reduction of semistable fatty acid hydroperoxides to their corresponding alcohols at "D". One, or possibly both, of these GSH-dependent factors may be activated by GSSG through thiol/disulfide exchange with a protein sulfhydryl group.

References

1. Pryor, W.A. (1973) *Fed. Proc.* **32**, 1862-186
2. Pryor, W.A. (1976) in *Free Radicals in Biology* (Pryor, W. A., ed) Vol. 1, pp. 1-49, Academic Press, New York
3. Aust, S.D., and Svingen, B.A. (1982) in *Free Radicals in Biology* (Pryor, W. A., ed) Vol. 5, pp. 1-28, Academic Press, New York
4. Tien, M., and Aust, S.D. (1982) in *Lipid Peroxides in Biology and Medicine* (Yagi, K., ed) pp. 23-39, Academic Press, New York
5. Halliwell, B., and Gutteridge, J.M.C. (1984) *Biochem. J.* **219**, 1-14
6. Sevanian, A., and Hockstein, P. (1985) *Ann. Rev. Nutr.* **5**, 365-390
7. Dahle, L.K., Hill, E.G., and Holman, R.T. (1962) *Arch. Biochem. Biophys.* **98**, 253-216.
8. Pryor, W.A., Stanley, J.P., and Blair, E. (1976) *Lipids* **11**, 370-379
9. McCay, P.B., and King, M.M. (1980) in *Vitamin E, A Comprehensive Treatise* (Machlin, L. J., ed) pp. 289-317, Marcel-Dekker, New York
10. Svingen, B.A., Buege, J.A., O'Neal, F.O., and Aust, S.D. (1979) *J. Biol. Chem.* **254**, 5892-5899
11. Tien, M., Svingen, B.A., and Aust, S.D. (1981) *Fed. Proc.* **40**, 179-182
12. Ernster, L., Nordenbrand, K., and Orrenius, S. (1982) in *Lipid Peroxides in Biology and Medicine* (Yagi, K., ed) pp. 55-79, Academic Press, New York
13. Borg, D.C., and Schaich, K.M. (1988) in *Oxygen Radicals and Tissue Injury* (Halliwell, B., ed) pp. 20-26. Allen Press, Kansas
14. Gutteridge, J.M.C. (1988) in *Oxygen Radicals and Tissue Injury* (Halliwell, B., ed) pp. 9-19, Allen Press, Kansas
15. Aruoma, O.I., Halliwell, B., Laughton, M.L., Quinlan, G.J., and Gutteridge, J.M.C. (1989) *Biochem. J.* **258**, 617-620
16. Plaa, G.L., and Witschi, H. (1976) *Ann. Rev. Pharmacol. Toxicol.* **16**, 125-141
17. Buege, J.A., and Aust, S.D. (1978) *Meth. Enzymol.* **52**, 302-310
18. Logani, M.K., and Davies, R.E. (1980) *Lipids* **15**, 485-495
19. Cadenas, E., Boveris, A., and Chance, B. (1984) in *Free Radicals in Biology* (Pryor, W. A., ed) pp. 211-242, Academic Press, New York
20. Stacey, N.H., and Kappus, H. (1982) *J. Toxicol. Environ. Health* **9**, 277-285
21. Riley, C.A., Cohen, G., and Liebermann, M. (1974) *Science* **183**, 208-210
22. Kappus, H. (1985) in *Oxidative Stress* (Sies, H., ed) pp. 273-310, Academic Press, New York
23. Smith, C.V., and Anderson, R.E. (1987) *Free Rad. Biol. Med.* **3**, 341-344
24. Svingen, B.A., O'Neal, F.O., and Aust, S.D. (1978) *Photochem. Photobiol.* **28**, 803-810
25. Gutteridge, J.M.C. (1987) in *Oxygen Radicals and Tissue Injury* (Halliwell, B., ed) pp. 9-19, FASEB, Bethesda, Maryland
26. Fridovich, I. (1975) *Ann. Rev. Biochem.* **44**, 147-159
27. Chance, B., Sies, H., and Boveris, A. (1979) *Physiol. Rev.* **59**, 527-605

28. Ganther, H.E., Lawrence, D.G., Serfass, R.E., and Hoekstra, W.G. (1976) in *Trace Elements in Human Health and Disease* (Prasad, A. S., ed) Vol. II, pp. 165-234, Academic Press, New York
29. Chow, C.K. (1979) *Amer. J. Clin. Nutr.* **32**, 1066-1081
30. Flohe, L. (1982) in *Free Radicals in Biology* (Pryor, W. A., ed) Vol. 5, pp. 223-254, Academic Press, New York
31. Chow, P., and Khan, U. (1983) *Biochem. Biophys. Res. Commun.* **115**, 932-937
32. Fesseden, R.W., and Verma, N.C. (1978) *Biophys. J.* **24**, 93-101
33. Bodannes, R.S., and Chan, P.C. (1979) *FEBS Lett.* **105**, 195-200
34. Ames, B.N., Cathcart, R., Schwiers, E., and Hochstein, P. (1981) *Proc. Nat. Acad. Sci. U.S.A.* **78**, 6858-6862
35. Niki, E., Yamamoto, Y., Kamiya, Y. (1985) Chem. Lett. 1267-1270
36. Willson, R.L. (1983) in *Biology of Vitamin E,* CIBA Foundation Symposium 101 pp. 19-44, Pitman Books, London
37. Girotti, A.W., Thomas, J.P., and Jordan, J.E. (1985) *Photochem. Photobiol.* **41**, 267-276
38. Burton, G.W., and Ingold, K.U. (1984) *Science* **224**, 569-573
39. Krinsky, N.I. (1989) *Free Rad. Biol. Med.* **7**, 617-635
40. Sugioka, K., Shimosegawa, Y., and Nakano, M. (1987) *FEBS Lett.* **210**, 37-39
41. Burton, G.W., Joyce, A., and Ingold, K.U. (1983) *Arch. Biochem. Biophys.* **221**, 281-290
42. Burton, G.W., Cheeseman, K.H., Doba, T., Ingold, K.U., and Slater, T.F. (1983) in *Biology of Vitamin E,* CIBA Foundation Symposium 101 pp. 4-18, Pitman Books, London
43. Burton, G.W., and Ingold, K.U. (1986) *Acc. Chem. Res.* **19**, 194-201
44. Witting, L.A. (1980) in *Free Radicals in Biology* (Pryor, W. A., ed) Vol. 4, pp. 295-319, Academic Press, New York
45. Chen, L.H., Lee, M.S., Hsing, W.F., and Chen, S.H. (1980) *Internat. J. Vit. Nutr. Res.* **50**, 156-162
46. McCay, P.B., Poyer, J.L., Pfeifer, P.M., May, H.E., and Gilliam, J.M. (1971) *Lipids* **6**, 297-306
47. Cohen, G. (1982) in *Lipid Peroxides in Biology and Medicine* (Yagi, K., ed) pp. 191-211, Academic Press, New York
48. Hill, K.E., and Burk, R.F. (1982) *J. Biol. Chem.* **257**, 10668-10672
49. Cadenas, E., Ginsberg, M., Rabe, U., and Sies, H. (1984) *Biochem. J.* **223**, 755-759
50. Kornbrust, D.J., and Mavis, R.D. (1980) *Lipids* **15**, 315-322
51. Lucy, J.A., (1972) *Ann. N.Y. Acad. Sci.* **203**, 4-11
52. Diplock, A.T., and Lucy, J.A. (1973) *FEBS Lett.* **29**, 205-210
53. Murphy, D.J., and Mavis, R.D. (1981) *Biochim. Biophys. Acta* **663**, 390-400
54. Diplock, A.T. (1983) in *Biology of Vitamin E,* CIBA Foundation Symposium 101 pp. 45-55, Pitman Books, London

55. Sunamato, J., Baba, Y., Iwamoto, K., and Kondo, H. (1985) *Biochim. Biophys. Acta* **833**, 144-150
56. Fukuzawa, K., Chida, H., Tokumura, A., and Tsukatani, H. (1981) *Arch. Biochem. Biophys.* **206**, 173-180
57. Niki, E., Kawakani, A., Saito, M., Yamamoto, Y., Tsuchiya, G., and Kamiya, Y. (1985) *J. Biol. Chem.* **260**, 2191-2196
58. Dam, H. (1957) *Pharmacol. Rev.* **9**, 1-16
59. Tappel, A.L. (1968) *Geriatrics* **23**, 97-105
60. Machlin, L.J., Brin, M. (1980) in *Human Nutrition-A Comprehensive Treatise* (Alfin-Slater, R., and Kritcheusky, D., eds) Vol. 3B, pp. 245-266, Plenum Publishing, New York
61. Winterle, J.D., Dalin, D., and Mill, T. (1984) *J. Org. Chem.* **49**, 491-495
62. Evarts, R.P., and Bieri, J.G. (1974) *Lipids* **9**, 860-864
63. Packer, J.E., Slater, T.F., and Willson, R.L. (1979) *Nature* **278**, 737-738
64. Niki, E., Tsuchiya, R., Tanimura, R., and Kamiya, Y. (1982) *Chem. Lett.* 789-792
65. McCay, P.B. (1985) *Ann. Rev. Nutr.* **5**, 323-340
66. Nishikimi, M., Yamada, H., and Yagi, K. (1980) *Biochim. Biophys. Acta* **627**, 101-103
67. Niki, E., Saito, T., and Kamiya, Y. (1983) Chem. Lett. 631-632
68. Niki, E., Tadatoshi, S., Kawakami, A., and Kamiya, Y. (1984) *J. Biol. Chem.* **259**, 4177-4182
69. Barclay, L.R.C., Locke, S.J., and MacNeil, J.M. (1983) *Can. J. Chem.* **61**, 1288-1290
70. Leung, H., Vang, M.J., and Mavis, R.D. (1981) *Biochim. Biophys. Acta* **664**, 266-272
71. Doba, T., Burton, G.W., and Ingold, K.U. (1985) *Biochim. Biophys. Acta* **835**, 298-303
72. Bascetta, E., Gunstone, F.D., and Walton, J.C. (1983) *Chem. Phys. Lipids* **33**, 207-210
73. Scarpa, M., Rigo, A., Maiorino, M., Ursini, F., and Gregolin, C. (1984) *Biochim. Biophys. Acta* **801**, 215-219
74. Lambelet, P., Saucy, F., and Loliger, J. (1985) *Experientia* **41**, 1384-1388
75. Ingold, K.U. (1990) in *Biological Oxidation Systems* (Reddy, C.C., Hamilton, G.A., and Madyastha, K.M. eds.) Vol.II, pp. 829, Academic Press, New York
76. McCay, P.B., Gibson, D.G., Fong, K.-L., and Hornbrook, K.R. (1976) *Biochim. Biophys. Acta* **431**, 459-468
77. Burk, R.F., Trumble, M.J., and Lawrence, R.A. (1980) *Biochim. Biophys. Acta* **618**, 35-41
78. McCay, P.B., Gibson, D.D., and Hornbrook, K.R. (1981) *Fed. Proc.* **40**, 199-205
79. Gibson, D.D., Hornbrook, K.R., and McCay, P.B. (1980) *Biochim. Biophys. Acta* **620**, 572-582

80. McCay, P.B., and Gibson, D.D. (1982) in *Lipid Peroxides in Biology and Medicine* (Yagi, K., ed) pp. 179-196, Academic Press, New York
81. Fairhurst, S., Barber, D.J., Clark, B., and Horton, A.A. (1982) *Toxicol.* **23**, 249-259
82. Fairhurst, S., Barber, D.J., Clark, B., and Horton, A.A. (1983) *Biochim. Biophys. Acta* **752**, 491-496
83. Gibson, D.D., Hawrylko, J., and McCay, P.B. (1985) *Lipids* **20**, 704-711
84. Christophersen, B.O. (1968) *Biochem. J.* **106**, 515-522
85. Reddy, C.C., Scholz, R.W., Thomas, C.E., and Massaro, E.J. (1982) *Ann. N.Y. Acad. Sci.* **393**, 193-195
86. McCay, P.B., Brueggemann, G., Lai, E.K., and Powell, S.R. (1989) *Ann. N.Y. Acad. Sci.* **570**, 32-45
87. Burk, R.F. (1982) *Biochem. Pharmacol.* **31**, 601-602
88. Reddy, C.C., Scholz, R.W., Thomas, C.E., and Massaro, E.J. (1982) *Life Sci.* **31**, 571-576
89. Burk, R.F. (1983) *Biochem. Biophys. Acta* **757**, 21-28
90. Haenen, G.R.M.M., and Bast, A. (1983) *FEBS Lett.* **159**, 24-28
91. McCay, P.B., Lai, E.K., Powell, S.R., and Breuggemann, G. (1986) *Fed. Proc.* **45**, 451 (abstract)
92. Hill, K.E., and Burk, R.F. (1984) *Biochem. Pharmacol.* **33**, 1065-1068
93. Wefers, H., and Sies, H. (1988) *Eur. J. Biochem.* **174**, 353-357
94. Liebler, D.C., Kling, D.S., and Reed, D.J. (1986) *J. Biol. Chem.* **261**, 12114-12119
95. Thirot, C., Durand, P.Jasseron, M.P., Kergonou, J.F., and Ducousso, R. (1987) *Biochem. Int.* **14**, 1-8
96. Robey, S., and Mavis, R. (1986) *Fed. Proc.* **45**, 1671 (abstract)
97. Murphy, M.E., and Kehrer, J.P. (1989) *Arch. Biochem. Biophys.* **268**, 585-593
98. Murphy, M.E., and Kehrer, J.P. (1987) *J. Chromatog.* **421**, 71-82
99. Graham, K.S., Reddy, C.C., and Scholz, R.W. (1989) *Lipids* **24**, 909-914
100. Jacoby, W.B. (1978) in *Functions of Glutathione in Liver and Kidney* (Sies, H., and Wendel, A., eds) pp. 157-163, Springer Verlag, Berlin
101. Friedberg, T., Bentley, P., Stasiecki, P., Glatt, H.R., Raphael, D., and Oesch, F. (1979) *J. Biol. Chem.* **254**, 12028-12033
102. Morgenstern, R., Meijer, J., DePierre, J.W., and Ernster, L. (1980) *Eur. J. Biochem.* **104**, 167-174
103. Morgenstern, R., Lundquist, G., Anderson, G., Balk, L., and DePierre, J.W. (1984) *Biochem. Pharmacol.* **33**, 3609-3614
104. Reed, D.J., and Tirmenstein, M.A. (1989) *Biochim. Biophys. Acta* **995**, 174-180
105. Reddy, C.C., Tu, C.-P.D., Burgess, J.R., Ho, C.-Y., Scholz, R.W., and Massaro, E.J. (1981) *Biochem. Biophys. Res. Commun.* **101**, 970-978
106. Morgenstern, R., and DePierre, J.W. (1983) *Eur. J. Biochem.* **134**, 591-597
107. Yonaha, M., and Tampo, Y. (1986) *Chem. Pharm. Bull.* **34**, 4195-4201

108. Morgenstern, R., and DePierre, J.W. (1985) in *Reviews in Biochemical Toxicology* (Hodgson, E., Bend, J. R., and Philpot, R. M., eds) Vol. 7, pp. 67-104, Elsevier, New York
109. Yonaha, M., and Tampo, Y. (1987) *Biochem. Pharmacol.* **36**, 2831-2837
110. Morgenstern, R., Lundquist, G., Jornvall, H., and DePierre, J.W. (1989) *Biochem. J.* **260**, 577-582
111. Morgenstern, R., and DePierre, J.W. (1988) in *Glutathione Conjugation, Its Mechanism and Biological Significance* (Ketterer, B., and Sies, H. V., eds) pp. 157-174, Academic Press, New York
112. Morgenstern, R., DePierre, J.W., and Ernster, L. (1979) *Biochem. Biophys. Res. Commun.* **87**, 657-663
113. Morgenstern, R., Guthenberg, C., and DePierre, J.W. (1982) *Eur. J. Biochem.* **128**, 243-248
114. Morgenstern, R., Guthenberg, C., Mannervik, B., and DePierre, J.W. (1983) *FEBS Lett.* **160**, 264-268
115. Nishino, H., and Ito, A. (1989) *Biochem. Int.* **19**, 731-735
116. Aniya, Y. (1989) *J. Pharmacobio-Dyn.* **12**, 235-240
117. Masukawa, T., and Iwata, H. (1986) *Biochem. Pharmacol.* **35**, 435-438
118. Aniya, Y., and Anders, M.W. (1989) *Arch. Biochem. Biophys.* **270**, 330-334
119. Aniya, Y., and Anders, M.W. (1989) *J. Biol. Chem.* **264**, 1998-2002
120. Scholz, R.W., Graham, K.S., Gumpricht, E., and Reddy, C.C. (1989) *Ann. N.Y. Acad. Sci.* **570**, 514-517
121. Mosialou, E., and Morgenstern, R. (1989) *Arch. Biochem. Biophys.* **275**, 289-294
122. Mannervik, B., and Danielson, U.H. (1985) *Biochem. J.* **231**, 263-267
123. Mannervik.B., Esterbauer, H., and Danielson, U.H. (1987) *Biochem. J.* **247**, 707-713
124. Haenen, G.R.M.M., Tsoi, J.N.L.T.T., Vermeulen, N.P.E., Timmerman, H., and Bast, A. (1987) *Arch. Biochem. Biophys.* **259**, 449-456
125. Sevanian, A., Muakkassah-Kelly, S.F., and Montestruque, S. (1983) *Arch. Biochem. Biophys.* **223**, 441-452
126. Tan, K.H., Meyer, D.J., Belin, J., and Ketterer, B. (1984) *Biochem. J.* **220**, 243-252
127. Nagasaka, Y., Fujii, S., and Kaneko, T. (1989) *Arch. Biochem. Biophys.* **274**, 82-86
128. Ursini, F., Maiorino, M., Valente, M., and Gregolin, C. (1982) *Biochim. Biophys. Acta* **710**, 197-211
129. Ursini, F., Maiorino, M., and Gregolin, C. (1985) *Biochim. Biophys. Acta* **839**, 62-70
130. Maiorino, M., Coassin, M., Roveri, A., and Ursini, F. (1989) *Lipids* **24**, 721-726
131. Maiorino, M., Ursini, F., Leonelli, M., Finato, N., and Gregolin, C. (1982) *Biochem. Int.* **5**, 575-583

132. Ursini, F., Maiorino, M., and Gregolin, C. (1983) in *Oxy Radicals and their Scavenger Systems* (Greenwall, R. A., and Cohen, G., eds) Vol. II, pp. 224-229, Elsevier, New York
133. Ursini, F., and Bindoli, A. (1987) *Chem. Phys. Lipids* **44**, 255-276
134. Tien, M., Bucher, J.R., and Aust, S.D. (1982) *Biochem. Biophys. Res. Commun.* **107**, 279-285
135. Bucher, J.R., Tien, M., and Aust, S.D. (1983) *Biochem. Biophys. Res. Commun.* **111**, 777-784
136. Tien, M., and Aust, S.D. (1982) in *Lipid Peroxides in Biology and Medicine* (Yagi, K., ed) pp. 23-39, Academic Press, New York
137. Vanekei, I. (1987) *Biochim. Biophys. Acta* **917**, 347-355
138. Fukuzawa, K., Chida, H., Tokumura, A., and Tsukatani, H. (1981) *Arch. Biochem. Biophys.* **206**, 173-180
139. Fukuzawa, K., Tokumura, A., Ouchi, S., and Tsukatani, A. (1982) *Lipids* **17**, 511-513
140. Burk, R.F., Patel, K., and Lane, J.M. (1983) *Biochem. J.* **215**, 441-445
141. Witting, L.A. (1965) *Fed. Proc.* **24**, 912-916
142. Scholz, R.W., Graham, K.S., and Reddy, C.C. (1990) *Biochem. Biophys. Res. Commun.* **166**, 960-966

Glutathione and Vitamin E and the Control of Lipid Peroxidation of Rat Liver Nuclei

Donald J. Reed and Mark A. Tirmenstein
Department of Biochemistry and Biophysics, Oregon State University, Corvallis, OR 97331 USA

Introduction

Lipid peroxidation is thought to be an important biological consequence of oxidative cellular damage (1). The destruction of unsaturated fatty acids, which occurs in lipid peroxidation, has been linked with altered membrane structure (2) and enzyme inactivation (3). In addition to lipid hydroperoxides and lipid radicals (4), lipid peroxidation may generate activated oxygen species such as hydroxyl radicals (5) and superoxide anions (6). The decomposition of peroxidized polyunsaturated fatty acids also generates reactive carbonyl compounds such as malondialdehyde (MDA) and hydroxyalkenals (7). Several studies indicate that such lipid peroxidation products can alter the structure and function of DNA (8-11).

Evidence strongly suggests that lipid peroxidation can cause extensive damage to subcellular organelles and biomembranes. Lipid peroxidation has been demonstrated to occur in isolated mitochondria, lysosomes and microsomes (12). Few studies, however, have focused on the susceptibility of the cell nucleus to lipid peroxidation. The nuclear membrane regulates the transport of mRNA into the cytoplasm and aids in the process of nuclear division. DNA is also frequently associated with certain regions of the nuclear membrane (13). It seems likely that nuclear membrane peroxidation may disrupt many of these critical functions. The proximity of the nuclear membrane to DNA could also contribute to the interaction of DNA with reactive compounds generated in lipid peroxidation. This fact is of importance since research indicates that hydroxyl radicals diffuse an average of only 60 angstroms before reacting with cellular components (14). Nuclear peroxidation may also increase interactions between more stable peroxidation products and DNA. The cytosolic enzymes aldehyde dehydrogenase (15), glutathione *S*-transferase (16) and glutathione peroxidase (17) have all been shown to metabolize various reactive lipid peroxidation products. Such cytosolic enzymes may metabolize peroxidation products generated

Biological Oxidation Systems, Volume 2

throughout the cell before they diffuse into the nucleus and interact with DNA.

The nuclear membrane system contains numerous sites at which outer and inner nuclear membranes are joined. These sites are referred to as nuclear pore complexes with an ultrastructure that is remarkably constant in a wide variety of eukaryotic cell types (13). Since the functional pore radius in isolated rat liver nuclei is 5.6-5.9 mm (18), these pores allow the free diffusion of small water-soluble constituents such as glutathione (GSH) between the nucleus and cytoplasm (19). This finding helps to confirm the semiporous nature of the nuclear envelope (20-22).

However, the confinement of GSH-synthesizing or metabolizing enzymes within the cytoplasm produces disparities between nuclear and cytoplasmic GSH levels under certain metabolic conditions. The relative concentration of nondiffusible GSH-binding macromolecules in the cytoplasm and nucleus may produce the distinctively lowered nuclear pool of GSH which is observed during inhibition of GSH synthesis by buthionine sulfoximine (19).

GSH is thought to play an important role in protecting cells against oxidative stress via the glutathione redox system. In addition, GSH protects microsomal incubations against both NADPH and ascorbate-induced lipid peroxidation (23-25). The mechanism involved in this protection remains uncertain, but results suggest that a microsomal protein is required for the GSH-dependent inhibition of lipid peroxidation (23).

The classical selenium dependent glutathione peroxidase (17) and certain glutathione *S*-transferases (26) have been shown to reduced lipid hydroperoxides to lipid alcohols and repair of DNA. Since lipid hydroperoxides have been shown to be involved in the propagation of lipid peroxidation (27), the reduction of these compounds would be expected to protect against lipid peroxidation.

Belouqui and Cederbaum showed that the addition of GSH produced a 40% reduction in NADPH-induced peroxidation of rat liver microsomes (28). However, the addition of the soluble enzymes glutathione peroxidase, glutathione reductase and glutathione *S*-transferase together with GSH did not significantly increase the protective capacity of GSH in these incubations.

In a similar set of experiments conducted by Gibson *et al.* (29), heat denatured hepatic microsomes initiated with ascorbate and ferric iron were not protected by the addition of soluble glutathione *S*-transferase or glutathione peroxidase fractions in the presence of GSH (29). Gibson *et al.* (29) concluded that lipid hydroperoxides must be free in solution if they are to be substrates for the soluble enzymes glutathione *S*-transferase and glutathione peroxidase. In this case, the lipophilic environment of membranes apparently restricts interactions between enzymes located in the aqueous phase and lipid hydroperoxides located in the lipid bilayer. The association of a glutathione-dependent peroxidase with membranes may encourage the reduction of lipid hydroperoxides located within lipid bilayers. The existence of a membrane-associated peroxidase activity has been proposed by Ursini *et al.* (30). These researchers purified a selenium-containing glutathione peroxidase from pig heart which is distinct from the classical glutathione peroxidase. Phospholipid hydroperoxides incorporated into liposomes were reduced by this enzyme but not by the classical glutathione peroxidase. Ursini *et al.* (30) concluded that this glutathione-dependent peroxidase is interfacial in character.

In addition to the peroxidase described by Ursini *et al.* (30), a membrane-bound glutathione *S*-transferase possessing glutathione peroxidase activity has been purified from rat liver microsomes (31). The solubilized enzyme possesses a monomer molecular weight of 14,000 and has been shown to reduce cumene hydroperoxide. This enzyme is unique with respect to the soluble glutathione *S*-transferases in that it is activated by N-ethylmaleimide (NEM) (32). At present, no research has investigated whether this enzyme can inhibit lipid peroxidation.

GSH, either alone (23) or in conjunction with added protein (33, 34), can protect microsomes against lipid peroxidation. Addition of GSH produced a 40% reduction in NADPH-induced peroxidation without addition of rat liver microsomes of other proteins (28). A microsomal glutathione *S*-transferase has been purified from human liver and found to have characteristics similar to those of the rat microsomal *S*-transferase as described by Morgenstern and De Pierre (31, 35).

Glutathione *S*-transferase activity present in rat liver microsomes may be regulated by reversible thiol/disulfide exchange. Activation of microsomal glutathione *S*-transferase by diamide or cystamine can be reversed by the addition of DTT whereas GSSG increased activity of only membrane-bound enzyme (36).

Mouse liver microsomal glutathione *S*-transferase has been purified in the NEM-activated as well as in an unactivated form. Inhibitor studies indicate a cooperative effect involving conversion of activated enzyme into the unactivated form (37). Lipid peroxidation inhibits microsomal glutathione *S*-transferase from rat and such inhibition appears modulated by dietary antioxidants (38).

Administration of allyl alcohol to rats leads to an activation of microsomal glutathione *S*-transferase and reduction of the glutathione dependent protection against lipid peroxidation. It is proposed that the activation of glutathione *S*-transferase *in vivo* by allyl alcohol proceeds via alkylation of the thiol group of the glutathione *S*-transferase by acrolein (39). Further, these workers speculate that acrolein decreases the GSH dependent protection against lipid peroxidation by alkylation of the thiol(s) group on a proposed vitamin E free radical reductase.

In this regard, Gibson *et al.*, (34) have reported that a cytosolic, GSH-dependent protein can protect microsomal membranes against peroxidation. GSH did not appear to be utilized in glutathione peroxidase activity but rather in the inhibition of the initiation of peroxidation possibly by serving in a reduction system for α-tocopherol (34).

A recent report provides indirect evidence for a rat liver free radical reductase in mitochondria and microsomal membranes that prevents vitamin E (chromanoxyl) radical accumulation (40). GSH increased the efficacy of NADPH in preventing the accumulation of the chromanoxyl radical, but was without effect in the absence of NADPH. NADH, NADPH and ascorbate prevent the accumulation leading these workers to speculate that microsomes and mitochondrial have both enzymatic and non-enzymatic mechanisms for reducing chromanoxyl radicals. These findings support the earlier speculation that GSH delays microsomal lipid peroxidation via the reduction of vitamin E radicals, which purportedly is catalyzed by a free radical reductase (41) and explains that lipoic acid is limited to non-enzymatic mechanism for protection against microsomal lipid peroxidation (42).

Little is known about the susceptibility of the cell nucleus to lipid peroxidation and possible damage to nuclear nucleic acids. The nuclear glutathione peroxidase activity appears to repair radical damage to DNA (43). Certain glutathione transferases of the rat liver nucleus possess glutathione peroxidase activity which reduces DNA hydroperoxides (43).

Since GSH and vitamin E are known to play an important role in protecting cells against oxidative stress, the ability of GSH and vitamin E to protect isolated rat liver nuclei against NADPH induced peroxidation was examined.

Methods and Materials

The *Methods and Materials* are as described in previous publications (44-46).

Results

Nuclei were isolated from rat liver homogenates by discontinuous sucrose gradient centrifugation, and lipid peroxidation was induced by 1.7 mM ADP, 0.11 mM EDTA, 0.1 mM $FeCl_3$, and either 1 mM NADPH or 0.5 mM ascorbate. The amount of lipid peroxidation was determined by measuring the formation of thiobarbituric acid-reactive products and the disappearance of lipid unsaturated fatty acid moieties. The addition of GSH (0.1 to 1.0 mM) produced a concentration-dependent lag period prior to the onset of lipid peroxidation (Figure 1). This GSH-induced lag period was abolished by pretreatment of nuclei with trypsin, thiol modifying reagents, disulfides (Table 1), or heating nuclei at 60°C for 15 min. Nuclei which were incubated with GSH also catalyzed the conversion of cumene hydroperoxide to cumyl alcohol (Table 2). Similarly, this activity was also inhibited by thiol modifying reagents, disulfides, and heating nuclei at 60°C for 15 min (44).

Glutathione was found to protect isolated rat liver nuclei against lipid peroxidation by inducing a lag period prior to the onset of peroxidation. This GSH-dependent protection was abolished by exposing isolated nuclei to the glutathione *S*-transferase inhibitor *S*-octylglutathione (Figure 2). In incubations containing 0.2 mM *S*-octylglutathione, the GSH-induced lag period was reduced from 30 to 5 minutes. *S*-octylglutathione (0.2 mM) also completely inhibited nuclear glutathione *S*-transferase activity and glutathione peroxidase activity by 85%. About 70% of the glutathione *S*-transferase activity associated with isolated nuclei was solubilized with 0.3% Triton X-100 (Table 3). This solubilized glutathione *S*-transferase activity was partially purified by utilizing a *S*-hexylglutathione affinity column (45). The partially purified nuclear glutathione *S*-transferase exhibited glutathione peroxidase activity towards lipid hydroperoxides in solution (Table 4). The data indicate that a glutathione *S*-transferase associated with the nucleus may contribute to glutathione dependent-protection of isolated nuclei against lipid peroxidation. This enzyme is distinct from the microsomal glutathione *S*-transferase.

Endogenous α-tocopherol levels in isolated rat liver nuclei were determined to be 0.045 mol% (mol α-tocopherol per mol phospholipid 100) (Table 5). This value corresponds to 970 polyunsaturated fatty acid (PUFA) moieties to one molecule of α-tocopherol in the nuclear membrane. Isolated nuclei, when incubated with various concentrations of exogenous α-tocopherol, took up only a small percent of initial levels of α-tocopherol present in the incubation media

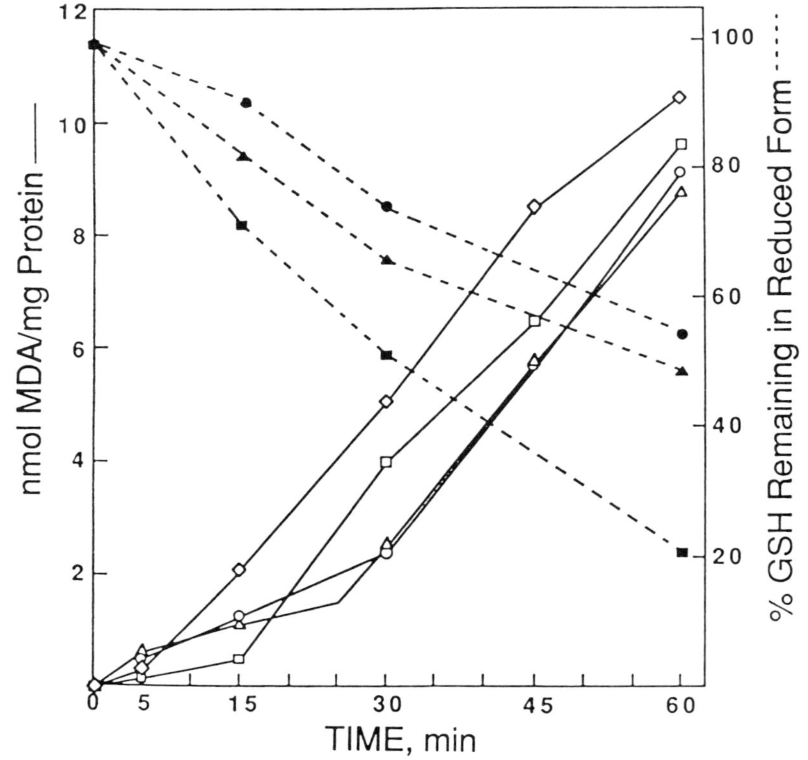

Figure 1. *The Effects of GSH on NADPH-Induced Nuclear Lipid Peroxidation.* Nuclear suspensions were incubated with the NADPH peroxidation system. Both nmol MDA/mg protein (—) and % GSH remaining in the reduced form (---) are plotted verses time of incubation. (\) No GSH, (□ , ■) 0.1 mM GSH, (△ , ▲) 0.5 mM GSH and (○ , ●) 1.0 mM GSH. Filled symbols represent data points for the % GSH remaining in the reduced form verses time plot. The incubation conditions and the NADPH peroxidation system are described in the experimental procedures section. All data points were means of n ≥ 3 experiments and S.D. ≥ 15%.

(46). Exogenous α-tocopherol, when incorporated in isolated nuclei above a threshold value of 0.085 mol%, effectively inhibited NADPH-induced lipid peroxidation. The addition of 1 mM glutathione lowered the threshold levels of α-tocopherol needed to inhibit lipid peroxidation to about 0.040 mol% (Table 6). The data indicate a glutathione-dependent enhancement of the ability of α-tocopherol to inhibit nuclear lipid peroxidation.

Table 1. *Effects of Disulfides and Thiol Modifying Reagents on GSH-Dependent Inhibition of Lipid Peroxidation*

Preincubation Conditions[a,b]	Concentration (mM)	Duration of lag prior to the onset of peroxidation (min)
Disulfides[c]		
Cystine	0.5	0
Cystine	0.1	10
Cystine	0.05	20
GSSG	10	0
GSSG	5	5
GSSG	1	15
GSSG	0.25	23
Cystamine	5	0
Cystamine	1	15
Cystamine	0.5	25
No Addition	--	30
Thiol modifying reagents[d]		
NEM	0.6	0
No Addition	--	25

[a] Preincubation conditions consisted of nucleic suspended in incubation buffer containing the disulfide or thiol modifying reagent and incubated at 37°C for 15 min. All values represent means of $n \geq 3$ experiments and SD $\leq$ 15%.

[b] Following preincubation, nuclei were pelleted and then resuspended in incubation buffer containing 1 mM GSH and either NADPH or ascorbate peroxidation system as indicated.

[c] NADPH peroxidation system.

[d] Ascorbate peroxidation system.

Discussion

NADPH-induced nuclear peroxidation was measured by monitoring the loss of unsaturated fatty acids and the formation of MDA. The results from these studies show that nuclear constituents are susceptible to lipid peroxidation. Analysis of fatty acid composition data indicated that the polyunsaturated arachidonic and docosahexaenoic acid fractions were primarily destroyed by peroxidation. Published reports indicate that MDA is formed from unsaturated fatty acids containing three or more double bonds (47). A correlation between loss of polyunsaturated fatty acids and the formation of MDA was found. GSH inhibited both MDA formation and loss of polyunsaturated fatty acids in isolated nuclei.

Table 2. *Cumene Hydroperoxide Levels Following Incubations with 1 mM GSH at 37 °C for 15 min: The Effects of Preincubations with Cystine, NEM, and Heat*

Preincubation Conditions[a]	Concentration (mM)	Cumene Hydroperoxide Levels Following Incubation[b] (μmol)
Controls[c]	---	0.40
Cystine	0.5	0.30
NEM	0.6	0.29
15 min at 60°C	---	0.39
No addition	---	N.D.[d]

[a] Preincubations consisted of exposing nuclear suspensions for 15 min to the designated treatment. Unless otherwise noted, all preincubations occurred at 37°C. All values are the means of three experiments and SD ≤ 15%.

[b] Following preincubations, nuclei were pelleted and then resuspended in incubation buffer containing 1 mM GSH. Nuclear suspensions were incubated at 37°C for 15 min. At time zero, 0.80 μmol cumene hydroperoxide was added.

[c] Control preincubations had no nuclei.

[d] Not detectable (less than 0.02 μmol).

Table 3. *Partial Purification of Nuclear Glutathione S-Transferase*

	Yield (%)	Specific Activity Milliunits (mU[a] per mg protein)	Purification Factor
Whole nuclei	100.0	9.4	---
Detergent solubilized fraction[b]	70.9	68.3	7.3
Affinity column[c]	35.6	679.0	72.5

[a] Milliunits of glutathione peroxidase activity is defined as the amount of enzyme catalyzing the conjugation of 1 nmol GSH with 1 nmol CDNB per min at 25°C.

[b] Nuclei were mixed with 0.3% Triton, 100 mM NaCl TM buffer containing 10% glycerol, and 1 mM EDTA. Nuclear suspensions were then centrifuged at 15,000 x g for 2 min. Supernatants were retained for glutathione *S*-transferase assays.

[c] *S*-hexylglutathione-agarose affinity column.

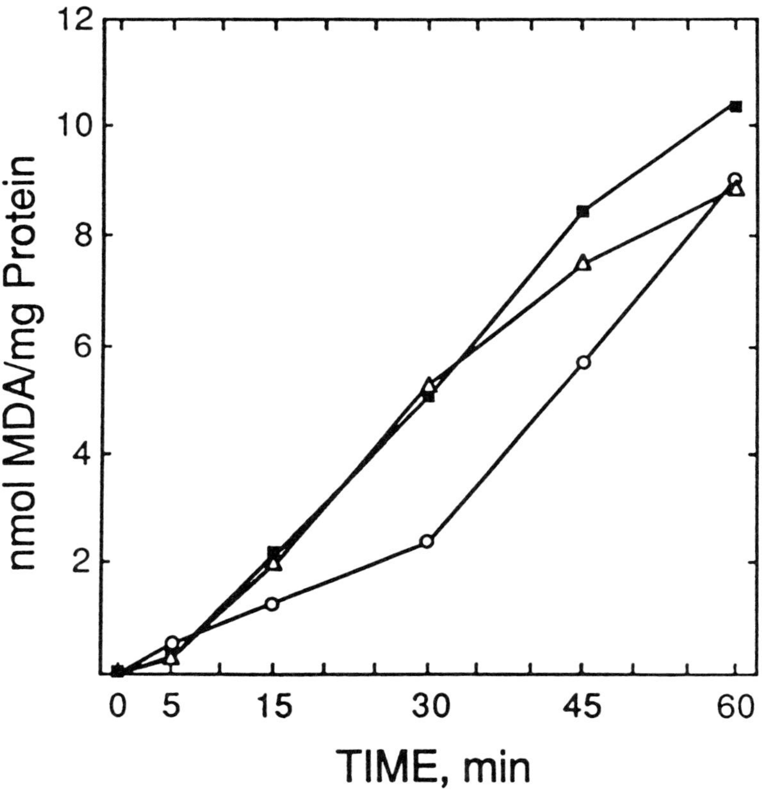

Figure 2. *Effects of S-Octylglutathione on the GSH Dependent Inhibition of NADPH-Induced Nuclear Lipid Peroxidation.* (■) No GSH, (○) 1 mM GSH and (△) 1 mM GSH and 0.2 mM S-octylglutathione. Nuclei were preincubated with NADPH and the indicated additions for 10 min prior to the initiation of peroxidation. The incubation conditions and the NADPH peroxidation system are described (46).

This is the first report that GSH can inhibit nuclear lipid peroxidation. The initial concentration of GSH determined the duration of the GSH-induced lag in peroxidation with higher concentrations of GSH extending the lag prior to the rapid peroxidation of unsaturated fatty acids. These effects occurred at concentrations of GSH below those typically found intracellularly, which range from 0.5 mM to 10 mM (48), depending upon the tissue examined. Interestingly enough, the ratio of GSH to GSSG was similar for all concentrations of GSH tested at the end of the lag period. About 70% of the GSH remained in the reduced form immediately prior to the onset of rapid peroxidation. It may be that the ratio of GSH to GSSG may be more important than the absolute concentration of GSH in determining whether the protein can inhibit lipid peroxidation. Although the mechanism for this effect is not known, GSSG may compete with GSH to bind at the protein and thereby slow the rate at which lipid peroxides are reduced and rapid peroxidation ensues.

Table 4. *GSH-Dependent Peroxidase Activity of Partially Purified Nuclear Glutathione S-Transferase*

Hydroperoxide	Specific Activity (U[a] per mg protein)
Hydrogen peroxide	N.D.[b]
Cumene hydroperoxide	0.567
Lipid hydroperoxide[c]	0.354
tert-Butyl hydroperoxide	0.080

[a] Units of glutathione peroxidase activity is defined as the amount of enzyme catalyzing the oxidation of 1 μmol GSH per min at 25°C.

[b] Not detectable (less than 0.028 units per mg protein).

[c] 14-Hydroperoxynonadecadienoic acid.

Table 5. *Time Course of α-Tocopherol (α-TH) Depletion and Induced Lipid Peroxidation in Isolated Rat Liver Nuclei*

	α-TH Supplemented Nuclei[a,b]		Control Nuclei[b]	
Time (min)	α-TH (mol%)	nmol MDA/mg Protein	α-TH (mol%)	nmol MDA/mg Protein
0	0.290	0.05	0.045	0.04
15	0.221	0.19	N.D.[c]	3.52
30	0.279	0.21	N.D.	6.50
45	0.244	0.27	N.D.	8.29
60	0.265	0.28	N.D.	9.74

[a] Nuclei were supplemented with α-TH according to the methods in the *Methods and Materials* section.

[b] One experiment typical of five separate experiments.

[c] Not detectable.

Table 6. *The Effects of GSH on the Time Course of α-Tocopherol (α-TH) Depletion and Induced Lipid Peroxidation in Isolated Rat Liver Nuclei Supplemented with α-TH*

	-GSH[a]		+GSH[a,b]	
Time (min)	α-TH (mol%)	nmol MDA/mg Protein	α-TH (mol%)	nmol MDA/mg Protein
0	0.107	0.08	0.099	0.17
15	0.118	0.37	0.082	0.23
30	0.087	1.15	0.078	0.22
45	0.034	3.72	0.077	0.23
60	N.D.[c]	6.51	0.099	0.28

[a] One experiment typical of four separate experiments.

[b] One mM GSH.

[c] Not detectable.

A protein is involved in the GSH-dependent inhibition of peroxidation protection afforded by GSH. Since GSH produced a lag prior to peroxidation, the GSH-dependent inhibition of peroxidation is not due to a nonspecific thiol effect. However, substrate specificity may be involved that causes a GSH dependence. Specificity for GSH was also demonstrated by Burk (23) in the inhibition of microsomal lipid peroxidation. No inhibition was seen when GSH was replaced by cysteine, DTT, or β-mercaptoethanol in this system (23). The cytosolic protein investigated by Gibson *et al.* (34) had much broader substrate specificity requirements. Cysteine, DTT, and β-mercaptoethanol all inhibited the lipid peroxidation of heated microsomes when incubated in the presence of this cytosolic protein (34). The GSH-dependent peroxidase purified by Ursini *et al.* (33) inhibited liposomal lipid peroxidation with either β-mercaptoethanol or GSH. However, concentrations of GSH above 1 mM were required to inhibit liposomal lipid peroxidation (33).

The GSH-dependent inhibition of lipid peroxidation was also affected by thiol modifying reagents and disulfides. Pretreatment with NEM abolished the protective effects of GSH. The disulfides GSSG, cystamine, and cystine produced similar inhibition but at varying concentrations. These data support the hypothesis that a labile protein is involved in the GSH-dependent inhibition of peroxidation and also suggests that an essential sulfhydryl group is required for its activity. The activity of many enzymes containing an essential sulfhydryl group can be modified by treatment with disulfides (49). This inhibition is mediated through the formation of protein-mixed disulfides (49).

Another potent inhibitor of the GSH-dependent protection against peroxidation is cumene hydroperoxide. Nuclei catalyze the reduction of cumene hydroperoxide to cumyl alcohol. Pretreatment of nuclei with 0.5 mM cystine, 0.6 mM NEM, or exposure to 60°C temperatures all inhibited the loss of cumene hydroperoxide. As previously demonstrated, these treatments also affected the ability of GSH to protect against lipid peroxidation. Therefore the peroxidase activity associated with nuclei appears to be responsible for the GSH dependent inhibition of lipid peroxidation.

Isolated nuclei catalyzed the GSH-dependent reduction of *tert*-butyl hydroperoxide and cumene hydroperoxide but failed to metabolize hydrogen peroxide. This substrate specificity pattern is consistent with those exhibited by the glutathione *S*-transferases (50, 51).

Incubation of isolated nuclei with *S*-octylglutathione abolished the GSH-dependent inhibition of peroxidation. In addition, *S*-octylglutathione inhibited the conjugation of CDNB with GSH (glutathione *S*-transferase activity) and the GSH-dependent reduction of cumene hydroperoxide (glutathione peroxidase activity) catalyzed by isolated nuclei. These data suggest that a nuclear glutathione *S*-transferase is responsible for the GSH-dependent inhibition of nuclear lipid peroxidation.

Approximately 70% of the nuclear glutathione *S*-transferase activity (as measured by CDNB conjugation) was solubilized by treating isolated nuclei with 0.3% Triton X-100. Partial purification of this detergent-solubilized glutathione *S*-transferase activity was achieved with a *S*-hexylglutathione affinity column. Following these procedures, specific activity measurements were conducted with the partially purified fraction and a variety of substrates. Activity was detected towards CDNB, cumene hydroperoxide, lipid hydroperoxides, and *tert*-butyl hydroperoxide. However, no activity was detected towards hydrogen peroxide.

Substrate specificities indicate the enzyme described by Ursini *et al.*, (30, 33) is distinct from the nuclear glutathione *S*-transferase. Our data also suggest that the microsomal enzyme which is activated by NEM (52) is distinct from the partially purified nuclear glutathione *S*-transferase described in this study. NEM inhibited rather than activated the nuclear enzyme, and no protein was detected below a molecular weight of 21,500 when the partially purified nuclear GSH *S*-transferase was analyzed by SDS-polyacrylamide gel electrophoresis.

Only slight glutathione-dependent peroxidase activity was detected when the nuclear glutathione *S*-transferase fraction or the selenium-containing glutathione peroxidase was incubated with peroxidized liposomes. However, this activity was increased by prior treatment of the peroxidized liposomes with phospholipase A_2. In the case of the selenium-containing glutathione peroxidase, phospholipase A_2 treatment increased the glutathione-dependent peroxidase activity sevenfold, whereas phospholipase A_2 treatment increased the activity of nuclear glutathione *S*-transferase threefold. The results with the selenium-containing glutathione peroxidase agree with those seen by Sevanian *et al.* (53). These researchers concluded that the selenium-containing glutathione peroxidase could not efficiently interact with lipid hydroperoxides incorporated into liposomes. Phospholipase A_2 treatment releases these lipid hydroperoxides from the liposomal membrane and allows the subsequent interaction of these compounds with the enzyme.

The nuclear glutathione *S*-transferase reduces lipid hydroperoxides in solution, but only low glutathione-dependent peroxidase activities were seen when peroxidized liposomes were incubated with the enzyme. It may be that the enzyme must be incorporated into membranes in order to act on lipid hydroperoxides still associated with the hydrophobic environment of the membrane. Experiments suggest that the partially purified glutathione *S*-transferase can inhibit the peroxidation of liposomes if it is reconstituted into liposomal membranes. Thus, some reduction of lipid hydroperoxides still associated with the membrane environment by the enzyme must occur.

Our findings suggest that a glutathione *S*-transferase associated with the nuclear membrane can utilize GSH to reduce lipid hydroperoxides. Since lipid hydroperoxides have been shown to be involved in the propagation of lipid peroxidation, this enzyme may be responsible for the GSH-dependent protection of nuclear membranes against lipid peroxidation and the protection of DNA against lipid peroxidation damage. Such protection would be complementary along with the vitamin E protection to provide a sparing effect for vitamin E.

Endogenous α-TH levels in isolated rat liver nuclei were determined to be 0.045 mol%. Calculations estimate that this level correlates with a ratio of one molecule of α-TH to 970 PUFA moieties in the nuclear membrane. We were unable to find other values in the literature for nuclear levels of α-TH in isolated rat liver nuclei. Other rat liver subcellular organelles have reported values of 3313 and 2100 for the ratio of PUFA moieties to α-TH in microsomes (54) and mitochondria (55) respectively.

Isolated rat liver nuclei incorporated exogenous α-TH, however, the percent of the initial α-TH incorporated in nuclei after washing is relatively low. Once taken up by membranes, the exogenous α-TH effectively inhibited NADPH-induced lipid peroxidation. Analogous studies assessing the ability of exogenous α-TH to protect rat liver microsomes were performed by Cadenas *et al.* (56) and Ohki *et al.* (57). Ohki *et al.* (57) demonstrated that exogenous α-TH inhibited lipid peroxidation even at the lowest concentration incorporated in washed microsomes.

In studies conducted by Cadenas *et al.* (56), virtually all added α-TH remained associated with microsomes following centrifugation. The exogenous α-TH inhibited lipid peroxidation but did so 50 times less effectively than endogenous α-TH. These researchers concluded that α-TH must be present at certain specific sites if it is to efficiently protect membranes against peroxidation. It is important to note that the microsomes supplemented with α-TH in this study were not washed prior to exposure to the peroxidation system. Unlike the preparations utilized in the present study and by Ohki *et al.* (57), washing may remove α-TH which is not fully incorporated or bound in the membrane and is therefore less effective in inhibiting lipid peroxidation. Support for this hypothesis is provided by experiments utilizing a model peroxidation system consisting of microsomal structural proteins in lipid micelles (58). Data from this research suggest that vitamin E bound to structural proteins inhibited lipid peroxidation while unbound vitamin E did not.

The present study supports the validity of the α-TH threshold effect in predicting the ability of α-TH to inhibit lipid peroxidation. Membrane α-TH levels above about 0.085 mol% inhibited nuclear lipid peroxidation, but values below this level were associated with high amounts of peroxidation. The reason for this threshold effect is unknown but may relate to the autocatalytic nature of lipid peroxidation. The initiation of lipid peroxidation starts a free radical chain reaction producing reactive chemical species such as lipid hydroperoxides which can then initiate further lipid peroxidation. The length of these chain reactions ultimately determines how many lipid peroxidation chain branching initiation sites are produced (59). α-TH inhibits lipid peroxidation by terminating free radical chain reactions. If enough α-TH is present in the membrane, the chain reaction is rapidly terminated and few chain branching initiation sites are produced. In addition, little α-TH is oxidized under these circumstances. If however, α-TH levels are not sufficient to quickly terminate these chain reactions, large amounts of chain branching initiation sites are produced which subsequently consume existing α-TH supplies and peroxidize PUFAs. The threshold effect for α-TH has been shown to occur in experiments utilizing liposomes (60, 61), but the present study is the first demonstration that the threshold effect also occurs in native membrane systems.

The addition of GSH to incubations was found to markedly lower the amount of α-TH required to inhibit NADPH-induced lipid peroxidation. α-TH threshold levels decreased from 0.085 mol% to 0.040 mol% following the addition of 1 mM GSH. GSH-dependent inhibition of nuclear peroxidation (44) may account for this GSH-dependent enhancement of the capacity of α-TH to inhibit lipid peroxidation. The enzymes glutathione *S*-transferase (26) and glutathione peroxidase (17) catalyze the GSH-dependent reduction of lipid hydroperoxides to lipid alcohols. Studies indicate that these enzyme activities can be associated with cellular membranes (33, 62). Since lipid hydroperoxides can initiate lipid peroxidation, the reduction of these compounds can contribute to the inhibition of peroxidation. The reduction of lipid hydroperoxides should decrease the number of chain branching initiation sites and thereby increase the effectiveness of existing levels of α-TH. Shlyapnikov and Miller (63), utilizing kinetic models, predicted that agents which reduce lipid hydroperoxides act to enhance the ability of antioxidants such as α-TH to inhibit lipid peroxidation.

Another mechanism which may account for the observed interaction between α-TH and GSH was proposed by Reddy *et al.* (25). They reported that the GSH-dependent inhibition of microsomal lipid peroxidation occurred in microsomal preparations from control rats but not in microsomes prepared from vitamin E-deficient rats. On the basis of these data, the existence of a microsomal protein which utilizes GSH to regenerate vitamin E was postulated.

In the nuclear peroxidation system, the addition of GSH reduced the threshold levels of α-TH required to protect against peroxidation. If GSH does regenerate α-TH levels through the actions of a nuclear protein, the rate of lipid peroxidation and α-TH consumption may be reduced, but the absolute amount of α-TH required for protection should remain unchanged. Although the data presented in this report do not discount either hypothesis, the evidence seems to suggest that GSH functions independently of α-TH to inhibit lipid peroxidation.

The levels of α-TH needed to protect against peroxidation in nuclear membranes were about five times less than those needed to inhibit peroxidation in liposomal membranes (60). In a study conducted by Liebler (60), liposomes prepared from egg phosphatidylcholine possessed an α-TH threshold value of about 100 nmol PUFA per nmol α-TH. An increased susceptibility of liposomal membranes as opposed to native membranes was also noted by Fukuzawa *et al.* (64). The explanation of this effect may relate to differences in membrane structure. Biological membranes contain such compounds as cholesterol which has been shown to suppress lipid peroxidation in liposomal membranes (65). Although cholesterol may be oxidized (66), it is more resistant than PUFAs to oxidation (66). Also, other membrane components such as proteins or other antioxidants may protect against lipid peroxidation.

In the presence of 1 mM GSH, 1 molecule of α-TH protected 1,100 PUFA moieties; while in the absence of GSH, α-TH protected only 510 PUFA moieties. It is interesting to note that a similar threshold value of α-TH was reported by Bieri and Poukka (67). In their experiments, a value of one molecule of α-TH to 1,100 PUFA moieties in rat erythrocytes was required to inhibit red cell hemolysis. Endogenous levels of α-TH in rat liver nuclei were slightly higher than the threshold levels of α-TH needed to protect nuclei in the presence of 1 mM GSH but below threshold values determined in the absence of GSH. Since the concentration of GSH utilized in this study is similar to those found intracellularly (48), this report suggests that GSH may be important in extending the ability of α-TH to inhibit lipid peroxidation *in vivo*.

Acknowledgments

We wish to thank Dr. Ann Brodie for her expertise with the SDS polyacrylamide gel electrophoresis studies, and Donna Kling for her excellent technical assistance. We would also thank Lynne Rogers and Cindy Spée for their help in preparing this manuscript. This work was supported in part by grant CH-109 from the American Cancer Society.

References

1. Plaa, G. L., and Witschi, H. (1976) *Ann. Rev. Pharmacol. Toxicol.* **16**, 125-141
2. Eichenberger, K., Bohni, P., Winterhalter, K. H., Kawato, S., and Richter, C. (1982) *FEBS Lett.* **142**, 59-62
3. Baker, S. P. and Hemsworth, B. A. (1978) *Biochem. Pharmacol.* **27**, 805-806
4. Porter, N. A. (1984) *Methods Enzymol.* **105**, 273-282
5. O'Brien, P. J., and Hawco, F. J. (1978) *Biochem. Soc. Trans.* **6**, 1169-1171
6. Svingen, B. A., O'Neal, F. O., and Aust, S. D. (1978) *Photochem. Photobiol.* **28**, 803-809
7. Brambilla, G., Sciaba, L., Faggin, P., Maura, A., Marinari, U. M., Ferro, M., and Esterbauer, H. (1986) *Mutat. Res.* **171**, 169-17

8. Akasaka, S. (1986) *Biochim. Biophys. Acta* **867**, 201-208
9. Ueda, K., Kobayashi, S., Morita, J., and Komano, T. (1985) *Biochim. Biophys. Acta* **824**, 341-348
10. Brawn, K., and Fridovich, I. (1981) *Arch. Biochem. Biophys.* **206**, 414-419
11. Mukai, F. H., and Goldstein, B. D. (1976) *Science* **191**, 868-869
12. Tappel, A. L. (1972) *Ann. N.Y. Acad. Sci.* **203**, 12-28
13. Franke, W. W. (1974) *Int. Rev. Cytol. Suppl.* **4**, 71-236
14. Roots, R. and Okada, S. (1975) *Radia. Res.* **64**, 306-320
15. Hjelle, J. J. and Peterson, D. R. (1983) *Toxicol. Appl. Pharmacol.* **70**, 57-66
16. Alin, P., Danielson, U. H., and Mannervik, B. (1985) *FEBS* **179**, 267-270
17. Christophersen, B. O. (1968) *Biochim. Biophys. Acta* **164**, 35-46
18. Peters, R. (1983) *J. Biol. Chem.* **258**, 11427-11429
19. Tirmenstein, M. A. and Reed, D. J. (1988) *Biochem. Biophys. Res. Commun.* **155**, 956-961
20. Reynolds, C. R. and Tedeschi, H. (1984) *J. Cell Sci.* **70**, 197-207
21. Lang, I., Scholz, M., and Peters, R. (1986) *J. Cell Biol.* **102**, 1183-1190
22. Paine, P. L., Moore, L. C., and Horowitz, S. B. (1975) *Nature* **254**, 109-114
23. Burk, R. F. (1983) *Biochim. Biophys. Acta* **757**, 21-28
24. Haenen, G. R. M. M. and Bast, A. (1983) *FEBS* **159**, 24-28
25. Reddy, C. C., Scholz, R. W., Thomas, C. E., and Massaro, E. J. (1982) *Life Sci.* **31**, 571-576
26. Ketterer, B. and Meyer D.J. (1989) *Mutation Res.* **214**, 33-40
27. Dahle, L. K., Hill, E. G., and Holman, R. T. (1962) *Arch. Biochem. Biophys.* **98**, 253-261
28. Belouqui, O. and Cederbaum, A. I. (1986) *Biochem. Pharmacol.* **35**, 2663-2669
29. Gibson, D. D., Hornbrook, K. R., and McKay, P. B. (1980) *Biochim. Biophys. Acta* **620**, 572-582
30. Ursini, F., Maiorino, M., and Gregolin, C. (1985) *Biochim. Biophys. Acta* **839**, 62-70
31. Morgenstern, R., and DePierre, J. W. (1983) *Eur. J. Biochem* **134**, 591-597
32. Morgenstern, R., DePeirre, J. W., and Ernster, L. (1979) *Biochem. Biophys. Res. Commun.* **87**, 657-663
33. Ursini, F., Maiorino, M., Valente, M., Ferri, L., and Gregolin, C. (1982) *Biochim. Biophys. Acta* **710**, 197-211
34. Gibson, D. D., Hawrylko, J., and McCay, P. B. (1985) *Lipids* **20**, 704-711
35. McLellan, L. I., Wolf, C. R., and Hayes, J. D. (1989) *Biochem. J.* **258**, 87-93
36. Aniya, Y., and Anders, M. W. (1989) *Arch. Biochem. Biophys.* **270**, 330-334
37. Anderson, C., Soderstrom, M., and Mannervik, B. (1988) *Biochem. J.* **249**, 819-823
38. Harris, C. M., and Stone, W. L. (1988) *Life Sci.* **42**, 415-420
39. Haenen, G. R., Vermeulen, N. P., Tai Tin Tsoi, J. N., Ragetli, H.M., Timmerman, H., and Blast, A. (1988) *Biochem. Pharmacol.* **37**, 1933-1938

40. Packer, L., Maguire, J. J., Mehlhorn, R. J., Serbinova, E., and Kagan, V. E. (1989). *Biochem. Biophys. Res. Commun.* **159**, 229-235
41. Haenen, G. R., Tai Tin Tsoi, J. N., Vermeulen, N. P., Timmerman, H., and Bast, A. (1987) *Arch. Biochem. Biophys.* **259**, 449-456
42. Bast, A., and Haenen, G. R. (1988) *Biochim. Biophys. Acta* **963**, 558-561
43. Tan, K.H., Meyer, D.J., Gilliet, N. and Ketterer, B. (1988) *Biochem. J.* **254**, 841-845
44. Tirmenstein, M. A. and Reed, D. J. (1988) *Arch. Biochem. Biophys.* **261**, 1-11
45. Tirmenstein, M. A. and Reed, D. J. (1989) *Biochem. Biophys. Acta* **995**, 174-180
46. Tirmenstein, M. A. and Reed, D. J. (1989) *J. Lipid Res.* **30**, 959-965
47. Pryor, W. A. and Stanley, J. P. (1975) *J. Org. Chem.* **24**, 3615-3617
48. Kosower, N. S. and Kosower, E. (1978) *Int. Rev. Cytol.* **54**, 109-160
49. Ziegler, D. M. (1985) *Ann. Rev. Biochem.* **54**, 305-329
50. Prohaska, J. R., and Ganther, H. E. (1977) *Biochem. Biophys. Res. Commun.* **76**, 437-445
51. Lawrence, R. A. and Burk, R. F. (1976) *Biochem. Biophys. Res. Commun.* **71**, 952-958
52. Morgenstern, R, Lundqvist, G., Andersson, G., Balk, L., and DePierre, J. W. (1984) *Biochem Pharmacol.* **33**, 3609-3614
53. Sevanian, A., Muakkassah-Kelly, S. F., and Montestruque, S. (1983) *Arch. Biochem. Biophys.* **223**, 441-452
54. Sevanian, A., Hacker, A. D., and Elsayed, N. (1982) *Lipids* **17**, 269-277
55. Gruger, E. H. and Tappel, A. L. (1971) *Lipids* **6**, 147-148
56. Cadenas, E., Ginsberg, M., Rabe, U., and Sies, H. (1984) *Biochem. J.* **223**, 755-759
57. Ohki, K., Takamura, T., and Nozawa, Y. (1984) *J. Nutr. Sci. Vitaminol.* **30**, 221-234
58. Tinberg, H. M. and Barber, A. A. (1970) *J. Nutr.* **100**, 413-418
59. Fukuzawa, K., Takase, S., and Tsukatani, H. (1985) *Arch. Biochem. Biophys.* **240**, 117-120
60. Liebler, D. C., Kling, D. S., and Reed, D. J. (1986) *J. Biol. Chem.* **261**, 12114-12119
61. Fukuzawa, K., Tokumura, A., Ouchi, S., and Tsukatani, H. (1982) *Lipids* **17**, 511-513
62. Friedberg, T., Bentley, P., Stasiecki, P., Glatt, H. R., Raphael, D., and Oesch, F. (1979) *J. Biol. Chem.* **254**, 12028-12033
63. Shlyapnikov, Y. A. and Miller, V. B. (1965) *Russ. J. Phys. Chem. Engl. Transl.* **39**, 1291-1294
64. Fukuzawa, K., Chida, H., Tokumura, A., and Tsukatani, H. (1981) *Arch. Biochem. Biophys.* **206**, 173-180
65. Mowri, H., Nojima, S., and Inoue, K. (1984) *J. Biochem.* **95**, 551-558

66. Terao, J., Sugino, K., and Matsushita, S. (1985) *J. Nutr. Sci. Vitaminol.* **31**, 499-508
67. Bieri, J. G. and Poukka, R. K. H. (1970) *J. Nutr.* **100**, 557-564

Mechanisms of Vitamin E Control of Lipid Peroxidation: Regeneration, Synergism, Asymmetry, Migration and Metal Chelation

V. Kagan[1], *L. Packer*[2], E. Serbinova, R. Bakalova, D. Stoyanovsky, Zh. Zhelev, M. Harfouf, S. Kitanova, D. Rangelova
[1] *Membrane Biostabilization Group, Institute of Physiology, Bulgarian Academy of Sciences, Acad. G. Bonchev Str. 23, Sofia 1113 Bulgaria*
[2] *Membrane Bioenergetics Group, Department of Molecular and Cell Biology 251 Life Science Addition, University of California, Berkeley, CA 94720 USA*

Abstract

The very high efficiency of vitamin E in preventing and inhibiting lipid peroxidation in biomembranes (where its molar concentration within the lipid bilayer is less than 0.1 mol%) is due to its multifunctional antioxidant properties. Data is presented which show that: (i) $\alpha-$tocopherol is non-uniformly distributed within the membrane lipid bilayer, being preferentially located in the regions enriched with polyenoic lipids (the main substrates of lipid peroxidation). (ii) $\alpha-$tocopherol local deficiency in membrane monolayers appearing in the course of oxidative stress may be replenished by means of its intermembrane exchange: this process is accelerated by accumulation of lipid peroxidation products and membrane disordering. (iii) Chromanoxyl radicals can be reduced to form chromanols by NAD(P)H-dependent enzymes in microsomes and mitochondria. In this regeneration, GSH acts synergistically; recycling of chromanoxyls also occurs non-enzymatically by ascorbate. (iv) Water-soluble antioxidants (*e.g.* carnosine) and protein-kinase C-dependent inhibitors of lipid peroxidation (activated by phorbol-12-myristate-13-acetate) synergistically increase $\alpha-$tocopherol antioxidant potency. (v) $\alpha-$tocopherol within the lipid bilayer chelates iron (necessary for propagation of lipid peroxidation), thus hydroperoxide cleavage sites are near to $\alpha-$tocopherol-rich membrane domains.

Introduction

More than 6 decades of studies with vitamin E have established its indisputable role in stabilizing biological membranes. This property is believed mainly to be due to its antioxidant action, but other factors may also be important viz ordering of lipids in the fluid regions of the bilayer, and formation of complexes with "chaotropic"

Biological Oxidation Systems, Volume 2

free fatty acids and lysophospholipids (1-3). Figure 1 is a scheme illustrating these effects of vitamin E. The antioxidant concept of vitamin E action is undisputed whereas the molecular mechanisms underlying the radical scavenging effects of tocopherols in membranes are still not fully understood. For many years its chain breaking activity due to lipid radical scavenging was thought to be solely responsible for the antioxidant activity of vitamin E (4,5). Recently however, evidence has been obtained that some other properties of tocopherols are also involved in its antioxidant action. Thus in Figure 1 the singlet oxygen quenching action of vitamin E is shown as is its effects on modulating the activity of important membrane associated enzymes such as phospholipase, cyclo-oxygenase, lipoxygenase, and protein kinase C. In this way vitamin E can exert a control on signal transduction and many important mediators of physiological function.

In this paper we shall discuss several of the mechanisms which may be important for the high antioxidant activity of vitamin E, namely: (i) radical scavenging; (ii) regeneration from chromanoxyl radicals; (iii) transition metal chelation; (iv) synergism with water soluble antioxidants, and (v) intermembrane exchange.

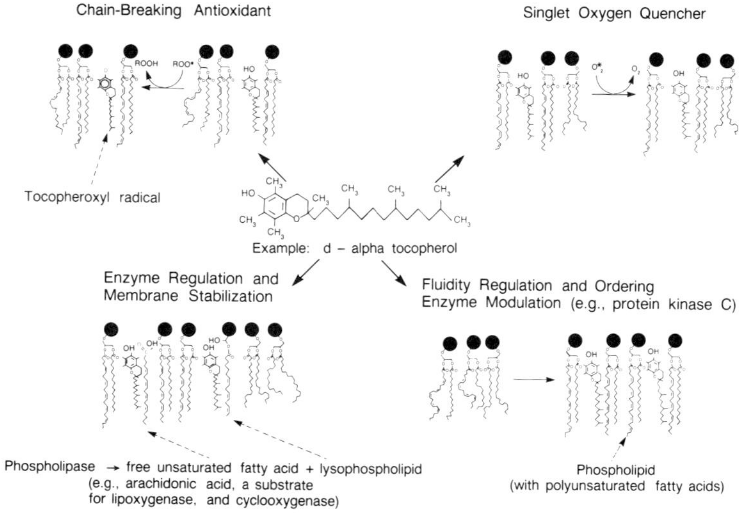

Figure 1. *Vitamin E as a Biological Response Modifier.* Scheme depicts mechanisms of the stabilizing effects of vitamin E in membranes.

Radical Scavenging Effects of Vitamin E

The essentially heterogeneous conditions of radical scavenging by tocopherols in biological and model lipid membranes may be attributed to restrictions in its mobility and accessibility of the lipid radicals to tocopherol molecules within the bilayer. Figure 2 illustrates this important aspect of vitamin E membrane organization, which is not the case when vitamin E is in solution.

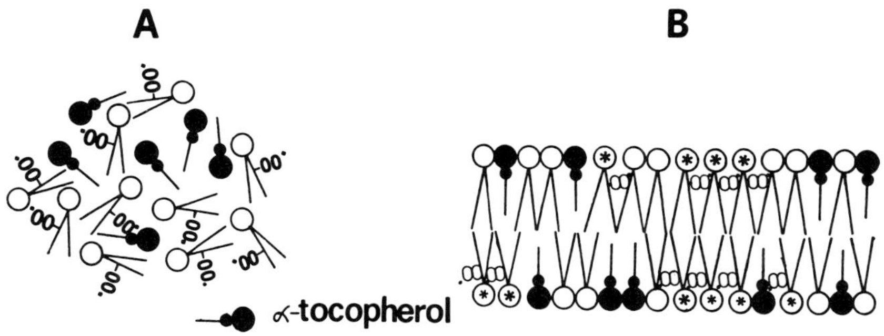

Figure 2. *Scheme of α-Tocopherol with Lipid Peroxy (ROO•) Radicals.* (A) in solution, (B) in the lipid bilayer.

The formation of clathrates of tocopherols and carbamides requires the penetration of the isoprenoid chain of α−tocopherol into the crystalline lattice of carbamide. Figure 3 shows the kinetics of clathrate formation when α−tocopherol is incubated in hexane or when tocopherol is in micellar form (in buffer), or when tocopherol is in the lipid bilayer (as in liposomes) during incubation with a carbamide derivative. The isoprenoid chain of α−tocopherol is readily accessible in solution and when it is in micelles. However, α−tocopherol is inaccessible after it is incorporated into liposomal bilayers.

It might be suggested that the specific membrane effects of tocopherols are predetermined by hydrophobic interaction of its hydrocarbon chain, even though its antioxidant function is localized in the chromanol ring (6,7). To evaluate the specific intermembrane interactions of α−tocopherol during radical scavenging, we decided to use α−tocopherol homologues of different chain lengths shown in Figure 4.

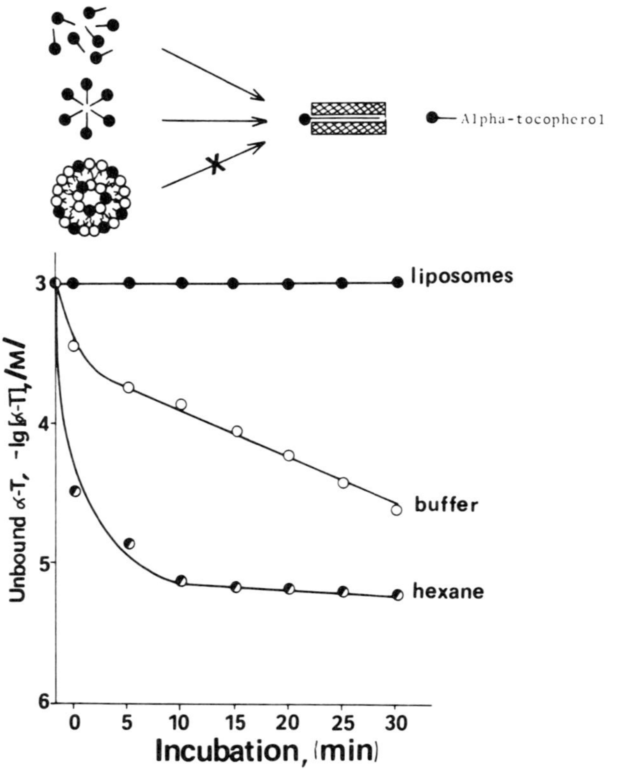

Figure 3. *Clathrate Formation with Tocopherol.* Upper - Scheme depicting clathrate formation which occurs during incubation of α-tocopherol with a carbamide derivative. Lower - Kinetics of tocopherol binding. Incubation conditions: 20 μM α-tocopherol, 0.6 mg lipids/ml, 0.1 M K, Na phosphate buffer at pH 7.4 and 20°C. Unbound tocopherol was measured by following fluorescence intensity (λ_{ex}=298nm, λ_{em}=325 nm) in suspensions or solution after removal by sedimentation (1000 g x 10 min) of the clathrates (tocopherol bound to carbamide derivative).

If we compare values of the reaction rate constants of $\alpha-$tocopherol and its homologues with cumene peroxide radicals in a homogeneous system no differences are found in activity (Table 1). However, this is not the case in ordered systems where in both model (liposomes) and natural membranes different efficiencies of lipid peroxidation inhibition are found. Inhibition of lipid peroxidation decreases in the order $C_1 > C_3 > C_6 > C_{11} > \alpha-$tocopherol for liposomes, microsomal and synaptosomal membranes. For $\alpha-$tocopherol and C_1, these differences exceed two orders of magnitude.

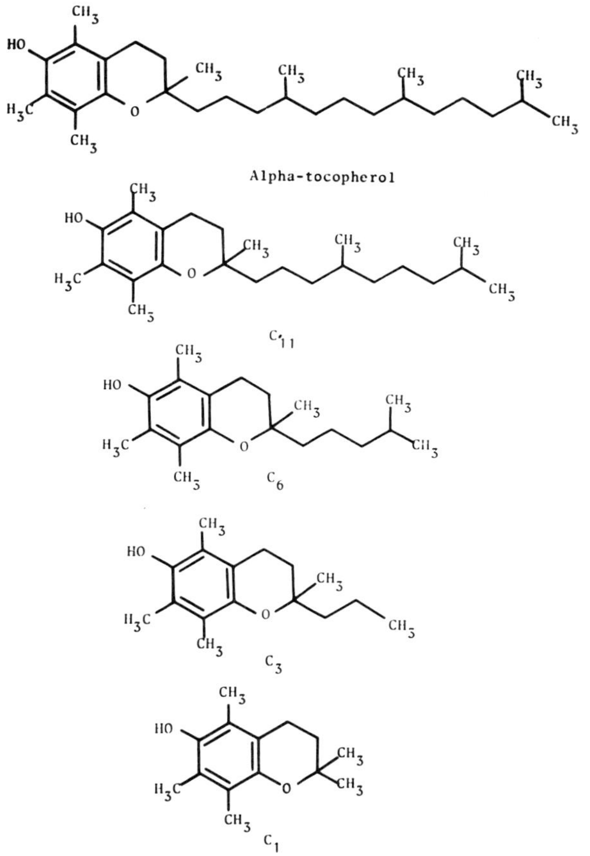

Figure 4. *Structural Formula of α-Tocopherol and Homologues with Different Chain Lengths*

Details of the mechanisms underlying these large differences in the antioxidant activities of a-tocopherol homologues in membranes are reported elsewhere (8). Here we shall only enumerate some of the most important mechanisms. The greater antioxidant activity of C_1 as compared to that of $\alpha-$tocopherol is not due either to its more efficient binding or incorporation into membranes. Results suggest that probably the higher intramembrane mobility, more efficient interaction with oxygen radicals, and a much less pronounced "clustering" of the vitamin E analogues into specific membrane domains determine the more efficient radical scavenging by $\alpha-$tocopherol homologues with shorter isoprenoid chains. Since $\alpha-$tocopherol more strongly interacts with membrane phospholipids both hydrophobically and by means of H-bonding, this restricts its motions in the membrane rendering it less efficient in scavenging of oxygen and lipid radicals (9,10).

Table 1. *Determination of Constants for 50% Inhibition (K_{50}) of Lipid Peroxidation by α-Tocopherol and Its Homologues*

α-Tocopherol Homologue	Lipid Peroxidation Inhibition[a] (K_{50}, M) Liposomes	Microsomes	Synaptosomes	Rate Constant for Interaction with Cumene Peroxide Radicals[b] ($\times 10^6\ M^{-1}s^{-1}$)
α-T	3.0×10^{-6}	2.0×10^{-6}	4.2×10^{-6}	3.3 ± 0.2
C_{11}	---	1.0×10^{-7}	1.0×10^{-6}	3.4 ± 0.2
C_6	---	5.0×10^{-8}	6.3×10^{-7}	3.3 ± 0.1
C_1	6.3×10^{-8}	1.2×10^{-8}	7.2×10^{-8}	3.5 ± 0.1

[a] Incubation conditions: 0.1 M K,Na-phosphate buffer, pH 7.4 (37°C), 20 μM Fe^{2+} for microsomes, 0.3 μM Fe^{2+} for synaptosomes, 40 μM Fe^{2+} for liposomes, 0.5 mM ascorbate for microsomes and liposomes, 0.01 mM ascorbate for synaptosomes, 1 mg of protein (lipids)/ml. Incubation time: 5 min, 37°C.

[b] Values given are the rate constants for the reactions of cumene peroxide radicals with chromanols at 60°C. According to Burlakova *et al.*, Biophyzika USSR, 1979, (3) 84-91.

The decreased antioxidant efficiency of α−tocopherol with its 16-carbon side chain, as compared with the shorter chain homologues is a feature which holds true not only for α−tocopherol homologues, but also for other natural and synthetic free radical scavengers. Some other examples are ubiquinols with different isoprenoid chain lengths, and butylated hydroxytoluene (BHT) and its alkylated derivatives. Table 2 shows that the rate constants for the reaction of cumene peroxide radicals with BHT derivatives in solution do not differ significantly, whereas lipid peroxidation inhibition in membranes by BHT is much more pronounced than by its alkylated homologues (11). With a ubiquinol series lipid peroxidation inhibition also decreases with the increase of the isoprenoid chain length and the ubiquinol components of natural membranes Q_9, Q_{10} possess low antioxidant activity (Table 3).

Thus we arrive at a paradoxical conclusion: the long hydrocarbon chain-containing compounds which are found in natural membranes (tocopherols, ubiquinols Q_9 and Q_{10}) are much less potent antioxidants in biomembranes *in vitro* than their homologues devoid of or containing shorter hydrocarbon chains.

Table 2. *Inhibition of Lipid Peroxidation by BHT and Its Derivatives in Rat Liver Microsomes and Determination of the Rate Constants for the Reactions of Cumene Peroxide Radicals with BHT and Its Derivatives.*

BHT derivative	$K_{50\%}$ Inhibition (M)	Rate Constants ($M^{-1}s^{-1}$)
R= C_4H_9, (BHT)	1.6 x 10^{-7}	4.9 x 10^{4}
R= −OH	1.0 x 10^{-8}	5.4 x 10^{4}
R= $-C_{16}H_{33}$	> 10^{-3}	6.9 x 10^{4}
R= $-C_{12}H_{25}$	> 10^{-3}	4.5 x 10^{4}

Conditions: as in Table 1.

Table 3. *Inhibition of Lipid Peroxidation by Ubiquinols in Rat Liver Microsomes.*

Ubiquinol	Inhibition of Lipid Peroxidation ($K_{50\%}$, M)	
	Fe^{2+} + ascorbate	Fe^{2+} + NADPH
Q_1	1.6 x 10^{-8}	4.8 x 10^{-8}
Q_3	1.6 x 10^{-7}	1.0 x 10^{-7}
Q_4	1.0 x 10^{-6}	8.3 x 10^{-7}
Q_5	5.7 x 10^{-6}	4.5 x 10^{-6}
Q_6	1.0 x 10^{-5}	8.4 x 10^{-6}
Q_7	4.8 x 10^{-5}	1.6 x 10^{-5}
Q_8	6.0 x 10^{-5}	2.4 x 10^{-5}
Q_9	7.6 x 10^{-5}	4.5 x 10^{-5}
Q_{10}	> 10^{-4}	1.0 x 10^{-4}

Conditions: 0.1 M K,Na-phosphate buffer, pH 7.4 (37°C), 20 lM Fe^{2+}, 0.5 mM ascorbate or NADPH, 1 mg protein/ml. Incubation: 5 min at 37°C. In these experiments ubiquinones were reduced by treatment with sodium borohydride to obtain the ubiquinols.

Was Nature mistaken in the choice of membrane antioxidants? Probably this is a hasty conclusion because "anchoring" of the antioxidant in the membrane is essential. α−Tocopherol fills in the spacing of the lipid bilayer, whereas its homologues, devoid of the hydrocarbon chain, like C_1, produce defects in the ordering of the membrane bilayer. Such defects lead to membrane damage (9). To illustrate this we investigated sarcoplasmic reticulum vesicles from rabbit skeletal muscle where the Ca^{2+}/ATP ratio is taken as a measure of membrane permeability to Ca^{2+}. The data in Table 4 show that neither $\alpha-$tocopherol, nor $\alpha-$tocopherol-acetate increases the Ca^{2+} permeability of sarcoplasmic reticulum membranes, whereas C_1 does increase Ca^{2+} permeability (12).

Table 4. *Effects of α-Tocopherol and Its Homologues on Ca^{2+} - Transporting Activity of Rabbit Skeletal Muscle Sarcoplasmic Reticulum Vesicles**

Concentration (M)	Ca^{2+}/ATP, (%) α-Tocopherol	α-Tocopherol-acetate	C_1
0	100	100	100
10^{-6}	100	100	100
$5x10^{-6}$	100	100	93
10^{-5}	100	100	86
$5x10^{-5}$	100	100	80
10^{-4}	100	100	67
$5x10^{-4}$	100	100	45

* Ca^{2+}/ATP ratio in the control preparation where no tocopherol was added was 1.7 which is taken as the 100% value.

Another important feature of the antioxidant compounds with long hydrocarbon chains is that they are not recognized by the cytochrome P-450 system, and thus are not metabolized in P-450-dependent reactions. However the short chain homologues will undergo oxidative transformations by the cytochrome P-450 system. Table 5 indicates that C_1 and C_6 give characteristic binding spectra of type I with microsomal cytochrome P-450 with values for spectral binding constants in the range of 0.4 mM to 0.5 mM. In contrast, C_{11} and $\alpha-$tocopherol do not bind to cytochrome P-450 even at concentrations as high as 10 mM. As a result the pharmacodynamics of the antioxidant action of alkylated and non-alkylated homologues is quite different. This may explain why *in vivo* the antioxidant compounds with shorter hydrocarbon chains do not exert a higher antioxidant activity than would be predicted by their antioxidant effects on membrane lipid peroxidation *in vitro*. Moreover, *in vivo* their antioxidant activity in the liver persisted for longer periods (Figure 5).

Table 5. *Spectral Binding Constants (K_S) of α-Tocopherol and Its Homologues with Cytochrome P-450 in Rat Liver Microsomal Membranes.*

α-Tocopherol homologue	K_S, (M)
C_1	4.0×10^{-4}
C_6	5.0×10^{-4}
C_{11}	$> 10^{-2}$
α-tocopherol	$> 10^{-2}$

Conditions: 0.1 M K, Na phosphate buffer, pH 7.6 (37°C), 0.2 mg microsomal protein in 1 ml.

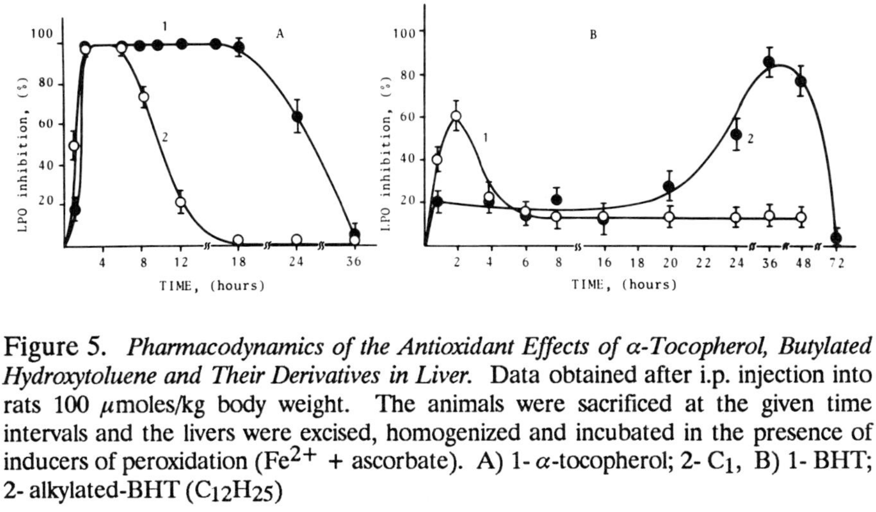

Figure 5. *Pharmacodynamics of the Antioxidant Effects of α-Tocopherol, Butylated Hydroxytoluene and Their Derivatives in Liver.* Data obtained after i.p. injection into rats 100 μmoles/kg body weight. The animals were sacrificed at the given time intervals and the livers were excised, homogenized and incubated in the presence of inducers of peroxidation (Fe^{2+} + ascorbate). A) 1- α-tocopherol; 2- C_1, B) 1- BHT; 2- alkylated-BHT ($C_{12}H_{25}$)

Transition Metal Chelation by α−Tocopherol

The high antioxidant activity of tocopherols is not determined solely by their radical scavenging effects. Transition metals such as iron and copper are known to play a very important role both in the initiation and propagation of free radical oxidation reactions (13). We have recently discovered (Figure 6A) that α−tocopherol containing liposomes bind iron; under the conditions used no iron is detected in liposomes in the absence of α−tocopherol. Moreover, the iron binding is dependent on α−tocopherol concentration in the liposomes (Figure 6B). Hence we suggest that in the presence of α−tocopherol at least in part, iron is not randomly distributed in the aqueous phase and on the membrane surface, and that some iron may partition into membrane domains where α–tocopherol is present (14). In other words the sites of generation and scavenging of lipid free radicals in membranes are in close proximity. This may facilitate more efficient regulation of lipid peroxidation (Figure 7).

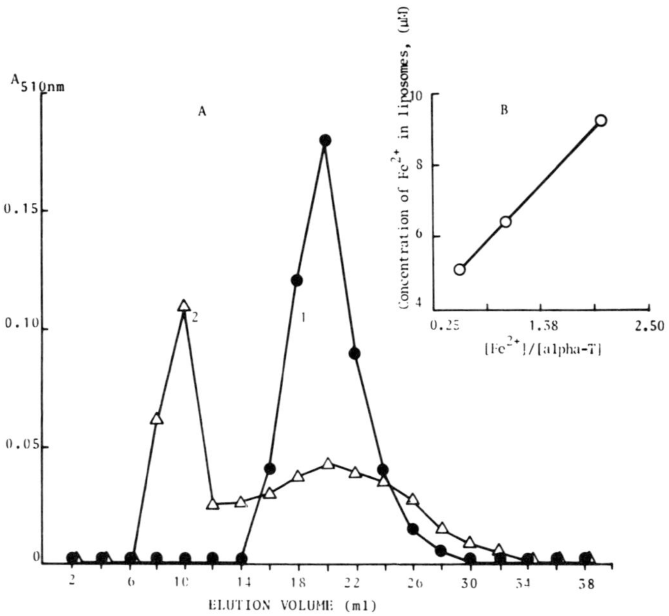

Figure 6. *Iron Binding by α-Tocopherol in Liposomes.* **A)** Elution profiles of iron from a Sephadex G-25 column filled with (1) ascorbate (10 mM) + Fe(II) ($4.6x10^{-4}$ M) (2) ascorbate (10 mM) + Fe(II) (4.6×10^{-4}M) + lecithin liposomes (3.4 mg/ml) with α-tocopherol incorporated at concentration 18.4 μM. Concentrations of Fe^{2+} in the fractions were determined using the 1,10-phenantroline procedure. **B)** Concentration dependence of Fe^{2+} binding on the tocopherol concentration in liposomes.

Intermembrane Exchange of α–Tocopherol

During lipid peroxidation membrane tocopherols become consumed. The resulting tocopherol-deficiency may be in specific membrane domains due to non-uniform distribution of tocopherol and radical initiating sites (15). Replenishment of such localized tocopherol deficiency may be accomplished by migration from "tocopherol-rich" to tocopherol-deficient membrane domains. Figure 8A is a scheme which explains the design of experiments in which tocopherol, incorporated into either saturated or unsaturated donor liposomes, can exert antioxidant effects in acceptor liposomes which

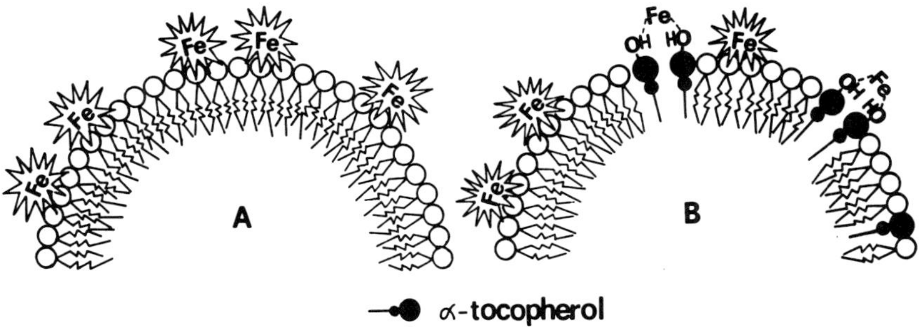

Figure 7. *Scheme Illustrating Binding of a-Tocopherol with Iron in Lipid Bilayers.* **A)** in the absence of tocopherol, **B)** after the incorporation of tocopherol. Sites of iron at the membrane surface where radicals are generated are shown schematically.

initially contain no tocopherol. We showed separately that no fusion of donor with acceptor liposomes occurs under the conditions used. Therefore our results can be explained only in terms of intermembrane exchange of tocopherol. The results of these experiments are given in Figure 8. The upper curve (8B) represents the accumulation of lipid peroxidation products in the mixture of donor and acceptor liposomes in the absence of antioxidants (controls). The dotted line (which is exactly half of the control one) is the theoretical curve when $\alpha-$tocopherol is present only in donor liposomes and does not exchange into the acceptor liposomes. The experimental data are illustrated by curves 2-4, which are below the theoretical curve. This indicates that membrane lipid peroxidation inhibition is due to $\alpha-$tocopherol transfer and exchange from donor liposomes into acceptor liposomes. In membranes containing preformed lipid peroxidation products the inhibition (*i.e.* $\alpha-$tocopherol exchange) is significantly higher than in liposomes without lipid oxidation products (Table 6). In contrast incorporation of cholesterol results in inhibition of $\alpha-$tocopherol intermembrane exchange probably due to ordering effects on the lipid bilayer.

Regeneration of Pentamethylchromanol from the Chromanoxyl Radical

Interaction of the tocopherol (C_1) with lipid radicals results in the formation of lipid hydroperoxides and the chromanoxyl radical.

$$ROO\cdot + \alpha\text{-T-OH} \rightarrow ROOH + \alpha\text{-T-O}\cdot$$

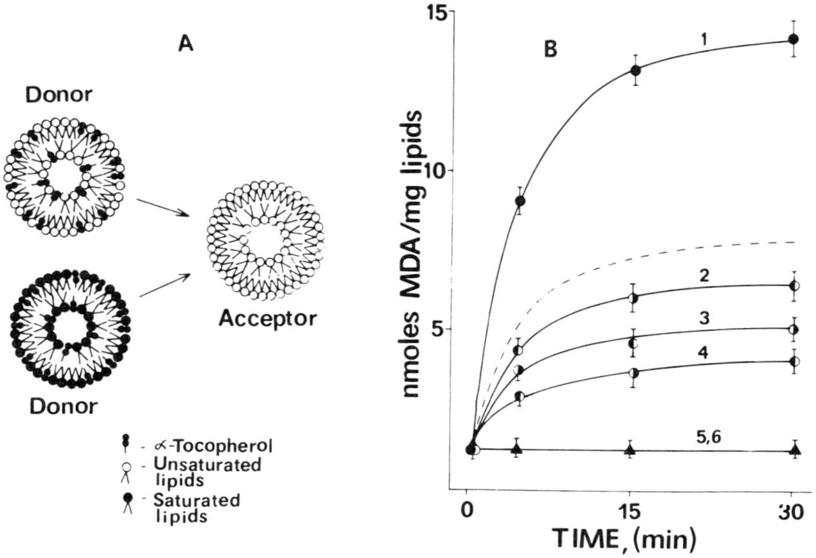

Figure 8. *Intermembrane Transfer of α-Tocopherol from Donor to Acceptor Liposomes.* **A)** Scheme illustrating α-tocopherol transfer from donor and exchange into acceptor liposomes. **B)** Accumulation of lipid peroxidation products in acceptor unilamellar liposomes. 1) absence, (2-4) in the presence of donor liposomes 12.5 nmoles/mg lipids α-tocopherol; The preincubation time of donor with acceptor liposomes were for: 2) 1 minute; 3) 30 minutes; 4) 60 minutes; 5) 30 minutes, but in this case with an α-tocopherol concentration of - 6.25 nmoles/mg lipids in both donor and acceptor liposomes. Dashed line - a theoretical curve for inhibition of lipid peroxidation if no α-tocopherol transfer occurs from the donor liposomes containing 6.25 mg/ml tocopherol into the acceptor liposomes having no tocopherol.

Chromanoxyl radicals can either be lost by reactions with other radicals *e.g.*

$$ROO\bullet \text{ (or } R\bullet\text{, or } \alpha\text{-T-O}\bullet) + \alpha\text{-T-O}\bullet \rightarrow \text{products}$$

or can be reduced thus providing for recycling of the chromanoxyl (tocopheroxyl) radical to chromanol (tocopherol).

$$\alpha\text{-T-O}\bullet + AH \rightarrow \alpha\text{-T-OH} + A$$

Packer *et al.* (16,17) were the first to demonstrate using pulse radiolysis that ascorbate could reduce tocopheroxyl radicals. However, until relatively recently it was still uncertain whether this kind of tocopherol regeneration occurs in natural membranes (18-21).

Figure 9 shows the ESR spectra of the chromanoxyl radical generated from C_1 after it was incorporated into rat liver microsomal membranes. The radical was generated by incubation in the presence

Table 6. *Lipid Peroxidation Inhibition Due to α-Tocopherol Transfer Between Donor and Acceptor Liposomes: Effects of Exogenously Added Cholesterol and Peroxidized Lipids*

Mixtures of Donor Plus Acceptor Liposomes (additions)	Lipid Peroxidation Inhibition, (%) Additions to Liposomes	
	Cholesterol[a]	Peroxidized Lipids[b]
None	58 ± 4	47 ± 2
To Donor Liposomes only	28 ± 2	60 ± 3
To Donor and Acceptor Liposomes	26 ± 2	71 ± 4

Conditions: 0.6 mg of lipids/ml, 0.1 M K, Na-phosphate buffer, pH 7.4 at 37°C, the reaction was started by adding 40 μM Fe^{2+} plus 0.5 mM ascorbate. Donor and acceptor liposomes were preincubated for 30 min at 37°C followed by incubation with inducers of lipid peroxidation inducers for 15 min at 37°C.
[a] Donor and acceptor liposomes were prepared from egg-yolk lecithin, where indicated cholesterol addition was 85 mole%.
[b] Donor and acceptor liposomes were prepared from rat cerebral cortex lipids, where indicated the amounts of peroxidized lipids added were equivalent to 20% by weight of the total lipids.

of an oxidation system comprised of soybean lipoxygenase and arachidonic acid. Addition of ascorbate, NADPH or NADH prevents the appearance of the chromanoxyl radical for a time period whose duration depends on the concentration of the reductant added. Both microsomal and submitochondrial membrane electron transport carriers are able to recycle chromanoxyl radicals in the presence of these reducing equivalents. NAD(P)H added to lipoxygenase + arachidonic acid in the presence of rat liver microsomes protects C_1 from oxidation. Separate HPLC analysis under conditions used indicate that in the presence of the reductants the loss of C_1 was much less, *i.e.* demonstrating a sparing effect. The amounts of C_1 remaining in the incubation medium were proportional to the concentration of NADPH added. GSH does not prevent the formation of chromanoxyl radical by itself, but exhibits a strong synergistic effect when combined with either NADH or NADPH (Figure 10). In this way the well known synergism in the antioxidant effects of a-tocopherol and GSH in membranes may be explained (18, 22-24).

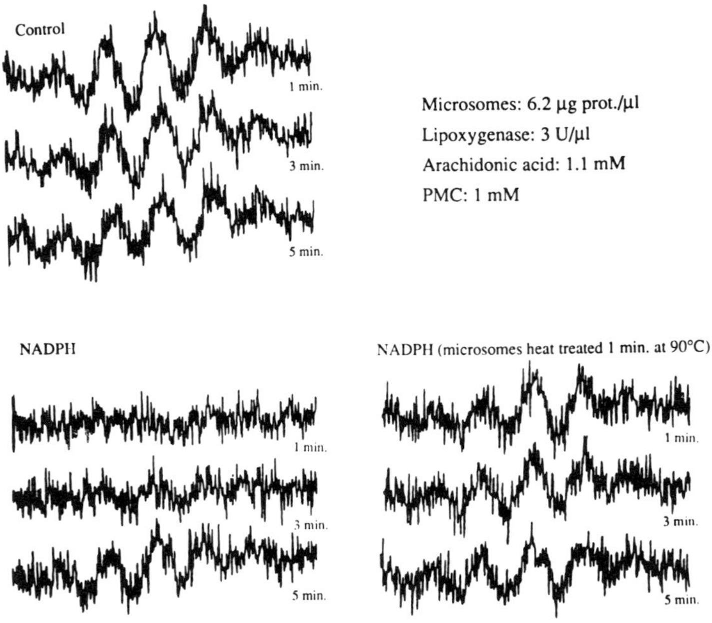

Figure 9. *Effect of NADPH on the Chromanoxyl Radical ESR Signal in Microsomes.* NADPH added was oxidized by microsomes after about 3 minutes after which the ESR signal appears. NADPH addition does not affect the ESR signal in heat denatured microsomes.

Similarly, synergism observed in the presence of α−tocopherol and other water-soluble antioxidants could also be due to the recycling of chromanoxyl tocopheroxyl radicals. Thus the data given in Figure 11 demonstrate that the water soluble histidine-containing dipeptide, carnosine, in the presence of α−tocopherol exerts a synergistic effect on lipid peroxidation induced in skeletal muscle sarcoplasmic reticulum or liver microsomal membranes (not shown). A similar synergistic effect on lipid peroxidation was observed when the carnosine homologue, cyclophenyl-alanyl-histidine was used in combination with vitamin E.

Synergistic phenomena of this type are probably common. Recently we have shown that phorbol-12-myristate-13-acetate (PMA) produces a concentration-dependent lipid peroxidation inhibition in brain, liver and skeletal muscle microsomes mediated by a protein-kinase C-supported mechanism (25, 26). In Figure 12 lipid peroxidation over 30 minutes was measured in microsomes isolated

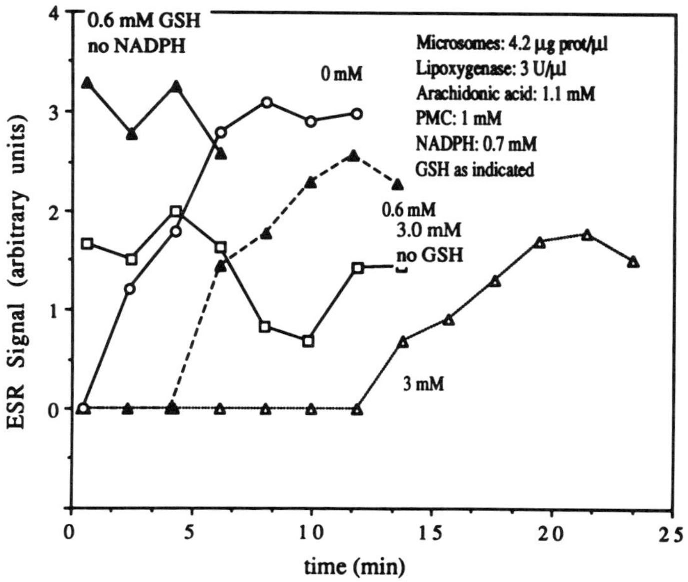

Figure 10. *Synergistic Effect of NADPH and GSH on the Time Course of Chromanoxyl Radicals in Microsomes.* Note that this synergistic effect is observed at different concentrations of GSH and at constant NADPH concentration.

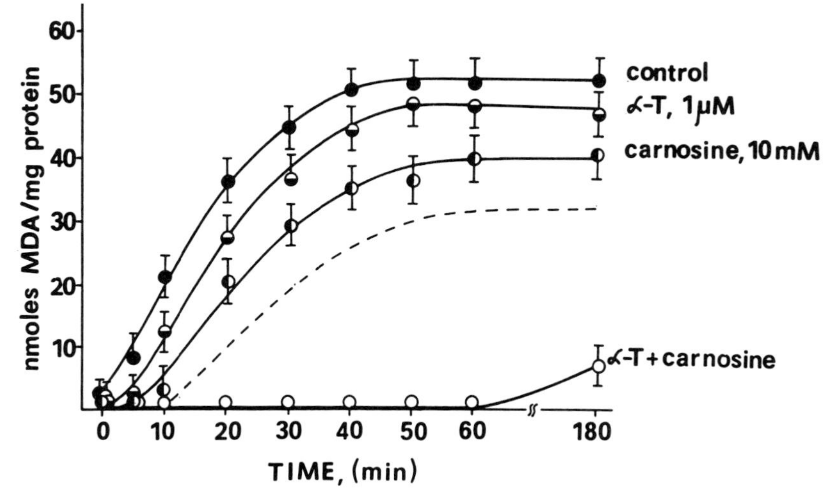

Figure 11. *Synergistic Effect of the Combination of α-Tocopherol and Carnosine on Lipid Peroxidation in Sarcoplasmic Reticulum Vesicles.* Broken lines- expected additive effect of α-tocopherol plus carnosine. Conditions - 1 mg protein/ml of sarcoplasmic reticulum vesicles in 0.1 M K, Na phosphate buffer, pH 7.4 at 37°C. The reaction was started by addition of 10 μM Fe^{2+} and 0.5 mM ascorbate.

from two types of homogenates, one incubated without (control) and one with 0.1 μM PMA present. In the PMA microsomes a pronounced transient inhibition of lipid peroxidation was observed. If these two types of microsomes were further treated with a relatively small amount of $\alpha-$tocopherol (0.1 mM) as expected the transient inhibition of lipid peroxidation is also observed in the control microsomes. In the PMA microsomes with tocopherol added a strong and prolonged synergistic inhibition of lipid peroxidation is observed. This synergistic effect of PMA (an activator of protein-kinase C) with $\alpha-$tocopherol (Figure 12) may be due to recycling of the tocopheroxyl radicals by some, as yet unknown mechanism.

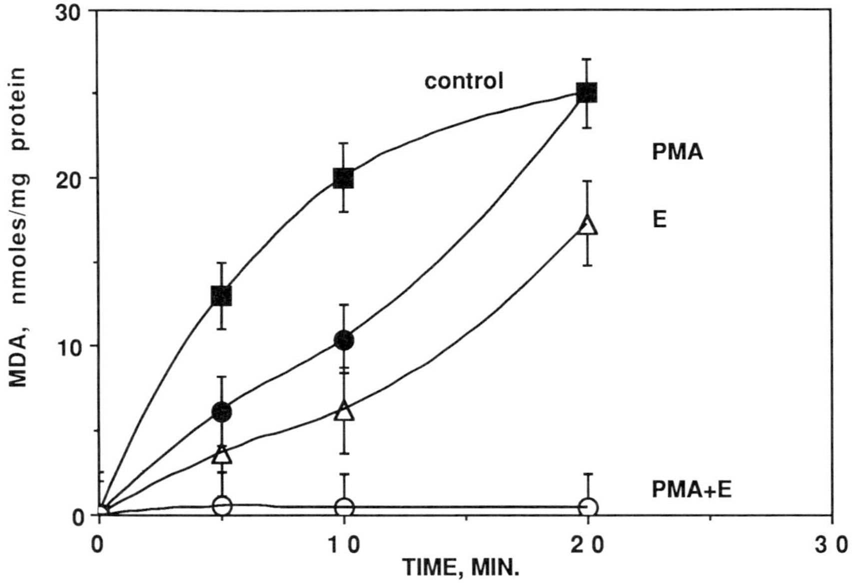

Figure 12. *Effect of Phorbol-12-Myristate-13-Acetate (PMA) on Fe^{2+} + Ascorbate-Induced Lipid Peroxidation in Rat Liver Microsomes.* Rat liver homogenates were prepared and incubated without (control) or with 0.1 μM PMA present (PMA microsomes) for 10 minutes at 30°C. Then the microsomal fractions were obtained from the two homogenates. α-tocopherol was added to both types of microsomes 5 minutes before addition of the inducers of lipid peroxidation (10 μM Fe^{2+} + 0.5 mM ascorbate) and then the MDA accumulation was assayed over a 20 minute period.

The interrelations between vitamin E and protein-kinase C suggest a regulatory process in which at normal or increased content of $\alpha-$tocopherol protein-kinase C is inhibited, *i.e.*, down regulation. Down regulation of protein kinase C activity has been reported by Mahoney and Azzi (27, 28). We speculate that the protein-kinase C-dependent inhibition of lipid peroxidation is not turned on when the concentration of vitamin E is high enough. As vitamin E decreases due to its consumption in free radical reactions, the vitamin E-dependent inhibition of protein-kinase C disappears, and the control of lipid peroxidation by protein kinase C may become increasingly important in the overall regulation of lipid peroxidation. Figure 13 illustrates the interrelationship between protein kinase C and vitamin E in regulation of lipid peroxidation. This PKC-dependent inhibition of lipid peroxidation can be prevented by activation of protein kinase A (PKA), a physiological antagonist of PKC. For example, forskolin acting via adenylate cyclase increases the activity of PKA and eliminates the PMA-dependent inhibition of lipid peroxidation. The same effect can be achieved under conditions of oxidative stress: the excessive accumulation of lipid peroxidation products stimulates adenylate or guanylate cyclase thus blocking the PKC dependent mechanism of regulation of lipid peroxidation. The details of this mechanism deserve further experimental study.

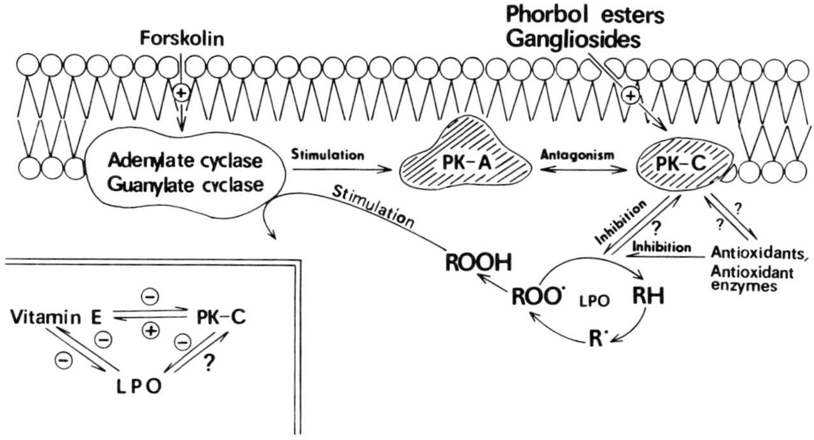

Figure 13. *Scheme Illustrating the Interrelation Between Protein-Kinase C (PKC) and Vitamin E in Regulation of Lipid Peroxidation.* By activating PKC phorbol-12-myristate-13-acetate (PMA) exerts an inhibitory effect on membrane lipid peroxidation. The PKC-dependent inhibition of lipid peroxidation is not turned on when the concentration of vitamin E is high. As vitamin E decreases due to its consumption in free radical reactions, the vitamin E - dependent inhibition of PKC disappears. This effect of PKC may be important in the overall regulation of lipid peroxidation.

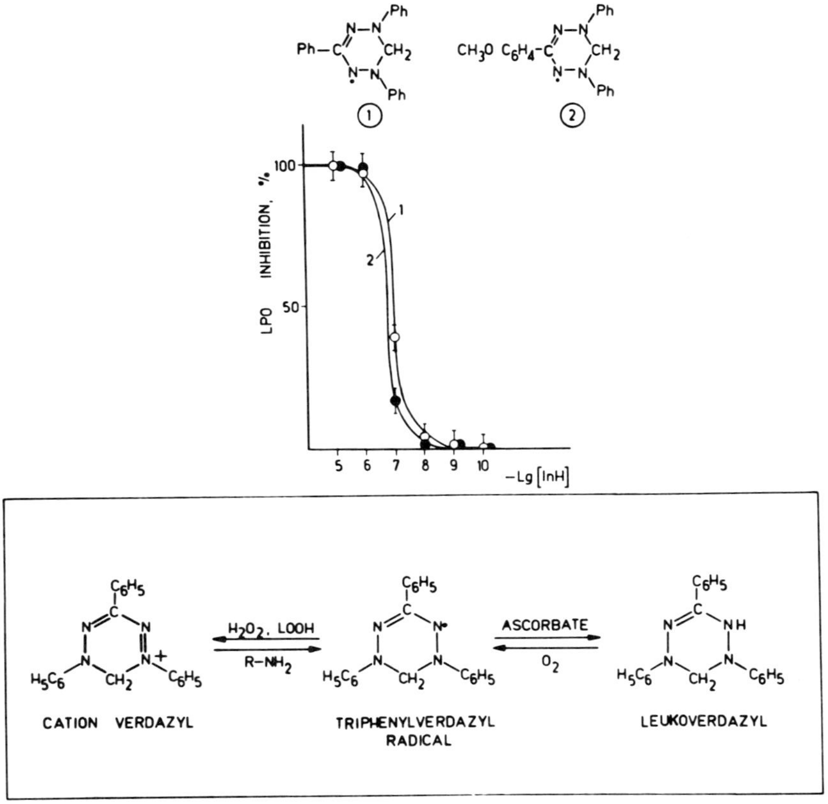

Figure 14. *Stable Verdazyl Radicals Inhibit Lipid Peroxidation in Rat Brain Synaptosomes.* 1) 1,3,5-triphenylverdazyl; 2) 3-methoxyphenyl-1,5-diphenylverdazyl. Incubation conditions: 1 mg protein/ml synaptosomes, 0.1 M K, Na phosphate buffer, pH 7.4 at 37°C. The reaction was started by addition of lipid peroxidation inducers 10 μM Fe^{2+} + 0.5 mM ascorbate. Insert: Shows reversible reduction and oxidation of verdazyl radicals to the leukoverdazyl and cation verdazyl species respectively.

The idea of reduction or recycling of free radicals with the subsequent formation of products possessing antioxidant properties might be exploited for the development of new potent inhibitors of lipid peroxidation in biomembranes. In this respect data on the antioxidant effects of stable verdazyl radicals are of interest. Verdazyl radicals appear to be very efficient inhibitors of enzymic and non-enzymic lipid peroxidation in liver microsomes, skeletal muscle, sarcoplasmic reticulum and brain synaptosomal membranes. Figure

14 illustrates the concentration dependence of lipid peroxidation inhibition in rat brain synaptosomes by 1,3,5-triphenylverdazyl and 3-methoxyphenyl-1,5-diphenylverdazyl radicals. The half-maximal inhibition is achieved at concentrations of the radicals as low as 10^{-8}- 10^{-7} M. We suggest this high efficiency of lipid peroxidation inhibition could be due to the reversible transformation of verdazyl radicals to leukoverdazyls possessing very high radical scavenging activity. We conclude that antioxidants which readily recycle their radical products before they are lost in radical reactions or in addition reactions are very promising candidates for developing new substances for preventing lipid peroxidation *in vivo.*

Acknowledgements

Research supported by the Bulgarian Academy of Sciences and NIH (CA47597).

References

1. Tappel, A.L., Brown, W.D., Zalkin, H., and Maier, V.P. (1961) *J. Am. Oil Chem. Soc.* **38**, 5-9
2. Machlin, L.J. (1980) *Vitamin E. A Comprehensive Treatise* (Machlin L.J. ed.), Marcel Dekker, New York
3. Witting, L.A. (1980) in *Free Radicals in Biology* (Pryor, W., ed.) pp 295-315, Academic Press, New York
4. Burton, G.W., and Ingold, K.U. (1986) *Acc. Chem. Res.* **19**, 194-201
5. Ingold, K.U., Webb, A.C., Witter, D., Burton, G.W., Metcalfe, T.A., and Muller, D.P.R. (1987) *Arch. Biochem. Biophys.* **259**, 224-225
6. Niki, E. (1987) *Chem. Phys. Lipids* **44**, 227-253
7. Diplock, A.T. (1982) *Acta Vitaminol. Enzymol.* **4**, 303-309
8. Kagan, V.E., Serbinova, E.A., Bakalova, R.A., Novikov, K.N., Skrypin, V.I., Evstigneeva, R.P., and Stoytchev, Ts.S. (1987) in *Free Radicals, Oxidant Stress and Drug Action* (Rice-Evans, C., ed) pp. 425-442. Richelieu Press, London
9. Kagan, V.E., Skrypin, V.I., Serbinova, E.A., Raikova, D.N., Tyurin, V.A., Buschev, B.N., Erin, A.N., and Stoytchev, Ts.S. (1986) *Dokl. Acad. Nauk USSR* **288**, 1242-1248 (in Russian)
10. Kagan, V.E. and Quinn, P.J. (1988) *Eur. J. Biochem.* **171**, 661-667
11. Harfouf, M., Serbinova, E., Ukhin, L., Komissarov, V.,Erin, A., Rakovsky, S., Savov, V., and Kagan, V. (1989) *Bull. Exp. Biol. Med. USSR* (in Press, in Russian)
12. Tyurin, V.A., Korol'kov, S.N., and Kagan, V.E. (1989) *Biochemistry USSR* **54**, 940-947 (in Russian)

13. Halliwell, B., and Gutteridge, M.C. (1985) *Free Radicals in Biology and Medicine* pp. 147-161, Clarendon Press, Oxford
14. Stoyanovsky, D.A., Kagan, V.E., and Packer, L. (1989) *Biochem. Biophys. Res. Commun.* **160**, 834-838
15. Packer, L. (1984) *Med. Biol.* **62**, 105-109
16. Packer, J.E., Slater, T.F., and Willson, R.F. (1979) *Nature* **278**, 737-739
17. Packer, J.E., Slater, T.F., and Willson, R.F. (1978), *Life Sci.* **23**, 2617-2620
18. McCay, P.B. (1985) *Ann. Rev. Nutr.* **5**, 323-340
19. Packer, L., Maguire, J.J., Mehlhorn, R.J., Serbinova, E.A. and Kagan, V.E. (1989) *Biochem. Biophys. Res. Comm* **159**, 229-235
20. Mehlhorn, R.J., Sumida, S., and Packer, L. (1989). *J. Biol. Chem.* **264**, 13778-13452
21. Maguire, J.J., Wilson, D.S. and Packer, L. (1989). *J. Biol. Chem.* **264**, in press
22. Guido, R.M.M. and Bast, A. (1983), *FEBS Letts.* **159**, 24-28
23. Liebler, D.C., Kling, D.S., and Reed, D.J. (1986) *J. Biol. Chem.* **261**, 12114-12119
24. Reddy, C.C., Scholtz, R.W., Thomas, C.E. and Massaro, E.J. (1982), *Life Sciences* **31**, 571-576
25. Kagan, V.E., Bakalova, R.A., Serbinova, E.A., Koynova, G.M., Baldenkov, G.N., Tkachuk, V.A. and Stoytchev,Ts.S. (1988) in *Free Radicals, Chemistry, Pathology and Medicine* (Rice-Evans, C., and Dormandy, T., eds) pp. 417-438. Richelieu Press, London
26. Baldenkov, G.N., Serbinova, E.A., Bakalova, R.A., Tkachuk, V.A., Kagan, V.E., and Stoytchev, Ts.S. (1988) *Free Rad. Res. Commun.* **4**, 277-281
27. Mahoney, C.W. and Azzi, A. (1988) Biochem. Biophys. Res. Comm. 154, 694-697
28. Mahoney, C.W. and Azzi, A. (1989) *Molecular Basis of Membrane-Associated Diseases* (Azzi, A., Drahota, Z. and Papa, S., eds.), pp.84-96. Springer-Verlag, Berlin

Vanadium as a Biological Hydrogen-Abstractive Agent

T. Ramasarma, S. Vijaya, Meera Rau, Lashmi Khandke, M. S. Patole, Sharada Gullapalli, Kalyani Penta, Mihir Chauduri and C. K. Ramakrishna Kurup
Department of Biochemistry, Indian Institute of Science, Bangalore 560 012 India

Abstract

Vanadate stimulates oxidation of NADH and NADPH non-enzymically and the rate of the reaction is enhanced by enzymes in plasma and microsomal membranes with NADH, and by xanthine oxidase with both NADH and NADPH. This H_2O_2 - generating reaction, with a stoichiometry of NADH: O_2: H_2O_2 of 1: 1: 1, is unusual in its inhibition by SOD which made its acceptance initially difficult. The first equivalent of NADH being used for a SOD-insensitive reduction of vanadate to a blue compound in microsomes, caution is suggested in interpretation of SOD effects. The activity is high with polymeric vanadate. The low activity with metavanadate increases on acidification/polymerization, chemical reduction to vanadyl and presence of H_2O_2, or converting into oxo-peroxo vanadate by treatment with H_2O_2. A common structural feature of O=V−O−V=O with infrared band of 985 cm^{-1} is noticed in all active vanadate forms. Mechanisms not involving superoxide need to be considered now in view of doubts on its formation in this reaction. Vanadate participates in several redox activities such as NADH oxidation, nitrogenase, lipid peroxidation and bromoperoxidase, with H-abstraction being the core reaction.

Vanadium is now considered to be essential in trace quantities, has therapeutic value in pharmacological doses and is toxic in excess. After the recognition of vanadate as the inhibitor of Na,K-ATPase, present as a contaminant in muscle-derived ATP samples (1), a large number physiological, metabolic and enzyme effects have been discovered for different forms of vanadate (see 2, 3, 4, for reviews). The plasma membranes, in addition to Na,K-ATPase and adenyl cyclase, have a powerful NADH dehydrogenase which is most active with ferricyanide as the electron acceptor, rather than with cytochrome *c* or oxygen and is also active with diferric transferrin

Biological Oxidation Systems, Volume 2

(5, 6). The electron flow of such an oxidation of NADH is opened to oxygen with the generation of highly active H_2O_2 in presence of vanadate. This report gives an overview of the developments of vanadium acting as a biological hydrogen-abstractive agent, particularly for oxidation of NADH and generation of H_2O_2 with membrane enzyme systems.

Vanadium Compounds as Catalysts in Chemical Redox Reactions

Vanadium (atomic wt. 50.94) is a transition element capable of forming compounds in valence states 3^+, 4^+ and 5^+, both cationic (vanadyl, V^{iv}) and anionic (meta- and ortho-vanadate, V^{v}) species. Vanadium chemistry is similar to that of phosphorus, with the bond lengths of V−O being about 0.1 Å longer than P-O. Vanadates profoundly influence reactions involving phosphate compounds. The interesting property of easily dehydrating and forming polymers distinguishes anionic vanadate from phosphate. At lower pH, orange-colored, stable form of decavanadate $(V_{10}O_{28})^{6-}$ predominates.

Vanadium salts are known for a long time to act as inorganic catalysts for oxidizing organic compounds such as aniline (7), cysteine (8), noradrenaline (9) and adrenaline (10). Reduced forms of vanadium are known to transfer electrons to nitrogen (11). Of particular interest is the stimulation of oxidation of NADH by vanadate compounds, first reported by Ramasarma *et al.* (12, date of receipt Jan. 11, 1980) and independently by Vyskocil *et al.* (13, date of receipt, Jan. 31, 1980). Their high rates in micromoles/min, the pH optimum of 7.0 and activity with metavanadate (0.8mM) are in contrast with results in our extensive studies (14, 15) of rates in nmoles/min, activity increasing in acid pH and high activity with polymeric vanadate at micromolar concentrations.

Vanadate Stimulates Oxidation of NADH to Generate Hydrogen Peroxide

We showed that stimulation by vanadate of oxidation of NADH is accompanied by oxygen uptake and formation of H_2O_2 with a stoichiometry of NADH: O_2: H_2O_2 of 1:1:1, and is inhibited by addition of Mn^{2+}, superoxide dismutase (SOD) and cytochrome *c* (15). These results and the identification of esr signals characteristic of DMPO−OH (16) led to belief that superoxide is involved in the reaction mechanism. The esr signals for reduced form of vanadate (V^{iv}) could not be picked up in this system, possibly because of small amounts, whereas Vyskocil *et al.* (17) were able to identify these signals as indicated in a footnote. In 1984, for the first time we proposed a reaction mechanism as given in scheme 1 in Figure 1 wherein O_2 receives the first electron from NAD radical, and the second electron from V^{iv} (15).

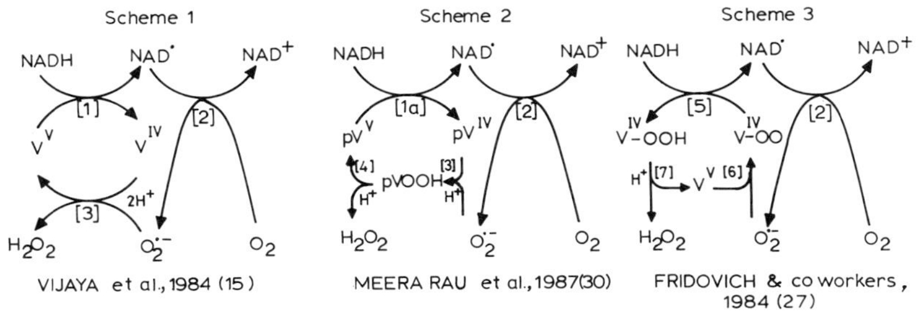

Figure 1. *Mechanism of NADH Oxidation Stimulated by Vanadate.* Scheme 1 shows that oxidation of NADH by V^V reduces it to V^{iv} and the NAD radical gives its electron to oxygen to generate superoxide which is further reduced by V^{iv} to H_2O_2 (15). In scheme 2, we included phosphovanadate as the electron carrier and the electron from reduced vanadate is donated to superoxide to form a vanadyl hydroperoxide species which then is protonated to from H_2O_2 regenerating V^V (33, 35). Scheme 3 represents the proposal of Fridovich and co-workers (19, 29) wherein the primary oxidant is $V^{iv}-OO$ formed from V^V and $O_2^{\cdot-}$ and H_2O_2 is obtained on hydrolysis of $V^{iv}-OOH$. In the light of the evidence for the formation of superoxide being now in doubt, these mechanisms need reevaluation.

We have repeatedly stressed that high rates are obtained for NADH oxidation with polymeric form of vanadate, but not ortho- or meta-forms. In addition we also detected the formation of hydroxyl radicals, by measuring oxidation of reduced cytochrome *c*, in the presence of H_2O_2 and deca-, meta- and ortho- forms of vanadate with relative efficiencies of 1.00, 0.75 and 0.50, respectively. The pH profile of these activities gave an interesting pattern (Figure 2). The activities of disappearance of NADH and oxygen uptake showed pH optimum near 5.0 and of cytochrome *c* redox reactions near 7.0 (2). We also clearly indicated in this article (2, page 289) that "H_2O_2 oxidizes vanadyl sulfate and yields a mixture of oxidated forms, including polymeric ones, capable of stimulating both nonenzymic and enzymic NADH oxidation." Evidence had been obtained with esr signals that hydroxyl radicals are produced during the reaction of vandyl with H_2O_2 (18). This reaction is now further explored by other workers.

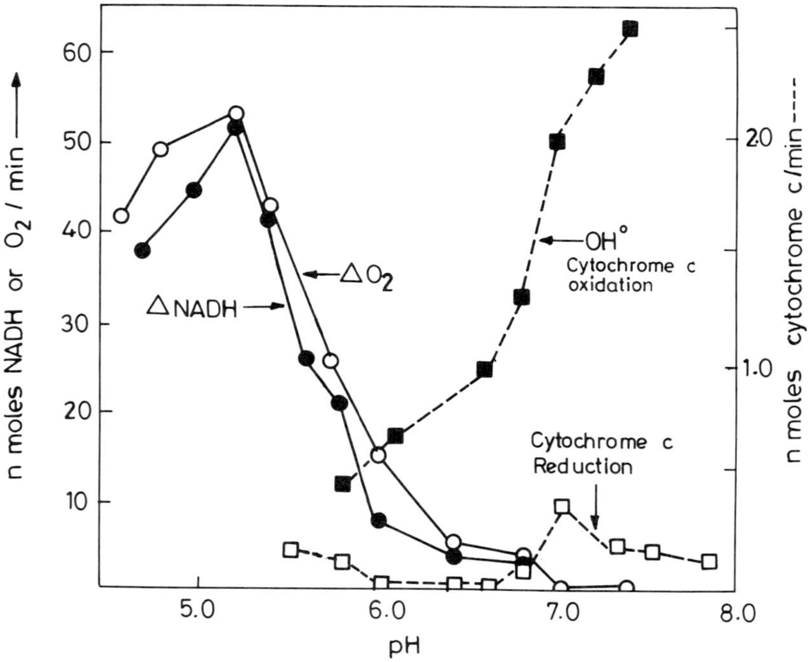

Figure 2. *The pH-Activity Profile of the Nonenzymic Oxidation of NADH (100 μM) by Polyvanadate (150 μM) in Phosphate Buffer (50 mM).* The following measurements were done by standard methods under identical conditions: decrease in absorbance at 340nm (Δ NADH) consumption of O_2 (Δ O_2), increase in absorbance at 550nm for reduction of added cytochrome *c* (Fe^{3+}) and decrease in absorbance at 550nm for oxidation of added cytochrome *c* (Fe^{2+}). (Adapted from ref. 2).

Metavanadate is Active Under Certain Conditions

One other development noticed in recent work is on the capability of metavanadate in stimulating NADH oxidation. To demonstrate this reaction, high concentrations of vanadate are required. The rates are usually low and a reduction of vanadate to vanadyl seems to be involved in these experiments. Thus, the reductants/reagents leading to active form of vanadate are: photochemically generated superoxide or KO_2 (19), sugars and sugar phosphates containing carbonyl function with produced yellow to blue-green products (20), ATP (Geetha Sharma, unpublished experiments), dithioerythritol, dithiothreitol and cysteine (21), and low concentration of ascorbate and phosphate (22). Addition of vanadyl sulfate alone (22) or in the presence of H_2O_2 (23) also showed increased NADH oxidation, confirming our brief report in 1981 (2) but without appearing to notice it.

The results now indicate that the systems of polyvanadate and of vanadyl-H_2O_2 are different in the following aspects: the concentration of vanadyl determines the extent of NADH oxidized; vanadyl and H_2O_2 seem to give a relatively unstable active form which sustains activity for a limited period; catalase inhibition indicates the need of H_2O_2 in the reaction; oxygen consumption accompanies but needs larger reducing equivalents. These results led to the view that this reaction represents an oxidase-peroxidase (Kalyani Penta, unpublished observations).

Vanadate-Stimulated NADH Oxidation in Biomembranes

Erdmann *et al.* (24) first showed that addition of vanadate in the reaction mixture containing calf cardiac cell plasma membranes resulted in disappearance of NADH. This was interpreted as an NADH-vanadate reductase activity. This activity was also found in mouse liver plasma membranes (25) and in rat liver microsomes wherein the rates with NADH were 50-fold higher compared to NADPH (26). For the first time, Ramasarma *et al.* (14) showed that vanadate-stimulated oxidation increased several-fold in acid pH and reached a maximum of a staggering value of 3200 moles/min per mg protein, that oxygen uptake accompanied oxidation of NADH, and that addition of Mn^{2+} or SOD inhibited the reaction. The paradox that a H_2O_2-generating oxidation of NADH by oxygen was inhibited by SOD, the enzyme known for dismutation of $O_2^{\cdot-}$ to H_2O_2 made acceptance of this work by the referees difficult and this was finally published in 1981 after three revisions and two years.

In the next major paper from our laboratory Vijaya *et al.* (27) first showed the stoichiometry of this reaction of 1:1:1 for NADH: O_2: H_2O_2 by measurement of each of these three reaction components, the specificity of decavanadate amongst the vanadate forms tested, the appearance of esr signals of 4-banded of DMPO-OH and 8-banded reduced form of vanadate (V^{iv}), as well as reconfirmed requirement of phosphate for maximum activity and inhibition by well-known superoxide quenchers. This paper ran into even greater difficulties and had the benefit of improvement from views of referee(s) of three journals before appearing in 1984. We interpreted the effect of vanadate as opening of the electron flow to oxygen from the NADH dehydrogenase which otherwise can give only poor rates with naturally occurring oxygen or cytochrome *c* as electron acceptors. It was found that addition of ferricyanide to vanadate-containing reactions mixtures (12) or cytochrome *c*, abolished oxygen uptake and then only the reductase activities manifested. In other words, the NADH dehydrogenase is converted to an oxidase in presence of vanadate, and further addition of ferricyanide or cytochrome *c* reverts it to reductase activity. The fixed ideas that a H_2O_2-generating system must not be inhibited by SOD and wherever superoxide is implicated cytochrome *c* reduction must be SOD sensitive led to initial disbelief of the originality of the effects of the vanadate system. We certainly do not think this is due either to the names of the authors or the laboratories it came from both in India and U.S.A. or the quality of the contents, as these two papers (14, 27) turned out to be most cited subsequently. This led us to formulate a theorem: "A good paper sent to the most appropriate referee will be rejected." These initial observations of ours attracted a number of workers to take up work on vanadate-stimulated NADH oxidation. They confirmed most and contradicted some effects, and in the process led to further developments. The properties of this system in biomembranes studied in our laboratory, especially the typical characteristics and some controversies, and our current understanding of the mechanism will be presented below.

Distribution of the Activity in Cell Components

The reaction is now found in variety of biomembranes: calf cardiac cell membranes (24), pig erythrocytes (25), mouse liver plasma membrane (14), rat liver microsomes (26), rat erythrocytes (27), sugar beet microsomes (28), bovine cream xanthine oxidase (29, 30), rat brain microsomes (31), and sarcoplasmic reticulum (32). In mouse liver tissue, smooth and rough endoplasmic reticulum were shown to be a little more effective than plasma membrane, whereas Golgi apparatus was less active (14).

We further showed that rat liver mitochondrial NADH oxidase was not similarly affected by vanadate when tested with intact mitochondria, digitonin particles (mitoplasts), submitochondrial particles (smp) (33). The rotenone and antimycin A-sensitive activity with exogenous NADH was present only with smp but not with whole mitochondria or mitoplasts, obviously because of non-permeability which our laboratory working on mitochondria for 3 decades is well aware. Small vanadate stimulation was obtained with mitochondria but not with smp and therefore we suggested that outermembranes may have this activity but not inner membrane. Liochev and Fridovich (34) confirmed this by testing with these membranes prepared by their special methods, but made a baffling comment that we "seemingly forgot the membrane of mitoplasts is impermeable to NADH" in the face of our clear data on the absence of NADH oxidation in mitoplasts, and good, vanadate-unaffected activity in smp (see Table 3 in ref. 33).

Active Form of Vanadate

From the early stages of this work, we have been claiming that a yellow-colored polymeric product, possibly decavanadate, obtained in acidic conditions is the active form of vanadate, and that meta- or ortho-vanadate was low in activity, if at all, and gains activity on acidification (14). We clearly showed concentration-dependent relative activities of different forms with erythrocyte membranes (27) and demonstrated that decavanadate has several-fold higher activity. This pattern is obtained in all membrane systems under study in our laboratory. Other laboratories continued to report activities with ortho- and meta-vanadate using high concentrations without clearly defining whether or not these solutions had gone through neutralization and had yellow color. We normalized the activities (assuming a linear basis) for 100 μM concentration of vanadate and NADH on which depends the activity (in addition to pH) and compared to results of approximate specific activities from different laboratories (Table 1). The high activity obtained with polyvanadate compared the metavanadate is obvious.

We reconfirmed that acidification and neutralization of metavanadate, known to dehydrate and polymerize readily, yielded a light yellow-colored solution which showed increased activity (Table 2). Addition of a trace of polyvanadate (1 μM) to a solution of metavanadate (500 μM), gave a very much higher rate, over and above the additive value, indicating a potentiation effect for metavanadate (38). Polyvanadate seems to act as a cofactor/catalyst and turns over several times its concentration of NADH. Fig. 3 gives the data at pH 5.0 with 100 μM decavanadate oxidizing 100 μM decavanadate NADH each in six successive additions to completion. Essentially similar results were obtained at pH 7.0. These results also point out that the active form is depleted in each cycle and the rate of reaction progressively decreased.

Phosphate is Required for Maximum Activity

One of the controversies was on the requirement of phosphate anions for maximum activity. The low activity increased to maximum with phosphate concentrations of 50-100 mM for membranes

Table 1. *Membrane-Dependent Vanadate-Stimulated NADH Oxidation*

Membrane	pH	Conc., μM			nmoles/min/mg		Authors (Country)	Ref.
		NADH	Meta V	Poly V	Expt.	Norm.		No.
Calf atrial pm	6.7	1500	1000	---	1260	8	Erdmann *et al.* 1979 (FRG/USA)	24
Pig Erythrocytes pm	7.0	100	---	300	197	66	Crane *et al.* 1980 (USA/India/Sweden)	25
Mouse Liver pm	7.4	100	---	100	360	72	Ramasarma *et al.* 1980 (USA/India)	12
Rat Liver pm	7.0	150	---	300	3200	710	Menon *et al.* 1980 (India/USA)	26
Mouse Liver pm	7.0	100	---	130	245	192	Ramasarma *et al.* 1981 (USA/India)	14
Rat Erythrocytes pm	7.0	100	---	100	70	70	Vijaya *et al.* (India/USA)	27
Sugar Beet mc	6.5	100	200	---	156	78	Briskin *et al.* 1985 (USA)	28
Xanthine Oxidase	7.0	100	300	---	2000	667	Darr & Fridovich 1984, 1985 (USA)	29
	7.8	100	200		1400	700		30
Xanthine Oxidase	6.6	200		150	2170	720	Khandke *et al.* 1986 (India)	35
Rat Liver mc	7.2	320	400	---	540	42	Liochev & Fridovich 1986 (USA)	41
Rat Liver mc	7.0	100	---	100	770	770	Patole *et al.* 1986 (India)	36
Rat Liver mc	7.0	100	---	150	1580	1053	Meera *et al.* 1987 (India)	33
Rat Liver mc	6.5	100	400	---	21	5	Coulombe *et al.* 1987 (USA)	37
Rat Liver mc	7.4	250	400	---	430	43	Liochev & Fridovich 1988 (USA/Bulgaria)	34
Rat Liver mc	7.0	100	---	100	600	600	Patole *et al.* 1988 (India)	38
		100	100		40	40		
Rat Liver mc	7.0	100	---	100	625	625	Patole *et al.* 1988 (India)	31
Sarco. Reticulum	---	---	---	1200	25	---	Molnar *et al.* 1988 (Hungary)	32
Rat Liver mc	7.2	100	---	308	507	133	Reif *et al.* 1989 (USA)	39
		100	308		46	11		

* Approximate specific activity normalized to concentration of 100 μM each of NADH and vanadate (not corrected for pH changes). pm, plasma membrane; mc, microsomes

Table 2. *Stimulation of NADH Oxidation by Vanadate Compounds*

Form of Vanadate	nmoles/min per mg protein		$NADH/O_2$
	Δ NADH	Δ O_2	
Decavanadate	444	498	0.90
Metavanadate (acid treated)	127	143	0.89
Metavandate	31	28	1.10
Monoperoxovandate	241	284	0.85
Diperoxovandate	63	71	0.89
Triperoxovandate	0	0	---

The reaction mixture consisted of rat liver microsomes (25 μg/ml), phosphate buffer (pH 7.0, 50 mM) vanadate compound (100 μM), and NADH (100 μM). The measurements of Δ NADH were done by decrease in absorbance at 340 nm in a spectrophotometer and Δ O_2 by uptake of oxygen in an oxygraph as described earlier (33). Treatment of metavanadate with acid was as described (38). Decavanadate was obtained in crystal form by extracting vanadium pentoxide with NaOH solution and adding ethanol to the orange-colored solution in cold. Monoperoxo-citrato-vanadate (54) and the di- and tri-peroxovanadates (53) were prepared as described.

tested in our laboratories. The stimulation of NADH oxidation by vanadate was "absent in Tris, acetate, borate and Hepes buffers," and addition of phosphate restored the activity (27). Phosphate and vanadate have a tendency to form heteropolymers (4) and this suggested that phosphovanadate complex is the reducible species, analogous to phosphomolybdate. Darr and Fridovich (30) maintained that phosphate is not needed and that this effect "can now be seen as a partial reversal by phosphate, of an inhibition imposed by Tris" in their system with xanthine oxidase which appears to behave different. It appears that this is likely to be due to changes in pH from unbuffered solutions, (possibly acidic which give higher activity) on addition of phosphate (pH 7.8). It is difficult to compare different experiments, as these workers did not give rates of NADH oxidation and also obtained varying values of absorbance change at 340nm in the presence (0.032, Table 1) and the absence (0.034, Table 3) of vanadate, with other conditions being the same (30). It is gratifying to see phosphate requirement was confirmed by other workers (22, 37). All the workers, including Fridovich's group, subsequently have been using phosphate buffer in the reaction mixtures.

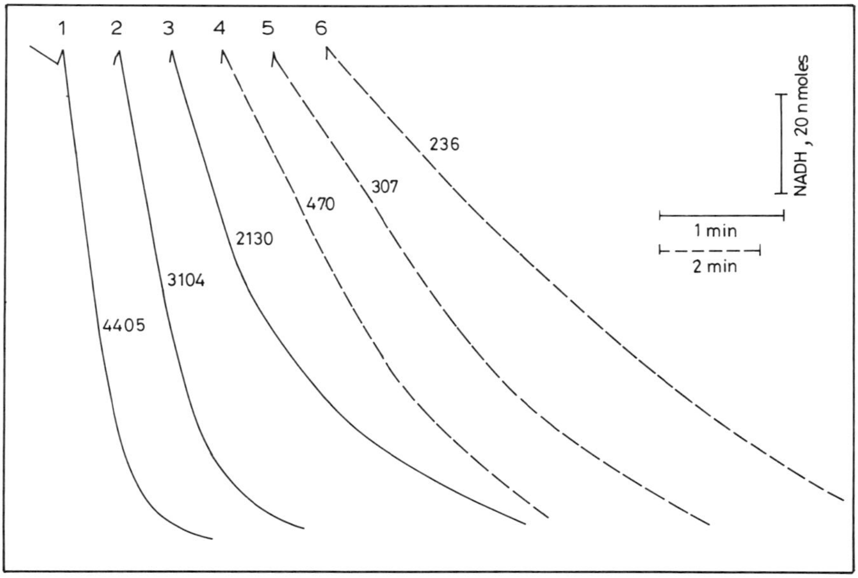

Figure 3. *Catalytic Nature of Polyvanadate in Stimulating NADH Oxidation.* The reaction mixture consisted of rat liver microsomes (25 μg protein/ml), phosphate buffer (pH 5.0, 50 mM), decavanadate (100 μM) and NADH (100 μM each, 6 additions). The reaction was monitored by decrease in absorbance at 340nm. The numbers on the lines are specific activity values (nmoles/min per mg protein).

Table 3. *Effect of SOD on Vanadate-Stimulated NADH Oxidation*

Measurement	SOD	Addition of NADH	
		FIRST	SECOND
Δ NADH	-	45	44
(nmoles NADH/min)	+	35	8
Δ O_2	-	5	48
(nmoles O_2/min)	+	2	2

The reaction mixture consisted of rat liver microsomes (25 μg/ml), phosphate buffer (pH 7.0, 50 mM), polyvanadate (100 μM), NADH (100 μM each addition) and SOD (10 μg/ml), where mentioned. The measurements were done as given under Table 2.

Activity is High in Acid pH

We have reported that the activity rose to high values in acid pH range (14). Some membrane preparations showed optimum activity around pH 5.0-6.0. This is confirmed by Coulombe *et al.* (37). Comparing the activities at neutral and acid ranges revealed interesting differences. For example, Golgi apparatus showed 2-fold higher values at pH 5.0 whereas at pH 7.0 it had ¼ activity of plasma membrane (14) for the oxidation of NADH, but not of NADPH. The instability of NADH on prolonged exposure to acid pH range did not affect this study as the rates were high and oxidation of added NADH was completed in a few minutes after addition of NADH. In pilot experiments we could show the product of oxidation was indeed NAD^+ under these conditions by the reappearance of 340nm absorbance on adding lactate, lactate dehydrogenase and Tris-glycine buffer (pH 8.8). We considered this effect of pH on activity, obtained also in nonenzymic system (see also Fig. 2), is due to formation and/or stabilization of active species of vanadate (15). Such an increased activity at acid pH was also found for the reduction of vanadate (40), but not of ferricyanide or of cytochrome *c* (12).

Stoichiometry of NADH Oxidation

We introduced into these studies the simultaneous measurements of NADH disappearance and oxygen uptake. The rate measurements of these gave NADH: O_2 ratio of 1:1 in non-enzymic (15) and membrane (27) systems indicating a 2e-reduction of molecular oxygen. The product obviously is H_2O_2. This was confirmed by release of ½ of oxygen consumed on adding excess catalase (15, 27, 33, 34), and also by measuring H_2O_2 formed (15, 27). It was also realized that the sensitiveness of the two measurements of NADH disappearance and oxygen uptake differed. Thus, the former is less sensitive to inhibition by Mn^{2+} and SOD than the latter (33).

Reduction of Vanadate

The reduction of polyvanadate added could readily be demonstrated first to a blue-colored, 650 nm-absorbing compound which is sensitive to SOD, and then to a colorless product which is insensitive to SOD, during the oxidation of NADH by rat liver microsomes (35). This reduction could also be obtained with ascorbate alone (40) under conditions similar to those used by Yoshino *et al.* (22). Polyvanadate, but not ortho- or meta- forms, was reduced and the rate and extent of this reaction depended on its concentration. The electron transfer from NADH to vanadate seems to involve participation of cytochrome b_5 (40). Reif *et al.* (39) later showed that purified NADH-b_5 reductase flavoprotein was sufficient to support polyvanadate-dependent NADH oxidation. These studies along with the identification of reduced form of vanadate by the esr in the reaction mixtures (27) established that vanadate can act as an electron acceptor.

It is important to take vanadate reduction reaction into account in the studies on SOD inhibition of NADH oxidation. In such experiments it is not possible to distinguish the electron flows for reduction of vanadate and of oxygen. This led to contradictory results on SOD effect. Thus, the first amount of added NADH (equivalent to vanadate present) seems to be used for reduction of polyvanadate which is SOD insensitive and the change in second such addition, used for oxygen uptake, is SOD sensitive. A typical set of results with rat liver microsomes, are given in Table 3. This will explain the purported lack of SOD inhibition in some experiments with polyvanadate (39).

Specificity of NADH

From the very beginning we found that the membrane enzymes used specifically NADH (14, 27), whereas non-enzymic system used both NADH and NADPH (15). With xanthine oxidase NMNH (29) and NADPH (34) could nearly substitute for NADH. Using mM concentration of metavanadate, NADPH was claimed to be the preferred substrate with rat liver microsomes but the rates were small (37). Using microsomal reductase proteins specific for NADH and NADPH, Reif *et al.* (39) could also show activity with NADPH, indicating that both can serve as electron donors under certain conditions.

Is Superoxide Involved?

The crucial finding in these studies is the marked inhibition by SOD of NADH oxidation supported by vanadate, but not by ferricyanide or by cytochrome *c* (14). The accompanying oxygen reduction to H_2O_2 was most sensitive (27, 33). This observation was repeatedly confirmed by all the laboratories working on this system. The active form of polyvanadate was not modified on pretreatment with SOD. Other scavengers of radicals, such as Mn^{2+}, DMPO and noradrenaline, also inhibited this reaction. No doubt participation of some form of radicals in the activation/reduction of oxygen is occurring. The unexpected inhibition by SOD, the enzyme that dismutates superoxide to H_2O_2, indicated strongly that this type of dismutation of superoxide was not responsible for H_2O_2 generated. It was not possible to measure superoxide in the reaction mixtures by the conventional SOD inhibitable reduction of cytochrome *c* because addition of cytochrome *c* abolished oxygen uptake and converted the apparent oxidase into a cytochrome *c* reductase, despite the presence of vanadate and oxygen. The importance of this experimental fact is yet to be recognized by the workers in the field.

We were led to believe that superoxide radicals are formed in this reaction on the support of esr data on DMPO-OH, provided first by us in 1984 (27). Recent evidence by Kuppuswamy and Zweier (4) that the DMPO−OH signal, the characteristic spectrum of 1:2:2:1 quartet, originates from the trapping hydroxyl radicals and not from breakdown of DMPO−OOH, purportedly formed by interaction of DMPO and superoxide. This, therefore, raises a doubt whether superoxide is indeed formed in the reaction mixtures of vanadate-stimulated NADH oxidation. As a consequence, the effect of SOD itself may be by a mechanism other than dismutating superoxide. In this connection, Ernster and co-workers (42) found an interesting alternative inhibitory action SOD on autooxidation of hydroquinone. This effect was "assessed in terms of an inhibition of NADPH oxidation, O_2 consumption and H_2O_2 formation" almost in parallel with present studies. This implies additional potential for SOD in dismutation of radicals of semiquinone variety, and possibly others. Indeed this may be an important function of SOD in cells, both aerobic and anaerobic (where it is present), in dismutating quinones of widely occurring menquinone and ubiquinone species and thereby detoxifying their radicals after univalent reductions, already proposed in 1976 by Winterbourn *et al.* (43)

Superoxide-Based Mechanisms

Fridovich and co-workers (19, 29) proposed a superoxide-based oxidant mechanism about the same time in 1984 as we proposed our two-source electron supply to oxygen (14). According to this the primary oxidant is V^{iv} −OO which is generated by reaction between V^v and $O_2^{\cdot-}$ and protonation and hydrolysis of V^{iv} −OOH yields H_2O_2 regenerating V^v as given in scheme 3, Fig. 1. Their proposal also includes a superoxide-dependent free radical chain reaction (19) and its termination when either $O_2^{\cdot-}$ or NAD dismutated. More intriguingly, by separately measuring $O_2^{\cdot-}$ and NADH disappearance, they calculated the chain lengths of 4 with xanthine oxidase (29) of 15 with the photochemical system (19), and of 30 and above with rat liver microsomes (44). In the absence of proof that superoxide is indeed produced in the actual reaction mixtures with vanadate as discussed in the previous section, these interpretations need to be reevaluated.

With the xanthine oxidase system we found that the oxidation of NADH by xanthine and vanadate are independent, and the latter activity, an intrinsic property, can occur without xanthine (35). We have not been able to appreciate a chain reaction in an overall stoichiometric utilization of oxygen, while we accepted the need to involve interaction of reduced oxygen with active vanadate and introduced reactions [3] and [4] into the mechanism (33, 34) as shown in scheme 2, Fig. 1. Also this offers an excellent explanation for the need of protons and therefore increased activity in acid pH range.

The reduction of oxygen to superoxide by NAD radical by reaction [2] has a high rate constant of $1.9 \times 10^9\ M^{-1}s^{-1}$ (45) and is the most accepted step. The least understood step is the one in which superoxide receives the second electron and gets reduced to H_2O_2, other than by SOD. Superoxide anion, however, is not a good oxidant of NADH. The rate constant of this reaction is $27\ M^{-1}s^{-1}$ (46) which increased to $1 \times 10^5\ M^{-1}s^{-1}$ when NADH was bound to lactate dehydrogenase (47) and to $2 \times 10^7\ M^{-1}s^{-1}$ when protonated superoxide anion reacted with NADH bound to glyceraldehyde-3-phosphate dehydrogenase (48). But increasing the rate constant is not a general property of all NADH-dependent dehydrogenases which should be expected if mere binding of NADH is implied. Several dehydrogenases show vanadate dependent NADH oxidation which seem to be due to their flavin content and not due to enzyme catalysis (Sharada Gullapalli and Kalyani Penta, unpublished data). In all the properties these systems resemble nonenzymic activity described earlier, including increased activity in acid pH (48). Radical-mediated chain reaction initiated by superoxide has been proposed as a mechanism, essentially based on the findings of Chan and Beilski (47, 48) who used a protein concentration of of 10.8 μM (1.3 mg/ml of pure lactate dehydrogenase), almost as a reagent, in the presence of xanthine oxidase or by pulse radiolysis. The proposed superoxide-dependent oxidation of NADH does not seem to be appropriate. Therefore the question what is the primary oxidant of reaction [1] in scheme 1 of Fig. 1 remains unanswered. A form of vanadate can fit this role. Our working basis (33) is shown in scheme 2, Fig. 1, which consists of three parts: first, reduction of V^v; second interaction of V^{iv} with superoxide to form VOOH; and third, its protonation and release of H_2O_2. The inhibitory effect of SOD is likely to be on reaction [1a] to stop generation of NAD radicals. The difference in the two proposals lies in reaction [1a] (scheme 2) and [5] (scheme 3) in which the primary oxidants are V^{v} and V^{iv} $-OO$, respectively. We now found that a form of peroxo-vanadate species, formed by interaction of V^{v} and H_2O_2, are active.

Peroxyvanadate Compounds Can Act as Primary Oxidants

In the hemoproteins iron-oxygen complexes (ferryl ions/perferryl) are known to be formed from O_2 or H_2O_2 and participate in the oxidations (49) and hydroxylations (50). Hamilton (51) postulated that peroxidic "oxenoid" species are capable of epoxidation and hydroxylation of aromatic hydrocarbons. An abstraction of hydrogen from C−H is implicit. Oxo-peroxo complexes of vanadium (V^{v}) are now found to act as biomimetic reagents for epoxidation of olefins and hydroxylation of alkanes and aromatic hydrocarbons (52). Collaborative work with Dr. Mihir Choudhuri, who synthesized a number of oxo-peroxo vanadium compounds (53), helped in our identifying these species as active in NADH oxidation. The data given in Table 2 show that oxo-monoperoxo-citrato-vanadate (54) was active, albeit less than decavanadate and the activity decreased on introducing a second peroxo-group. The activity with both these compounds was sensitive to addition of SOD, again indicating that SOD action is more complex than merely preventing formation of VOO complex. The compound with three peroxo-groups and on oxo-groups was devoid of activity. These peroxo-vanadates, obtained by treating metavanadate solution with H_2O_2, occur as dimeric vanadates (54) and have the peroxo-groups in triangular bidendate form. In addition to a band around 950 cm^{-1} of O=V (55), polymeric-, deca-, acidified meta-, oxo-monoperoxo and oxo-diperoxo-vanadate species show a typical infrared absorption at 980-985 cm^{-1} ascribed to V−O−V (56). This band is absent in all inactive species including metavanadate, which has only O=V, and triperoxo-vanadate, which has no O=V, and therefore is ascribed to O=V−O or more precisely

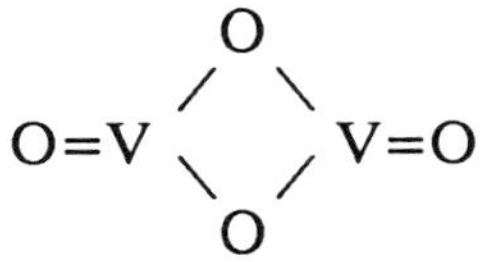

present in active species of polyvanadates and peroxovanadates. This structure is indeed present in the two most active compounds tested oxo-monoperoxovanadate (54) and decavanadate (57) as seen from the X-ray crystal structure data (Fig. 4). It is instructive to notice that the distance between the two end oxygens in both cases is 5.5 Å.

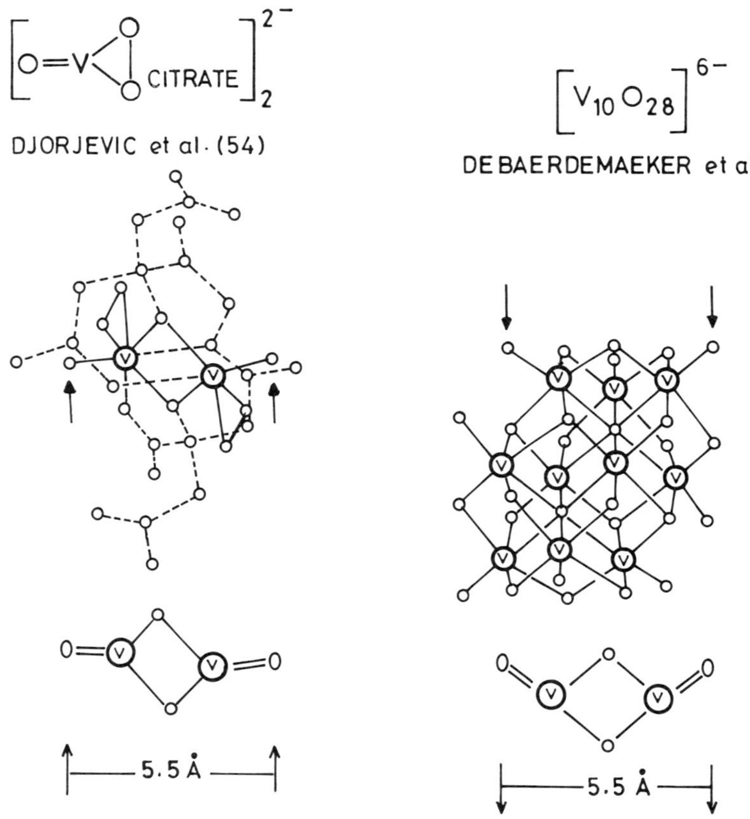

Figure 4. *The X-Ray Crystal Structures of Active Vanadate Forms.* The structures are adapted for oxo-monoperoxo-citrato vanadate dimer from Djorjevric *et al.* (54) and for decavanadate from Debaerdemaeker *et al.* (57). The bonds not belonging to vanadate are shown as broken lines. Two end oxygens are marked by arrows and the corresponding structures, common to both the active forms give a O−V−O−V−O distance of 5.5 Å.

Biological H-Abstracting Activities

Vanadate is now found to have wide range of biological activities. In addition to NADH oxidation vanadate compounds can also participate in three other types of redox systems which can be considered as H-abstraction from NAD(P)H or C−H of fatty acids, and reducing or fixing gaseous substrates, oxygen, nitrogen or bromine, into organic products (Table 4).

Table 4. *Vanadate-Dependent H-Abstration Systems*

Enzyme System	Reductant	Electron Acceptor	Product
Oxidase	NADH	O_2	H_2O_2
Nitrogenase	NADPH	N_2	NH_3
Lipid Peroxidation	PUFA	O_2	RC-OOH
Bromoperoxidase	RC-H	Br_2	RC-Br

An alternative nitrogenase present in a mutant of *Azotobacter croococcum* is now found to be a vanadium-containing enzyme (58) with 2 atoms of vanadium present in the AcI protein (Mr 210K), but not Ac 2 (Mr 60K). The putative active site is considered to contain VFeS cluster and participates in the transfer of 6 electrons from 3 NADPH to N_2 with simultaneous hydrolysis of about 12 Mg ATP.

During 1988 a series of publications appeared reconfirming that vanadate supports lipid peroxidation. In our studies on brain microsomes, we found that more oxygen was consumed over that needed for generation of H_2O_2 and this was traced to polyvanadate and NADH specific lipid peroxidation (31). This activity was inhibited by dihydroxyphenolates and by hemin and enhanced by traces of Fe^{2+} in a concentration-dependent fashion. Indeed in this tissue the common NADPH-Fe^{2+} dependent lipid peroxidation occurs to a small extent. Liochev and co-workers (59) showed vanadyl (V^{iv}) can peroxidize mitochondrial lipids and phosphatidyl choline suspensions. Keller and co-workers (60) found by spectral evidence that vanadyl and H_2O_2 together produced hydroxyl radicals and these then cause peroxidation of fatty acids (18:3 and 20:4). Human placental tissue microsomes were also found to have vanadate-dependent lipid peroxidation (61). In all these reactions some vanadate-derived active species is abstracting a hydrogen from a fatty acid carbon.

Another example of naturally occurring vanadate-containing proteins with biological activity is the bromoperoxidase reaction involving H_2O_2-dependent introduction to a Br atom into organic compounds of a large variety. Vanadium containing bromo-perioxidases from brown algae, red algae and lichen have been isolated and the mechanism of peroxidative bromination is considered to be obtained by enzyme-bound bromonium (Br^+) ion intermediate

(62) but this was contradicted (63). The specific activity had increased with increase in concentration of vanadium up to 1 atom per mole of enzyme (Mr 64K), present in the pentavalent form which can be reduced to V^{iv} by dithionate as judged by esr (64). It is not clear what is the active form and how vanadate functions in the reaction although H-abstraction by a form of peroxyvanadate is a possibility.

With its ability to form homo and hetero polymers and complexes with oxygen, and possibly sulfur, thereby providing specialized structures, vanadium will be able to interact with biologically active compounds and exert a regulatory role in cellular activities.

Acknowledgement

Financial assistance from the Department of Science and Technology, Government of India, New Delhi is acknowledged.

References

1. Cantley L.C. Jr., Josephson, L., Warner, R., Yanagasiwa, M., Lechne, C. and Guidotti, G. (1977) *J. Biol. Chem.* **252**, 7421-7423
2. Ramasarma, T. and Crane, F.L. (1981) *Current Topics in Cellular Regulation* **20**, 247-301
3. Jandhyala, B.S. and Hom, G.J. (1983) *Life Sci.* **33**, 1325-1340
4. Boyd, D.W. and Kustin, K. (1984) *Adv. Inorg. Biochem.* **6**, 311-365
5. Crane, F.L., Goldenberg, H., Morre, D.J. and Low, H. (1979) *Subcell. Biochem.* **6**, 345-399
6. Sun, I.L., Navas, P., Crane, F.L., Morre, D.J., and Low, H. (1987) *J. Biol. Chem.* **262**, 15915-15921
7. Witz, G. and Osmond, L. (1976) *C.R. Hebd. Seances Acad. Sci.* **83**, 348-350
8. Bergl., F., Bray, R.C., and Harrap, K.R. (1958) *Nature* (Lond) **181**, 1654-1655
9. Cantley, L.C. Jr., Fergnson, J.H. and Kustin, K. (1978) *J. Am. Chem. Soc.* **100**, 5210-5212
10. Quist, E.E. and Hokin, L.E. (1978) *Biochem. Biophys. Acta.* **511**, 202-212
11. Volpin, M.E., Shur, V.B., Latyaeva, V.N., Vyshinskaya, L.I., and Shulgaitser (1966) *Izvest. Akad. Nauk SSBR, Ser. Khim.* **385**
12. Ramasarma, T., Mackellar, W. and Crane, F.L. (1980) *Indian J. Biochem. Biophys.* **17**, 163-168
13. Vyskocil, F., Teisinger, J. and Dlouha, H. (1980) *Nature* (Lond) **786**, 516-517
14. Ramasarma, T., Mackellar, W. and Crane F.L. (1981) *Biochim. Biophys. Acta* **646**, 88-98
15. Vijaya, S. and Ramasarma, T. (1984) *J. Inorg. Biochem.* **20**, 247-254

16. Finkelstein, E., Rosen, G.M. and Rauckman, E.J. (1982) *Mol. Pharmacol.* **21**, 262-265
17. Vyskocil, F., Dlouha, H. and Teisinger, J. (1981) *Nature (Lond)* **294**, 288
18. Ozawa, T. and Hanaki, A. (1989) *Chem. Pharm. Bull.* **37**, 1407-1409
19. Liochev, S. and Fridovich, I. (1986) *J. Free Rad. Biol. Med.* **1**, 287-292
20. Liochev, S. and Fridovich, I. (1987) *Biochim. Biophys. Acta* **924**, 319-322
21. Keller, R.J., Coulombe, R.A., Sharma, R.P., Grover, T.A. and Pietta, L.H. (1989) *Arch. Biochem. Biophys.* **217**, 40-48
22. Yoshino, S., Sullivan, S.G. and Stern, A. (1989) *Arch. Biochem. Biophys.* **272**, 76-80
23. Liochev, S. and Fridovich, I. (1987) *Arch. Biochem. Biophys.* **255**, 274-278
24. Erdmann, E., Kraweitz, W., Phillip, G., Hackbarth, Schmitz, W., Scholz, H. and Crane, F.L. (1979) *Nature (Lond)* **282**, 335-336
25. Crane, F.L., Mackellar, W.C., Morre, D.J., Ramasarma, T., Goldenberg, H., Grebing, C. and Low, H. (1980) *Biochem. Biophys. Res. Communs.* **93**, 746-754
26. Menon, A.S., Rau, M., Ramasarma, T. and Crane, F.L. (1980) *FEBS Lett.* **114**, 139-142
27. Vijaya, S., Crane, F.L. and Ramasarma, T. (1984) *Mol. Cell. Biochem.* **62**, 175-185
28. Briskin, D.P., Thornley, W.R. and Poole, R.J. (1985) *Arch. Biochem. Biophys.* **236**, 228-237
29. Darr, D. and Fridovich, I. (1984) *Arch. Biochem. Biophys.* **232**, 562-565
30. Darr, D. and Fridovich, I. (1984) *Arch. Biochem. Biophys.* **243**, 220-227
31. Patole, M.S. and Ramasarma, T. (1988) *J. Neurochem.* **51**, 491-496
32. Molnar, E., Kiss, Z., Dux, L. and Guba, F. (1988) *Acta Biochim. Biophys. Hung.* **23**, 63-74
33. Meera Rau, Patole, M.S., Vijaya, S., Ramakrishna Kurup, C.K. and Ramasarma, T. (1987) *Mol. Cell. Biochem.* **75**, 151-159
34. Liochev, S. and Fridovich, I. (1988) *Arch. Biochem. Biophys.* **263**, 299-304
35. Khandke, L., Gullapalli, S., Patole, M.S. and Ramasarma, T. (1986) *Arch. Biochem. Biophys.* **244**, 742-749
36. Patole, M.S., Ramakrishna Kurup, C.K. and Ramasarma, T. (1986) *Biochem. Biophys. Res. Communs.* **141**, 171-175
37. Coulombe, Jr., R.A., Briskin, D.P., Keller, R.J., Thornley, W.R. and Sharma, R.P. (1987) *Arch. Biochem. Biophys.* **255**, 267-273
38. Patole, M.S., Gullapalli, S. and Ramasarma, T. (1988) *Free Rad. Res. Communs.* 4, 201-207
39. Reif, D.W., Coulombe, Jr., R.A. and Aust, S.D. (1989) *Arch. Biochem. Biophys.* **270**, 137-143
40. Patole, M.S., Ramakrishna Kurup, C.K. and Ramasarma, T. (1987) *Mol. Cell. Biochem.* **75**, 161-167
41. Kuppusamy, P. and Zweier, J.L. (1989) *J. Biol. Chem.* **264**, 9880-9884

42. Cadenas, E., Diego, M., Anders, B., Christina, L., Juan, S. and Ernster, L. (1988) *Free Rad. Biol. Med.* **5**, 71-80
43. Winterbourn, C.C., French, J.K. and Claridge, R.F.C. (1976) *FEBS Lett.* **94**, 269-272
44. Liochev. S. and Fridovich, I. (1986) *Arch. Biochem. Biophys.* **250**, 139-145
45. Willson, R.L. (1970) *J. Chem. Soc. D. Chem. Commun.* 1005
46. Land, E.J. and Swallow, A.J. (1971) *Biochim. Biophys. Acta.* **234**, 34-42
47. Beilski, B.H.J. and Chan, P.C. (1976) *J. Biol. Chem.* **251**, 3841-3844
48. Chan, P.C. and Belski, B.H.J. (1980) *J. Biol. Chem.* **255**, 874-876
49. Dunford, H.B. (1982) *Adv. Inorg. Biochem.* **4**, 41-68
50. Groves, J.T., McClusky, G.A., White, R.E. and Coon, M.J. (1978) *Biochem. Biophys. Res. Communs.* **81**, 154-160
51. Hamilton, G.A. (1974) in "*Molecular Mechanisms of Oxygen Activation*" Hayaishi, O. ed. Acad. Press, N.Y. pp 405-451
52. Mimoun, H., Saussine, L., Daire, E., Postel., M., Fischev, J. and Weiss, R. (1983) *J. Am. Chem. Soc.* **105**, 3101-3110
53. Chaudhuri, M.K. and Ghosh, S.K. (1984) *Inorg. Chem.* **23**, 534-537
54. Djorljevic, C., Lee, M. and Sinn, E. (1989) *J. Am. Chem. Soc.* **28**, 719-723
55. Frederickson, J.L.D. and Hausen, D.M. (1963) *Anal. Chem.* **35**, 818-827
56. Yuchi, A., Muranaka, H., Yamada, S. and Tanaka, M. (1980) *Bull. Chem. Soc. Jpn.* **53**, 1560-1563
57. Debaerdemaeker, T., Arrieta, J.M. and Amigo, J.M. (1982) *Acta Cryst.* **838**, 2465-2468
58. Robson, R.L., Eady, R.R., Richardson, T.H., Miller, R.W., Hawkins, M. and Postgate, J.R. (1986) *Nature* **322**, 358-360
59. Liochev, S., Ivancheva, E. and Russanov, E. (1988) *Free Red. Res. Comms.* **4**, 317-323
60. Byczkowski, J.Z., Wan, B. and Kulkarni, A.P. (1988) *Env. Contam. Toxicol.* **41**, 696-703
61. Itoh, N., Izumi, Y. and Yamada, H. (1987) *Biochemistry* **26**, 282-289
62. DeBoer, E. and Wever, R. (1988) *J. Biol. Chem.* **263**, 12326-12332
63. Yu, H. and Whittaker, J.W. (1986) *Biochem. Biophys. Res. Commun.* **160**, 87-92

Constitutive and Inducible Repair Systems in Oxidative Stress

Stefan W. Ryter, Robert E. Pacifici and *Kelvin J.A. Davies*
Institute for Toxicology, Department of Biochemistry, University of Southern California Health Science Campus, 1985 Zonal Ave. PSC 614-616, Los Angeles, CA 90033 USA

Abstract

To defend against an onslaught of oxidant insults, prokaryotic and eukaryotic cells maintain enzymatic and chemical systems which scavange reactive species or arrest the propagation of damage. Despite these antioxidant defenses, oxidative damage to biomolecules can be demonstrated *in vivo*. The resistance of prokaryotes and eukaryotes to high degrees of oxidative stress has implied the existence of enzyme systems to remove and/or repair oxidatively damaged regions of biomolecules, and subsequently restore the molecules to their native functional states or replace them completely. The present review will examine evidence in support of such "repair systems". Furthermore, changes in gene expression patterns have been associated with transient adaptation to oxidative stress. The mechanisms by which inducible systems may confer adaption to oxidative stress are currently being investigated in several laboratories.

Introduction

Oxidative Stress

Virtually all biochemical entities, including lipids, proteins, carbohydrates, and nucleic acids, can sustain structural alterations as a consequence of free-radical mediated reactions (1-3). In living organisms, damage to biomolecules by free radicals and oxidants may translate into a variety of disease states, including ischemia-reperfusion injury, cancer, and aging (1). The toxic effects of many xenobiotics, such as quinonoid antitumor drugs, bipyridyl herbicides, and elements of air pollution and smoke, also involve free-radical mechanisms (1). Furthermore, aerobic organisms endogenously produce reactive oxygen intermediates as by-products of mitochondrial respiration, or "spillage" from excessively activated phagocytes. The condition known as "oxidative -stress" arises when such exogenous or endogenous oxidant insults supercede cellular

Biological Oxidation Systems, Volume 2

defenses and repair capacities, resulting in a debilitation of normal function, and ultimately, cell death. Both elevated oxidant exposure and deficiency in antioxidant defenses may evoke oxidative stress.

Enzymatic Antioxidant Defenses

The identification of superoxide dismutases (SOD) in 1968 first demonstrated that cells possess systems to cope with free radicals occuring *in vivo* (4). SOD's catalyze the dismutation of two superoxide anion radicals (O_2^-) to hydrogen peroxide and oxygen. Since 1968, a number of other enzymatic and chemical systems have been identified which scavange intracellular radicals and arrest the propagation of free radical damage (1-3). Scheme one, part A, lists these known systems, collectivly refered to as cellular antioxidant defences. Glutathione peroxidase utilizes glutathione as an electron donor to reduce toxic hydrogen peroxide or lipid hydroperoxides, to water or lipid alcohols, respectively. The phospholipid hydroperoxide glutathione peroxidase (PHGPX), may directly reduce membrane associated lipid hydroperoxides (5). Glutathione reductase regenerates reduced glutathione from glutathione disulfide, at the expense of NADPH from the hexose monophosphate shunt. Catalase also reduces H_2O_2 to water and oxygen, although its higher apparant k_m for H_2O_2 probably restricts its importance to conditions of high H_2O_2 concentration, as found in peroxisomes. It has been postulated that the combined action of superoxide dismutases and glutathione peroxidases serve to remove O_2^- and H_2O_2, lest these species participate in the metal catalyzed Haber-Weiss reaction to produce an even more reactive oxidant, the hydroxyl radical (•OH) (2). DT Diaphorase (NADPH-quinone oxidoreductase) catalyzes the divalent reduction of quinonoid compounds to more polar hydroquinones, bypassing their univalent reduction to (oxygen-reducing) semi-quinone radicals.

Antioxidant Compounds

In addition to enzymatic antioxidant systems, cells also possess both water- soluble and lipid-soluble compounds which readily oxidize, upon reaction with free radicals, to derivatives with relatively little reactivity. Vitamin E and β-carotene terminate the propagation of peroxidative chain reactions in lipid bilayers. Ascorbic acid acts as a cytoplasmic antioxidant, and may reduce vitamin E oxidation products. Uric acid and proteins which chelate transition metals, (cerruloplasmin, transferrin, albumin) function as extracellular antioxidants in plasma, by restricting the ability of iron and copper to function as catalysts in free-radical generating reactions (1).

Scheme 1. ANTIOXIDANT DEFENSES AND REPAIR SYSTEMS

A. Antioxidant Systems which Remove Primary Reactive Species or Arrest the Propagation of Free-Radical Damage.

1. Enzymes: Superoxide Dismutases, Glutathione Peroxidases/Glutathione Reductase, Alkyl Hydroperoxide Reductase, Catalases, DT Diaphorase.

2. Antioxidant Compounds: α-tocopherol, β-carotene, Ascorbic acid, Uric acid, Ubiquinone, Retinoic acid, Transition Metal Chelators, Glutathione.

B. Enzymatic Repair Systems, Which Remove, Repair, or Replace Oxidatively Damaged Biomolecules.

1. Protein Degradation/Repair Enzymes: Proteinases, Proteases, Peptidases, (*eg.* Macroxyproteinase, or M.O.P).

2. Lipid Degradation/Repair Enzymes: Phospholipase A_2, Phospholipase C, Acyl Transferases, Glutathione Peroxidase/Transferase/Reductase, Alkyl Hydroperoxide Reductase, Phospholipid Hydroperoxide Glutathione Peroxidase (PHGPX).

3. DNA Degradation/Repair Enzymes: Exonuclease III, Endonuclease III, Endonuclease IV, Glycosylases, DNA Polymerases, DNA Ligase.

Repair of Oxidative Damage

Despite the presence of antioxidant enzymes and compounds, considerable evidence exists that oxidative modification to cellular constituants occurs *in vivo.* Oxidation products of DNA bases have been detected in the urine of mammals, and increase after exposure to oxidants and radiation. Based on such studies, Ames *et al.* estimate that each human cell sustains 10^3 oxidative "hits" per day to thymine DNA bases alone (1). The occurance *in vivo* of chemical species identical to •OH reaction products with biomolecules in pulse radiolysis model studies, implies that the antioxidant defenses do not safeguard against such reactions with absolute efficiency. Prokaryotic and eukaryotic cells, however, display a high degree of resistance to oxidative stress, as exemplified by minimal accumulation of damaged biomolecules over time.

To explain these apparant discrepancies, Davies (3) hypothesized that cells must have the capacity to process oxidant damage, that is, to remove, degrade, repair or replace oxidatively modified biomolecules. This chapter will focus on evidence that has accumulated in support of such "repair systems" (originally described as "secondary antioxidant defences" (3,6) Part B of scheme one depicts the known degradative and repair activities that cope with oxidatively damaged proteins, lipids, and DNA. Repair systems for DNA and phospholipids imply the removal and degradation of individually damaged componants, such as altered nucleotides or oxidized fatty acyl chains, followed by the restoration of "proper " or native chemical moieties to the greater molecule. The repair of proteins denatured during oxidative stress, however, refers to the complete degradation of those proteins to free amino acids, followed by their *de novo* synthesis from amino acid pools. Thus, used in this context, the process of "repair" may refer to a particular biomolecule, or to the cell as a whole.

A growing area of research involves the characterization of transient adaptive responses to oxidative stress, and associated changes in patterns of gene expression, resulting in the selective induction of distinct protein families. The present review will also discuss these so-called "global" or "stress inducible responses".

Degradation of Oxidatively Modified Proteins

Mechanisms of Intracellular Protein Degradation

The discovery of cytoplasmic proteolytic enzymes has eclipsed the old view that cytosolic proteins undergo degradation primarily in the lysosomes, by intra-lysosomal proteases. Lysosomes have now gained distinction as the sites of organelle decomposition, whereas cytoplasmic proteolytic systems largely attend to the degradation of cytoplasmic proteins (9). In general, proteins destined for complete degradation undergo sequential processing by proteinases, proteases, and peptidases. Proteinases initially recognize and cleave cellular proteins that bear signals calling for their degradation, whether they be attached biochemical markers or alterations in physical properties. Proteases hydrolyze the intermediate-sized polypeptide products of proteinase activity into smaller peptides. Peptidases finish the task of protein degradation by hydrolyzing small peptide remnants into free amino acids which can be reused for protein synthesis. The terms "protein degradation", "proteolysis", and "protein cleavage" will herein be used interchangeably in reference to the enzymatic hydrolysis of peptide bonds. In bacteria, the proteolytic degradation of some abnormal proteins requires ATP, which directly stimulates the cytoplasmic proteases "La " and "Ti" (10-11). In eukaryotes, energy dependent proteolysis involves the conjugation of the peptide ubiquitin to certain malfunctional proteins, which serves as a signal for degradation. In this case, both the conjugation of ubiquitin and the stimulation of proteolysis require ATP (12-14). The degradation of truncated polypeptides and products of nonsense/missense mutations generally proceeds through these energy dependent pathways (10-15).

The degradation of oxidatively modified proteins in both prokaryotes and eukaryotes, however, proceeds through ATP and ubiquitin independent cytoplasmic proteolytic pathways, the discussion of which will dominate the present chapter (3,16-19, 20-51).

Intracellular Protein Oxidation

Proteins can sustain intracellular oxidation by two mechanisms: attack of free-radicals and oxidants,(3,15-19,20-29, 41-43) or enzyme catalyzed oxidation by mixed function oxidases (40,44-51). The oxidation of proteins by mixed function oxidases, may constitute a mechanism which marks proteins for degradation, whereas oxidation by free-radicals represents a random degenerative process. Proteins oxidized by both mechanisms show similar increases in proteolytic susceptibility (3,15-19,20-51). Furthermore, both mechanisms share similar characteristics, in that they proceed through oxygen intermediates,often involve transition metal catalysts, and produce similar product profiles (See Fig. 1).

Free-radical mediated protein oxidation, may have several deleterous consequences in addition to the loss of enzyme function. Hydroxyl radical potentially reacts with the α-carbon of any amino acid, with slightly higher rate constants for residues with sulfhydryl, aromatic, or heterocyclic R groups. •OH attack on proteins may result in protein fragmentation, which can occur at sites other than peptide bonds (3,6-9,16-19). Oxidative modification of proteins can also disrupt secondary and tertiary structure, leading to the exposure of hydrophobic regions normally shielded in the interior of the protein. Increases in hydrophobicity promote the non-covalent aggregation of proteins and decrease overall water solubility (3,16-39). Oxidative modification also changes amino acid charges, shifting the isoelectric point of the protein. Tyrosyl and thiyl radicals produced by •OH mediated oxidation of amino acid residues, may react with one another to form amino acid cross links, and thus, covalently aggregate proteins. The accumulation of such insoluble protein aggregates has been implicated in the formation of lipofuscin (age pigment)(31), as well as cataract formation in the lens (22).

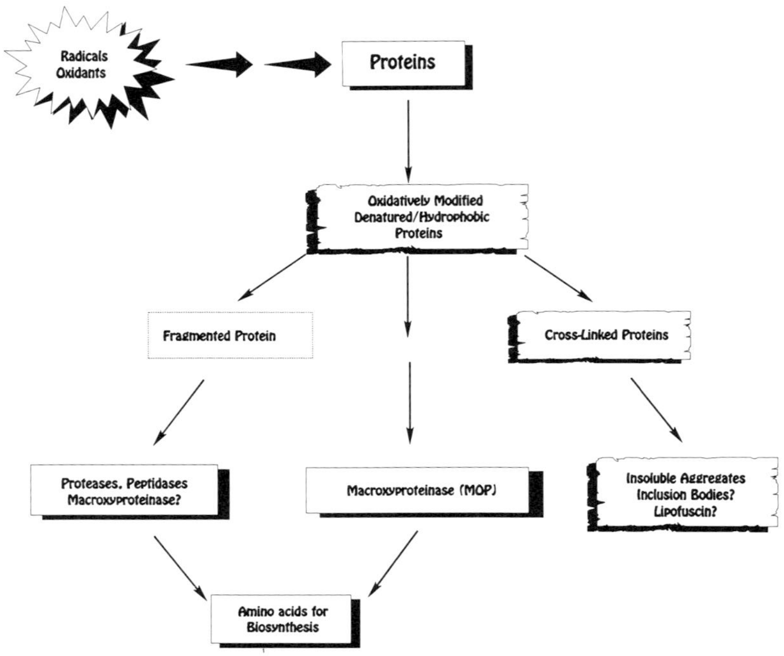

Figure 1. *Degradation and Repair of Oxidatively Modified Proteins.* Proteins are susceptible to reactions with radicals and oxidants. Oxidative modification of proteins disrupts secondary and tertiary structure, increasing protein hydrophobicity by exposing interior hydrophobic residues. Exposed hydrophobic residues become substrates for proteinase cleavage by Macroxyproteinase (M.O.P.). Under aerobic conditions radical attack leads to protein fragmentation. Protein fragments are also more susceptible to degradation by cytosolic proteases, and peptidases. M.O.P may also participate in the degradation of fragments. Under severe oxidizing conditions, or in the absence of oxygen, radical-radical reactions predominate, leading to the formation of insoluble cross-linked protein aggregates. Such aggregates are poor substrates for degradation, and have been implicated in the formation of lipofuscin, inclusion bodies in the red cell, and cataracts in the lens.

The free amino acids produced from the complete degradation of oxidatively modified proteins can be reused for protein synthesis.

Oxidatively Modified Proteins as Preferential Substrates for Proteolysis

Both eukaryotes and prokaryotes display a burst of proteolytic activity following exposure to oxidants (3,6-9,11-14,16,20-47). Furthermore, when pure proteins are subjected to oxidation *in vitro*, [utilizing either hydrogen peroxide, oxygen radicals, (3,15-19,22-39) or mixed function oxidase systems (40,44-51)], and subsequently incubated with cell free extracts from untreated cells, the oxidatively modified proteins undergo rapid and selective degradation (3,15-19,22-39). These phenomena have been demonstrated using whole cells and extracts from erythrocytes, reticulocytes, hepatocytes, skeletal muscle, cardiac muscle, *Esherichia coli*, and *Salmonella typhimurium*. Oxidatively modified proteins,(3, 15-19.22-51) as well as other proteins,(9) undergo complete degradation to free amino acids in intact cells and cell-free extracts, indicating co-ordination between proteinase, protease, and peptidase activities. The lack of accumulation of partially degraded peptides implies excess peptidase activity over rate-limiting proteinase activity.

Unlike many other proteolytic substrates, the degradation of oxidatively modified proteins does not require the presence of ATP or ubiquitin (3,15-19). In fact, ATP typically inhibits somewhat the degradation of oxidized proteins in cell extracts. These observations pointed the way toward the characterization of energy independent proteolytic pathways for oxidatively modified proteins.

Bacterial Proteolysis

Rapid degradation of endogenous or exogenous oxidatively damaged proteins has been observed in intact bacteria and cell-free extracts of *E. coli.*, respectively (16,19,23-24).

The proteolytic enzymes of *E. coli* which require ATP for activity, "La" and "Ti" (10-11) exhibit no preferential hydrolysis of oxidized substrates. *E. coli lon*$^-$ mutants, which lack the ability to make the ATP dependent protease "La", display a proteolytic capacity equal to that of wild type *E. coli*, with respect to the degradation of oxidatively modified proteins (24). The ATP independent series of proteases "Do", "Re", "Me", "Fa", and "So", however, preferentially degrade oxidized substrates (3,6,23-24). The importance of each of these enzymes appears to vary with the substrates tested. This energy-independent series of proteolytic enzymes may be the evolutionary precursors for the mammalian high molecular weight complexes discussed in the next section. In *E. coli*, ATP independent proteolytic activity has also been detected against oxidatively modified glutamine synthasE (40-41).

Eukaryotic Proteolytic Systems, (Macroxyproteinase)

A single high molecular weight proteinase complex accounts for 80-90% of the total proteolytic activity toward oxidized proteins in erythrocytes (25,36,37). Similar complexes, typically weighing 600-700 kD and composed of 8-12 distinct proteins, exist in all mamalian cells studied thus far, although their functions had previously been unknown (52). The name recently given to such a complex isolated from red blood cells, "Macroxyproteinase" or "M.O.P", short for "Macro-Oxy-Proteinase", refers to its large molecular weight, its preference for oxidized substrates, and its multicatalytic sites. Sulfhydryl reagents, serine reagents, and transition metal chelators all inhibit macroxyproteinase activity, implying the existence of cysteine, serine, and metallo active sites (25,36,37). The effectiveness of these inhibitors, as well as the relative importance of each active site, varies with the proteolytic substrates tested, including oxidatively modified hemoglobin, superoxide dismutase, "foreign proteins", and fluorogenic peptides. These observations resolved conflicts between previously reported inhibitor profiles for similar 600-700 kD complexes from other sources (52). The M.O.P complex appears to degrade its substrates completly to free amino acids, implying that M.O.P sequentially applies proteinase, protease, and peptidase activities in a co-ordinated fashion (25,36-37). Ubiquitin has no stimulatory effect on M.O.P, whereas ATP actually inhibits its activity slightly. ATP and ubiquitin independence has also been demonstrated for all such 600-700 kD complexes previously reported (52). Indeed such complexes may represent different forms of M.O.P. in various mammalian species. The 300 kD proteinase complex partially purified from liver against an activity towards oxidized glutamine synthase (48,52) may actually constitute a fragment of the M.O.P complex.

In reticulocytes, a 1500 kD proteinase complex ("Megapain", "Megasin", or "UCDEN") which catalyzes most of the ATP and ubiquitin dependant proteolytic activity, (53-54) can account for no more for than 10-15% of the proteolytic activity toward oxidatively damaged hemoglobin (25,36-37). Furthermore, ATP and ubiquitin stimulate the ability of this complex to degrade casein, but do not stimulate any activity toward oxidatively damaged substrates (25,36, 37).

Hydrophobicity as a Recognition Signal for the Degradation of Oxidized Proteins.

The degree of oxidant induced protein denaturation and hydrophobicity correlates strongly with proteolytic susceptibility (3, 16-19,20-39). Oxidative modification disrupts protein secondary and tertiary structure, (16-19) allowing interior hydrophobic residues to come into contact with the solvent, thus decreasing overall solubility. Under mild oxidizing conditions, hydrophobicity increases linearly with the degree of oxidative modification. As the strength of the oxidizing conditions increases, (especially under anaerobic conditions) termination reactions between amino-acid radicals occur more frequently. The resulting inter- and intra-protein amino acid cross-links begin to limit further protein unfolding, and associated increases in hydrophobicity. In highly oxygenated systems, free radical attack will cause protein fragmentation, rather than covalent aggregation, therefore the extent of denaturation/ hydrophobicity continues to increase even with severe oxidizing conditions (3,16-19). It follows that proteolytic susceptibility also correlates with the degree of denaturation and hydrophobicity, (linear regression co-efficients r=0.98 for erythrocytes,(3) 0.96-0.98 for *E. coli*) (23) until aggregation events begin to interfere with further denaturation and proteolysis.

These observations led to the proposition that changes in protein hydrophobicity resulting from oxidative modification may mark a protein for intracellular degradation (6,19). Studies of M.O.P (25,36-37) and other reported 600-700 kD complexes (54) have shown a proteolytic preference for sequences rich in hydrophobic/bulky amino acid residues. M.O.P likely exists in an active state within cells, displaying little activity towards native proteins which shield their hydrophobic residues within their interiors. Following oxidative modification, the resulting perturbations in protein structure expose hydrophobic residues, (3,16-19) making them available for recognition and cleavage by M.O.P.

Structural analysis of M.O.P has depicted it as cylindrical in shape, with a central cavity or hole (49). The entrance to the M.O.P cylinder possibly recognizes the exposed hydrophobic residues of oxidatively modified proteins.

De Novo Synthesis Concludes the Repair of Oxidatively Damaged Proteins

The degradation of oxidatively damaged proteins serves to rid the cell of aberrant and useless polypeptides, and prevent these species from aggregating and crosslinking to form insoluble, deleterious material. Intact free amino acids released from the complete degration of such polypeptides, can be reutilized for the *de novo* synthesis of proteins. Since oxidative stress can trigger the elevated synthesis of large amounts of "stress proteins" as well as enhanced proteolytic activity, it follows that the amino acids liberated through the proteolytic pathways would be directly reused in the synthesis of the "stress proteins". CuZn superoxide dismutase (CuZnSOD), an oxidative stress protein, can sustain irreversible damage from its end product hydrogen peroxide. Subsequently, CuZnSOD becomes highly susceptible to proteolytic degradation (3,25,36-37). Intact free amino acids released from damaged CuZnSOD likely can be used for the immediate resynthesis of this enzyme. In this sense, the complete turnover of CuZnSOD may be considered to be controlled by reactive oxygen intermediates (O_2^- and/or H_2O_2).

Degradation and Repair of Oxidized Membrane Lipids

Consequences of Lipid Peroxidation in Biological Membranes

The toxic effects of radiation, certain xenobiotics, oxidizing gases and reactive oxygen intermediates has been associated with their capacity to initiate lipid peroxidation in membranes. These agents can abstract a labile hydrogen atom from the methylenic carbons of polyunsaturated fatty acids, producing alkyl radicals. In the presence of oxygen, alkyl radicals form lipid peroxyl radicals which participate in further hydrogen abstraction events, producing fatty acid hydroperoxides and more alkyl radicals. Such autocatalytic chain reactions multiply the damaging effects of a single initiation event. The presence of phospholipids containing polyunsaturated fatty acids renders biological membranes vulnerable to disruption by peroxidative chain reactions (56). Free radical initiated lipid peroxidation may underlie the tissue damage observed after ischemia/ reperfusion incidents (57).

Following membrane peroxidation, the presence of phospholipids with hydroxy-, epoxy-, carbonyl-, and peroxy- substituted acyl chains may cause perturbations in membrane structure. The formation of such polar species within the non-polar membrane matrix may lead to increased membrane viscosity, rigidity, (58,59) and increased occurance of intermembrane and transbilayer phospholipid exchange (fusion and flip-flop) (60-61). A raising of the gel to liquid-crystalline transition temperature by conjugated dienes and other peroxidation products may create phase separations in membranes, disrupting the orderly packing of lipids and proteins, and creating regions of membrane microheterogeneity (60-62). Furthermore, many peroxidation end products can also react to exacerbate membrane damage (56). For example, malondialdehyde may form shiff bases with amino groups, leading to protein and phospholipid crosslinking. Hydroperoxides may paticipate in metal catalyzed multiplication reactions which produce highly reactive alkoxyl radicals. To cope with the oxidation of phospholipids *in vivo*, cells recruit enzymatic componants of phospholipid biosynthetic pathways.

It has been proposed that the complete repair of oxidatively modified membrane constituants occurs in three distinct phases: the selective hydrolysis of peroxidized fatty acyl chains from the glycerophosphate backbone of phospholipids, the detoxification of free fatty acid hydroperoxides, and finally, the energy dependant reacylation of the lysophospholipids which remain in the membrane (63). (See fig. 2)

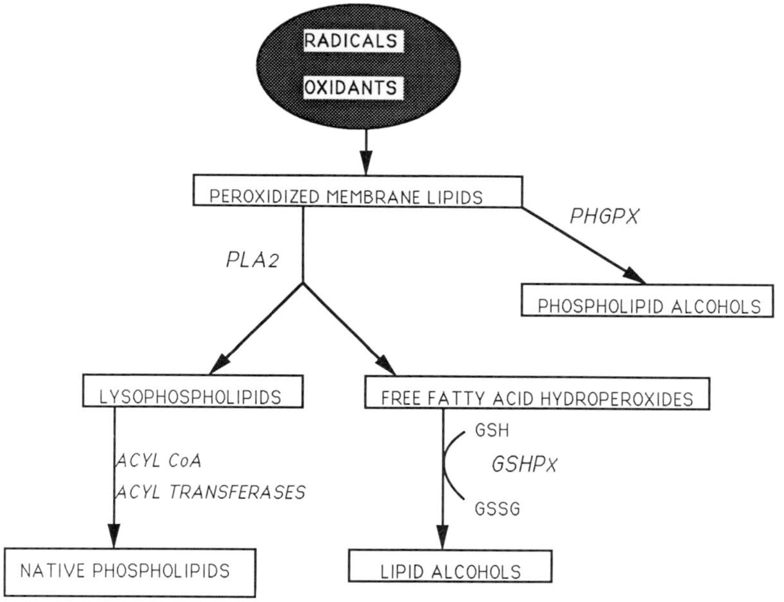

Figure 2. ***Repair of Peroxidized Membrane Lipids.*** Free radicals and oxidants may initiate the peroxidation of membrane phospholipids. Fatty acid hydroperoxides may be repaired/removed by two possible mechanisms. In one model, Phospholipase A_2 (PLA_2) cleaves phopspholipid hydroperoxides into free fatty acid hydroperoxide and lysophospholipid. Cytosolic Glutathione Peroxidase (GSHPX) reduces free fatty acid hydroperoxides to fatty acid alcohols, utilizing Glutathione (GSH) as a cofacter which it oxidizes to Glutathione Disulfide (GSSG). Acyltransferases reacylate the lysophospholipids which remain in the membrane, using acyl CoA.

Alternatively, Phospholipid Hydroperoxide Glutathione Peroxidase (PHGPX) may directly reduce membrane phospholipid hydroperoxides to phospholipid alcohols.

Removal and Degradation of Oxidized Fatty Acids from Lipid Membranes

Phospholipase A_2 isozymes, ubiquitous in prokaryotes and eukaryotes, appear to function in the first step of the phospholipid repair processs. Phospholipase A_2 cleaves fatty acyl chains from the Sn-2 position of phospholipids, producing equimolar 1-acyl lysophospholipid and free fatty acid (64). In general, this activity catalyzes the deacylation steps necessary during phospholipid biosynthesis and turnover (64). By mobilizing free arachidonic acid from the membrane, Phospholipase A_2 also limits the rate of eicosanoid biosynthesis through the cyclooxygenase and lipoxygenase pathways,(65) and thus helps to modify many hormone-regulated physiological responses (66-67). The selective hydrolysis of oxidized phospholipids by phospholipase A_2 would serve the dual protective function of relieving the membrane from the disruptive effects of polar and reactive species, and facilitating the detoxification of hydroperoxides by cytosolic peroxidases. Evidence exists which demonstrates preferential phospholipase activity towards lipids containing polyunsaturated fatty acids at the Sn-2 position, and a high degree of specificity toward oxidized phospholipids and oxidized membranes (68).

In ischemia-reperfusion systems, the amplification of phospholipase A_2 activity in conjunction with oxidative stress may contribute to the expression of organ damage. Sharp increases in free-fatty acid concentrations have been recorded in ischemic tissues (69). The activation of phospholipase A_2 in ischemic heart (69) and brain (70) probably accounts for the accumulation of free fatty acids observable in those tissues. Upon reoxygenation of these organs, free unsaturated fatty acids become more susceptible to peroxidation than those which remain in phospholipids (57).

Possible signals for increased phospholipase A_2 activity may include aberrations in membrane structure (71). In response to perturbed regions, increased phospholipid turnover mediated by phospholipase A_2, may adjust the fatty acid composition of phospholipids in an attempt to equilibrate membrane characteristics in accordance with environmental or physiochemical conditions (72). This could be the underlying mechanism for the phenomenon known as "homeoviscous adaptation" of membranes (73). Phospholipase A_2 displays a preferential activity towards phospholipids in multiphasic systems, (74) and this may explain its selective hydrolysis of oxidized substrates.

By liberating fatty acid hydroperoxides from membranes, phospholipase A_2 renders such species available for degradation. The cytosolic glutathione peroxidase,(75) which utilizes glutathione to reduce lipid hydroperoxides to less deleterious lipid alcohols, can hardly exert this activity if its substrates remain associated with the membrane (63). The effectiveness of glutathione peroxidase as a primary antioxidant defense may well depend on the activity of phospholipase A_2 (63).

Phospholipase C, an enzyme which hydrolyzes glycero-phosphate esters to release phospholipid head groups from membranes, also displays a preferential activity towards oxidized lipids (76). Phospholipase C releases inositol tris-phosphate and diacylglycerol as second messangers in receptor-mediated signal transduction pathways,(77) and also provides for the formation of specific phospholipids by remodeling in lieu of *de novo* synthesis. In addition to phospholipase A_2 mediated hydrolysis, oxidatively damaged phospholipids may alternatively undergo degradation by the combined action of phospholipase C and diacylglycerol lipase (76).

Repair of Phospholipids Following Hydrolysis by Phospholipase A_2

The reacylation phase of phospholipid repair, can be considered analagous to acylation events necessary for the the synthesis and turnover of native phospholipids, and may be the rate limiting step of phospholipid repair (78). Reacylation requires the ATP dependant activation of fatty acids with coenzyme A by fatty acid:CoA ligase, followed by the action of Acyl-CoA:1-acyl-sn-glycerolysophospholipid acyltransferases. Lysophospholipids produced by phospholipase A_2 require reacylation lest they exert deletereous effects on membrane function. Accumulation of lysophospholipids to high concentrations may promote intermembrane phospholipid exchange. At low concentrations, lysophospholipids may regulate insulin secretion from rat islets of Langerhans, by facilitating the release of membrane associated calcium (79).

Both increased phospholipase and acyltransferase activity have been observed in membranes subjected to oxidizing conditions (80). In rod outer segment membranes, the stimulated reacylation at the sn2 position did not exhibit specificity toward any particular fatty acyl CoA species (78).

In erythrocytes, oxidative stress generated by tert- butyl hydroperoxide induces the incorporation of free fatty acids into phosphatidylethanolamine (81). This enhanced acylation may occur in response to elevated concentrations of lysophospholipids, peroxidation products, or free fatty acids; or in response to perturbations in membrane structure. In contrast, exogenous arachidonic acid hydroperoxide (82) or hyperbaric hyperoxia (83) inhibit phospholipid acylation in synaptosomes and whole erythrocytes respectively. One possible explanation for this discrepancy is that acylation may be stimulated during lipid peroxidation only prior to the overt accumulation of fatty acid hydroperoxides (82).

Such results suggest that the extent of lipid peroxidation influences the cellular capacity for phospholipid repair, until the accumulation of hydroperoxides overwhelms the cells ability to detoxify these species.

Direct Reduction of Membrane Lipid Hydroperoxides by PHGPX

An alternative to the above scheme of lipid hydroperoxide removal, cytosolic reduction, and subsequent lysophospholipid reacylation, has been proposed by Ursini, Maiorino, and their associates (5,84,85). In this model, a novel glutathione peroxidase, named "Phospholipid Hydroperoxide Glutathione Peroxidase" (PHGPX) may catalyze the direct reduction of membrane bound lipid hydroperoxides to their corresponding alcohols. The fate of phospholipid alcohols generated by this enzyme, and their effects on membrane structure and function, have not yet been explained.

The Repair of Oxidatively Damaged DNA

Consequences of Free Radical Reactions with DNA

Extensive pulse radiolysis studies of nucleic acids and free nucleotides in aqueous solution have elucidated the nature of free-radical induced DNA damage (86-88). Hydroxyl radical (•OH) mediated addition and hydrogen-abstraction reactions with bases and deoxyribose sugars, generate a panoply of oxidative modifications, DNA strand breaks, cross-links, and abasic lesions. In radiolysis models, •OH preferentially adds across the double bonds of pyrimidine bases. Typical end products of pyrimidine oxidation include hydroxyhydroperoxides, ring saturation and fragmentation products, thymine glycols, urea derivatives, and hydroxymethyl uracil. Free radical attack on purine bases in DNA produces hydroxylated bases, or immidazole ring-opening; yielding formamidopyrimidine derivatives. 8-hydroxydeoxyguanosine represents a characteristic product of hydroxyl radical initiated purine oxidation. Radical initiated oxidation of the deoxyribose moiety produces lactone or ring-opened aldehyde derivatives, which may facilitate the release of the attached base from the DNA (87).

Certain oxidative DNA base modifications labilize the N-glycosydic bond between the base and its corresponding deoxyribose sugar, liberating the damaged base from the DNA phosphodiester backbone. The resulting apurinic/apyrimidinic lesions (AP sites), if not promptly repaired, interfere with DNA replication by blocking the progress of DNA polymerases (89). AP sites arise as a common endpoint in a number of other DNA damaging processes, including spontaneous depurination, and modification by alkylating agents and electrophillic carcinogens. In bacteria, (following induction of the SOS lesion-bypass system) AP sites may be mutagenic if bypassed by DNA polymerase, which indiscriminately incorporates deoxyadenosine opposite the AP site in the nascent DNA chain (89).

Radiation, H_2O_2, and other oxidants, produce single strand breaks with unusual termini. The 5' end most often terminates with a phosphate group. The 3' side of the break may either terminate in a phosphate group or a sugar fragment such as 3'-phosphoglycolate. The removal of a 3' sugar fragment to generate a 3' hydroxyl termini, precludes the repair of such strand breaks (90,91).

Evidence has accumulated for the existence of enzymes which specifically recognize and remove oxidative DNA modifications in a multi-step excision repair process (See Fig 3). The present review will encompass only those activities which participate in the removal of oxidative DNA damage, excluding a discussion of the repair of ultraviolet photodamage and adducts of chemical carcinogens (Fig. 3).

Enzymatic Repair of Oxidatively Damaged Prokaryotic DNA

In *Escherichia coli*, (*E. coli*) exonuclease III, endonuclease III, and endonuclease IV appear to be the major enzymes in the recognition step of excision repair (92). These enzymes act in concert to remove oxidatively modified bases, abasic sugars and sugar fragments, ultimately creating nucleotide-free gaps with 3' hydroxy termini. A free 3' hydroxyl serves as a primer for DNA polymerase in the next step of repair: the reinsertion of missing nucleotides. DNA ligase completes excision repair by sealing the phosphodiester bond between the last newly polymerized nucleotide and the native DNA.

Exonuclease III (the xth gene product) possesses 3' phosphatase, 3'-5' exonucleolytic, and 5'-3' endonucleolytic activities (92). The 3'-5' exonucleolytic activity removes 3' phosphoglycolate residues in order to render single strand breaks with blocked 3' termini repairable by DNA polymerase and ligase (93). The 5'-3' endonuclease activity functions as a class II AP endonuclease, cleaving phosphodiester bonds on the 5' side of apurinic/apyrimidinic lesions. This activity also recognizes the 5' side of urea derivatives, but not other thymine

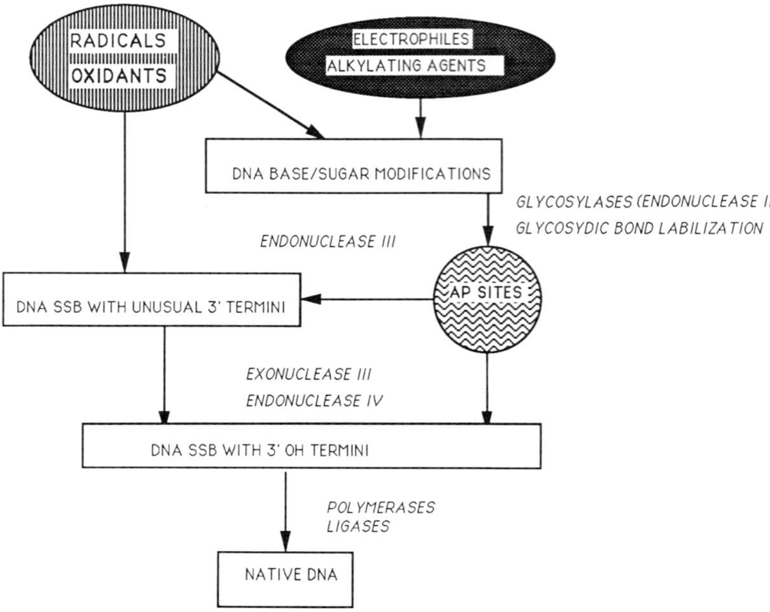

Figure 3. *Repair of Oxidatively Damaged DNA.* Radicals and oxidants react with DNA to generate oxidative DNA base or sugar modifications, and single strand breaks (SSB). Modified bases may either labilize their glycosydic bonds or undergo enzymatic removal by damage specific glycosylases. The resulting Apurinic/Apyridinic lesions (AP Sites) can be repaired by AP endonucleases. Endonuclease III cleaves on the 3' side of AP sites, creating breaks with blocked 3' termini which must be further repaired by Endonuclease IV. Exonuclease III and Endonuclease IV cleave on the 5' side of AP sites, generating breaks with 3' hydroxy termini, which can be used to prime repolymerization by DNA polymerases. DNA Ligase completes excision repair by resealing nicked DNA. Exonuclease III and Endonuclease IV may also co-operate to repair single strand breaks with blocked 3' termini generated by oxidants such as H_2O_2.

oxidation products (94). The endonucleolytic activity of exonuclease III accounts for some 85% of the total measurable AP endonuclease activity in *E. coli*. An analagous enzyme occurs in *H. influenzae* (92).

Endonuclease III (the *nth* gene product) acts as a class I apurinic endonuclease, catalyzing the β-elimination of phosphate groups on the 3' side of AP sites, leaving a saturated abasic sugar on the 3' terminus of the break (92). The blocked 3' end cannot serve as a substrate for DNA polymerase I, but can undergo further processing by endonuclease IV (95). Endonuclease III also carries an associated glycosylase activity, which recognizes and cleaves the glycosydic bond of glycol, hydro, and ring-opened derivatives of thymine, generating an AP site as a substrate for AP endonucleolytic activities.

Endonuclease IV (the *nfo* gene product) operates as a double-strand specific class II apurinic endonuclease. Endonuclease IV incises the phosphodiester backbone on the 5' side of AP sites. creating a free 3'-OH terminus. This process leads to the spontaneous loss of the 5' abasic deoxyribose terminus. Baily and Verly have shown that Endonuclease IV also cleaves on the 5' side of 3'-terminal abasic 2',3' unsaturated dideoxyribonucleotides produced by class I apurinic endonucleases, suggesting that endonuclease III and endonuclease IV co-operate in the complete repair of certain lesions (95). Together, endonuclease lll and IV comprise 10% of the endonucleolytic activity in *E. coli* extracts. Furthermore, Endonuclease IV, also bears 5' phosphatase and phosphoglycollatase capabilities, suggesting that endonuclease IV shares a role with exonuclease lll in the repair of single strand breaks containing various sugar fragments on the 3' end (93).

The isolation of *xth*$^-$, *nth*$^-$, and *nfo*$^-$ *E. coli* mutants, (deficient in exonuclease III, endonuclease III, and endonuclease IV, respectively, has contributed greatly to our understanding of the relative importance of these activities in the repair of oxidant-induced DNA damage (96,97). The *xth*$^-$ mutants show increased sensitivity to hydrogen peroxide and tert-butyl hydroperoxide, but do not show a marked increse in sensitivity to γ-irradiation or bleomycin. Mutants carrying the *nfo* deletion display even greater susceptibility to tert-butyl hydroperoxide than do *xth*$^-$ mutants, although no increase in H_2O_2 sensitivity is observed.

The combination of *xth*$^-$ and *nfo*$^-$ mutations, however, greatly potentiates hydrogen peroxide and organic hydroperoxide toxicity (97). This supports the proposition that exonuclease III and endonuclease IV co-operate in the repair of single-strand breaks with blocked 3'-ends, a radiomimetic lesion produced by H_2O_2 (98). No increase in susceptibilty to oxidants has been observed in *nth*$^-$ *E. coli*. Two minor endonuclease activities, endonuclease VIII and endonuclease IX have been isolated from *nth*$^-$ mutants. These enzymes share similar activities with endonuclease III, except that they exhibit preference for relaxed DNA. The presence of these two enzymes may compensate for the loss of endonuclease III in nth$^-$ *E. coli*, with respect to oxidative damage (92). Endonuclease Vlll operates on DNA containing AP sites and thymine-radiation products, and carries thymine- glycol glycosylase activity. Endonuclease lX also possesss a class I AP endonuclease activity similar to endonuclease III, and a related urea-specific glycosylase activity.

Other N-glycosylases with specificity for oxidative DNA modification have been purified, including hydroxymethyluracil glycosylase, (from mice) and formamidopyrimidine glycosylase (from *E. coli*) (92).

Eukaryotic DNA Repair Systems

Our understanding of DNA repair capabilities in eukaryotic organisms has not yet matured. Specificities of known eukaryotic DNA repair enzymes with respect to oxidant inflicted damage require further characterization. The presence of thymine glycol and thymidine glycol in human and rat urine suggests that these lesions occur in mammalian DNA, and experience a removal/ repair process (99). The time dependant dissapearance of 8-hydroxydeoxyguanosine in DNA extracts from irradiated mouse livers also implies the existence of a eukaryotic removal/repair system for this modified DNA base (100).

Class II AP endonucleases have been purified to varying degrees from *S. Cerevisiae*, *Drosophilia Melanogaster*, mice epidermis, calf thymus, rat liver, and Hela cells (92). The major yeast AP endonuclease also acts as a 3' diesterase which recognizes 3' phosphoglycoaldehyde esters and other 3' sugar fragments characteristic of oxidant damage (101). A mammalian "redoxyendonuclease" similar to endonuclease III of *E. coli* has been purified against its activity toward oxidatively modified DNA (102). A novel enzyme purified from *Neurospora crassa* acts as a single - strand -specific endonuclease as well as a double-strand specific exonuclease, and can repair DNA damage generated by 4-nitroquinolone-1 oxide (103).

Enzymes analagous to endonuclease III of *E. coli*, containing both class I AP endonuclease and pyrimidine glycosylase activities, occur in bovine and human derived cell lines. The identical sequence specificity of the bacterial and mamalian endonucleases toward various oxidized DNA fragments suggests an evolutionary conservation of certain DNA repair functions between prokaryotes and eukaryotes (104).

In addition to excision repair, certain oxidative DNA base modifications might undergo direct repair. Glutathione S-Transferases and peroxidases have been reported which can utilize thymidine hydroperoxide as substrates (105). Since DNA can sustain oxidative demethylation, DNA methylases may function to restore methylation patterns and maintain epigenetic phenomenon.

Global Responses to Oxidative Stress (Inducible Enzyme Systems)

The heat shock response, ubiquitous in prokaryotes and eukaryotes, represents the best characterized of the known stress-inducible responses. Heat stress triggers the elevated synthesis of a series of proteins (heat shock proteins, HSPs) far above constitutive levels, in conjunction with a decline in overall protein synthesis. Such induced gene expression has been associated with transient thermotolerance.

Inducible Prokaryotic Responses

E. Coli and *S. Typhimurium* exposed to sublethal doses of hydrogen peroxide develop a tolerance toward subsequent lethal doses of hydrogen peroxide or γ-irradiation. Demple and Halbrook have shown that H_2O_2 stimulates the repair of damaged viral DNA transfected into *E. coli* (106).

Sublethal doses of H_2O_2 induce the synthesis of thirty proteins in *S. typhimurium*, five of which belong to the heat-shock protein family (107). Nine proteins induced, are constitutively overexpressed in dominant *Oxy* R mutants indicating that these genes respond to positive regulation by the *Oxy* R gene product (108). Deletions of the *Oxy* R locus destroy this inducibility. Two of these genes have been identified as *Kat* G and SODa, encoding catalase and mangano-superoxide dismutase, respectively. Such induced proteins may reasonably be labeled "oxidative stress proteins" or OSPs.

Ames *et al.* have recently identified an NAD(P)H-dependant alkyl hydroperoxide reductase, whose expression also falls under the control of the *Oxy* R regulon in *E. coli* and *S. typhimurium*, which may function in a similar fashion as the glutathione peroxidase/reductase system (198,110).

A distinct response occurs in bacteria following exposure to quinones such as paraquat and plumbagin, which generate superoxide through a redox cycling mechanism (111-113). In addition to all the *Oxy* R regulated genes, this response induces 33 other proteins not common to the peroxide induced response, including glucose 6-phosphate dehydrogenase and additional heat shock proteins (112). Endonuclease IV has recently been identified as one of the proteins whose elevated synthesis occurs in the global response to redox cycling drugs, but not in the hydrogen peroxide mediated response (112,114). Endonuclease IV represents the only example of an inducible DNA repair enzyme known to date.

Inducible Eukaryotic Responses

In eukaryotes the synthesis of the glucose regulated protein family (GRPs), responds to different classes of stress: glucose starvation or stimulation by sulfhydryl reagents, and calcium ionophores. One member of this family, GRP78 is believed to function by sequestering denatured or misfolded proteins in the endoplasmic reticulum (115). Work currently proceeds toward the characterization of a global response to singlet oxygen-mediated photooxidative stress in mammalian cells [Gomer, *et al.* (116)] Light activated dihematoporphyrin ethers induces the 78 kd glucose regulated protein (Grp-78), the 70 kd heat shock protein, metalothionein, and heme oxygenase. The mechanisms by which the elevated synthesis of such proteins may confer an adaptive to oxidative stress, however, remains an enigma.

Other work currently in progress (Salo, D.C., Davies, K.J.A. unpublished results) involves the overexpression of OSPs, HSPs, and GRPs, during the oxidative stress, heat shock, and glucose depletion experienced by excercising rats. These studies have clearly demonstrated the overexpression of a variety of stress/shock proteins, including HSP70, which may be involved in the mitochondrial biogenesis (117) which accompanies excercise training (118,119).

Acknowledgements

This work was supported by grant # ES03598 for the NIH/NIEHS to K.J.A. Davies.

References

1. Ames, B.N. (1987) *Ann. Intern. Med.* **107**, 526-545
2. Fridovich, I. (1976) in *Free Radicals in Biology* (Pryor, W.A., ed.) Vol.1, pp 239 Academic Press, New York
3. Davies, K.J.A., (1986) *J. Free Rad. Biol. Med.* **2**, 155
4. McCord, J.M., Fridovitch, J., (1969) *J. Biol. Chem.* **244**, 6049
5. Ursini, F., Maiorino, M., Zalenti, M.Serri, L.Gregolin, C., (1982) *Biochem. Biophys. Acta.* **710**, 197-211.
6. Davies, K.J.A. (1988) in *Cellular Antioxidant Defense Mechanisms*, (Chow, C.K. ed) Vol **2**, pg 22, CRC press, Boca Raton
7. Davies, K.J.A., (1985) in *Cellular and Molecular Aspects of Aging: The Red Cell as a Model* (Eaton, J.W., Konzen, D. K., White, J. G., eds) p 15, Alan R. Liss, New York
8. Davies K.J.A. (1986) in *Superoxide and Superoxide Dismutase in Chemistry, Biology, and Medicine*. (Rotillio, ed) p. 443, Elsevier, Amsterdam

9. Bond, J.S., Butler, P.E., (1987) *Ann. Rev. Biochem.,* **56**, 333
10. Goldberg, A.L., Waxman, L. (1985) *J. Biol. Chem.* **260,** 12029
11. Hwang, B.J., Park, W.J., Chung, C.H., Goldberg, A.L., (1987) *Proc. Natl. Acad. Sci. USA* **8 4,** 5550
12. Hershko, A., Ciechanover, A. (1982) *Annu. Rev. Biochem.* **51**, 335
13. Goldberg, A.L.Dice, F.J. (1974) *Annu. Rev Biochem* **43**, 835
14. Goldberg, A.L., St.John, C. (1976) *Annu. Rev. Biochem*. **45**, 747
15. Wolff, S.P., Garner, A., Dean, R.T. (1986) *Trends Biol. Sci.* **11,** 27
16. Davies, K.J.A., (1987) *J. Biol.Chem.* **262** , 9895
17. Davies, K.J.A., Delsignore, M.E., Lin, S.W. (1987) *J.Biol. Chem* **262** 9902
18. Davies, K.J.A., Delsignore, M.E. (1987) *J. Biol. Chem* **262** , 9908
19. Davies, K.J.A., Lin, S.W., Pacifici, R.E. (1987) *J. Biol Chem.* **262,** 9914
20. Davies, K.J.A., Goldberg, A.L. (1987) *J. Biol. Chem* **262** ,8220
21. Davies, K.J.A., Goldberg, A.L. (1987) *J. Biol. Chem* **262,** 8227
22. Taylor, A., Davies, K.J.A., (1987) *J. Free Rad. Biol. Med.* **3** , 371
23. Davies, K.J.A., Lin, S.W., (1988) *J. Free Rad. Biol. Med.* **5,** 215
24. Davies, K.J.A., Lin, S.W. (1988) *J. Free Rad. Biol. Med.* **5,** 225
25. Salo, D.C., LIn, S.W., Pacifici, R.E., Davies, K.J.A. (1988) *J. Free Rad. Biol. Med.* **5,** 335
26. Marcillat, O., Zhang, Y., Lin, S.W., Davies, K.J.A. (1988) *Biochem. J.* **254** , 677
27. Marcillat, O., Zhang, Y., Lin, S.W., Davies, K.J.A. (1989) *Biochem J.* **259,** 181
28. Davies K.J.A. (1988) in *Cellular Antioxidant Defense Mechanisms*, (Chow, C. K. ed) Vol. 2, p. 25, CRC Press, Boca Raton
29. Davies, K.J.A. (1985) in *Cellular and Molecular Aspects of Aging, the Red Cell as a Model* p. 15 (Eaton, J. W., Konzen, D. K., White, J. G. (eds) New York, Alan R. Liss
30. Davies, K.J.A. (1986) in *Superoxide and Superoxide Dismutase in Chemistry, Biology, and Medicine*, (Rottilio, ed.) p 443, Elsevier, Amsterdam
31. Davies, K.J.A. (1988) in *Lipofuscin 1987, State of the Art*, pg. 109 (Zs-Nagy, ed.) Elsevier, Amsterdam
32. Davies, K.J.A. (1988) in *Lipid Peroxidation in Biological Systems*, (Sevanian, A. ed.) pg. 100 Am. Oil Chem. Soc.
33. Davies, K.J.A. (1988) in *The Role of Oxygen Radicals in Cardiovascular Disease* (L'Abbate, A., Ursini, F.) p.143, Kluwer Academic Publishers, Dortrecht
34. Pacifici, R.E., Lin, S.W., Davies, K.J.A. (1988) in *Oxygen Radicals in Biology and Medicine* (Simic, M .G., Taylor, K. A., eds.) p 575, Plenum Press, New York
35. Davies, K.J.A. (1988) in *Oxygen Radicals in Biology and Medicine*, (Simic, M.G.,Taylor, K.A., eds.) p. 575, Plenum Press, New York,
36. Pacifici, R.E., Davies, K.J.A. (1988) *FASEB J.* **2** , A 1007
37. Davies, K.J.A., Lin, S.W., Pacifici, R.E. (1988) (I.C.O.P.Newsletter) Aug, p.3
38. Taylor, A., Blondin, B., Davies, K.J.A., Davies, K.J.A., Murakami, K. (1987) *Abstr. 4th Int. Cong. Oxygen Radicals* W-20

39. Murakami, K., Jahngen, J.H., Lin, S.W., Davies, K.J.A., Taylor, A., (1989) *Free Rad. Biol. Med.*
40. Levine, R.L., Oliver, C.N., Fulkes, R.M., Stadtman E.R., (1981) *Proc. Natl. Acad. Sci. U.S.A.* **78,** 2120
41. Dean, R.T., Pollack, J.K., (1985) *Biochem. Biophys. Res. Commun.* **126**, 1082
42. Dean, R.T., Thomas S.M., Garner, A., (1986) *Biochem J.* **240,** 489
43. Wolff, S.P., Dean, R.T. (1986) *Biochem J.* **234** , 399
44. Fucci, L., Oliver, C.N., Coon, M.J.Stadtman, E.R., (1983) *Proc. Natl. Acad. Sci. USA,* **80**, 1521
45. Levine, R.L., *J. Biol. Chem.* **258**, 11823
46. Levine, R. L., *J. Biol. Chem.* **258**, 11828
47. Nakamura, K., Stadtman, E.R., (1984) *Proc. Natl. Acad. Sci. U.S.A.* **81**, 2011
48. Rivett, A.J., (1985) *J. Biol. Chem.* **260** 300
49. Stadtman, E.R., Wittenberger, M.E. (1985) *Arch . Biochem. BiophyS.* **239** 379
50. Stadtman, E.R., (1986) *Trend Biochem. Sci.* **11,** 11
51. Roseman, J.E., Levine, R.L., (1987) *J. Biol. Chem.* **262** , 2451
52. Rivett, A.J., (1989) *Arch. Biochem. Biophys .* **268** , 1
53. Waxman, L., Fagen, J.M., Goldberg, A.L., (1987) *J. Biol. Chem.* **262**, 8303
54. Hough, R., Pratt, G.Reichsteiner, M., (1987) *J. Biol. Chem.* **262** 2451
55. Cervera, J., Levine, R.L., (1988) *FASEB J.* **2** , 2591
56. Mead, J.F., (1976) in *Free Radicals in Biology and Medicine*, (W.A. Pryor ed.) p 51, Academic Press, New York
57. Majewska, M.C., Strosznajder, J., Lazarewicz, J.W., (1978) *Brain. Res.* **158**, 423
58. Galeotti, T. Borrello, S. Palombini, G., Masott, L., Ferrari, M. B. Cavatorta, P, Arcioni, A., Stremmenos, C., Zannoni, C., *FEBS Lett.* **84**, 125
59. Dobretsov, G.E.Borschevskaya, T.A., Petrov, V.A., Vladimirov, Y.A. (1977) *FEBS Lett.* **169,** 169
60. Coolbear, K.P., Keough, K.M.W. (1983) *Biochim, Biophys. Acta.* **732**, 531
61. Kornburg, R.D., McConnell, H.M. (1971) *Biochem.* **10**, 1111
62. Borsukov, L.I., Victorov, A.V., Vasilenko, I.A., Evstigneva, R.P., Bergelson, L.D., (1980) *Biochem. Biophys. Acta.* **598**, 153
63. Van Kujik, F.J.G.M., Sevanian, A., Handelman, G.J., Dratz, E.A. (1987) *Trends Biochem. Sci.* **12**, 31
64. Van Den Bosch, H., (1980) *Biochim. Biophys. Acta.* **604**, 191-246
65. Axelrod, J., Burch, R. M., Jelsema, C., *Trends. Neuro. Sci.* **11,** 117-123
66. Slivka, S.R., Insel, P.A. (1987) *J. Biol. Chem.* **262** 4200-4207
67. Nolan, R.D., Dusting, G.J., Martin, T.J., (1981) *J. Pharmacol,* **30**, 2121-2125
68. Sevanian, A.Kim, E., (1987) *J. Free Rad. Biol. Med.* **1**, 263
69. Chan, P.H., Fishman, R.A., Schmidley, J.W., Chen, S.F., (1984) *J. Neurosci Res.* **12** , 595
70. Bentham, J.M., Higgens, A.F.Woodward, B. (1987) *Basic Res. Cardiol.* **82**, 127
71. Sevanian, A., Wratten, M.L., Mcleod, L.L., Kim, E. (1988) *Biochim. Biophys. Acta,* **961** 16

72. Dawson, R.M.C. (1982) *J. Am. Oil Chem. Soc.* **59** 401- 406
73. Cossins, A. R., Slnensky, M. (1984) (1984) in *Shinitzky*, M. Boca Raton CRC press, Inc., p 7
74. Sevanian, A., (1988) in Chow, C. K. (ed): Boca Raton: CRC Press, 78
75. Flohe, L., (1982) in *Free Radicals in Biology* (Pryor, W.A. ed.) New York, Academic Press, 223
76. Gamache, D.A., Fawzey, A.A., Franson, R.C. (1988) *Biochem. Biophys. Acta.* **958**, 116
77. Hotkin, L.E. (1985) *Ann Rev. Biochem.* **54**, 205
78. Zimmerman, W.F., Keys, S. (1988) *Exp. Eye. Res.* **47**, 247
79. Metz, S.A., (1988) *Biochem. Biophys. Acta.* **968** 239
80. Lubin, B.H., Shohet, S.B., Nathan, D.G., (1972) *J.Clin. Invest.* **51** 338
81. Dise, C.A., Goodman, D.B.P. (1986) *Biochem. Biophys. Acta.* **859** 69
82. Zaleska, M.M., Wilson, D.F. (1989) *J. Neurochem*. **52**, 255
83. Dise, C.A., Clark, J.M., Lambersten, C.J., Goodman, D.B.P. (1987) *J. Appl. Physiol.* **62** 533
84. Ursini, F., Maiorino, M., Gregolin, C. (1985) *Biochem. Biophys. Acta.* **839,** 62-70
85. Thomas, J. P., Maiorino, M., Ursini, F., Girotti, A. W. *J. Biol. Chem.* **265** , 454-461
86. Von Sonntag, C., Schuchmann, H.P. (1986) *Int. J. Radiat. Biol.* **49**, 1-34
87. Teoule, R. (1987) *Int. J. Radiat. Biol.* **59**, 573-589
88. Teebor, G.W., Boorstein, R.J., Cadet, J. (1988) *Int. J . Radiat. Biol.* **54**, 131-150
89. Loeb, L.A., (1985) *Cell,* **40**, 483-484
90. Hutchinson, F. (1985) *Prog. Nuc. Acid. Res. Mol. Biol.* **32**, 115-154
91. Von Sonntag, C., Hagen, U., Schon-Bopp, A., Shulte-Frohlinde, D. (1981) *Adv. Radiat. Biol.* **9,** 109-142
92. Wallace, S.S. (1988) *Env. Mol. Mutag.* **12,** 431-477
93. Demple, B., Johnson, A., Fung, D. (1986) *Proc. Natl. Acad. Sci. USA* **83,** 7731-7735
94. Kow, Y.K., Wallace, S.S. (1985) *Proc. Natl. Acad. Sci. USA* **82,** 8354-8358
95. Bailly, V., Verly, W.G. (1989) *Biochem J.* **259**, 761-768
96. Weiss, B., Cunningham, R.P. (1985) *J. Bact.* **162,** 607-610
97. Cunningham, R.P., Saporito, S.M., Spitzer, S .G., Weiss, B. (1986) *J. Bact.* **168,** 1120-1127
98. Bernelot-Moens, C., Demple, B. (1989) *Nuc. Acid. Res. 17, 587-600*
99. Cathcart, R., Schwiers, E., Saul, R. L., Ames, B. N. *Proc. Natl. Acad. Sci. USA* **81,** 5633-5637
100. Kasai, H., Crain, P.E., Kuchino, Y., Nishimura, S., Ootsuyama, A., Tanooka, H. (1986) *Carcinogenesis* **7**, 1849-1851
101. Johnson, A., Demple, B. (1988) *J. Biol. Chem.* **263,** 18017-18022

102. Doetsch P.W., Henner, W.D., Cunningham, R.P., Toney J.H., Helland, D.E. (1987) *Mol. Cell. Biol*. **7** 26-32
103. Ramotar, D., Auchincloss, A.H., Fraser, M.J. (1987) *J. Biol. Chem.* **262**, 425-431
104. Doetsch, P.W., Henner, W.D., Cunningham, R.P., Toney, J.H., Helland, D.E. (1987) *Mol. Cell. Biol.* **7**, 26-32
105. Ketterer, B., Mayer, D.J. (1989) *Mutat. Res.* **214** 33-40
106. Demple, B., Halbrook, J. (1983) *Nature* **304**, 466-468
107. Morgan, R.W., Christman, M.F., Jacobson, F.S., Storz, G., Ames, B.N. (1986) *Proc. Natl. Acad. Sci. USA* **83**, 8059-8063
108. Christman, M.F., Morgan, R.W., Jacobson, F.S., Ames, B.N. (1985) *Cell* **41**, 753-762
109. Jacobson, F., S., Morgan, R.W., Christman, M.F., Ames, B.N (1989) *J. Biol. Chem.* **264** 1488-1496
110. Storz, G., Jacobson, F.S., Tartaglia, L .A., Morgan, R.W., Silveira, L.A., Ames, B.N. (1989) *J. Bact.* **171**, 2049-2055
111. Farr, S.B., Natvig, D.O., Kogoma, T. (1985) *J. Bact.* **164**, 1309-1316
112. Walkup, L.K.B., Kogoma, T. (1989) *J. Bact,.* **171**, 1476-1484
113. Greenburg, J. T., Demple, B., *J. Bact*. **171**, 3933-3939
114. Chan, E., Weiss, B. (1987) *Proc. Natl. Acad. Sci. USA* **84**, 3189-3193
115. Watowitch, S.S., Morimoto, R.S. (1988) *Mol Cell Biol.* **8** 393-405
116. Gomer, C.J., Ferrario, A., Hayashi, N., Rucker, N., Szirth, B.C., Murphree, A.L., (1988) *Las. Sur. Med.* **8.** 450-463
117. Deshaies, R.J., Koch, B.D., Schekman, R., (1988) *Trends Biochem. Sci.* **13,** 235-246
118. Deshaies, R.J., Koch, B.D., Werner-Washburne, M., Craig, E., Schekman, R. (1988) *Nature* **332,** 800-805
119. Davies, K.J.A., Packer, L., Brooks, G.A., (1981) *Arch. Biochem. Biophys.* **209,** 539-554

Oxidative Damage to Proteins Mediated by Singlet Oxygen

D. Balasubramanian
Centre for Cellular and Molecular Biology, Hyderabad, 500 007 India

Summary

Singlet oxygen oxidizes *trp* residues in proteins into N-formylkynurenine and related products, just as peroxide does. It oxidizes *tyr* residues as well, but not to bityrosine or dihydroxyphenylalanine, as peroxide is able to do. *His* residues are oxidized by singlet oxygen, and the products appear to be more acidic. *His* oxidation in proteins also leads to covalent cross-linking of the protein, as monitored by the presence of higher molecular weight products in electrophoregrams. Blocking of *his* by adduct formation with diethylpyrocarbonate leads to inhibition of cross-linking.

The oxygen molecule normally occurs in its ground electronic state with a spin multiplicity of three, or in the triplet state. From here, it can be excited to the singlet state, either to the $^1\Delta_g$ state or the higher $^1\Sigma_g$ state, as shown in Figure 1. Table 1 highlights some facts about the singlet state of molecular oxygen.

Table 1. *Some Facts about Singlet Oxygen*

	$^1\Delta_g$	$^1\Sigma_g^+$
Energy Above Ground State	22.5 kcal/m	37.5 kcal/m
Radiative Lifetime	45 min	7 sec
Bond Length*	1.218 Å	1.223 Å
Half-Life at 1 Atmosphere	50 millisec	--
Gas Collisions Survived	10^8	10^2
Solution State Lifetime	1 μs	100 ps
Light Emission	1.27 μ	762 nm

* *cf.* 1.20 Å in the ground state.
Most data was compiled from reference 1.

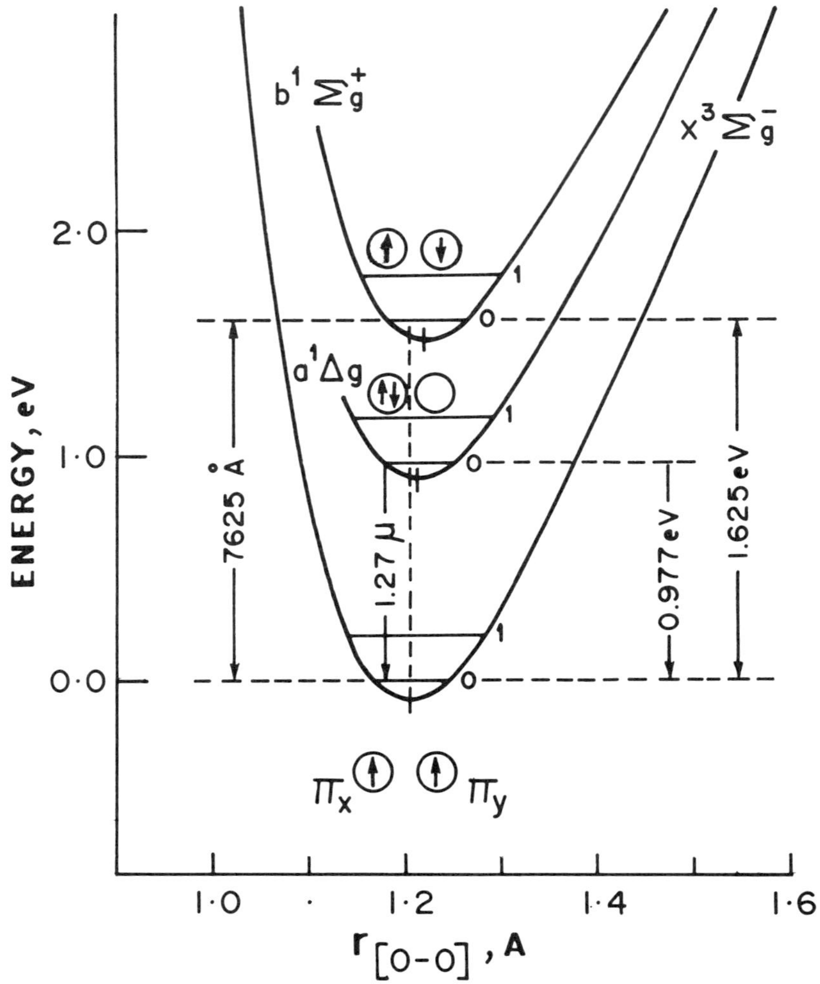

Figure 1. *Potential Energy Curves of the Oxygen molecule in Its Ground Triplet and Excited Singlet States.* The electronic configurations of the relevant orbitals are also indicated.

Besides its higher energy, singlet oxygen possesses another advantage which enables it to be more reactive. The substrate molecule with which reaction occurs is invariably in the singlet electronic state, and spin selection rules inhibit a ready reaction between singlets and triplets, which makes triplet oxygen somewhat more sluggish than singlet oxygen. Of the two singlet oxygens described above, $^1\Delta_g$ O_2 is longer lived and stabler than $^1\Sigma_g$ O_2 and is thus the more common species with which most investigations are done. From the chemist's point of view, $^1\Delta_g$ O_2 resembles ethylene electronically, and is expected to be electrophilic in nature, and to be involved in two electron reactions. In contrast, $^1\Sigma_g$ O_2 resembles ground state oxygen electronically and is expected to be involved in one electron reactions.

Singlet oxygen is produced from the ground state molecule in two ways. In the direct excitation method, gaseous oxygen is subjected to microwave discharge or to Nd-YAG laser light (1.06 micron) illumination so as to populate the molecule in its singlet state (1,2,3). In the second method, a dye molecule is excited to its higher electronic level by light and allowed to transfer its energy (usually from its long lived triplet state) to ground state molecular oxygen. Indeed it was through such a photosensitization mechanism that singlet oxygen was first realized and studied. The earliest report is that of Raab (4) in 1900 who discovered that microbes stained with certain dyes are killed in air when exposed to light but not in the dark, and he termed this the photodynamic effect. Kautsky and de Bruijn (5) explained this photodynamics as due to the formation of $^1O_2$ by photosensitization by the dye and the reaction of the singlet oxygen with acceptor or substrate molecules. They described a technique in which the sensitizer dye was deposited on silica and irradiated in air to produce singlet oxygen. In recent years, an improved version of this method has been suggested by Midden and Wang (6) for the clean and simple generation of singlet oxygen for solution state studies, termed the surface-separated-sensitizer or SSS method (7). The studies reported in the present paper were made using the Midden-Wang SSS procedure for the generation of singlet oxygen.

Photosensitization and photodynamic effects in biological systems are subjects of considerable current attention. While photodynamics by externally added dyes (acridine orange, acriflavin, fluorescein, methylene blue, rose bengal, thionine and hematoporphyrin) is more commonly studied, we have been interested in identifying sensitizers endogenous to cells and tissues and their role in photodamage and oxidation. These are relevant to tissues such as the eye lens, which do not display any significant metabolic activity and which have turnover periods of the order of years and decades. Even low level chronic damage tends to accumulate over time in these systems, and photodynamic action by endogenous chromophores, particularly those that absorb in the visible and near-ultraviolet region of light becomes important. Some of these are kynurenine and its derivatives that occur as oxidation products of tryptophan (*trp*), Maillard reaction products that arise from the glycation of some of the residues in the eye lens proteins crystallins, and clinically abnormal concentrations of coenzymes, metabolites or metal ions.

Besides photosensitization pathways, singlet oxygen and related active oxygen species are produced in biological systems through dark reactions as well. Kanofsky (8) has critically reviewed singlet oxygen production in these systems. Polymorphonuclear leukocytes are phagocytes which generate several oxidants using the metabolic pathway called the respiratory burst (9). The coenzyme NADPH is oxidized by molecular oxygen in the presence of the enzyme NADPH Oxidase and generates the superoxide anion, which in turn yields H_2O_2.

$$2\,O_2 + NADPH \rightarrow NADP^+ + O_2^-$$

$$2\,O_2^- + 2\,H^+ \rightarrow H_2O_2 + O_2$$

The hydrogen peroxide reacts with halide ions to produce singlet oxygen, and the reaction is catalyzed by the peroxidase family of enzymes.

$$H_2O_2 + X^- \rightarrow HOX + OH^-$$

$$H_2O_2 + HOX \rightarrow H_2O + X^- + {}^1\Delta_g\,O_2$$

The halide ion could be Cl^-, Br^-, or I^- and the coenzyme that effects the reaction can be lactoperoxidase, chloroperoxidase or myeloperoxidase. In addition, peroxidase catalyzed oxidation of indole-3-acetic acid also generates singlet oxygen (10); also, the dismutation of the superoxide anion itself is reported to produce singlet oxygen (11, 12).

Eosinophils are another type of polymorphonuclear leukocytes which aid in the killing of invading parasitic worm larvae. These cells contain a large concentration of the enzyme eosinophil peroxidase, which is distinct from myeloperoxidase and which generates singlet oxygen (13). Kanofsky and co-workers (14) have shown that intact human eosinophils generate singlet oxygen upon stimulation with phorbol esters.

Besides these leukocytes, lipoxygenases, the enzymes which catalyze the oxidation of long chain unsaturated fatty acids, appear to be able to generate singlet oxygen, as monitored by its chemiluminescence at 1268 nm (15, 16). The reaction is thought to occur by the Russell mechanism (17).

$$2\,RR'HCO2 \rightarrow RR'CO + RR'CHOH + {}^{1}\Delta_g\,O_2$$

Thus singlet oxygen appears to be a significant intermediate in several biological processes. Figure 2 is a summary clock diagram (18) of the generation of singlet oxygen from various biological reactions and processes.

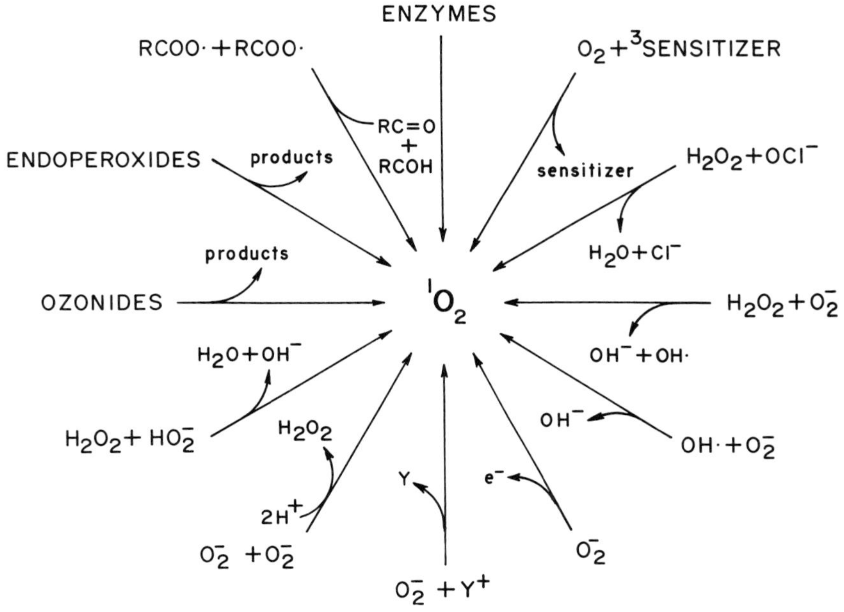

Figure 2. *A Clock Diagram of the Production of Singlet Oxygen by Photochemical, Chemical and Biological Systems* (18)

What are the biomolecules that are vulnerable for oxidation by singlet oxygen? Clearly lipids, particularly unsaturated ones, some coenzymes, several amino acid sidechains, and nucleic acid bases would be the targets. With the advent of the SSS method for the convenient production of singlet oxygen, studies on several of these systems and compounds are becoming increasingly frequent. In our laboratory, we have been concerned with the oxidation of proteins by active oxygen species, particularly the covalent modification of the eye lens crystallins in relation to aging and cataract formation. As mentioned earlier, the eye lens is a metabolically inert and slow turn-over tissue, and thus chemical damage to its constituent molecules tends to accumulate with time and lead to malfunction, namely opacification of the eye lens. One well-recognized cause of lens opacity is oxidative modification and protein cross-linking of the constituent crystallins.

Among the twenty amino acid residues present in proteins, those with inert alkyl and aryl chains, namely alanine (*ala*), valine (*val*), leucine (*leu*), isoleucine (*ile*), proline (*pro*), phenylalanine (*phe*), and glycine (*gly*) are expected not to react with singlet oxygen. Nor is any oxidation expected with the side chains of glutamic acid (*glu*), aspartic acid (*asp*), lysine (*lys*), arginine (*arg*), or even asparagine (*asp*), glutamine (*gln*), serine (*ser*), or threonine (*thr*). Matheson and Lee (3) have studied the chemical reaction rates of amino acids with singlet oxygen and reported that histidine (*his*) reacts fastest (rate constant $k=1x10^8$ $mole^{-1}s^{-1}$), followed by tryptophan (*trp*) ($k=3x10^7$), methionine (*met*) ($k=1.6x10^7$) and tyrosine (*tyr*) ($k=0.8x10^7$). Cysteine (*cys*) also reacts readily with singlet oxygen to produce cystine (19), which itself is only very slowly photo-oxidized (20). Hence, within the lifetime of singlet oxygen in solution, the only residues that tend to be oxidized would be *his*, *trp*, *met*, *cys*, and *tyr*. Our aim is to attempt to analyze the consequence of the oxidation of each of these residues on the structure and properties of the protein molecule. We have done this analysis in some detail on several control proteins and on the lens crystallins, particularly in regard to their reaction with singlet oxygen (21). We summarize some of that work in this review.

Materials and Methods

We have used the Midden-Wang method (6), as modified by Murali Krishna *et al.* (7), to generate singlet oxygen by photo-sensitization using rose bengal as the dye. 6 mg of rose bengal (Aldrich Chem. Co.) was dissolved in 6 ml methanol, and 2 g control pore glass (CPG 1460, pore diameter 460 Å, particle size 37-74 micron, Pierce Chem. Co.) were added to the solution and the solvent removed in a rotary evaporator. The dye-coated particles were sprinkled on the top of a glass plate (75 x 50 mm, Fisher) that had double stick Scotch tape attached to it, until the plate surface was completely covered. After blowing off any loose particles from the surface, the sensitizer coated plate, or the SSS plate, was ready for use. "Boerner" microscope slides (Thomas Scientific Co.) were used to hold the sample solutions. Each Boerner slide has 8 wells, each of 0.2 μl volume. The reaction conditions, precautions and the set-up have been described elsewhere (7, 21). The proteins were obtained from Sigma Chemical Co., St. Louis, MO, and were of the highest purity available.

The oxidation of *trp* and *tyr* residues is conveniently monitored by following the change in their fluorescence spectral intensity and the appearance of new bands arising from the oxidation products such as N-formylkynurenine (NFK), dihydroxyphenylalanine (*dopa*), bityrosine and so on. Fluorescence spectra were recorded using Aminco SPF-500 and Hitachi 650-10S spectrofluorimeters and microcells of 0.15 ml capacity. The histidine content of the proteins was estimated using the diethylpyrocarbonate (DEPC) adduct spectrophotometric assay (22). Protein charge modification, cross-linking and the formation of high molecular weight products were detected by running polyacrylamide gel eletrophoresis (PAGE), using a Pharmacia Phastsystem or a Bio-Rad Mighty-Small assembly.

Results and Discussion

trp Oxidation

In our studies, we have attempted to assess the consequences of the oxidation of each type of amino acid residues on the protein structure and properties. To this end, we have chosen some "control" proteins in order to isolate and highlight the effect of the oxidation of specific sidechains. The first such control is the 26-residue long peptide of the bee venom, melittin, which has the primary structural sequence GIGAVLKVLTTGLPALISWIKRKRQQ-NH_2. The only amino acid in this sequence that would readily react with singlet oxygen is the *trp* (or W) residue in position 19 of the chain. Melittin thus serves as the *trp* control. Reaction of melittin in pH 7.4 phosphate buffered saline (PBS) solution with singlet oxygen generated in the SSS set-up led to a loss of the *trp* fluorescence emission near 350 nm and the concomitant appearance of a new emission band near 435 nm (excitation 320 nm). Based on its excitation and emission maxima and the fluorescence intensity profile, this new band is assigned to N-formylkynurenine (NFK) and closely related products of the oxidation of *trp* (21, 23). Singlet oxygen is thus able to oxidize the *trp* residue of melittin to NFK and related products. That singlet oxygen is involved in the reaction was confirmed in two ways: (i) addition of the singlet oxygen quencher sodium azide (24) in millimolar concentrations to the protein solution prevented the oxidation, and (ii) when the reaction was conducted with the protein dissolved in PBS solution in D_2O rather than in H_2O, the loss in *trp* emission was much higher. Singlet oxygen is known to have a far longer lifetime in deuterated solvents (25).

NFK has been suspected to be a photosensitizer moiety which might promote photodynamic damage to proteins through the generation of active oxygen species (23). Such photodynamic effects are thought to cause protein covalent cross-linking (26, 27). However, when we analyzed the molecular weights of the reaction mixture by PAGE runs in the presence of sodium dodecylsulfate (SDS), no higher molecular weight products were detected (Figure 3). *Trp* oxidation and NFK formation *per se* does not lead to melittin cross-linking.

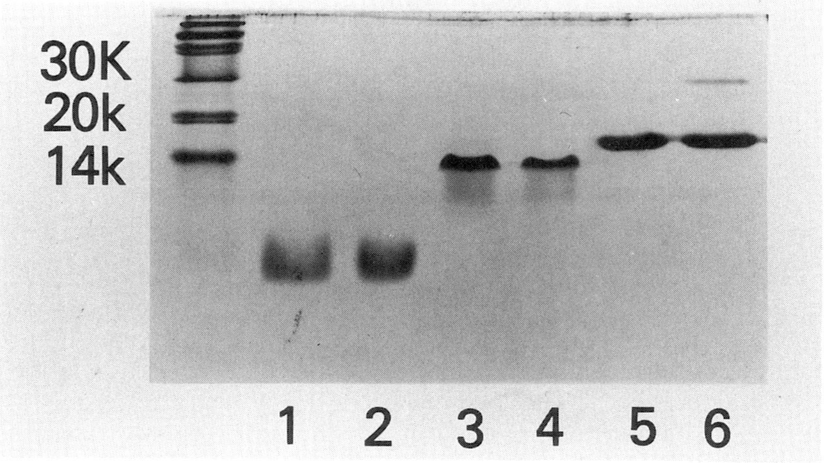

Figure 3. *Effects of Singlet Oxygen Oxidation of Melittin, BPTI and RNase A as Seen by the SDS Electrophoresis on a 20% Polyacrylamide Slab Gel.* Samples were reacted in the SSS set-up for 2 hours and loaded on the gel after boiling at 373° K in the presence of SDS and mercaptoethanol. Lane 1: melittin control; lane 2: melittin exposed to singlet oxygen; lane 3: BPTI control; lane 4: BPTI exposed; lane 5: RNase A control; lane 6: RNase A exposed to singlet oxygen, and extreme left lane: molecular weight markers (Figure from ref. 21).

tyr Oxidation

As control, we chose two proteins - bovine pancreatic trypsin inhibitor (BPTI) and ribonuclease a (RNase A). BPTI is a 58-residue protein devoid of *trp* and *his*, but containing 4 *tyr*, 6 *cys* and one *met* residues that could be oxidized by singlet oxygen. Reaction of BPTI in PBS with singlet oxygen in the SSS set-up led to a drop in the *tyr* fluorescence intensity, but no concomitant appearance of any new emission bands.

We have elaborated this point elsewhere, and we note here that though *tyr* is oxidized, it does not appear to produce *dopa*, dopachrome or dityrosine - compounds that display fluorescence at longer wavelengths. Joshi *et al.* (28) have suggested that singlet oxygen might not be involved in the transformation of *tyr* to *dopa* or to dopachrome. It is also known that *tyr* is converted to dityrosine by H_2O_2 in the presence of the enzyme horseradish peroxidase, presumably by a radical mechanism (29). The oxidation product of *tyr* reaction with singlet oxygen remains yet to be characterized. That this oxidation does not lead to cross-linking of BPTI is also evident from the SDS-PAGE shown in Figure 3. Though we have not monitored whether the *met* residue in BPTI is oxidized or not, we may conclude that such oxidation does not lead to protein cross-linking or higher molecular weight products of BPTI. It is worth noting here that the SDS-PAGE runs were made in the presence of mercaptoethanol, so that we detect only non-disulfide cross-links in the gels.

his Oxidation

The next control protein that we chose, RNase A, has an amino acid composition similar to BPTI, including the absence of *trp*, but with one difference. RNase A has four *his* residues in its sequence while BPTI has none. Singlet oxygen reaction with RNase A revealed fluorescence changes similar to those seen in BPTI, *i.e.*, the *tyr* emission decreases but with no concomitant generation of any bands indicative of *dopa* or dityrosine formation. However, Figure 3 reveals that the oxidation of RNase A by singlet oxygen does lead to the formation of non-disulfide cross-linked higher molecular weight products. Note that as far as the amino acid compositions are concerned, RNase A differs from BPTI in that it has *his* residues. It is thus tempting to ascribe the cross-linking of RNase A to the reaction of its *his* residues.

In an effort to assess this implication, we estimated the *his* content of RNase A before and after reaction with singlet oxygen for various time periods. Using the DEPC-adduct assay (22), we estimated 4 *his* per molecule of unreacted RNase A, and found this number to decrease to three after exposure to singlet oxygen in the SSS set-up for three hours, and to two after four hours. Similar reduction in the *his* content of the eye lens crystallins has been noted by us (21). That *his* residues are involved in the protein cross-linking reaction is supported by inhibition experiments. In these, we protected the available *his* residues of RNase A by reacting the protein with DEPC for a short time and purifying the modified protein by eluting it off a Sephadex G25 column. As Figure 4 shows, the modified protein failed to produce any significant amount of the higher molecular weight product upon reaction with singlet oxygen.

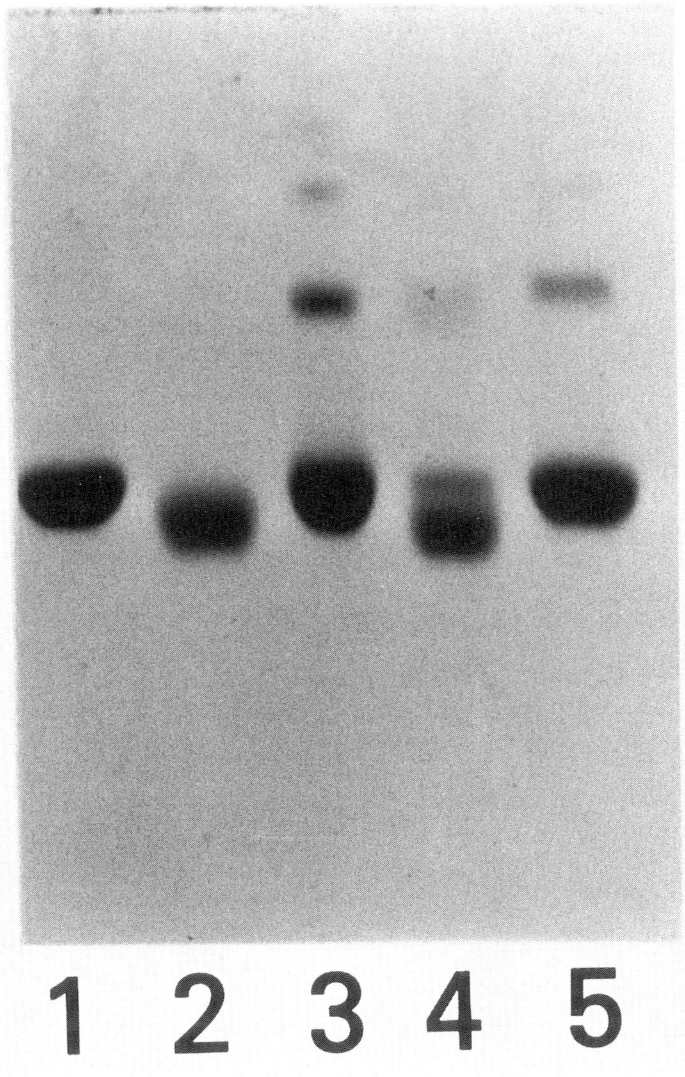

Figure 4. *SDS PAGE (15% gel) Patterns Showing the Effects of Pre-Treatment with DEPC or L-Lysine on the Cross-Linking of RNase A in the SSS Set-Up.* Lanes 1: control protein; 2: protein modified with DEPC prior to oxidation; 3: control protein exposed to singlet oxygen for 90 min.; 4: DEPC-modified protein exposed to singlet oxygen for 90 min.; and 5: control protein oxidized by singlet oxygen for 90 min. in the presence of L-lysine.

Tomita *et al.* (30) and, more recently, Uchida and Kawakishi (31) have shown that a major product of the oxidation of *his* is *asp* and closely related products. Furthermore, it has also been suggested that photodynamic cross-linking of proteins might occur through amide bonds involving photo-oxidized *his* residues and free amino groups (32). While this remains to be established in the present cases, it is worth noting form Figure 4, lane 5, that the formation of higher molecular weight product is inhibited when RNase A is oxidized in the presence of 5 mM free L-lysine in the buffer.

Conclusions

1. Singlet oxygen oxidizes *trp* residues in proteins to produce NFK and related products.
2. It also oxidizes tyr residues but not to dopa or bityrosine, as peroxide does in the presence of enzymes. The nature of the oxidized products needs to be clarified.
3. *His* residues in the protein are oxidized by singlet oxygen. Such oxidation ultimately leads to protein cross-links and high molecular weight products, whereas the oxidation of the *tyr*, *trp*, *met* or *cys* residues do not appear to do so.
4. Such oxidative damage to the proteins of metabolically inert tissues, notably the crystallins of the eye lens, could accumulate over long periods of time and cause irreversible precipitation and phase separation (21).

Acknowledgments

I thank the NIH/NEI for a visiting scientist position during 1988-1989. Some of this work was done during that time with the collaboration of Dr. J. S. Zigler and Dr. X. Du there. I also thank Dr. Murali Krishna Cherukuri for advice and help with the SSS set-up.

References

1. Chadha, M.S. (1976) *Singlet Oxygen - Proceedings of a Symposium*, Bhabha Atomic Research Centre, Bombay, India
2. Wasserman, H.H., and Murray, R.W. (1979) *Singlet Oxygen* Academic Press, New York
3. Matheson, I.B.C., and Lee, J. (1979) *Photochem. Photobiol.* **29**, 879-881
4. Raab, O. (1900) *Z. Biol.* **39**, 524

5. Kautsky, H., and de Bruijn H. (1931) *Naturwiss.* **19**, 1043
6. Midden, W.R., and Wang, S.Y. (1983) *J. Am. Chem. Soc.* **105**, 4129-4135
7. Murali Krishna, C., Lion, Y., and Riesz,P. (1987) *Photochem. Photobiol.* **45**, 1-6
8. Kanofsky, J.R. (1989) *Chem. - Biol. Interactions* **70**, 1-28
9. Babior, B.M. (1978) *New Eng. J. Med.* **298**, 659-668
10. Kanofsky, J.R. (1988) *J. Biol. Chem.* **263**, 14171-14175
11. Khan, A.U. (1970) *Science* **168**, 476-477
12. Corey, E.J., Mehrotra, M.M., and Khan, A.U. (1987) *Biochem. Biophys. Res. Commun.* **145**, 842-846
13. Weiss, S.J., Test, S.T., Eckmann, C.M., Roos, D., and Regiani, S. (1986) *Science* **234**, 200-203
14. Kanofsky, J.R., Hoogland, H., Wever, R., and Weiss, S.J. (1988) *J. Biol. Chem.* **263**, 9692-9696
15. Kanofsky, J.R., and Axelrod, B. (1986) *J. Biol. Chem.* **261**, 1099-1104
16. Khan, A.U. (1987) *ACS Symp. Ser.* **339**, 58-75
17. Russell, G.A. (1957) *J. Am. Chem. Soc.* **79**, 3871-3877
18. Krinsky, N.I. (1977) *Trends Biochem. Sci.* **2**, 35-38
19. Gennari, G., Gauzo, G., and Jori, G. (1974) *Photochem. Photobiol.* **20**, 497-500
20. Weil, L. (1965) *Arch. Biochem. Biophys.* **110**, 57-68
21. Balasubramanian, D., Du, X., and Zigler, J.S., Jr. (1990) *Photochem. Photobiol.* submitted
22. Miles, E.W. (1977) *Methods Enzymol.* **47**, part E, 431-444
23. Walrant, P., and Santus, R. (1974) *Photochem. Photobiol.* **20**, 411-417
24. Foote, C.S. (1968) *Acc. Chem. Res.* **1**, 104-110
25. Merkel, P.B., and Kearns, D.R. (1972) *J. Am. Chem. Soc.* **74**, 7244-7253
26. Verweij, H., Dubbelman, T.M.A.R., and van Stevenninck, J. (1981) *Biochim. Biophys. Acta* **647**, 87-94
27. Zigman, S. (1981) in *Mechanisms of Cataract Formation in the Human Lens* (Duncan, G., ed) pp.117-149, Academic Press, New York
28. Joshi, P.C., Carraro, C., and Pathak, M.A. (1987) *Biochem. Biophys. Res. Commun.* **142**, 265-274
29. Gross, A.J., and Sizer, I.W. (1959) *J. Biol. Chem.* **234**, 1611-1614
30. Tomita, M., Irie, M., and Ukita, T. (1969) *Biochem.* **8**, 5149-5160
31. Uchida, K., and Kawakishi, S. (1989) *Bioorg. Chem.* **17**, 330-343
32. Dubbelman, T.M.A.R., De Goeij, A.F.P.M., and van Stevenninck, J. (1978) *Biochim. Biophys. Acta* **511**, 141-151

Molecular Mechanisms of Oxidative Cell Damage

Sten Orrenius, George E.N. Kass, Steven K. Duddy, and Pierluigi Nicotera
Department of Toxicology, Karolinska Institutet
Box 60400 S-104 01 Stockholm Sweden

Abstract

Exposure of mammalian cells to oxidative stress induced by redox-active quinones and other prooxidants results in the depletion of intracellular glutathione, followed by the modification of protein thiols and the loss of cell viability. Protein thiol modification during oxidative stress is normally associated with an impairment of various cell functions, including inhibition of agonist-stimulated phosphoinositide metabolism, disruption of intracellular Ca^{2+} homeostasis, and perturbation of normal cytoskeletal organization. The latter effect appears to be responsible for the formation of the numerous plasma membrane blebs, typically seen in cells exposed to cytotoxic concentrations of prooxidants. Following the disruption of thiol homeostasis in prooxidant-treated cells, there is an impairment of Ca^{2+} transport and a subsequent perturbation of intracellular Ca^{2+} homeostasis, resulting in a sustained increase in cytosolic Ca^{2+} concentration. This Ca^{2+} increase can cause activation of various Ca^{2+}-dependent degradative enzymes (phospholipases, proteases, endonucleases) which may contribute to cell killing. In contrast with the cytotoxic effects of excessive oxidative damage, low levels of oxidative stress can lead to the activation of a number of enzymes. In particular, the activity of protein kinase C is markedly increased by redox-cycling quinones through a thiol/disulfide exchange mechanism, and this may represent a mechanism by which prooxidants can modulate cell growth and differentiation.

Introduction

In aerobically growing eukaryotic cells, ATP synthesis is coupled to the enzymatic reduction of molecular oxygen (O_2) to H_2O via a sequential, four electron transfer reaction. However, one- and two-electron reduction of O_2 can also occur under physiological conditions in mitochondria and other cellular compartments, generating superoxide anion free radicals ($O_2^{\cdot-}$) and other active oxygen species, including hydrogen peroxide (H_2O_2) and the hydroxyl radical ($\cdot OH$)

(1). Although several intracellular mechanisms exist that normally prevent the accumulation of active oxygen species during cell metabolism, these protective systems can become overwhelmed and toxic damage to cells may follow. Thus, accumulating evidence implicates oxygen radicals and other oxygen-derived species as causative agents in ageing and a variety of human diseases, including cancer (2).

The term 'oxidative stress' is generally applied to conditions in which the intracellular prooxidant-antioxidant ratio favors the former (3). Recent studies by our laboratory and others have examined the effects of oxidative stress on various cellular functions. Experimental systems commonly used to expose various types of cells and tissues to oxidative stress include enzymatic generation of active oxygen species (*e.g.*, using xanthine/xanthine oxidase) (4), direct exposure to peroxides or prooxidants (5) and intracellular generation of active oxygen species by treatment with redox-cycling chemicals such as quinones or bipyridilium compounds (6,7). Notable biochemical consequences of cellular oxidative stress include marked alterations in the regulation of intermediary metabolism (5,8), disruption of intracellular thiol and Ca^{2+} homeostasis (9,10), modification of signal transduction pathways (11-13) and tumor promotion (2).

Role of Glutathione and Protein Thiol Modification During Oxidative Stress

Glutathione (GSH) plays a unique role in the cellular defense against active oxygen species and reactive intermediates. GSH functions both as a reductant in the metabolism of hydrogen peroxide and organic hydroperoxides and as a nucleophile which can conjugate electrophilic molecules (14). During glutathione peroxidase-catalyzed metabolism of hydroperoxides, GSH serves as an electron donor, and the glutathione disulfide (GSSG) formed in the reaction is subsequently reduced back to GSH by glutathione reductase, at the expense of NADPH. Under conditions of oxidative stress, when the cell must cope with large amounts of H_2O_2 or organic hydroperoxides, the rate of glutathione oxidation exceeds the slower rate of GSSG reduction by glutathione reductase, and GSSG accumulates. To avoid the detrimental effects of increased intracellular levels of GSSG (*e.g.*, formation of mixed disulfides with protein thiols), the cell actively excretes GSSG, which can lead to depletion of the intracellular glutathione pool (Fig. 1).

Although it is now well established that cell killing caused by conditions of oxidative stress is preceded by depletion of intracellular

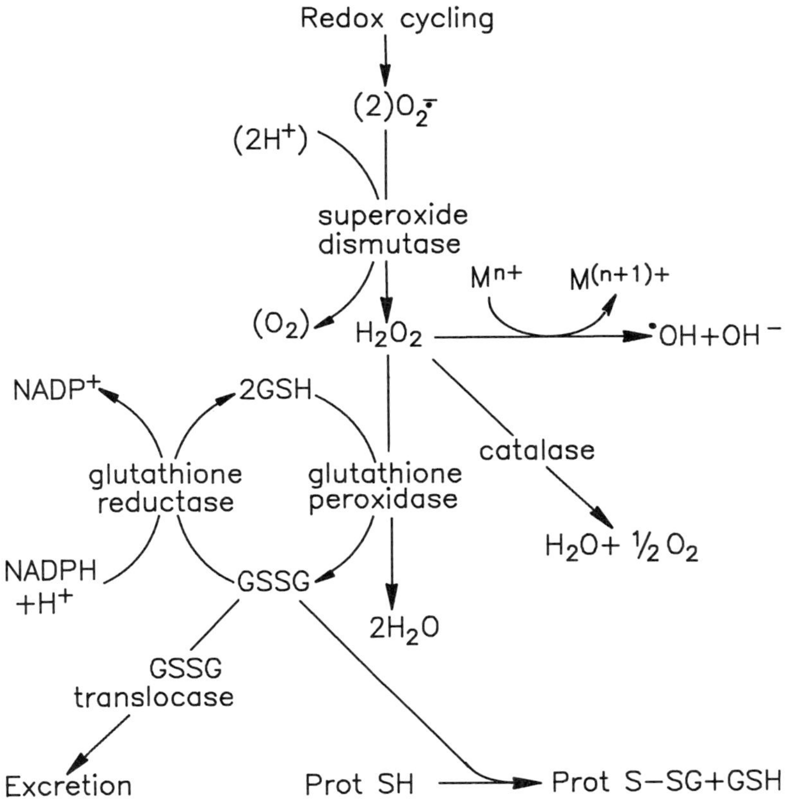

Figure 1. *Formation and Metabolism of Active Oxygen Species*

GSH (15), the exact relationship between GSH depletion and cell death has not yet been clarified. Sulfhydryl groups are, in general, highly reactive; thus, the sulfhydryl groups contained in molecules other than GSH, for example cellular protein thiols, may represent critical targets for oxidative stress-mediated cell killing. Indeed, the generation of conditions of oxidative stress during the metabolism of menadione (2-methyl-1,4-naphthoquinone) in isolated rat hepatocytes results in the loss of protein thiols. This loss of protein thiols follows glutathione depletion and precedes the onset of cell death. Moreover, pretreatments which deplete intracellular GSH (*e.g.*, diethylmaleate), potentiate menadione-induced protein thiol depletion and cytotoxicity (16).

Oxidative Stress Disrupts the Cytoskeleton

Exposing hepatocytes to menadione to generate conditions of oxidative stress results in an alteration of surface morphology characterized by the loss of microvilli and the appearance of multiple blebs on the surface of the hepatocytes (17). Many other toxic agents, including 1-methyl-4-phenyl-1,2,3,6-tetrahydropyridine (MPTP) (18) and the Ca^{2+} ionophore A23187 (19), cause similar alterations in surface structure, indicating that plasma membrane blebbing is a common event in the progression of toxic injury. Such blebs appear before alterations of the plasma membrane permeability characteristics and seem to be initially reversible.

From the findings that hepatocyte blebbing can also be induced by the microfilament modifying drugs cytochalasin D, phalloidin, and microcystin-LR (20,21), it appears that the blebbing caused by menadione and other cytotoxic agents may be due to microfilament disruption. The direct relationship between perturbation of the cytoskeleton during oxidative stress and the resulting cell killing is at present unclear. Many cellular processes such as cell division, intracellular transport, cell to cell contact, and the control of cell motility and shape are critically dependent on an intact cytoskeleton; disruption of such processes by oxidative stress would clearly be detrimental to cell survival. Furthermore, blebs have been observed to rupture, compromising the integrity of the plasma membrane.

Mechanistically, perturbation of normal cytoskeletal organization by oxidative stress may result from the disruption of intracellular thiol and Ca^{2+} homeostasis. Thiol oxidation perturbs microfilament organization through formation of disulfide cross-links between actin molecules (22). A change in intracellular Ca^{2+} distribution can affect cytoskeletal structure because calcium ions and cytoskeleton-associated Ca^{2+}-binding proteins play a pivotal role in the regulation of the microfilament network (23). Alternatively, Ca^{2+}-activated catabolic enzymes, such as the calpain family of cytosolic proteases, can cleave cytoskeletally-associated proteins, and thereby cause alterations of the microfilament network and blebbing (24,25). It is likely that both thiol modification and Ca^{2+}-dependent processes (see below) may combine to produce cytoskeletal alterations during oxidative stress, and that this may play a determinant role in the onset of cell killing.

Oxidative Stress Interferes with Ca^{2+} Homeostasis

Under physiological conditions, the cytosolic free Ca^{2+} concentration ($[Ca^{2+}]_i$) in mammalian cells is very low (0.1 μM) compared with the concentration of Ca^{2+} in the extracellular fluids (1-2 mM). Such a low $[Ca^{2+}]_i$ is maintained through the concerted action of Ca^{2+}-translocases present in mitochondria, the endoplasmic reticulum, the nucleus, and the plasma membrane (26,27). The plasma membrane translocases (which are Ca^{2+}-ATPases in liver cells and erythrocytes, and a Na^+/Ca^{2+} exchanger in excitable tissues) balance the passive influx of Ca^{2+} from the extracellular environment, and the Ca^{2+}-translocating activites of mitochondria, endoplasmic reticulum, and perhaps also liver cell nuclei, contribute to the low $[Ca^{2+}]_i$, by active Ca^{2+} sequestration into their respective compartments.

All of these Ca^{2+} transport systems are highly sensitive to agents which affect the redox status of pyridine nucleotides and the oxidation state of protein sulfhydryl groups. Consequently, a perturbation of the intracellular redox balance could result in the disruption of intracellular Ca^{2+} homeostasis (Fig. 2). This hypothesis has received support from experiments conducted *in vivo* (28), or utilizing intact cells and subcellular fractions (10,29-31). Work from our laboratory and others has shown that during the metabolism of menadione, oxidation of pyridine nucleotides and protein thiols leads to an impairment of the Ca^{2+}-sequestering capacity of both the mitochondria and the endoplasmic reticulum. The endoplasmic reticulum is the major intracellular store of mobilizable Ca^{2+} in cells (32), and we have recently demonstrated that inhibition of the endoplasmic reticular Ca^{2+} translocase rapidly releases the entire endoplasmic reticular Ca^{2+} pool into the cytosol (33). The Ca^{2+} released by inhibition of the endoplasmic reticular Ca^{2+} translocase is only very slowly removed from the cytosol by the plasma membrane Ca^{2+} pump. In contrast with the endoplasmic reticulum, mitochondria have a low affinity for Ca^{2+}, and it is now clear that in the presence of physiological Ca^{2+} concentrations, they take up little Ca^{2+}. However, when cytosolic $[Ca^{2+}]$ is high [$>1\,\mu$M], mitochondria are extremely efficient Ca^{2+} buffers, thus becoming a crucial line of defense against pathological elevations of cytosolic Ca^{2+} concentrations. During conditions of oxidative stress, the loss of both reduced pyridine nucleotides and protein thiols inhibits mitochondrial Ca^{2+} sequestration resulting in the loss of this vital defence mechanism against elevated $[Ca^{2+}]_i$, while at the same time releasing Ca^{2+} into the cytosol. If, as is often the case, the plasma membrane

Ca^{2+}-extruding system is also inhibited by thiol modification, the continuous influx of the extracellular Ca^{2+} down its concentration gradient will also contribute to the sustained elevation of cytosolic Ca^{2+} concentration observed in cells undergoing oxidative stress.

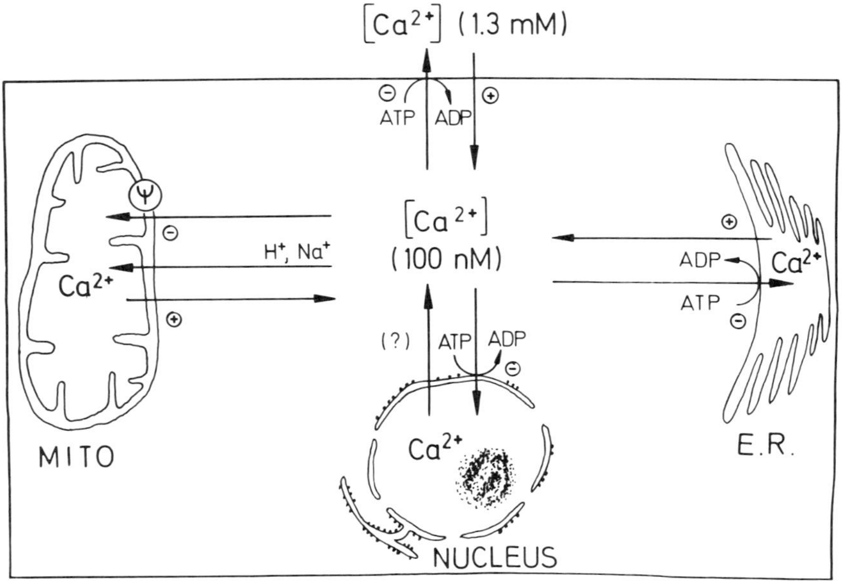

Figure 2. *Ca^{2+}-Transport Systems in Hepatocytes and How Each of Them Can Be Inhibited (-) or Stimulated (+) During Exposure to Oxidative Stress*

Mechanisms of Cell Damage and Cell Killing during Oxidative Stress

Pathological elevations of $[Ca^{2+}]_i$ following inhibition of the cellular Ca^{2+} transport systems by oxidative stress, has been shown to activate several Ca^{2+}-dependent cytotoxic mechanisms. These mechanisms include, in addition to disruption of the cytoskeletal network (see above), the pathological activation of Ca^{2+}-stimulated catabolic enzymes, such as proteases (*e.g.*, the calpains), phospholipases and endonucleases (Fig. 3) (34). A further notable consequence of the disruption of thiol homeostasis during exposure to cytotoxic concentrations of prooxidants is the impairment of hormone-stimulated phosphoinositide metabolism (13), which will result in the isolation of the damaged cell from the hormonal control of the body.

Modulation of Signal Transduction Pathways by Oxidative Stress

As opposed to the cytotoxic effects of excessive concentrations of prooxidants, recent investigations and extensive indirect evidence indicate that chemical oxidants and other agents that induce conditions of mild oxidative stress in cells can modulate cellular growth and differentiation (35). In particular, oxidative stress can lead to tumor promotion and tumor progression (2). Tumor promotional activity has been demonstrated for reactive oxygen species such as H_2O_2, superoxide anion radicals, ozone, and hyperbaric oxygen, and organic peroxides including benzoylperoxide (36,37). The mechanism by which agents that induce oxidative stress function as tumor promoters remains, however, unresolved.

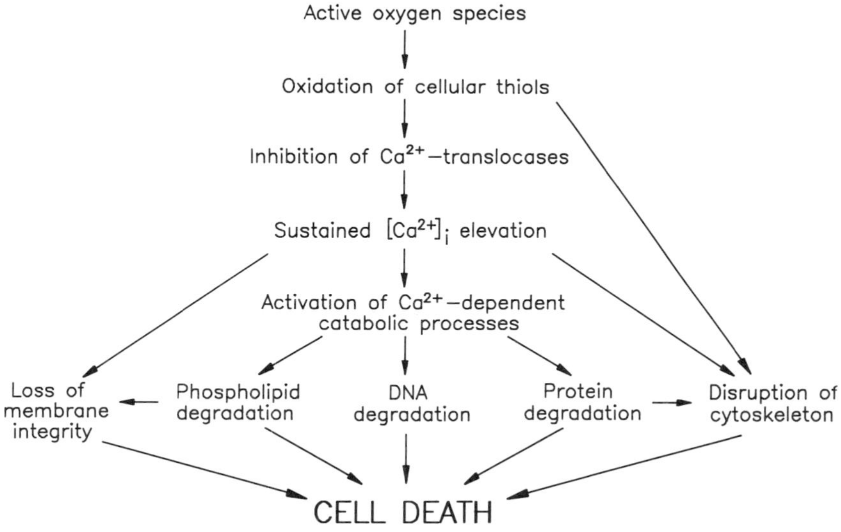

Figure 3. *Mechanisms Involved in Oxidative Stress-Mediated Cell Killing*

It has been shown that conditions of oxidative stress can result in the formation of DNA strand breaks (38). Interestingly, the potent skin tumor promoter, 12-O-tetradecanoylphorbol 13-acetate (TPA), also stimulates generation of reactive oxygen species in cell types such as polymorphonuclear leukocytes (39) and primary mouse epidermal cells (40), thereby inducing DNA strand breakage (41,42). Based on the observations that DNA strand breaks stimulate polyADP-ribosylation of chromatin proteins, Cerutti and co-workers have proposed that alterations of chromatin conformation and function through polyADP-ribosylation result in modification of the expression of genes involved in the control of cell growth and differentiation, such as c-*fos* and c-*myc* (43). Since the expression of these genes is also modulated by protein kinase C, our laboratory has recently started to investigate the potential role of protein kinase C in oxidative stress-mediated tumor promotion.

Exposure of rat hepatocytes to low levels of oxidative stress results in a rapid, 2-3-fold increase in the specific activity of protein kinase C (11). Only the cytosolic protein kinase C displayed the increase in specific activity (Table 1) whereas the activity of the membrane bound form remained unchanged. The increase in protein kinase C specific activity is not due to an increase in protein kinase C copy number. Rather, oxidative modification of the protein kinase C enzyme itself, most likely through oxidative stress-induced modification of the thiol/disulfide balance of the enzyme, is responsible for the activition. The involvement of thiol residues in the activation phenomenon was confirmed when we found that partially pure protein kinase C from rat brain could be activated using low concentrations of GSSG in a glutathione redox buffer.

Table 1. *Activation of Hepatocyte Protein Kinase C During Exposure to Oxidative Stress*

Treatment[a]	Protein Kinase C Activity[b]
Vehicle (DMSO)	11.2
50 μM Menadione	32.4
250 μM DMNQ[c]	27.8

[a] Isolated rat hepatocytes were treated with redox-cycling quinones for 30 min before isolating and measuring the cytosolic protein kinase C activity, as described in (11).
[b] Expressed as pmol ^{32}P incorporated into histone H1/min/mg protein.
[c] DMNQ, 2,3-dimethoxy-1,4-napthoquinone

It is not yet established whether the oxidative modification of protein kinase C results in the stimulation of substrate phosphorylation *in situ*. However, the oxidatively modified enzyme may overexpress responses to various physiological stimuli, and this may represent a potential mechanism for oxidative stress-mediated tumor promotion. Indeed, the induction of the enzyme ornithine decarboxylase (which is controlled by protein kinase C) and skin tumor promotion by TPA has been reported to be enhanced by adriamycin, a redox-cycling anthracycline antibiotic (44). It should be expected that overexpression of protein kinase C-dependent responses would have a significant impact on the control of cellular growth and differentiation. Weinberg and colleagues have recently reported that overproduction of protein kinase C enzyme and overexpression of its enzyme activity are sufficient for inducing dramatic growth abnormalities, such as growth to a higher confluence density, shorter exponential doubling time, and development of foci of cells displaying anchorage-independent growth in soft agar (45).

In addition to protein kinase C, other protein kinases have been shown to become activated under conditions of oxidative stress. A protein kinase which phosphorylates the α-subunit of eukaryotic initiation factor 2 is activated by low concentrations of GSSG (46). Several laboratories have also reported that reactive oxygen species can activate the rat liver insulin receptor tyrosine kinase (47,48). More recently, induction of oxidative stress in mouse epidermal cells has been found to stimulate phosphorylation of the ribosomal protein S6 (49) through an apparently Ca^{2+}-dependent event. Analysis of the primary structure of a number of protein kinases (serine, threonine and tyrosine kinases) has revealed a substantial sequence homology in the catalytic (kinase) domains (50). Thus, a common structural feature may predispose different protein kinases to activation through alteration(s) of there thiol/disulfide status. Consequently, modulation of kinase activities involved in signal transduction and cellular metabolism could constitute a mechanism by which a non-cytotoxic state of oxidative stress is able to have profound implications on cell division and differentiation.

References

1. Fridovich, I. (1983) *Ann. Rev. Pharmacol. Toxicol.* **23,** 239-257
2. Cerutti, P.A. (1985) *Science* **227,** 375-381
3. Sies, H. (ed) (1985) in *Oxidative Stress* pp. 507, Academic Press, London
4. Muehlematter, D., Larsson, R., and Cerutti, P. (1988) *Carcinogenesis* **9,** 239-245
5. Sies, H., Brigelius, R., Graf, P. (1987) in *Advances in Enzyme Regulation* (Weber, G., ed) Vol. 26, pp. 175-189, Pergamon Press, Oxford
6. Smith, M.T., Evans, C.G., Thor, H., and Orrenius, S. (1985) in *Oxidative Stress* (Sies, H., ed) pp. 91-113, Academic Press, London
7. Orrenius, S., Rossi, L., Eklöw-Låstbom, L., and Thor, H. (1986) in *Free Radicals in Liver Injury* (Poli, G., and Cheesman, K.H., eds) pp. 99-105, IRL Press Ltd., Oxford

8. Ziegler, D.M. (1985) *Ann. Rev. Biochem.* **54,** 305-329
9. Di Monte, D., Ross, D., Bellomo, G., Eklöw, L., and Orrenius, S. (1984) *Arch. Biochem. Biophys.* **235,** 334-342
10. Nicotera, P., McConkey, D., Svensson, S.-Å., Bellomo, G., and Orrenius, S. (1988) *Toxicology* **52,** 55-63
11. Kass, G.E.N., Duddy, S.K., and Orrenius, S. (1989) *Biochem. J.* **260,** 499-507
12. Gopalakrishna, R., and Anderson, W.B. (1989) *Proc. Natl. Acad. Sci. USA* **86,** 6758-6762
13. Bellomo, G., Thor, H., and Orrenius, S. (1987) *J. Biol. Chem.* **262,** 1530-1534
14. Orrenius, S., and Moldéus, P., (1984) *Trends Pharmacol. Sci* **5,** 432-435
15. Nicotera, P., and Orrenius, S. (1986) in *Biological Reactive Intermediates III* (Kocsis, J.J., Jollow, D.J., Witmer, C.M., Nelson, J.O., and Snyder, R., eds) pp. 41-51, Plenum, New York
16. Di Monte, D, Bellomo, G., Thor, H., Nicotera, P., and Orrenius, S. (1984) *Arch. Biochem. Biophys.* **235,** 343-350
17. Thor, H., Smith, M.T., Hartzell, P., Bellomo, G., Jewell, S.A., and Orrenius, S. (1982) *J. Biol. Chem.* **257,** 12419-12425
18. Kass, G.E.N., Wright, J.M., Nicotera, P., and Orrenius, S. (1988) *Arch. Biochem. Biophys.* **260,** 789-797
19. Jewell, S.A., Bellomo, G., Thor, H., Orrenius, S., and Smith, M.T. (1982) *Science* **217,** 1257-1259
20. Prentki, M., Chaponnier, C., Jeanrenaud, B., and Gabbiani, G. (1979) *J. Cell Biol.* **81,** 592-607
21. Eriksson, J.E., Paatero, G.I.L., Meriluoto, J.A.O., Codd, G.A., Kass, G.E.N., Nicotera, P., and Orrenius, S. (1989) *Exp. Cell Res.* **185,** 86-100
22. Mirabelli, F., Salis, A., Marinoni, V., Finardi, G., Bellomo, G., Thor, H., and Orrenius, S. (1988) *Arch. Biochem. Biophys.* **264,** 261-269
23. Weeds, A. (1982) *Nature* **296,** 811-816
24. Nicotera, P., Hartzell, P., Davis, G., and Orrenius, S. (1986) *FEBS Lett.* **209,** 139-144
25. Mirabelli, F., Salis, A., Vairetti, M., Bellomo, G., Thor, H., and Orrenius, S. (1989) *Arch. Biochem. Biophys.* **270,** 478-488
26. Carafoli, E. (1987) *Ann. Rev. Biochem.* **56,** 395-433
27. Nicotera, P., McConkey, D.J., Jones, D.P., and Orrenius, S. (1989) *Proc. Natl. Acad. Sci. USA* **86,** 453-457
28. Tsokos-Kuhn, J.O., Smith, C.V., Hughes, H., and Mitchell, J.R. (1988) *Mol. Pharmacol.* **34,** 209-214
29. Moore, G.A., O'Brien, P.J., and Orrenius, S. (1986) *Xenobiotica* **16,** 873-882
30. Thor, H., Hartzell, P., Svensson, S.-Å., Orrenius, S., Mirabelli, F., Marinoni, V., and Bellomo, G. (1985) *Biochem. Pharmacol.* **34,** 3717-3723
31. Nicotera, P., Moore, M., Mirabelli, F., Bellomo, G., and Orrenius, S. (1985) *FEBS Lett.* **181,** 149-153
32. Somlyo, A.P., Bond, M., Somlyo, A.V. (1985) *Nature* **314,** 622-625

33. Kass, G.E.N., Duddy, S.K., Moore, G.A., and Orrenius, S. (1989) *J. Biol. Chem.* **264,** 15192-15198
34. Orrenius, S., McConkey, D.J., Bellomo, G., and Nicotera, P. (1989) *Trends Pharmacol. Sci.* **10,** 281-285
35. Sohal, R.S., Allen, R.G., and Nations, C. (1986) *J. Free Rad. Biol. Med.* **2,** 175-181
36. Zimmerman, R., and Cerutti, P. (1984) *Proc. Natl. Acad. Sci. USA* **81,** 2085-2087
37. Slaga, T.J., Klein-Szanto, A.J.P., Triplett, L.L., Yotti, L.P., and Trosko, J.E. (1981) *Science* **213,** 1023-1024
38. Imlay, J.A., and Linn, S. (1988) *Science* **240,** 1302-1309
39. Goldstein, B.D., Witz, G., Amoruso, M., Stone, D.S., and Troll, W. (1981) *Cancer Lett.* **11,** 257-262
40. Fischer, S.M., and Adams, L.M. (1985) *Cancer Res.* **45,** 3130-3136
41. Emerit, I., and Cerutti, P. (1981) *Nature* **293,** 144-146
42. Birnboim, H.C. (1982) *Science* **215,** 1247-1249
43. Crawford, D., Zbinden, I., Amstad, P., and Cerutti, P. (1988) *Oncogene* **3,** 27-32
44. Perchellet, J.P., Kishore, G.S., and Perchellet, E.M. (1985) *Cancer Lett.* **29,** 127-137
45. Housey, G.M., Johnson, M.D., Hsiao, W.-L., O'Brian, C.A., Murphy, J.P., Kirschmeier, P., and Weinstein, I.B. (1988) *Cell* **52,** 343-354
46. Ernst, V., Levin, D.H., London, I.M. (1978) *Proc. Natl. Acad. Sci. USA* **75,** 4110-4114
47. Chan, T.M., Chen, E., Tatoyan, A., Shargill, N.S., Pleta, M., Hochstein, P. (1986) *Biochem. Biophys. Res. Commun.* **139,** 439-445
48. Hayes, G.R., and Lockwood, D.H. (1987) *Proc. Natl. Acad. Sci. USA* **84,** 8115-8119
49. Larsson, R., and Cerutti, P. (1988) *J. Biol. Chem.* **263,** 17452-17458
50. Parker, P.J., Coussens, L., Totty, N., Rhee, L., Young, S., Chen, E., Stabel, S., Waterfield, M.D., and Ullrich, A. (1986) *Science* **233,** 853-859

Oxidative Myocardial Injury and Cardiac-Derived Experimental Systems

David R. Janero
Research Department, Pharmaceuticals Division, CIBA-GEIGY Corporation
Summit, NJ 07901 USA

Abstract

Partially-reduced oxygen, either in the form of short-lived inorganic species with an unpaired electron(s) (i.e., "oxy-radicals") or molecules with a more stable electronic configuration (e.g., hydrogen peroxide), can incite and exacerbate tissue injury. A cardinal degenerative feature of this "oxygen toxicity, " particularly in the presence of transition metals, is the non-enzymatic, non-metabolically-coupled oxidation of membrane phospholipid polyunsaturated fatty-acyl moieties to form hydroperoxides. Partially-reduced oxygen is injurious to the heart and may directly mediate myocardial ischemic pathophysiology (especially in the post-ischemic reperfusion phase) as a cytotoxic promoter of heart muscle-cell (cardiomyocyte) membrane peroxidation. These considerations, along with pressing medical concerns associated with the high mortality due to ischemic heart disease, have helped make cardiac-derived experimental systems attractive and therapeutically relevant. Such systems are currently employed for studying tissue injury due to oxygen toxicity and evaluating potential pharmacologic targets and interdictions.

Introduction

The aim of this chapter is two-fold: (a) to provide an overview of the potential pathogenic roles of partially-reduced oxygen in, especially, the heart; and (b) to illustrate the applicability, utility, and limitations of cardiac-derived experimental systems in the study of peroxidative tissue injury. The primary literature as well as recent reviews will be cited with the hope of stimulating the reader to delve further into the peroxidative aspects of biological oxidation.

Partially-Reduced Oxygen as a Pathophysiologic Agent

For chemotrophs, molecular oxygen (O_2) is a bioenergetic substrate essential to efficient energy conservation *via* oxidative

Biological Oxidation Systems, Volume 2

phosphorylation (1). In this process, O_2 serves as the terminal acceptor of electrons from energy-rich pyridine-nucleotide and flavin products of glycolysis, β-oxidation, and the citric acid cycle to form the nontoxic, tetravalent reduction product, water (H_2O).

$$O_2 + 4H^+/e^- \rightarrow 2H_2O$$

The severe toxicity of O_2 at a concentration only slightly greater than that in air (2) led Gerschman *et al.*, in 1954, to theorize that tissue injury could be mediated by highly-reactive, partially-reduced inorganic species derived from O_2 and having one or more unpaired electrons -- *i.e.* oxygen-containing free radicals ("oxy-radicals") (3). Subsequent studies employing, for instance, chemical antioxidants (4) or spin-trapping (5) have established that the major pathogenic oxy-radicals in biologic systems are the superoxide anion radical ($O_2^{\cdot-}$) and the hydroxyl radical (•OH), the respective one- and three-electron reduction products of O_2. Two-electron O_2 reduction yields a non-radical species, hydrogen peroxide (H_2O_2). Thus, the tetravalent reduction of O_2 to H_2O may be expressed as a series of one-electron, partial-reduction intermediates:

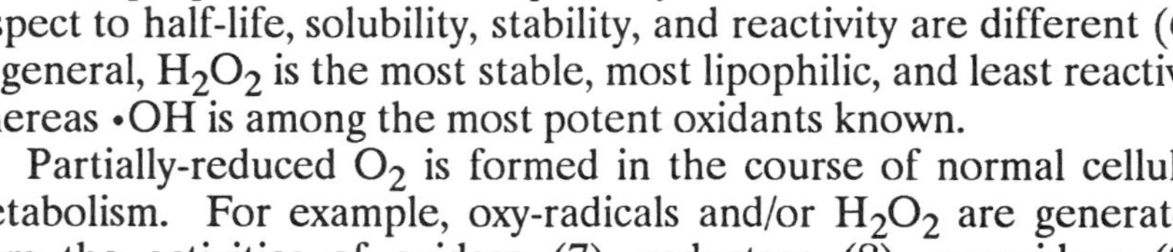

The properties of these partially-reduced intermediates with respect to half-life, solubility, stability, and reactivity are different (6). In general, H_2O_2 is the most stable, most lipophilic, and least reactive, whereas •OH is among the most potent oxidants known.

Partially-reduced O_2 is formed in the course of normal cellular metabolism. For example, oxy-radicals and/or H_2O_2 are generated from the activities of oxidase (7), reductase (8), peroxidase (9), hydroxylase (10), and oxygenase (11) enzymes and through oxidation-reduction reactions involving quinones (12) and flavins (13). Perhaps the most extensively studied intracellular source of partially-reduced O_2 is the metalloflavoprotein xanthine oxidase (XOD) (xanthine: O_2 oxidoreductase, EC 1.2.3.2), which catalyzes the degradation of purines to uric acid and, in the process, passes reducing equivalents to O_2, forming H_2O_2, $O_2^{\cdot-}$, and •OH (14). Extracellular sources of partially-reduced O_2 include environmental pollutants (15, 16), pesticides and herbicides (17), and various organics (18). Oxy-radicals and H_2O_2 produced within a cell may reach the extracellular compartment (19). Radiation (20) and certain drugs (21, 22) stimulate both intra- and extra-cellular oxy-radical formation.

Generation of oxy-radicals and H_2O_2 in living systems is not invariably detrimental to cells, nor does it lead vectorially to tissue injury (2). Indeed, partially-reduced O_2 may serve a beneficial role in maintaining tissue homeostasis, as illustrated by the bactericidal action of the $O_2^{\bar{\cdot}}$ generated through the phagocytic "respiratory burst" (19). These considerations support a fundamental tenet of O_2-toxicity: partially-reduced O_2 species can be pathophysiologic only when their net production exceeds the capacity of cellular defense mechanisms to detoxify them (23). The major cellular detoxification mechanisms which interact directly with partially-reduced O_2 include the enzymes superoxide dismutase (SOD) ($O_2^{\bar{\cdot}}:O_2^{\bar{\cdot}}$ oxidoreductase, EC 1.15.1.1) (24) and catalase (CAT) ($H_2O_2:H_2O_2$ oxidoreductase, EC 1.11.1.6) (25).

$$\text{SOD: } 2O_2^{\bar{\cdot}} + 2H^+ \rightarrow H_2O_2 + O_2$$

$$\text{CAT: } 2H_2O_2 \rightarrow 2H_2O + O_2$$

Three essential dietary nutrients, α-tocopherol (vitamin E), ascorbic acid (vitamin C), and β-carotene, can directly detoxify oxy-radicals by chemically interacting with ("scavenging") them (26). Through redox mechanisms, a variety of small molecules convert oxy-radicals and H_2O_2 to H_2O, *e.g.*, the peroxidase-catalyzed reaction of H_2O_2 and reduced glutathione (GSH) to form oxidized glutathione disulfide (GSSG) (27).

$$H_2O_2 + 2GSH \xrightarrow{\text{GSH peroxidase}} GSSG + 2H_2O$$

Transition metals, principally iron in biologic systems, support the transformation of partially-reduced O_2 species (28, 29). Thus, in a mixture of ferrous iron (Fe^{2+}) and H_2O_2 ["Fenton's reagent" (30)], Fe^{2+} promotes H_2O_2 decomposition *via* the iron-catalyzed Haber-Weiss reaction to generate the highly-reactive $\cdot OH$.

$$Fe^{2+} + H_2O_2 \rightarrow Fe^{3+} + {}^{-}OH + \cdot OH$$

Redox reactions among H_2O_2 and oxy-radicals themselves may result in their transformation to other species, *e.g.*:

$$H_2O_2 + \cdot OH \rightarrow H_2O + H^+ + O_2^{\bar{\cdot}}$$

$$H_2O_2 + O_2^{\cdot -} \rightarrow O_2 + {}^-OH + \cdot OH$$

Similarly, oxy-radical protonation may occur at low pH, yielding, for instance, the hydroperoxyl radical ($HO_2\cdot$) from $O_2^{\cdot -}$.

$$O_2^{\cdot -} + H^+ \rightarrow HO_2\cdot$$

Complexation between iron and partially-reduced O_2 has been theorized to produce reactive ferryl adducts (31).

$$Fe^{2+} + H_2O_2 \rightarrow FeOH^{3+} \text{ (or } FeO^{2+}\text{)} + {}^-OH$$

The extent to which these and other such transformation or spontaneous-dismutation reactions occur in biologic, as opposed to strictly chemical, systems is a matter of some controversy (32).

The variegated nature of the reactions between tissue components and partially-reduced O_2 has engendered particularly diverse possibilities as to how oxy-radicals and H_2O_2 might promote tissue injury. Destruction of extracellular matrix constituents, such as collagen (33) and glycosaminoglycans (34), by oxy-radicals has been shown to compromise tissue function. The effects of partially-reduced O_2 on proteins include oxidation of amino acid residues (35) and sulfhydryl moieties (36), alteration of primary (and higher-order) structure *via* fragmentation (37) and proteolytic cleavage (38), and covalent peptide modification *via* ADP-ribosylation (39, 40). Partially-reduced O_2 damages non-protein macromolecules as well, inciting, for instance, DNA base-modification (41) and strand-breakage (42). Sugars, either free molecules or covalently-bound carbohydrate moieties, are degraded by oxy-radicals and H_2O_2 (43).

The principal pathogenic means by which partially-reduced O_2 promotes tissue injury is considered to be the non-enzymatic, non-metabolically-coupled introduction of O_2 into polyunsaturated fatty acid (PUFA) ("PUFA autoxidation") to form hydroperoxides ("PUFA peroxidation"). Since rather detailed chemical treatments of PUFA autooxidation have appeared recently (44-49), current thinking concerning the process in biologic systems will be conceptualized here.

PUFA autooxidation takes place in three stages: initiation, propagation, and termination. Autoxidation of PUFA may be initiated upon hydrogen-atom abstraction from allylic or *bis*-allylic positions by an oxy-radical (e.g., $\cdot$OH) to generate a fatty alkyl radical with a conjugated-diene structure ($\cdot$PUFA).

$$\text{PUFA} + \cdot\text{OH} \rightarrow \cdot\text{PUFA} + \text{H}_2\text{O}$$

Oxygenation of the fatty alkyl radical yields a fatty peroxyl radical (PUFA-OO•).

$$\cdot\text{PUFA} + \text{O}_2 \rightarrow \text{PUFA-OO}\cdot$$

In biologic systems, which contain PUFA's as good hydrogen donors, the fatty peroxyl radical tends to abstract a hydrogen atom from another PUFA to furnish a fatty hydroperoxide (PUFA-OOH), the primary product of non-enzymatic PUFA peroxidation, and a new fatty peroxyl radical.

$$\text{PUFA-OO}\cdot + \text{PUFA} + \text{O}_2 \rightarrow \text{PUFA-OOH} + \text{PUFA-OO}\cdot$$

Repetition of this reaction in an autocatalytic cycle propagates peroxidation at the expense of PUFA substrate. This "chain-reaction" continues until terminated by, for instance, PUFA depletion, oligomerization of radical species to form non-radical products, or the interdiction of "chain-breaking" antioxidants such as α-tocopherol.

Depending upon the local tissue conditions (*e.g.*, pH, antioxidant content) and the presence and chemical nature of transition metals such as iron, fatty hydroperoxides may participate in oxy- and lipid-radical-mediated reactions to afford either additional fatty peroxyl radicals or fatty alkoxyl radicals (PUFA-O•), the latter tending to epoxidize (49). Since low levels of fatty hydroperoxides exist in tissues (50), generation of PUFA peroxyl radicals through iron-catalyzed fatty hydroperoxide decomposition can serve to initiate PUFA peroxidation. Such hydroperoxide-dependent PUFA peroxidation may require redox reactions between oxy-radicals and iron to establish a suitable ferric iron (Fe^{3+}):Fe^{2+} ratio for optimal initiation (46, 51). In biologic systems, perhaps the most important transformations of fatty hydroperoxides are intramolecular rearrangements to generate cyclic peroxides and prostaglandin-like endoperoxides, the latter a prime source of secondary aldehydic end-products such as malondialdehyde upon thermal and metal-catalyzed decomposition (47, 48, 52).

The ready autoxidation of molecules having allylic hydrogens, such as PUFA's, to hydroperoxides (53) and the high concentration of PUFA's esterified to the phospholipids of cellular membranes (54) make these supramolecular assemblies particularly susceptible to peroxidative injury from partially-reduced O_2. Since the solubility of O_2 in oils and, by inference, in the membrane phospholipid bilayer is greater than in the aqueous cytosol or in tissue fluids (55), the two key elements in lipid peroxidation, PUFA's and O_2, share an intimate environmental relationship. Likewise, membranous organelles are principal cellular sites of abnormally-reduced O_2 production (56) and contain redox-active iron which can help initiate PUFA peroxidation (57). Another factor making biomembranes prime targets of O_2-toxicity is the autopropagatory nature of PUFA autoxidation (*vide supra*), which serves as an intramembranous amplifier of the initial peroxidative stimulus. Membrane damage from phospholipid peroxidation has been attributed to the loss of PUFA's themselves (58), the peroxides formed (59), and the secondary end-products of lipid peroxide decomposition (60). Specific membrane effects of phospholipid PUFA peroxidation include changes in the activities of receptors (61), ion channels/pumps (62), and enzymes (63, 64); and biophysical derangements involving fluidity and permeability (65, 66). Therefore, PUFA peroxidation compromises the structural and physiologic integrity of cellular membranes and jeopardizes their roles as continuous, selectively-permeable hydrophobic barriers and as matrices which organize and compartmentalize cellular metabolism for maximal efficiency. Disruption of the homeostatic mechanisms responsible for cell integrity results. Since the cell is the basic unit of structure and function in all living things (67), O_2-toxicity, initially manifest as membrane phospholipid peroxidation, readily expresses itself in terms of tissue and organ pathophysiology.

Membrane Peroxidation and Myocardial Ischemic-Reperfusion Injury

Four major experimental preoccupations exist for researchers now in the field of cardiovascular O_2-toxicity: (a) pro-inflammatory cells such as neutrophils and platelets (68, 69); (b) dysfunction of the vascular endothelium (70); (c) oxidative lipoprotein modification and atherogenesis (71); and (d) ischemic-reperfusion damage to heart muscle (myocardium) (72). Although these four themes are not exclusive from the standpoints of either organismic physiology or the biology of cardiovascular disease, this section is focused primarily on myocardial ischemic-reperfusion injury, a leading cause of mortality in the industrialized world (73). Other reviews may be consulted for alternative treatments of this topic (72, 74-85).

Myocardial ischemia may be defined as the transient or progressive reduction in the oxygenated, coronary blood supply to the myocardium below that required to sustain normal aerobic metabolism in the heart-muscle cell (cardiomyocyte) (86). The cellular and physiologic derangements of the ischemic-reperfused heart are heterogeneous and, at best, only incompletely understood. Initially, ischemic cardiomyocytes evidence changes which are reversible with restoration of nutritive coronary flow, but prolonged ischemia results in irreversible, severe cardiomyocyte injury, formation of an infarcted zone of left ventricular necrosis, and compromised pump function (87-90). Major cardiomyocyte abnormalities with ischemia-reperfusion include depletion of high-energy phosphates (91); sodium, calcium, and water influx (92); protease inactivation (93); phospholipase activation (94); defective lipid acylation (95); amphiphile formation (96, 97) and release (98); and liberation of metabolic products [e.g., purine catabolites (99)]. Many of these derangements are associated with disturbances in cardiomyocyte bioenergetics (100) and membranes (101, 102). Functional abnormalities of the ischemic-reperfused myocardium include non-uniform blood-flow restoration to the ischemic tissue ("no-reflow phenomenon") (103); depressed contractility ("myocardial hybernation") (104); prolonged contractile impairment ("myocardial stunning") (105); and electrophysiological disturbances ("reperfusion arrhythmias") (106).

The most direct therapeutic means of limiting myocardial ischemic necrosis, prompt reestablishment of coronary flow, is being aggressively pursued as the fundamental clinical anti-ischemic strategy (107). Although survival is enhanced with efficient thrombolytic recanalization (108), reperfusion can extend myocardial necrosis by destroying cardiomyocytes expected to have remained viable (109, 110), a paradoxical phenomenon termed "reperfusion injury" (111). The severity of the ischemic period preceding reperfusion appears to be the primary prognostic indicator of the degree of reperfusion injury (112), and it remains unclear to what extent reperfusion injury constitutes the expression of cardiomyocyte abnormalities established during ischemia (113).

The precise mechanistic determinants of cardiac ischemic-reperfusion injury are unknown. A causal role for partially-reduced O_2 as a pathogenic agent in this disease process has been postulated. The so-called "free-radical hypothesis" of cardiac ischemic-reperfusion injury rests on the premise that reperfusion supports extensive formation of partially-reduced O_2 species which are injurious to the cardiomyocyte, especially as a result of their pro-peroxidative nature (*vide supra*). The net toxic effect of oxy-radicals/H_2O_2 on the ischemic-reperfused myocardium would likely be modulated by many factors *in vivo*: the O_2 content of the ischemic heart tissue itself and the degree of "hyperoxia" established with reperfusion (114), the duration and intensity of the ischemic insult

(115), the acuity of reperfusion (116), qualitative and quantitative variations in partially-reduced O_2 production over the course of injury (117, 118), and temporal changes in the vulnerability of ischemic cardiomyocytes to partially-reduced O_2 (119). Indeed, amphiphiles (120), calcium (121), and acidosis (122) are known amplifiers of O_2-toxicity to cardiac tissue.

Potential sources of oxy-radicals / H_2O_2 in the post-ischemic myocardium include enzymatic arachidonic acid oxygenation (123), catecholamine oxidation (124, 125), and the mitochondrial respiratory chain (56). Perhaps the two most intensively-investigated sources of partially-reduced O_2 are XOD and neutrophils. It has been thought that the relative lack of O_2 during ischemia limits XOD activity which, upon reperfusion and reintroduction of O_2, accelerates in the presence of accumulated purine substrates from cardiomyocyte nucleotide catabolism to yield, enzymatically, $O_2^{\bar{\cdot}}$, H_2O_2, and •OH (126). Activated neutrophils release copious amounts of partially-reduced O_2, and neutrophil accumulation is temporally, regionally, and quantitatively associated with myocardial ischemic-reperfusion injury (127, 128). The proposition that cardiomyocyte XOD is a primary contributor to myocardial reperfusion injury has been questioned (129, 130); resolution of the associated controversy may reside in the rich XOD content of the (micro)vascular endothelium (131). Likewise, demonstration that oxy-radical scavengers reduce infarction in leukopenic dogs (132) has been taken as evidence against neutrophils being a primary contributor to ischemic-reperfusion injury; yet this finding may be rationalized simply in terms of multiple oxy-radical sources in cardiac ischemic-reperfusion damage, some being neutrophil-independent.

Partially-reduced O_2 fulfills the four basic criteria of Dale (133), as elaborated by Mullane *et al.* (127), for a primary chemical mediator which induces a biologic response -- in this case, peroxidative myocardial ischemic-reperfusion damage:

a) *Presence.* Partially-reduced O_2 is present in ischemic-reperfused myocardium (117, 118, 134-141).
b) *Association.* Production of partially-reduced O_2 in ischemic myocardium is associated with reperfusion (118, 138, 140, 141).
c) *Action.* Administration of oxy-radical generating systems to various *in vitro* cardiac preparations incites changes reminiscent of ischemic-reperfusion injury (81, 143-145). Partially-reduced O_2 may be toxic to the cardiomyocyte through a variety of potential mechanisms, most particularly by supporting membrane-phospholipid PUFA peroxidation (*vide supra*).

Myocardial membranes, rich in PUFA's, are peroxidized with reperfusion (136, 138, 139, 142, 146-152).

d) *Pharmacologic manipulation.* A variety of agents which dismutate/scavenge partially-reduced O_2 (153-157) or chelate iron into a form not catalytic to PUFA peroxidation (158, 159) reduce infarct size, improve post-ischemic contractility, and/or act as anti-arrhythmics. Indeed, the chain-breaking antiperoxidant α-tocopherol has been used clinically as adjunctive, cardioprotective therapy against reperfusion damage (160).

Although available data clearly implicate partially-reduced O_2 as a primary mediator of myocardial ischemic-reperfusion injury *via* its pro-peroxidative activity, conclusive evidence awaits elucidation of the cause-and-effect interrelationships among oxy-radicals/H_2O_2, the ischemic-reperfused cardiomyocyte, impaired pump function, and infarction. The following section will illustrate the major cardiac-derived experimental systems being used to gain improved understanding of myocardial (and, more generally, tissue) oxidative injury.

Application of Cardiac-Derived Systems to Study Peroxidative Injury

This presentation of cardiac-derived experimental systems follows the levels of increasing structural complexity by which living matter is organized.

Molecular Systems. Since PUFA-esters of membrane phospholipids are the prime tissue substrates for peroxidation, liposomes reconstituted from membrane lipids have met with wide acceptance in the study of peroxidative injury (161). Liposomal peroxidation systems can be constructed with defined lipid compositions and have the unique advantage of eliminating interference due to non-PUFA oxidative transformations (162), but are inherently limited by the prior removal of the phospholipids from their biomembrane milieu. Liposomes prepared from myocardial-membrane phospholipids have been utilized to help elucidate the chemistry of (cardiac) lipid peroxide formation and decomposition (48) and the mechanisms by which various agents/pharmaceuticals inhibit (myocardial) PUFA peroxidation (163-167).

Membrane Systems. Membranes and membranous organelles can be isolated from either myocardial tissue homogenates (168) or cardiomyocytes (169) by differential/gradient centrifugation or by more elaborate means (170). Preparations enriched in one of the three major myocardial-membrane types, sarcolemma, sarcoplasmic reticulum, or mitochondria, have been employed to study the response of membrane-associated enzymes (63), ion channels/pumps (62), phospholipids (162), and metabolic processes (56) to perturbation by partially-reduced O_2. Isolated myocardial membranes have also been examined for their endogenous antiperoxidant tone (171, 172).

The use of cardiac membranes in oxidative-injury systems is limited mainly by existing technology, which does not permit rapid, selective isolation of the major myocardial-membrane types in completely pure form with high yields (170, 171). Particularly problematic are the definitive characterization of membrane subtypes, the quantification of cross-contamination, and the applicability of myocardial subfractionation protocols developed for the normal heart to injured tissue. Protocols involving salt extraction of myocardial contractile proteins may alter the topology of membrane constituents (173). As technology advances, it is anticipated that cardiac-membrane peroxidative systems will aid precise assessment of the organellar compartmentalization of peroxidative-injury stimuli and uncover site-specific intramembrane effects of partially-reduced O_2.

Cellular (Cardiomyocyte) Systems. Since the basic unit of cardiac structure and function is the heart-muscle cell (86), isolated cardiomyocytes have proven to be a very useful system for probing quantitatively the mechanisms by which partially-reduced O_2 might affect the heart. Cardiomyocyte systems are free of vascular, neurological, humoral, and mechanical influences characteristic of intact hearts, and the phenomena observed therein are representative of a single population of cells, without interference from non-muscle cells. The physiologic integrity of the isolated cardiomyocyte is also compelling, especially when metabolic parameters are being evaluated.

Two general cellular systems have been used most frequently to study oxy-radical/H_2O_2-induced heart-cell injury: monolayers of beating, neonatal rat cardiomyocytes (97) and suspensions of quiescent rat cardiomyocytes (174). The former can be maintained in primary culture, whereas adult cardiomyocyte suspensions must be utilized within a few hours after preparation. Neonatal cardiomyocytes have a higher glycolytic capacity than do adult cells (175), and it remains to be determined whether endogenous cardiomyocyte antioxidant systems change with time *ex vivo*. General experimental strategies employed in the study of cardiomyocyte injury by partially-reduced O_2 include exposing the cells to injury stimuli such as hypoxia/anoxia (97, 176), to various sources of oxy-radicals/H_2O_2 (177), or to compounds which are transformed/metabolized to yield these species (178, 179); incubating the cardiomyocytes with organic peroxides (180); and inducing changes in cellular antioxidant levels (181).

Isolated cardiomyocytes represent a particularly attractive system for addressing and integrating the various cellular and subcellular consequences of oxidative stress, especially since a variety of biochemical, metabolic, physiologic, and (ultra)structural parameters may be evaluated, often simultaneously (97, 182). With the increasing sophistication of tissue-culture methodology, especially in regard to adult heart-muscle cells (183), it would be predicted that cardiomyocyte-based model systems will gain importance in the identification and mechanistic assessment of novel cardioactive therapeutics for ischemic-reperfusion injury (184).

Tissue Systems. Although tissue homogenates of myocardium are easily prepared (168), they have rather limited diagnostic use, for biochemical and physiological integrity is lacking. Tissue slices have been employed to evaluate O_2-toxicity in brain, lung, and liver (185), but comparable myocardial tissue-slice systems have not been developed. Since tissue slices retain a major portion of whole-tissue organization, and many samples can be prepared from a single organ, they would appear to represent an exploitable approach to bridge cardiomyocyte and whole-organ systems. The rabbit interventricular septal preparation of Burton *et al.* exemplifies a tissue-based myocardial system in which biochemical, metabolic, and fine-structural parameters have been examined during oxy-radical exposure or with changes in flow/oxygenation (143). Rat (144) and rabbit (186) papillary muscles have been used in a similar manner. The key advantage of these tissue systems rests with the ability to monitor cardiac-muscle contractility and mechanical function. Neither the septal nor papillary preparations are homogeneous with respect to cell type, and the potential participation of, for example, vascular endothelium in the response of tissue-based experimental systems to O_2-toxicity must be considered (81).

Whole-Organ Systems. Mechanistic information on the complex process of ischemic-reperfusion heart injury and its mediation by partially-reduced O_2 is often difficult to obtain from the intact organ, either *in vivo* or *ex vivo*, although whole-organ systems allow investigation/manipulation of hemodynamics so that true ischemia may be achieved experimentally. Such systems have generally been used as occlusion-reperfusion models to assess the effect of anti(per)oxidant interventions in terms of infarct size (187), left-ventricular function (188), electrophysiology (137, 156), or cardiomyocyte enzyme/metabolite release (188). Ischemic-reperfused hearts *ex vivo* are often employed to measure and characterize, by physical means such as electron spin resonance spectroscopy, oxy- and lipid-radical production (117, 135, 138, 141). Improved diagnostic use of whole-organ systems for the study of myocardial ischemic-reperfusion injury awaits better measurements of infarction (189, 190) and more detailed *in situ* analysis of cardiomyocyte injury dynamics in as noninvasive a manner as possible. Tomographic imaging of infarct-avid radioactive compounds/nuclides (191) and magnetic resonance techniques (192) are now available to help provide direct analyses of ischemic-reperfusion injury and the role of oxy-radicals/H_2O_2 in mediating the associated organ abnormalities at the whole-heart level.

Concluding Remarks

The information presented provides a conceptual context for the hypothesis that partially-reduced O_2 species play an important pathogenic role in the cardiomyocyte derangements and myocardial dysfunction associated with ischemia-reperfusion. Examples of cardiac-derived experimental systems have been cited to illustrate their use in meeting the growing need to elucidate the mechanisms responsible for the adverse sequelae of post-ischemic myocardial reperfusion. Several considerations make this need a pressing one: the continued high mortality due to myocardial ischemia (73), the rather common occurrence of "spontaneous" ischemic-reperfusion events (193), the identification of a "silent ischemic" patient population (107), and the increasing use of therapeutic recanalization (107-110).

The involvement of oxy-radicals/H_2O_2 in post-ischemic reperfusion injury remains to be defined conclusively. The extremely complex pathophysiology of this disease process often obscures the distinction between coincidence and causality and makes it naive to assume that partially-reduced O_2 is the sole, let alone essential, pathogenic agent. Given this complexity, it is likely that, under at least some conditions of ischemia-reperfusion, cardiomyocyte injury may be neither directly nor primarily dependent upon oxy-radicals/H_2O_2. Consideration must be given to the possibility that the cytotoxic actions of partially-reduced O_2 in some circumstances link a multi-factorial chain of events leading to post-ischemic myocardial damage and in other circumstances initiate directly or even obviate such a cascade. Recent reports of cross-talk between phospholipid peroxides and phospholipases (194) exemplify such thinking. Consequently, as discussed elsewhere (81, 195), negative studies on the effects of antioxidants in occlusion-reperfusion models do not serve as general proof that partially-reduced O_2 has no pathogenic role in post-ischemic cardiac dysfunction.

The pro-peroxidant property of oxy-radicals/H_2O_2 serves as an attractive mechanistic rationalization of the diverse pathologic effects that partially-reduced O_2 has on myocardial physiology in the context of cardiomyocyte-membrane perturbation. The detrimental, non-peroxidative cellular effects of these species should not be ignored. For example, studies demonstrating that $O_2^{\bar{\cdot}}$ mediates the sodium/hydrogen-ion antiporter (196) and that H_2O_2 triggers covalent protein modification (i.e., ADP-ribosylation) (39, 40) offer an alternative, non-peroxidative outlook on O_2-toxicity in terms of oxidant messengers and abnormal signal-transduction. Peroxidative and non-peroxidative modes of cellular oxidative injury may represent two distinct expressions of O_2-toxicity which necessitate different therapeutic strategies: for the former, an antiperoxidant inhibitor of non-enzymatic PUFA oxidation; for the latter, a means to modulate or reset normal cellular signalling mechanisms which have gone awry. Clearly, although randomized placebo-controlled clinical trials of recombinant human SOD have been initiated in patients with acute myocardial infarction undergoing early reperfusion (82), immediate concern regarding the therapeutic amelioration of O_2-toxicity in myocardial ischemia-reperfusion must rest with careful description and dissection of oxy-radical/H_2O_2-mediated mechanisms underlying this complex disease process through employment of appropriate experimental systems.

Acknowledgments

The author thanks M. Worcel for support; M. Adamczk, B. Burghardt, C. Burghardt, M. Cardell, D. Feldman, M. Garber, W. Lee, R. Lopez, and J. Pittman for technical assistance in his cited studies; C. Laspina for secretarial assistance; and D. Feldman, J. Thakkar, D. Hrenivk, and C. Yarwood for comments.

References

1. Ernster, L., and Schatz, G. (1981) *J. Cell. Biol.* **91**, 227s-255s
2. Jamieson, D. (1989) *Free Radic. Biol. Med.* **7**, 87-108

3. Gerschman, R., Gilbert, D.L., Nye, S.W., Dwyer, P., and Fenn, W.O. (1954) *Science* **119**, 623-626
4. Gutteridge, J.M.C. (1984) *Biochem. J.* **224**, 761-767
5. Perkins, M.J. (1980) *Adv. Phys. Org. Chem.* **17**, 1-64
6. Pryor, W.A. (1986) *Annu. Rev. Physiol.* **48**, 657-667
7. Fernandez, V., Barrientos, X., Kipreos, K., Valenzuela, A., and Videla, L.A. (1985) *Endocrinology* **117**, 496-501
8. Lai, C.-S., Grover, T.A., and Piette, L.H. (1979) *Arch. Biochem. Biophys.* **193**, 373-378
9. Halliwell, B. (1977) *Biochem. J.* **163**, 441-448
10. Hill, M.A., Marota, J.A.J., and Shiman, R. (1988) *J. Biol. Chem.* **263**, 5646-5655
11. Kukreja, R.C., Kontos, H.A., and Hess, M.L. (1988) *FASEB J.* **2**, A835
12. Sugioka, K., Nakano, M., Totsune-Nakano, H., Minakami, H., Tero-Kubota, S., and Ikegami, Y. (1988) *Biochim. Biophys. Acta* **936**, 377-385
13. Muller, F. (1987) *Free Radic. Biol. Med.* **3**, 215-230
14. Kuppusamy, P., and Zweier, J. (1989) *J. Biol. Chem.* **264**, 9880-9884
15. Cosgrove, J.P., Borish, E.T., Church, D.F., and Pryor, W.A. (1985) *Biochem. Biophys. Res. Commun.* **132**, 390-396
16. Mehlhorn, H., Cottam, D.A., Lucas, P.W., and Wellburn, A.R. (1987) *Free Radic. Res. Commun.* **3**, 193-197
17. Sutton, H.C., and Winterbourn, C.C. (1984) *Arch. Biochem. Biophys.* **235**, 106-115
18. Shaw, S., and Jayatilleke, E. (1987) *Biochem. Biophys. Res. Commun.* **143**, 984-990
19. Ward, P.A., Warren, J.S., and Johnson, K.J. (1988) *Free Radic. Biol. Med.* **5**, 403-408
20. Raleigh, J.A. (1988) *Pharmacol. Ther.* **39**, 109-113
21. Augusto, O., Weingrill, C.L.V., Schreier, S., and Amemiya, H. (1986) *Arch. Biochem. Biophys.* **244**, 147-155
22. Muller, K., and Kappus, H. (1988) *Biochem. Pharmacol.* **37**, 4277-4280
23. Cotgreave, I.A., Moldeus, P., and Orrenius, S. (1988) *Annu. Rev. Pharmacol. Toxicol.* **28**, 189-212
24. Fridovich, I. (1989) *J. Biol. Chem.* **264**, 7761-7764
25. Percy, M.E. (1984) *Can. J. Biochem. Cell Biol.* **62**, 1006-1014
26. Machlin, L.J., and Bendich, A. (1987) *FASEB J.* **1**, 441-445
27. Ross, D. (1988) *Pharmacol. Ther.* **37**, 231-249
28. Aust, S.D., Morehouse, L.A., and Thomas, C.E. (1985) *Free Radic. Biol. Med.* **1**, 3-25
29. Bacon, B.R., and Britton, R.S. (1989) *Chem. Biol. Interact.* **70**, 183-226
30. Fenton, H.J. (1896) *J. Chem. Soc. Lond.* **69**, 546-562
31. Koppenol, W.H., and Liebman, J.F. (1984) *J. Phys. Chem.* **88**, 99-101
32. Minotti, G., and Aust, S.D. (1987) *J. Biol. Chem.* **262**, 1098-1104

33. Greenwald, R.A., and Moy, W.W. (1979) *Arthritis Rheum.* **22**, 251-259
34. Greenwald, R.A., and Moy, W.W. (1980) *Arthritis Rheum.* **23**, 455-463
35. Maier, K.L., Matejkova, E., Hinze, H., Leuschel, L., Weber, H., and Beck-Speier, I. (1989) *FEBS Lett.* **250**, 221-226
36. Harman, L.S., Mottley, C., and Mason, R.P. (1984) *J. Biol. Chem.* **259**, 5606-5611
37. Wolff, S.P., and Dean, R.T. (1986) *Biochem. J.* **234**, 399-403
38. Marcillat, O., Zhang, Y., Lin, S.W., and Davies, K.J.A. (1988) *Biochem. J.* **254**, 677-683
39. Lotscher, H., Winterhalter, K.H., Carafoli, E., and Richter, C. (1979) *Proc. Natl. Acad. Sci. USA* **76**, 4340-4344
40. Schraufstatter, I.U., Hyslop, P.A., Hinshaw, D.B., Spragg, R.G., Sklar, L.A., and Cochrane, C.G. (1986) *Proc. Natl. Acad. Sci. USA* **83**, 4908-4912
41. Aruoma, O.I., Halliwell, B., and Dizdaroglu, M. (1989) *J. Biol. Chem.* **264**, 13024-13028
42. Imlay, J.A., and Linn, S. (1988) *Science* **240**, 1302-1309
43. Cheeseman, K.H., Beavis, A., and Esterbauer, H. (1988) *Biochem. J.* **252**, 649-653
44. Sevanian, A., and Hochstein, P. (1985) *Annu. Rev. Nutr.* **5**, 365-390
45. Girotti, A.W. (1985) *Free Radic. Biol. Med.* **1**, 87-95
46. Braughler, J.M., Duncan, L.A., and Chase, R.L. (1986) *J. Biol. Chem.* **261**, 10282-10289
47. Frankel, E.N. (1987) *Free Radic. Res. Commun.* **3**, 213-226
48. Janero, D.R., and Burghardt, B. (1989) *Lipids* **24**, 125-131
49. Gardner, H.W. (1989) *Free Radic. Biol. Med.* **7**, 65-86
50. Marshall, P.J., Warso, M.A., and Lands, W.E.M. (1985) *Anal. Biochem.* **145**, 192-199
51. Aruoma, O.I., Halliwell, B., Laughton, M.J., Quinlan, G.J., and Gutteridge, J.M.C. (1989) *Biochem. J.* **258**, 617-620
52. Pryor, W.A., and Stanley, J.P. (1975) *J. Org. Chem.* **40**, 3615-3617
53. Pryor, W.A. (1976) in *Free Radicals in Biology* (Pryor, W.A., ed) pp. 1-49. Academic Press, New York
54. White, D.A. (1973) in *Form and Function of Phospholipids* (Ansell, G.B., Hawthorne, J.N., and Dawson, R.M.C., eds), 2nd Ed., pp. 441-482. Elsevier, Amsterdam
55. Lawrence, J.H., Loomis, W.F., Tobias, C.A., and Turpin, F.H. (1946) *J. Physiol.* **105**, 197-204
56. Bindoli, A. (1988) *Free Radic. Biol. Med.* **5**, 247-261
57. Schaich, K.M., and Borg, D.C. (1988) *Lipids* **23**, 570-579
58. Gamache, D.A., Fawzy, A.A., and Franson, R.C. (1988) *Biochim. Biophys. Acta* **958**, 116-124
59. Girotti, A.W., Thomas, J.P., and Jordan, J.E. (1985) *Arch. Biochem. Biophys.* **236**, 238-251

60. Jain, S.K. (1984) *J. Biol. Chem.* **259**, 3391-3394
61. Muakkassah-Kelly, S.F., Andresen, J.W., Shih, J.C., and Hochstein, P. (1983) *J. Neurochem.* **41**, 1429-1439
62. Kaneko, M., Beamish, R.E., and Dhalla, N. (1989) *Am. J. Physiol.* **256**, H368-H374
63. Kukreja, R.C., Okabe, E., Schrier, G.M., and Hess, M.L. (1988) *Arch. Biochem. Biophys.* **261**, 447-457
64. Kramer, J.H., Mak, I.T., and Weglicki, W.B. (1984) *Circ. Res.* **55**, 120-124
65. Ohyashiki, T., Ohtsuka, T., and Mohri, T. (1988) *Biochim. Biophys. Acta* **939**, 383-392
66. Arduini, A., Stern, A., Storto, S., Belfiglio, M., Mancinelli, G., Scurti, R., and Federici, G. (1989) *Arch. Biochem. Biophys.* **273**, 112-120
67. Palade, G.E. (1964) *Proc. Natl. Acad. Sci. USA* **52**, 613-634
68. Engler, R.L. (1989) *Am. J. Cardiol.* **63**, 19E-23E
69. Janero, D.R., Burghardt, B., and Burghardt, C. (1988) *Thrombosis Res.* **50**, 789-802.
70. Hennig, B., and Chow, C.K. (1988) *Free Radic. Biol. Med.* **4**, 99-106
71. Palinski, W., Rosenfeld, M.E., Yla-Herttuala, S., Gurtner, G.C., Socher, S.S., Butler, S.W., Parthasarathy, S., Carew, T.E., Steinberg, D., and Witztum, J.L. (1989) *Proc. Natl. Acad. Sci. USA* **86**, 1372-1376
72. Del Maestro, R.F. (1980) *Acta. Physiol. Scand.* **492**, 153-168
73. Roberts, R., and Marmor, A.T. (1983) *Annu. Rev. Med.* **34**, 377-390
74. Meerson, F.Z., Kagan, V.E., Kozlov, P., Belkina, L.M., and Arkhipenko, Y.V. (1982) *Basic Res. Cardiol.* **77**, 465-485
75. McCord, J.M., and Roy, R.S. (1982) *Can. J. Physiol. Pharmacol.* **60**, 1346-1352
76. Hammond, B., and Hess, M.L. (1985) *J. Am. Coll. Cardiol.* **6**, 215-220
77. Werns, S.W., Shea, M.J., and Lucchesi, B.R. (1985) *Free Radic. Biol. Med.* **1**, 103-110
78. Ferrari, R., Ceconi, C., Curello, S., Cargnoni, A., Agnoletti, G., Boffa, G.M., and Visioli, O. (1986) *Eur. Heart J.* **7A**, 3-12
79. Simpson, P.J., and Lucchesi, B.R. (1987) *J. Lab. Clin. Med.* **110**, 13-30
80. Becker, L.C., and Ambrosio, G. (1987) *Prog. Cardiovasc. Dis.* **30**, 23-44
81. Burton, K.P. (1988) *Free Radic. Biol. Med.* **4**, 15-24
82. Flaherty, J.T., and Weisfeldt, M.L. (1988) *Free Radic. Biol. Med.* **5**, 409-419
83. Bolli, R. (1988) *J. Am. Coll. Cardiol.* **12**, 239-249
84. Marklund, S.L. (1988) *J. Mol. Cell. Cardiol.* **20(Suppl. II)**, 23-30
85. Burrell, C.J., and Blake, D.R. (1989) *Br. Heart J.* **61**, 4-8
86. Hjalmarson, A. (1980) *Cardiology* **65**, 226-247
87. Hearse, D.J., Baimbridge, M.V., and Jynge, P. (1981) in *Protection of the Ischemic Myocardium: Cardioplegia* , pp. 21-49. Raven Press, New York
88. Spann, J.F. (1983) *Am. J. Med.* **74**, 877-886
89. Carmeliet, E. (1984) *Circulation* **70**, 149-151
90. Buckberg, G.D. (1986) *J. Thorac. Cardiovasc. Surg.* **92**, 483-487

91. Kupriyanov, V.V., Lakomkin, V.L., Steinschneider, A.Y., Severina, M.Y., Kapelko, V.I., Ruuge, E.K., and Saks, V.A. (1988) *J. Mol. Cell. Cardiol.* **20**, 1151-1162
92. Jennings, R.B., Schaper, J., Hill, M.L., Steenberger, C., and Reimer, K.A. (1985) *Circ. Res.* **56**, 262-278
93. Bolli, R., Davenport, N.J., Goldstein, R.E., and Epstein, S.E. (1983) *Cardiovasc. Res.* **17**, 274-281
94. Chiariello, M., Ambrosio, G., Cappelli-Bigazzi, M., Nevola, E., Perrone-Filardi, P., Marone, G., and Condorelli, M. (1987) *J. Pharmacol. Exp. Ther.* **241**, 560-568
95. Chien, K.R., Han, A., Sen, A., Buja, L.M., and Willerson, J.T. (1984) *Circ. Res.* **54**, 313-322
96. Liedtke, A.J. (1988) *J. Mol. Cell. Cardiol.* **20** (Suppl. II), 65-74
97. Janero, D.R., and Burghardt, C. (1988) *J. Cell. Physiol.* **137**, 1-13
98. Janero, D.R., and Burghardt, C. (1989) *J. Cell. Physiol.* **140**, 150-160
99. Van Belle, H., Goossens, F., and Wynants, J. (1987) *Am. J. Physiol.* **252**, H886-H893
100. Pisarenko, O., Studneva, I., Khlopkov, V., Solomatina, E., and Ruuge, E. (1988) *Biochim. Biophys. Acta* **934**, 55-63
101. Das, D.K., Engelman, R.M., Rousou, J.A., Breyer, R.H., Otani, H., and Lemeshow, S. (1986) *Am. J. Physiol.* **251**, H71-H79
102. Post, J.A., Ruigrok, T.J.C., and Verkeij, A.J. (1988) *J. Mol. Cell. Cardiol.* **20** (Suppl. II), 107-111
103. Kloner, R.A., Ganote, C.E., and Jennings, R.B. (1974) *J. Clin. Invest.* **54**, 1496-1508
104. Kloner, R.A., Przyklenk, K., and Patel, B. (1989) *Am. J. Med.* **86** (Suppl. 1A), 14-22
105. Braunwald, E., and Kloner, R.A. (1982) *Circulation* **66**, 1146-1149
106. Podzuweit, T., Binz, K.-H., Nennstiel, P., and Flaig, W. (1989) *Cardiovasc. Res.* **23**, 81-90
107. Braunwald, E. (1988) *J. Am. Coll. Cardiol.* **12**, 85A-92A
108. Wilcox, R.G., von der Lippe, G., Olsson, C.G., Jensen, G., Skene, A.M., and Hampton, J.R. (1988) *Lancet* **2**, 525-530
109. Braunwald, E., and Kloner, R.A. (1985) *J. Clin. Invest.* **76**, 1713-1719
110. Nayler, W.G., and Elz, J.S. (1988) *Circulation* **74**, 215-221
111. Hearse, D.J. (1977) *J. Mol. Cell. Cardiol.* **9**, 605-616
112. Califf, R.M. (1989) *J. Am. Coll. Cardiol.* **13**, 1130-1132
113. Sobel, B.E. (1974) *Circ. Res.* **34** (Suppl. III), 173-181
114. Allen, B.S., Rosenkranz, E.R., Buckberg, G.D., Vinten-Johansen, J., Okamoto, F., and Leaf, J. (1986) *J. Thorac. Cardiovasc. Surg.* **92**, 543-552
115. Hugenholtz, P.G. (1988) *J. Mol. Cell. Cardiol.* **20**, 367-369
116. Okamoto, F., Allen, B.S., Buckberg, G.D., Bugyi, H., and Leaf, J. (1986) *J. Thorac. Cardiovasc. Surg.* **92**, 613-620

117. Kramer, J.H., Arroyo, C.M., Dickens, B.F., and Weglicki, W.B. (1987) *Free Radic. Biol. Med.* **3**, 153-159
118. Bolli, R., Patel, B.S., Jeroudi, M.O., Lai, E.K., and McCay, P.B. (1988). *J. Clin. Invest.* **82**, 476-485
119. Arduini, A., Mezzetti, A., Porreca, E., Lapenna, D., DeJulia, J., Marzio, L., Polidoro, G., and Cuccurullo, F. (1988) *Biochim. Biophys. Acta* **970**, 113-121
120. Mak, I.T., Kramer, J.H., and Weglicki, W.B. (1986) *J. Biol. Chem.* **261**, 1153-1157
121. Hearse, D.J., and Tosaki, A. (1988) *J. Mol. Cell. Cardiol.* **20**, 213-223
122. Wolbarsht, M.L., and Fridovich, I. (1989) *Free Radic. Biol. Med.* **6**, 61-62
123. Kim, R.S., Sukhu, B., and LaBella, F.S. (1988) *Biochim. Biophys. Acta* **961**, 270-277
124. Singal, P.K., Beamish, R.E., and Dhalla, N.S. (1983) *Adv. Exp. Med. Biol.* **161**, 391-401
125. Jewett, S.L., Eddy, L.J., and Hochstein, P. (1989) *Free Radic. Biol. Med.* **6**, 185-188
126. Downey, J.M., Hearse, D.J. and Yellon, D.M. (1988) *J. Mol. Cell. Cardiol.* **20** (Suppl. II), 55-63
127. Mullane, K.M., Westlin, W., and Kraemer, R. (1988) *Ann. N.Y. Acad. Sci.* **524**, 103-121
128. Kukreja, R.C., Weaver, A.B., and Hess, M.L. (1989) *Biochim. Biophys. Acta.* **990**, 198-205
129. Downey, J.M., Miura, T., Eddy, L.J., Chambers, D.E., Mellert, T., Hearse, D.J., and Yellon, D.M. (1987) *J. Mol. Cell. Cardiol.* **19**, 1053-1060
130. Eddy, L.J., Stewart, J.M., Jones, H.P., Engerson, T.D., McCord, J.M., and Downey, J.M. (1987) *Am. J. Physiol.* **253**, H709-H711
131. Simpson, K., Kolce, J., Chudej, L., Christodoulides, N., Senn, G., and Bittar, N. (1988) *FASEB J* . **2**, A487
132. Mitsos, S.E., Askew, T.E., Fantone, J.C., Kunkel, S.L., Abrams, G.D., Schork, A., and Lucchesi, B.R. (1986) *Circulation* **73**, 1077-1086
133. Dale, H.H. (1933) *Bull. Johns Hopkins Hosp.* **53**, 297-347
134. Zweier, J.L., Rayburn, B.K., Flaherty, J.T., and Weisfeldt, M.L. (1987) *J. Clin. Invest.* **80**, 1728-1734
135. Zweier, J.L., Flaherty, J.T., and Weisfeldt, M.L. (1987) *Proc. Natl. Acad. Sci. USA* **84**, 1404-1407
136. Arroyo, C.M., Kramer, J.H., Dickens, B.F., and Weglicki, W.B. (1987) *FEBS Lett.* **221**, 101-104
137. Hearse, D.J., and Tosaki, A. (1987) *Circ. Res.* **60**, 375-383
138. Garlick, P.B., Davies, M.J., Hearse, D.J., and Slater, T.F. (1987) *Circ. Res.* **61**, 757-760
139. Zweier, J.L. (1988) *J. Biol. Chem.* **263**, 1353-1357
140. Baker, J.E., Felix, C.C., Olinger, G.N., and Kalyanaraman, B. (1988) *Proc. Natl. Acad. Sci. USA* **85**, 2786-2789

141. Zweier, J.L., and Kuppusamy, P. (1988) *Proc. Natl. Acad. Sci. USA* **85**, 5703-5707
142. Barsacchi, R., Coassin, M., Maiorino, M., Peolsi, G., Simonelli, C., and Ursini, F. (1989) *Free Radic. Biol. Med.* **6**, 573-579
143. Burton, K.P., McCord, J.M., and Ghai, G. (1984) *Am. J. Physiol.* **246**, H776-H873
144. Blaustein, A.S., Schine, L., Brooks, W.W., Fanburg, B.L., and Bing, O.H.L. (1986) *Am. J. Physiol.* **250**, H595-H599
145. Ytrehus, K., Myklebust, R., and Mjos, O.D. (1986) *Cardiovasc. Res.* **20**, 597-603
146. Rao, P.S., Cohen, M.V., and Mueller, H.S. (1983) *J. Mol. Cell. Cardiol.* **15**, 713-716
147. Roth, E., Torok, B., Zsoldos, T., and Matkovics, B. (1985) *Basic Res. Cardiol.* **80**, 530-536
148. Ferrari, R., Ceconi, C., Curello, S., Guarnieri, C., Caldarera, C.M., Albertini, A., and Visioli, O. (1985) *J. Mol. Cell. Cardiol.* **17**, 937-945
149. Lesnefsky, E.J., VanBenthuysen, K., McMurtry, I., Fennessey, D.M., Travis, V.L., and Horwitz, L.D. (1986) *Circulation* **74** (Suppl. II), 347
150. Romaschin, A.D., Rebeyka, I., Wilson, G.J., and Mickle, D.A.G. (1987) *J. Mol. Cell. Cardiol.* **19**, 289-302
151. Herbaczynska-Cedro, K., and Gordon-Majszak, W. (1989) *Cardiovasc. Res.* **23**, 98-103
152. Lesnefsky, E.J., VanBenthuysen, K.M., McMurtry, I., Shikes, R.H., Johnston, R.B., and Horwitz, L.D. (1989) *J. Cardiovasc. Pharmacol.* **13**, 895-901
153. Przyklenk, K., and Kloner, R.A. (1989) *Circ. Res.* **64**, 86-96
154. Chi, L., Tamura, Y., Hoff, P.T., Macha, M., Gallagher, K.P., Schork, M.A., and Lucchesi, B. (1989) *Circ. Res.* **64**, 665-675
155. Kuzuya, T., Hoshida, S., Nishida, M., Kim, Y., Fuji, H., Kitabatake, A., Kamada, T., and Tada, M. (1989) *Cardiovasc. Res.* **23**, 323-330
156. Bernier, M., Manning, A.S., and Hearse, D.J. (1989) *Am. J. Physiol.* **256**, H1344-H1352
157. Nejima, J., Knight, D.R., Fallon, J.T., Vemura, N., Manders, W.T., Canfield, D.R., Cohen, M.V., and Vatner, S.F. (1989) *Circulation* **79**, 143-153
158. Van der Kraaij, A.M.M., van Eijk, H.G., and Koster, J.F. (1989) *Circulation* **80**, 158-164
159. Reddy, B.R., Kloner, R.A., and Przyklenk, K. (1989) *Free Radic. Biol. Med.* **7**, 45-52
160. Cavarocchi, N.C., England, M.D., O'Brien, J.F., Solis, E., Russo, P., Schaff, H.V., Orszulak, T.A., Pluth, J.R., and Kaye, M.P. (1986) *J. Surg. Res.* **40**, 519-527
161. Chatterjee, S.N., and Agarwal, S. (1988) *Free Radic. Biol. Med.* **4**, 51-72
162. Janero, D.R., and Burghardt, B. (1988) *Lipids* **23**, 452-458
163. Janero, D.R., and Burghardt, B. (1988) *Biochem. Pharmacol.* **37**, 3335-3342

164. Janero, D.R., Burghardt, B., and Lopez, R. (1988) *Biochem. Pharmacol.* **37**, 4197-4203
165. Janero, D.R., and Burghardt, B. (1989) *Res. Commun. Chem. Pathol. Pharmacol.* **63**, 163-173
166. Janero, D.R., Lopez, R., Pittman, J., and Burghardt, B. (1989) *Life Sci.* **44**, 1579-1588
167. Janero, D.R., Burghardt, B., Lopez, R., and Cardell, M. (1989) *Biochem. Pharmacol.* **38**, 4381-4387
168. Inui, M., Wang, S., Sarto, A., and Fleischer, S. (1988) *J. Biol. Chem.* **263**, 10843-10850
169. Weglicki, W.B., Owens, K., Kennett, F.F., Kessner, A., Harris, L., Wise, R.M., and Vahouny, G.V. (1980) *J. Biol. Chem.* **255**, 3605-3609
170. Kako, K.J. (1987) *Med. Res. Rev.* **7**, 507-522
171. Janero, D.R., and Burghardt, B. (1989) *J. Mol. Cell. Cardiol.* **21**, 1111-1124
172. Janero, D.R., and Burghardt, B. (1989) *Lipids* **24**, 33-38
173. Barchi, R.L., Bonilla, E., and Wong, M. (1977) *Proc. Natl. Acad. Sci. USA* **74**, 34-38
174. Severs, N.J., Slade, A.M., Powel, T., Twist, V.W., and Warren, R.L. (1982) *J. Ultrastruct. Res.* **81**, 222-239
175. Ross, P.D., and McCarl, R.L. (1984) *Am. J. Physiol.* **246**, H389-H397
176. Hohl, C., Oestreich, R., Rosen, P., Wiesner, R., and Grieshaber, M. (1987) *Arch. Biochem. Biophys.* **259**, 527-535
177. Scott, J.A., Fischman, A.J., Khaw, B.-A., Homcy, C.J. , and Rabito, C.A. (1987) *Biochim. Biophys. Acta.* **899**, 76-82
178. Julicher, R.H.M., van der Laarse, A., Sterrenberg, L., van Treslong, C.H.F.B., Bast, A., and Noordhoek, J. (1985) *Res. Commun. Chem. Pathol. Pharmacol.* **47**, 35-47
179. Donck, L.V., Van Reempts, J., Vandeplassche, G., and Bogers, M. (1988) *J. Mol. Cell. Cardiol.* **20**, 811-823
180. Van der Laarse, A., Julicher, R.H.M., Sterrenberg, L., and van Treslong, C.H.F.B. (1986) *Cardiovasc. Res.* **20**, 20-25
181. Lewko, W.M. (1987) *Biochem. Pharmacol.* **36**, 219-223
182. Scott, J.A., Kaw, B.A., Locke, E., Haber, E., and Homcy, C. (1985) *Circ. Res.* **56**, 72-77
183. Piper, H.M., Probst, I., Schwartz, P., Hutter, J.F., and Spieckermann, P.G. (1982) *J. Mol. Cell. Cardiol.* **14**, 397-412
184. Sen, A., Miller, J.C., Reynolds, R., Willerson, J.T., Buja, L.M., and Chien, K.R. (1988) *J. Clin. Invest.* **82**, 1333-1338
185. Fraga, C.G., Leibovitz, B.E., and Tappel, A.L. (1988) *Free Radic. Biol. Med.* **4**, 155-161
186. Wood, M.A., and Hess, M.L. (1987) *Mol. Cell. Biochem.* **78**, 161-167
187. Nakazawa, H., Ban, K., Ichimori, K., Minezaki, K., Okino, H., Masuda, T., Aoki, N., and Hori, S. (1988) *Jpn. Circ. J.* **52**, 646-654

188. Zucchi, R., Poddighe, R., Limbruno, U., Mariani, M., Ronca-Testoni, S., and Ronca, G. (1989) *J. Mol. Cell. Cardiol.* **21**, 67-73
189. Barnard, R.J., Okamoto, F., Buckberg, G.D., Sjostrand, F., Rosenkranz, E.R., Vinten-Johansen, J., Allen, B.S., and Leaf, J. (1986) *J. Thorac. Cardiovasc. Surg.* **92**, 502-512
190. Smith, H.J. (1986) *J. Mol. Cell. Cardiol.* **18** (Suppl. 4), 105-107
191. Mahmarian, J.T., Pratt, C.M., Borges-Neto, S., Cashion, W.R., Roberts, R., and Verani, M.S. (1988) *Circulation* **78**, 831-839
192. Schaefer, S., Gober, J., Valenza, M., Karczman, G.S., Matson, G.B., Camacho, S.A., Botvinick, E.H., Massie, B., and Weiner, M.W. (1988) *J. Am. Coll. Cardiol.* **12**, 1449-1455
193. Conti, C.R., Feldman, R.L., and Pepine, C.J. (1982) *Am. Heart J.* **103**, 584-588
194. Borowitz, S.M., and Montgomery, C. (1989) *Biochem. Biophys. Res. Commun.* **158**, 1021-1028
195. Richard, V.J., Murry, C.E., Jennings, R.B., and Reimer, K.A. (1988) *Circulation* **78**, 473-480
196. Shibanuma, M., Kuroki, T., and Nose, K. (1988) *J. Cell. Physiol.* **136**, 379-383

Mitochondrial Generation of Hydroxyl Radical During Reperfusion of Ischemic Myocardium

Dipak K. Das, Richard M. Engelman,
Xuekun Liu and Anna George
Cardiovascular Division, Department of Surgery, University of Connecticut School of Medicine, Farmington, Connecticut 06032 USA

Abstract

Although the presence of free radicals has been indicated in ischemic-reperfused heart, the exact nature and source of these free radicals are not known. The present study utilized a chemical trap, salicylic acid, to trap hydroxyl radical which could be detected as hydroxylated benzoic acid using high pressure liquid chromatography. Since the hydroxylated product is extremely stable, heart was subjected to subcellular fractionation after ischemia and reperfusion, and each fraction was separately examined for the presence of hydroxyl radical. When each of the subcellular fractions was examined for the presence of •OH signal, a small peak appeared in the mitochondrial fraction, even in the preischemic heart. However, the intensity of this peak remained unaltered when the heart was treated with DMSO, or DMTU, or SOD plus catalase, suggesting that the peak was not related to •OH. The same peak also appeared in the ischemic tissue samples. In contrast, a larger peak emerged for •OH in the mitochondria from 1 min reperfused heart. This signal was scavenged completely by DMSO, DMTU, or SOD plus catalase, suggesting that the signal indeed originated from •OH.

To further confirm the identity of the •OH signals found in the mitochondrial fraction of the reperfused heart, we produced pure •OH radical from the reaction of hypoxanthine (100 μM), xanthine oxidase (8 mU), $FeCl_3$ (100 μM), and EDTA (100 μM). When these reactions were carried out in the presence of salicylate (2 mM) and injected onto HPLC, a peak for •OH emerged at the identical position. Since •OH reacts with salicylic acid and produces 2,3- and 2,5-dihydroxybenzoic acids, we also analyzed the authentic standards of these two hydroxylated benzoic acids. These two standards had identical retention times to those obtained from reperfused heart. These results indicated for the first time the presence of •OH in the

Biological Oxidation Systems Volume 2

mitochondrial fraction during early reperfusion, which decreased in intensity as the reperfusion progressed.

Introduction

Reperfusion of ischemic myocardium is associated with a number of undesirable, but unavoidable, events such as transient rise in resting tension [1], reduction in contractile functions [2], massive influx of calcium ions [3,4], release of several intracellular enzymes [5,6], degradation of membrane phospholipids [7,8], and generation of oxygen-derived free radicals [9-11]. These interrelated factors, either alone or in combination, are responsible for the amplification of cellular injury which is developed during ischemic insult and give rise to the phenomenon of "reperfusion injury." Among the aforementioned factors, generation of free radicals is thought to play a major role in the development of reperfusion injury [12-14].

Although the deleterious effects of reperfusion of ischemic myocardium have been linked to the production of oxygen-derived free radicals, the exact nature and source of these free radicals remain unknown. A number of potential sources have been described which include xanthine oxidase [15], polymorphonuclear leukocytes (PMN) [16], oxygen-carrying proteins [17], phospholipid degradation pathway [18], catecholamines [19], and mitochondria [20]. The fact that xanthine oxidase-xanthine reaction produces superoxide anion (O_2^-) [21] and allopurinol, a potent inhibitor of xanthine oxidase, reduces infarct size in post-ischemic heart [22,23] has led many investigators to believe that xanthine oxidase is a potential source of free radicals in the ischemic-reperfused heart. Recent studies, however, did not support the idea, since (with the exception of rat heart) hearts of many animals, including man, do not contain any xanthine oxidase activity [24,25]. The beneficial effects of allopurinol have been ascribed to be due to the ability of this drug to directly scavenge hydroxyl radical (•OH) and hypohalite radicals (OCl•) [25,26]. Activated PMN which generates O_2^- during respiratory burst represents another potential source of free radicals [27]. Studies supporting the role of PMN in reperfusion injury are mostly indirect, *e.g.*, a nonsteroidal anti-inflammatory agent, ibuprofen, has been found to inhibit the PMN influx into infarcted tissue, with a corresponding reduction of ultimate infarct size [28]. Similarly, PMN depletion was associated with a significant reduction in ultimate infarct size in experimental animals [29]. Other studies, however, showed neither a decrease in infarct size in leukopenic dogs after 21 h of reperfusion [29], nor did they find any reduction in infarct size and functional recovery after ibuprofen treatment [30]. In addition, the fact that reperfusion injury occurs even when ischemic heart is reperfused with blood-free system (buffer) [31] makes the hypothesis of PMN being a source of free radical questionable.

Recently, evidences have been presented to indicate that free radicals generated from the mitochondrial respiratory chain may be a potential source of free radicals associated with reperfusion of ischemic myocardium [32-34]. Mitochondria are likely to be most susceptible to ischemic injury among the cellular components, and it appears that reperfusion has some deleterious effects on mitochondrial functions [35]. It has also been known for a long time that mitochondrial respiration, when inhibited, generates a significant amount of activated O_2 consisting of O_2^-, •OH, hydrogen peroxide (H_2O_2), and singlet oxygen ($^1O_2$) [36].

There is no direct evidence to support the hypothesis of mitochondria being a source of free radicals in the ischemic-reperfused heart. The major problem is that free radicals are extremely short-lived and, even when they are spin-trapped and examined by an electron paramagnetic resonance spectrometer (EPR), they must be handled in a very special manner, viz, they must be instantly frozen in liquid N_2 and processed at liquid N_2 temperature within a very short period of time [37,38]. To overcome this problem, we have used a stable chemical trap, salicylic acid, which reacts with •OH to form hydroxylated benzoic acid. Since this product is extremely stable, subcellular organelles could be isolated and examined individually. Using this technique, we are able to directly demonstrate the presence of •OH in the mitochondria of ischemic-reperfused heart.

Materials and Methods

Preparation of Heart for Ischemia and Reperfusion

Male Sprague-Dawley rats weighing 250-350 g were anesthetized with an intravenous injection of sodium pentobarbital. The hearts were removed and perfused with Krebs-Henseleit bicarbonate buffer equilibrated with a gas mixture of 95 percent O_2 and 5 percent CO_2 (pH 7.4), using Langendorff's non-recirculating technique [39]. The heart was perfused for 15 minutes in the presence of 2 mM salicylate at 100 g H_2O perfusion pressure and at 37°C for stabilization. Flow rate was maintained at 15 ml/min. Retrograde aortic flow was then reduced to about 0.1 ml/min, and the heart was made ischemic for 30 min. Reperfusion was performed for 30 min, following 30 min ischemia with nonrecirculating Krebs-Henseleit bicarbonate buffer containing 2 mM salicylate.

To examine whether the generated radical was indeed •OH, parallel experiments were performed by simultaneously adding specific scavengers of •OH, either DMSO or DMTU along with salicylate. The experiments were terminated at different points: prior to ischemia, after 30 min of ischemia, and after 1, 2, 5, 15, and 30 min of reperfusion. At least ten experiments were performed for each time frame. Perfusates and hearts were processed as described below.

Subcellular Fractionation of Heart

Hearts were homogenized in Krebs-Henseleit bicarbonate buffer using a Polytron homogenizer (Brinkman, NY). The homogenate was centrifuged at 100 g to remove nuclear debris. The homogenate was then centrifuged at 10,000 g to collect mitochondria. Mitochondria-free homogenate was further centrifuged at 100,000 g to precipitate microsomal fraction. Mitochondria (500 g - 6,500 g) and microsomes (10,000 g - 100,000 g) were further purified by repeated precipitation and washing as described elsewhere [40]. Marker enzymes were assayed to judge the purity of the preparation. Nuclear fraction was purified by sucrose density gradient as described previously [40].

Analysis of •OH by HPLC

Each subcellular fraction was suspended in 1 ml tris-sucrose buffer and 800 μl treated with 40 μl 70% $HClO_4$. The precipitate was removed by centrifugation, and the supernatant was filtered through Rainin Nylon-66 membrane filter (0.45 μ pore size). Twenty μl of the sample was injected onto an Altex ultrasphere 3 μ ODS (75 mm x 4.6 mm) equipped in a Water Associates HPLC unit consisting of a Model 510 pump and Model 460 electrochemical detector. 2,5-and 2,3-dihydroxybenzoic acid (hydroxylated product of salicylic acid and •OH) were eluted by buffer containing 0.03 M sodium acetate and 0.05 M sodium citrate (pH 4.5) at a flow rate of 1.0 ml/min [41]. The detection potential was maintained at 0.6 V, employing Ag/AgCl reference electrode. Peaks were identified by authentic standards as well as by injection of pure •OH generated chemically. •OH peak emerged at the same position as the mitochondrial samples.

Results

Generation of •OH During Reperfusion of Ischemic Heart

When each of the subcellular fractions was examined for the presence of •OH signal, a small peak appeared in the mitochondrial fraction, even in the preischemic heart (Fig. 1). However, the intensity of this peak remained unaltered when the heart was treated with DMSO, or DMTU, or SOD plus catalase, suggesting that the peak was not related to •OH. The same peak also appeared in the

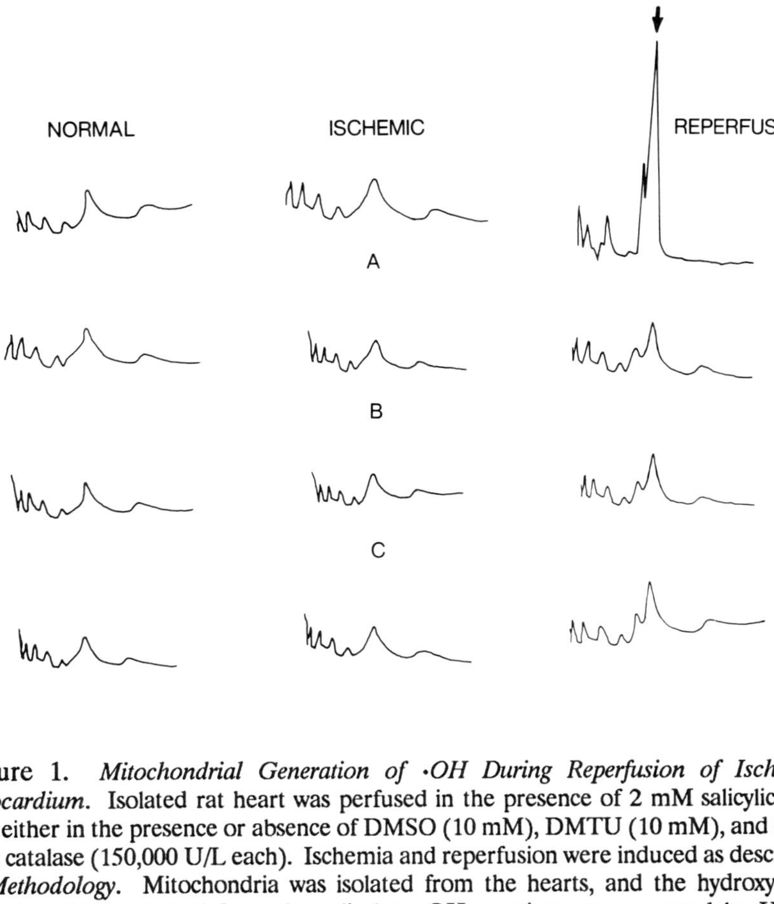

Figure 1. *Mitochondrial Generation of •OH During Reperfusion of Ischemic Myocardium.* Isolated rat heart was perfused in the presence of 2 mM salicylic acid and either in the presence or absence of DMSO (10 mM), DMTU (10 mM), and SOD plus catalase (150,000 U/L each). Ischemia and reperfusion were induced as described in *Methodology*. Mitochondria was isolated from the hearts, and the hydroxylated benzoic acids generated from the salicylate-•OH reactions were assayed by HPLC using an electrochemical detector. The detection potential was maintained at 0.6 V employing Ag/AgCl reference electrode.
A. Control; **B.** 10 mM DMSO; **C.** 10 mM DMTU; **D.** 150,000 U/L SOD + catalase.

ischemic tissue samples. In contrast, a larger peak emerged for •OH in the mitochondria from 1 min reperfused heart. This signal was scavenged completely by DMSO, DMTU, or SOD plus catalase, suggesting that the signal indeed originated from •OH.

Analysis of •OH-Salicylate Reaction Products by HPLC

To further confirm the identity of the •OH signals found in the mitochondrial fraction of the reperfused heart, we produced pure •OH radical from the reaction of hypoxanthine (100 μM), xanthine oxidase (8 mU), $FeCl_3$ (100 μM), and EDTA (100 μM). When these reactions were carried out in the presence of salicylate (2 mM) and injected onto HPLC, a peak for •OH emerged at the identical position (Fig. 2). Since •OH reacts with salicylic acid and produces 2,3- and 2,5-dihydroxybenzoic acids, we also analyzed the authentic standards of these two hydroxylated benzoic acids. As shown in Figure 2, these two standards also had identical retention times.

Generation of •OH as a Function of Duration of Reperfusion

Since oxygen free radicals are known to appear at the onset of reperfusion, we examined the presence of •OH as a function of reperfusion time. As expected, •OH was reduced as reperfusion progressed (Table 1). A slight reduction of •OH was noticed after 2 min of reperfusion. The reduction of •OH signals was significant after 15 min and 30 min of reperfusion, confirming the earlier observations that oxygen free radicals are generated during the early reperfusion phase.

Table 1. *Mitochondrial Generation of •OH as a Function of Duration of Reperfusion*

	Duration of Reperfusion (min)				
	1	2	5	15	30
	•OH Peak Response (mm)				
Control	60 ± 7	65 ± 9	21. ± 7	11. ± 4	9. ± 3
+ DMSO (10 mM)	10 ± 6	14 ± 8	1.2 ± 1.0	0	0
+ DMTU (10 mM)	16 ± 7	17 ± 5	2.1 ± 1.5	0.5 ± 0.3	0.6 ± 0.2
+ SOD + Catalase (150,000 U/L each)	18 ± 4	19 ± 8	2.5 ± 1.2	0.2 ± 0.1	0.3 ± 0.1

Presence of •OH signal in the cytosolic fraction

We also examined other subcellular fractions for the presence of •OH. Nuclear fractions and microsomes did not exhibit any signal for •OH when examined by HPLC. However, cell-free supernatant fraction showed a small •OH signal in the reperfused heart (Fig. 3). The intensity of the signal was relatively small compared to the mitochondrial signal (only 5% as compared to mitochondria), but this peak was scavenged by DMSO, DMTU, or SOD plus catalase.

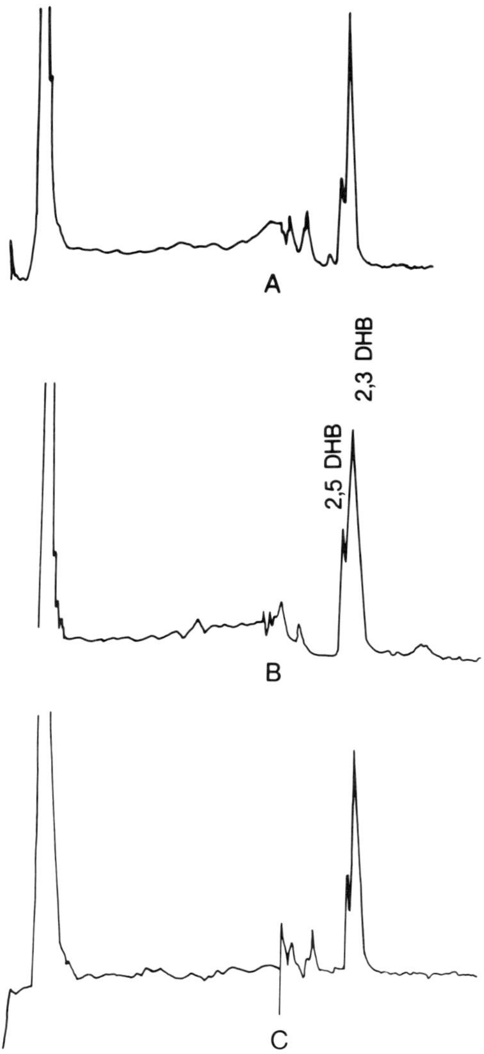

Figure 2. *Analysis of •OH-Salicylate Reaction Products by HPLC.* **A.** •OH was generated chemically by the reaction of hypoxanthine, and allowed to react with salicylate, XO, $FeCl_3$, and EDTA; **B.** A mixture of authentic standards of 2,3-DHB and 2,5-DHB; **C.** •OH-salicylate product extracted from reperfused heart as described in the legends of Figure 1.

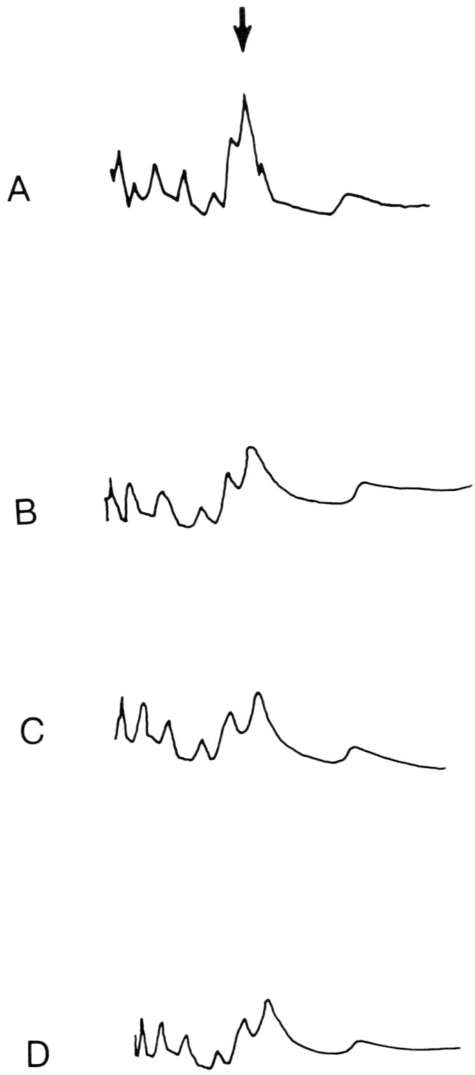

Figure 3. *Cytosolic Generation of ·OH During Reperfusion of Ischemic Myocardium.* Isolated rat heart was perfused in the presence of 2 mM salicylic acid and either in the presence or absence of DMSO (10 mM), DMTU (10 mM), and SOD plus catalase (150,000 U/L each). Ischemia and reperfusion were induced as described in *Methodology*. Cytosol was separated from the hearts, and the hydroxylated benzoic acids generated from the salicylate-·OH reactions were assayed by HPLC using an electrochemical detector. The detection potential was maintained at 0.6 V employing Ag/AgCl reference electrode. **A.** Control; **B.** DMSO; **C.** DMTU; **D.** SOD + catalase.

Discussion

The results reported in this study demonstrated for the first time a direct evidence for the presence of •OH in the mitochondria of ischemic-reperfused heart. Although previous studies suggested mitochondria could be a potential source of free radicals, the evidence was only indirect and was based on *in vitro* studies. The present study utilized a physiologic model using isolated perfusion heart preparations.

A large number of reducing equivalents are accumulated in the mitochondrial respiratory chain during ischemia. As mentioned earlier, respiring mitochondria release a variety of reactive oxygen species including O_2^-, •OH, H_2O_2, and $^1O_2$. Under normal conditions, the primary oxygen free radical O_2^- is effectively removed by the SOD present in the tissue [42]. However, ischemia reduces the activity of SOD, preventing its availability for inactivating the O_2^- during reperfusion [9]. O_2^- production by mitochondria is thought to occur at the NADH dehydrogenase step, near to the ubiquinone component [43]. Additional research supporting mitochondrial generation of free radicals has appeared in these studies, which has documented a reduction in coenzyme Q_{10} values after an ischemic insult, with concomitant increases in the •OH signal. Supporting this is a recent study of coenzyme Q_{10}-treated animals showing a reduction of NADH-diaphorase activity with simultaneous reduction of lipid peroxidation [39]. Finally, electron paramagnetic resonance spectroscopy has been utilized to directly analyze free radicals. Using this modality, Baker and associates have demonstrated that the locus of ischemic reperfusion injury may well involve the mitochondrial respiratory chain [10]. Our study confirms this hypothesis and further identifies the free radical as •OH.

A substantial body of evidence now suggests that O_2^- produced initially from molecular O_2 undergoes dismutation, forming H_2O_2 as indicated below:

$$O_2 + e^- \rightarrow O_2^- \qquad \text{(i)}$$

$$2O_2^- + 2H^+ \rightarrow O_2 + H_2O_2 \qquad \text{(ii)}$$

The O_2^- was thought to undergo a Haber-Weiss reaction in the presence of H_2O_2, forming •OH:

$$O_2^- + H_2O_2 \rightarrow O_2 + OH^- + \bullet OH \qquad \text{(iii)}$$

Although this reaction is thermodynamically feasible, and the overall stoichiometry is widely accepted under *in vitro* conditions, the reaction is kinetically very slow or even negligible. The modified reaction has been suggested to occur in two steps by iron-catalyzed Fenton-type reactions, as described below:

$$Fe^{+++} + O_2^{-} + H_2O_2 \rightarrow O_2 + Fe^{++} \qquad \text{(iv)}$$

$$Fe^{++} + H_2O_2 \rightarrow Fe^{+++} + OH^{-} + \cdot OH \qquad \text{(v)}$$

$$O_2^{-} + H_2O_2 \rightarrow O_2 + OH^{-} + \cdot OH \qquad \text{(iv) + (v)}$$

With present technology, it is virtually impossible to identify O_2^- in an *in vivo* system, even if it may be produced in mitochondria. O_2^- has an extremely short half-life which rapidly undergoes dismutation, yielding H_2O_2. O_2^- then undergoes a Fenton-type reaction catalyzed by as transient a metal such as iron. The exact source of iron to catalyze this reaction remains highly speculative [44]. A recent study from our laboratory suggested oxygen-carrying heme proteins as a potential source of iron [17].

•OH is extremely reactive and virtually reacts with a number of compounds, including lipids and proteins. Breakdown of phospholipids and its peroxidation has been indicated in ischemic-reperfused myocardium [7]. This ultimately leads to myocardial dysfunction causing reperfusion injury. In this regard, the presence of •OH in the mitochondria should be an important facet to understanding the pathophysiology of myocardial reperfusion injury.

Summary

Although the presence of free radicals has been indicated in ischemic-reperfused heart, the exact nature and source of these free radicals are not known. The present study utilized a chemical trap, salicylic acid, to trap hydroxyl radical which could be detected as hydroxylated benzoic acid using high pressure liquid chromatography. Since the hydroxylated product is extremely stable, heart was subjected to subcellular fractionation after ischemia and reperfusion, and each fraction was separately examined for the presence of hydroxyl radical. The results indicated for the first time the presence of hydroxyl radical in the mitochondrial fraction during early reperfusion, which decreased in intensity as the reperfusion progressed.

Acknowledgements

The study was supported by USPHS grants NIH HL22559, HL34360, and HL33889. The authors gratefully acknowledge the secretarial assistance of Mrs. Joanne D'Aprile.

References

1. Jennings, R.B., Reimer, K.A. and Steenbergen, C. (1986) *J. Mol. Cell Cardiol.* **18**, 769-780
2. Karmazyn, M. (1986) *Am. J. Physiol.* **251**, H133-H140
3. Harding, D.P. and Poole-Wilson, P.A. (1980) *Cardiovasc. Res.* **14**, 435-445
4. Das, D.K., Engelman, R.M., Prasad, M.R., Rousou, S.A., Breyer, R.H., Jones, R., Young, H. and Cordis, G.A. (1989) *Biochem. Pharmacol.* **38**, 465-471
5. Ganote, C.E. and Kaltenbach, J.P. (1979) *J. Mol. Cell. Cardiol.* **11**, 389-406
6. Hearse, D.J. and Humphrey, S. (1975) *J. Mol. Cell. Cardiol.* **7**, 463-482
7. Das, D.K., Engelman, R.M., Rousou, J.A., Breyer, R.H. and Otani, H. (1986) *Am. J. Physiol.* **251**, H71-H79
8. Otani, H. (1988) *Circ. Res.* **63**, 930-936
9. Das, D.K., Engelman, R.M., Rousou, J.A., Breyer, R.H., Otani, H. and Lemeshow, S. (1986) *Basic. Res. Cardiol.* **81**, 155-166
10. Baker, J.E., Felix, C.C., Olinger, G.N. and Kalyanaraman, B. (1988) *Proc. Natl. Acad. Sci. (USA)*. **85**, 2786-2789
11. Arroyo, C.M., Kramer, J.H., Dickens, B.F. and Weglicki, W.B. (1987) *FEBS Lett.* **221**, 101-104
12. Granger, D.N., Höllwarth, M.E. and Parks, D.A. (1986) *Acta. Physiol. Scand.* **548** (suppl), 47-63
13. Jolly, S.R., Kane, W.J., Bailie, M.B., Abrams, G.D. and Lucchesi, B.R. (1984) *Circ. Res.* **54**, 277-285
14. Das, D.K. and Engelman, R.M. (1989) in *Oxygen Radicals, Systemic Events and Disease Processes* (Das D.K. and Essman W.B., eds.) Krager, Basel, pp 97-128
15. McCord, J.M. (1985) *N. Engl. J. Med.* **312**, 159-162
16. Romson, J.L., Hook, B.G., Rijot, V.H., Schork, M.A., Swanson, D.P. and Lucchesi, B.R. (1982) *Circulation* **66**, 1002-1011
17. Prasad, M.R., Engelman, R.M., Jones, R.M. and Das, D.K., *Biochem. J.* (in press)
18. Otani, H., Engelman, R.M., Rousou, J.A., Breyer, R.H. and Das, D.K. (1986) *J. Mol. Cell. Cardiol.* **18**, 953-961
19. Rao, P.S., Cohen, M.V. and Mueller, H.A. (1983) *J. Mol. Cell. Cardiol.* **15**, 713-716
20. Otani, H., Prasad, M.R., Jones, R.M. and Das, D.K. (1984) *Am. J. Physiol.* **257**, H252-H258

21. Chambers, D.K., Parks, D.A., Patterson, G., Roy, R., McCord, J.M., Yoshida, S., Parmley, L.F. and Downey, J.M. (1985) *J. Mol. Cell. Cardiol.* **17**, 145-152
22. Werns, S.W., Shea, M.J., Mitsos, S.E., Dysko, R.C., Fantone, J.C., Schork, M.A., Abrams, G.D., Bertram, P.H. and Lucchesi, B.R.. *Circulation* **73**, 518-524
23. McCord, J.M. and Roy, R.S. (1982) *Can. J. Physiol. Pharmacol.* **60**, 1346-1352
24. Das, D.K., Engelman, R.M., Clement, R., Otani, H., Prasad, M.R. and Rao, P.S. (1987) *Biochem. Biophys. Res. Commun.* **148**, 314-319
25. Downey, J.M., Hearse, D.J. and Yellon, D.M. (1988) *J. Mol. Cell. Cardiol.* **20** (suppl II), 55-63
26. Hoey, B.M., Butler, J. and Halliwell, B. (1988). *Free. Rad. Res. Commun.* **4**, 259-263
27. Werns, S.W. and Lucchesi, B.R. (1988) *Free. Rad. Biol. Med.* **4**, 31-37
28. Romson, J.L., Hook, B.G., Kunkel, S.L., Abrams, G.D., Schork, M.A. and Lucchesi, B.R. (1983) *Circulation* **67**, 1016-1023
29. Chatelain, P., Latour, J.G., Tran, D., DeLorgoni, M., Dupras, G. and Bourassa, M. (1987) *Circulation* **75**, 1083-1090
30. Reimer, K.A., Jennings, R.B., Cobb FR, Murdoch, R.H., Greenfield, J.C., Becker, L.C., Buckley, B.H., Hutchins, G.H., Schwartz, R.P., Bailey, K.R. and Passamani, E.R. (1985) *Circ. Res.* **56**, 651-665
31. Poole-Wilson, P.A. (1987) *Clin. Physiol.* **7**, 439-453
32. Genko, S., Sugiyam, M., Shoji, T., Kagiyama, A., Sato, H. and Ogura, R. (1989) *J. Mol. Cell. Cardiol.* **21**, 1029-1036
33. Guarnieri, C., Muscari, C., Ventura, C. and Marelli, I. (1985) *Free. Rad. Res. Commun.* **1**, 123-128
34. Nohl, H. (1987) *FEBS Lett.* **214**, 269-273
35. Kane, J.L., Murphy, M.L., Bissett, J.K., DeSoyza, N., Doherty, J.E. and Straub, K,.D. (1975) *Am. J. Cardiol.* **36**, 218-224
36. Sugiyama, S., Ozawa, T., Kato, T. and Shohachi, S. (1980) *Am. Heart J.* **100**, 829-837
37. Rosen, G.M. and Rauckman, E.J. (1984) in *Methods in Enzymology* (Packer, L., ed), Academic Press, NY, **105**, 198-209
38. Bolli, R., Patel, B,.S., Jasodi, M.O., Lai, E.K. and McCay, P.B. (1988) *J. Clin. Invest.* **82**, 476-485
39. Otani, H., Tanaka, H., Inoue, T., Umemoto, M., Omoto, K., Tanaka, K., Sato, T., Osako, T., Masuda, A., Nonoyama, A. and Kagawa, T. (1984) *Circ. Res.* **55**, 168-175
40. Das, D.K. and Steinberg, H. (1985) *Am. J. Physiol.* **248**, E58-E63
41. Grootveld, M. and Halliwell, B. (1986) *Biochem. J.* **237**, 499-504
42. Babior, B.M. (1978) *N. Engl. J. Med.* **198**, 659-688
43. Cadenas, E., Boveris, A., Ragan, C.I., Stopani, A.O.M. (1977) *Arch. Biochem. Biophys.* **180**, 248-257
44. Halliwell, B. and Gutteridge, J.M.C. (1984) *Biochem. J.* **219**, 1-14

Role of Lipid Peroxides in Aging

Kunio Yagi and Sadaaki Komura
Institute of Applied Biochemistry, Yagi Memorial Park, Mitake, Gifu 505-01 Japan

Abstract

Increases in lipid peroxide levels in the serum, liver, and skin of the male senescence-accelerated mouse (SAM-P/1) were found to occur prior to the appearance of age-related clinical signs, suggesting that lipid peroxides increased in the early period of life may initiate or promote the aging process.

A simple and reliable method [1] to determine the lipid peroxide level in human serum made it possible to demonstrate that the serum lipid peroxide level in humans increases with age [2,3]. In addition, an increase in the level is thought to be causative of degenerative diseases such as atherosclerosis [4-6] and cataract [7], both of which are recognized as age-related diseases. From these facts, we proposed that an increase in the lipid peroxide level in the blood may be one of the direct causes of aging itself. However, little experimental evidence is available for assessment of the possible relationship between lipid peroxides in the blood and the aging process. For investigation of this problem, animal models of aging would be useful.

The senescence-accelerated mouse (SAM), developed by Takeda *et al.* [8], is a murine model of aging. The life span of accelerated senescence-prone (SAM-P/1) mice is only about two-thirds of that of accelerated senescence-resistant (SAM-R/1) mice. SAM-P/1 mice are characterized by senescence-associated clinical signs that are observable with advancing age; these signs include loss of passivity and reactivity, loss of glossiness of the hair, increased coarseness of the skin, hair loss, periophthalmic lesions, increased lordokyphosis, and systemic amyloidosis [8-10]. Hosokawa *et al.* [11] devised a grading score system to evaluate the degree of senescence with respect to the symptoms described above and confirmed that a steady and irreversible increase in the score was seen with advancing age.

Using these animals, we studied chronological changes in lipid peroxide levels in the serum, liver, and skin of SAM in relation to the appearance of the age-related symptoms [12,13]. Our data suggest that an increase in the lipid peroxide level is not the result of the progress of aging, but possibly the cause of it.

Changes in Grading Score of SAM with Age

The mean values of grading scores of SAM-P/1 and SAM-R/1 mice are shown in Figure 1. Senescence-associated scores of mice of both strains began to rise slightly at 3 months of age. From that age, the scores of SAM-P/1 mice continued to rise until 10 months of age, reaching the value of 4.4, while those of SAM-R/1 mice were always below 0.4 at all ages examined in this study.

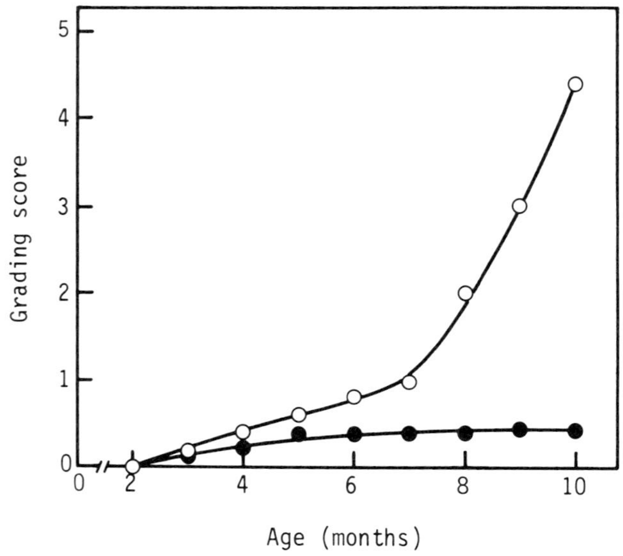

Figure 1. *Changes in Grading Score in Male SAM with Age.* Grading score of senescence in SAM was measured by the following criteria according to Hosokawa *et al.* [11]: A, skin and hair (glossiness, coarseness, and skin ulcer); B, eyes (periophthalmic lesions, corneal opacity, ulcer of the cornea, and cataract); C, spine (lordokyphosis). ○ , SAM-P/1 mice; ● , SAM-R/1 mice. Number of mice in each group was 5 at each age. Mean value of grading scores is given.

The above chronological changes observed in the degree of senescence of SAM-P/1 and SAM-R/1 mice inbred in our institute were quite consistent with those reported by Hosokawa *et al.* [11]; they described that for young mice of both strains, there was no difference in grading score, whereas a difference between the two became obvious after 6 months of age. These results indicate that SAM-P/1 and SAM-R/1 mice used in the present study have well retained their original nature.

Changes in Serum Lipid Peroxide Levels of SAM with Age

The pattern of change in the serum lipid peroxide level was different between SAM-P/1 and SAM-R/1 mice, as shown in Figure 2. The serum lipid peroxide level of SAM-R/1 mice tended to increase gradually with advancing age. This increase seems to reflect the progress of aging in SAM-R/1 mice. On the contrary, the level of SAM-P/1 mice was increased markedly at 2 months of age and remained at a high level until 8 months of age. Thus, serum lipid peroxide levels of SAM-P/1 mice at 2 and 3 months of age were significantly higher than those of SAM-R/1 mice. At 9 and 10 months of age, serum lipid peroxide levels of SAM-P/1 mice decreased rapidly and were lower than those of SAM-R/1 mice of the same age. This decrease seen in old SAM-P/1 mice is, in principle, in accord with that observed in the case of human serum. As reported in our previous paper [2], lipid peroxide levels in human sera from subjects over 71 years of age were slightly lower than those from subjects in the 61-70 year category. The elevation of the serum lipid peroxide level in SAM-P/1 mice obviously preceded the appearance of age-related symptoms, suggesting that this elevation may have a role as a trigger for the accelerated aging.

Changes in Liver Lipid Peroxide Levels of SAM with Age

The pattern of change in the liver lipid peroxide level of SAM-P/1 mice and that of SAM-R/1 mice are shown in Figure 3. The liver lipid peroxide level of SAM-R/1 mice did not change significantly during the 5-month period after birth and decreased thereafter. In contrast, the level of SAM-P/1 mice changed with a similar pattern as found in their serum. The level at 2 months of age in SAM-P/1 mice was significantly higher than that in SAM-R/1 mice. Accordingly, we consider that lipid peroxides increased in the liver possibly initiate, and cause acceleration of, aging *via* their release into the bloodstream.

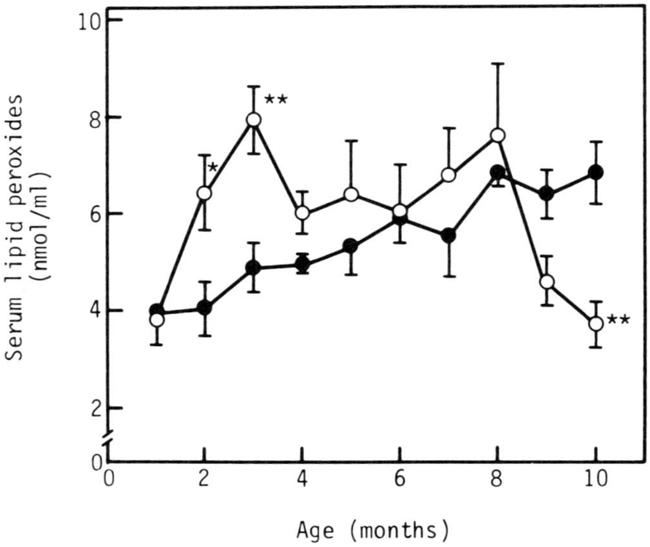

Figure 2. *Changes in Serum Lipid Peroxide Levels of Male SAM Over a Period from 1 to 10 Months of Age.* ○, SAM-P/1 mice; ●, SAM-R/1 mice. Number of mice in each group was 5 at each age. Mean value ± SE is given. Significantly different from the values of SAM-R/1 mice: *$p<0.05$, **$p<0.01$.

Relationship Between the Grading Score and Serum or Liver Lipid Peroxide Levels in SAM

The serum lipid peroxide level of SAM-P/1 mice in each grading score category was calculated and compared with that of SAM-R/1 mice having a grading score of 0. The mean value of serum lipid peroxide levels of SAM-P/1 mice having a grading score of 0 tended to be higher, and that of the mice having a score of 1.0 was significantly higher, as compared with that of SAM-R/1 mice. On the contrary, the value of SAM-P/1 mice having a grading score of 2.0 was significantly lower than that of SAM-R/1 mice. Among SAM-P/1 mice with various degrees of senescence, the maximum value of the lipid peroxide level was found in the mice having a grading score of 1.0. Although three SAM-P/1 mice were observed to possess high grading scores of 5.0, 6.0, and 8.0, their serum lipid peroxide levels were not extremely high, i.e., 3.35, 3.53, and 5.34 nmol/ml, respectively.

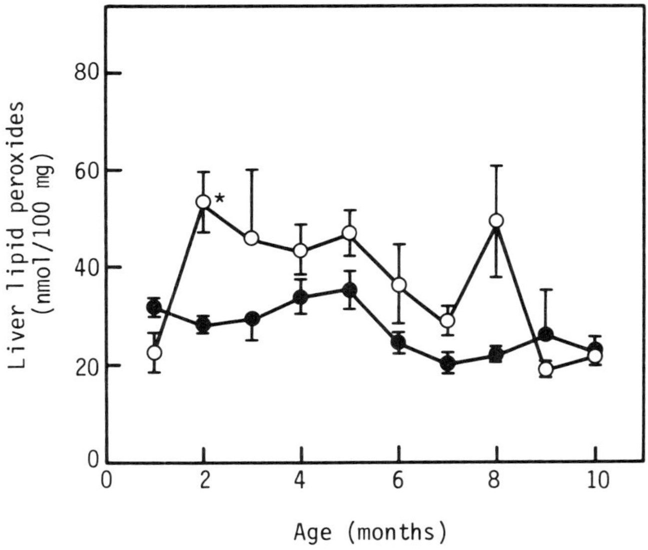

Figure 3. *Changes in Liver Lipid Peroxide Levels of Male SAM Over a Period from 1 to 10 Months of Age.* ○, SAM-P/1 mice; ●, SAM-R/1 mice. Number of mice in each group was 5 at each age. Mean value ± SE is given. Significantly different from the values of SAM-R/1 mice: $*p<0.05$.

The liver lipid peroxide level of SAM-P/1 mice in each grading score category was also calculated and compared with that of SAM-R/1 mice having a grading score of 0. The mean value of liver lipid peroxide levels of SAM-P/1 mice having a grading score of 0 or 1.0 was significantly higher than that of SAM-R/1 mice having a score of 0. The levels of three SAM-P/1 mice having high grading scores of 5.0, 6.0, and 8.0 were rather low, being 15.6, 11.2, and 21.8 nmol/100 mg wet weight, respectively. These results clearly indicate that the increase in serum and liver lipid peroxide levels preceded the appearance of age-related symptoms.

Symptoms Appearing in the Skin of SAM

The occurrence of symptoms in the skin of SAM-P/1 and SAM-R/1 mice was examined with the naked eye. In both SAM-P/1 and R/1 mice of 3~4 months of age, no symptoms were seen. At 5~6 months of age, however, 3 out of 5 SAM-P/1 mice examined had a decrease in hair glossiness in the head skin and the dorsal neck skin, and 2 of these 3 mice also had ulcers in the dorsal neck skin; while SAM-R/1 mice of the same age showed no skin symptoms. At 12~14 months of age, all of SAM-P/1 mice examined had severe changes in the skin such as periophthalmic lesions, alopecia, and skin ulcers. In SAM-R/1 mice of 12~14 months of age, only one out of 5 mice showed slight coarseness of the hair on its head. No changes were seen in the skin of the hip of any of the animals.

Lipid Peroxide Levels in the Skin of SAM

Lipid peroxide levels in four regions of the skin (below eyes, head, dorsal neck, and hip) of both strains of SAM of 3~4, 5~6, and 12~14 months of age are shown in Figure 4. In both 3~4- and 5~6-month-old animals, the levels in SAM-P/1 mice tended to be higher than those of age-matched SAM-R/1 mice. In particular, the levels in the skin regions, except for below the eyes, of 3~4-month-old SAM-P/1 mice were significantly increased as compared with the levels in the respective regions of the control SAM-R/1 mice. In the case of 12~14-month-old mice, no significant difference was seen between two strains.

Among the lipid peroxide levels in the four skin areas examined in SAM-P/1 mice, the highest level was found in the dorsal neck skin of the 3~4-month-old mice and in the head skin of the mice of 5~6 months of age. The highest value in each age category was significantly different from the levels in the skin below the eyes and in the skin of the hip.

It can also be seen from Figure 4 that the skin lipid peroxide levels in SAM-P/1 mice of 3~4 and 5~6 months of age were higher than those of 12~14-month-old SAM-P/1 mice in all regions of the skin examined. In the skin of the dorsal neck of SAM-P/1 mice, the lipid peroxide level was maximum at 3~4 months of age and thereafter decreased. In the head skin of SAM-P/1 mice, the skin lipid peroxide level was maximum at 5~6 months of age, at which age no symptoms except for a decrease in hair glossiness were observed in this region of the skin. These results indicate that the increase in the skin lipid peroxide level preceded the appearance of the symptoms in

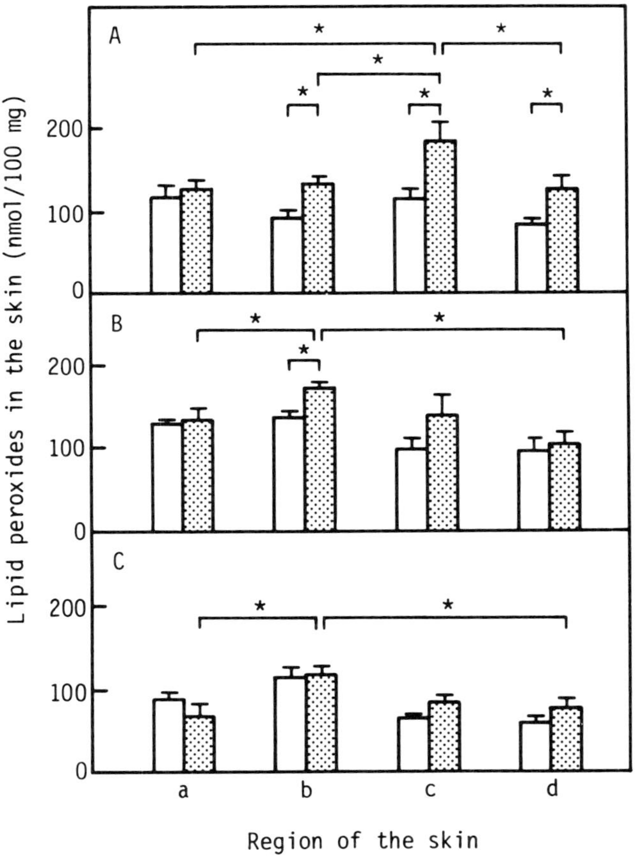

Figure 4. *Lipid Peroxide Levels in the Skin of Male SAM of Three Different Ages.* [Open Box], SAM-R/1 mice; [Stippled Box], SAM-P/1 mice. A, 3~4 months; B, 5~6 months; C, 12~14 months of ages. a, b, c, and d indicate the region of the skin: a, below eyes; b, head; c, dorsal neck; d, hip. Mean value ± SE is given. Number of mice in each group was 5 at each age. Significant difference between the two groups in the same age category: $*p<0.05$.

the skin. The lipid peroxides increased in the skin would be the most probable cause of the symptoms, since it is well known that lipid peroxides provoke injury to cells [14-16]. Supporting this view are the data showing that the lipid peroxide levels in the skin of the head and of the dorsal area of the neck of SAM-P/1 mice, where symptoms in the skin are most severe, were increased markedly. In relation to these facts, it should be recalled that lipid peroxides attached to the skin bring about many histological changes in the tissue, such as hyperkeratosis, dyskeratosis, hypergranulosis, intraepidermal edema, and spongiosis in the epidermis [17,18]. In the case of the skin below the eyes, lipid peroxide levels were not increased, and so the lesions observed in this area might be caused by irritation due to excessive lacrimation.

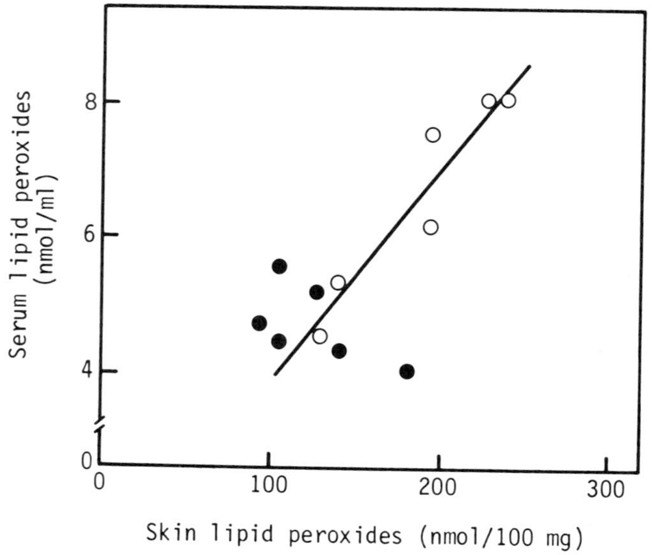

Figure 5. *Correlation Between Lipid Peroxide Levels in the Dorsal Neck Skin and those in the Serum in Male SAM of 3~4 Months of Age.* ○, SAM-P/1 mice; ●, SAM-R/1 mice. Regression line for SAM-P/1 mice was $y = 0.0308x + 0.802$ ($r = 0.934$, $p<0.05$).

Relationship Between Lipid Peroxide Levels in the Skin and those in the Serum of SAM

In Figure 5, the correlation between lipid peroxide levels in the dorsal neck skin and those in the serum of 3~4-month-old SAM-P/1 and SAM-R/1 mice is shown. In SAM-P/1 mice, a positive correlation was found between lipid peroxide levels in the dorsal neck skin and those in the blood, whereas no such correlation was found in SAM-R/1 mice. From this result, we consider that in SAM-P/1 mice lipid peroxides increased in the serum are transferred through the vascular wall to the skin tissue where they provoke the skin symptoms involved in aging.

References

1. Yagi, K. (1976) *Biochem. Med.* **15**, 212-216
2. Suematsu, T., Kamada, T., Abe, H., Kikuchi, S., and Yagi, K. (1977) *Clin. Chim. Acta* **79**, 267-270
3. Hagihara, M., Nishigaki, I., Maseki, M., and Yagi, K. (1984) *J. Gerontol.* **39**, 269-272
4. Yagi, K. (1982) in *Lipid Peroxides in Biology and Medicine* (Yagi, K., ed) pp. 223-242, Academic Press, New York
5. Yagi, K. (1984) *BioEssays* **1**, 58-60
6. Yagi, K. (1986) *Trends Biochem. Sci.* **11**, 18-19
7. Yagi, K., Komura, S., Ihara, N., Abe, H., Konishi, H., and Arichi, S. (1985) *J. Appl. Biochem.* **7**, 202-206
8. Takeda, T., Hosokawa, M., Takeshita, S., Irino, M., Higuchi, K., Matsushita, T., Tomita, Y., Yasuhira, K., Hamamoto, H., Shimizu, K., Ishii, M., and Yamamuro, T. (1981) *Mech. Ageing Dev.* **17**, 183-194
9. Hosokawa, M., Takeshita, S., Higuchi, K., Shimizu, K., Irino, M., Toda, K., Honma, A., Matsumura, A., Yasuhira, K., and Takeda, T. (1984) *Exp. Eye Res.* **38**, 105-114
10. Takeshita, S., Hosokawa, M., Irino, M., Higuchi, K., Shimizu, K., Yasuhira, K., and Takeda, T. (1982) *Mech. Ageing Dev.* **20**, 13-23
11. Hosokawa, M., Kasai, R., Higuchi, K., Takeshita, S., Shimizu, K., Hamamoto, H., Honma, A., Irino, M., Toda, K., Matsumura, A., Matsushita, M., and Takeda, T. (1984) *Mech. Ageing Dev.* **26**, 91-102
12. Yagi, K., Yoshino, K., Komura, S., Kondo, K., and Ohishi, N. (1988) *J. Clin. Biochem. Nutr.* **5**, 21-27
13. Komura, S., Yoshino, K., Kondo, K., and Yagi, K. (1988) *J. Clin. Biochem. Nutr.* **5**, 255-260
14. Tappel, A. L. (1973) *Fed. Proc.* **32**, 1870-1874

15. Sasaguri, Y., Nakashima, T., Morimatsu, M., and Yagi, K. (1984) *J. Appl. Biochem.* **6**, 144-150
16. Yagi, K. (1987) in *Advances in Membrane Biochemistry and Bioenergetics* (Kim, C. H., Tedeschi, H., Diwan, J. J., and Salerno, J. C., eds) pp. 553-560, Plenum Publishing Corporation, New York
17. Tanaka, T., and Hayakawa, R. (1986) *J. Clin. Biochem. Nutr.* **1**, 201-207
18. Yagi, K. (1987) in *The Biological Role of Reactive Oxygen Species in Skin* (Hayaishi, O., Imamura, S., and Miyachi, Y., eds) pp. 109-116, University of Tokyo Press, Tokyo

Hepatocellular Regeneration as a Determinant of Halomethane Hepatotoxicity

Harihara M. Mehendale
Departments of Pharmacology & Toxicology, University of Mississippi Medical Center
2500 North State Street, Jackson, Mississippi 39216-4505 USA

The possibility of unusual toxicity due to interaction of toxic chemicals upon environmental or occupational exposures to two or more chemicals, particularly when such exposures involve levels ordinarily considered harmless individually is an important toxicological concern. Progress in this area of environmental toxicology has suffered for want of a model where the two interactants are individually nontoxic. Models where toxic doses of chemicals are employed are not very representative of low level, environmental exposure. Prior exposure to nontoxic levels of the pesticide Kepone^{R} (chlordecone) results in a 67-fold amplification of CCl_4 lethality in experimental animals. This propensity for chlordecone to potentiate hepatotoxicity of halomethanes such as CCl_4, $CHCl_3$, and $BrCCl_3$ has been a subject of intense inquiry to unravel the underlying mechanism. The biological effects of this interaction include extensive hepatotoxicity characterized by histopathological alterations, hepatic dysfunction, and perturbation of related biochemical parameters. Mechanisms such as induction of microsomal cytochrome P-450 by chlordecone and greater lipid peroxidation are inadequate to explain the remarkably powerful potentiation of toxicity and lethality. Compelling experimental data support the hypothesis that hepatocellular division during early time points after the administration of CCl_4 is an important determinant of the progression (or repair of it) of the liver injury and consequent destruction (or restoration) of the hepatolobular architecture and function. This paper advances a hypothesis for the mechanism of hepatotoxic and lethal effect of CCl_4 as being primarily related to the accelerated progression of liver injury due to suppressed hepatocellular regeneration and hepatolobular restoration. This is in contrast to the widely accepted putative mechanism, one which invokes only bioactivation followed by runaway lipid peroxidation as the events determining the course of the progressive phase of liver injury. The concept being advanced here accepts bioactivation and (perhaps lipid peroxidation) as the primary initiating events of cell injury, but proposes that they

Biological Oxidation Systems, Volume 2

are not the determinants of the progressive phase of liver injury. Incapacitation of the liver cells from regeneration is the determinant of the progression of liver injury which leads to the ultimate outcome of hepatotoxicity and lethality.

Introduction

The propensity and the specificity with which chlordecone (CD; Fig. 1) amplifies the hepatoxic actions of a nontoxic dose of halomethane solvents has been well established (1-22). Dietary exposure of rats to 10 ppm CD for a total dose of roughly 10 mg/kg, results in a greatly potentiated hepatotoxic response to ordinarily nontoxic dose of CCl_4 as determined by functional, biochemical and histopathological parameters (1-16). CCl_4 lethality is increased 67-fold (Table 1) by CD in male (5) and 25-fold in female rats (6). The specificity of CD in potentiating halomethane hepatotoxicity is illustrated by the ineffectiveness of mirex (3,17,20,22) or photomirex (2) or phenobarbital (22). Exposure to phenobarbital (PB) at 50 times greater levels (on a molar basis) caused less than 50% of the potentiation of CCl_4 hepatotoxicity caused by CD (3) and the lethality was increased only by a statistically insignificant 1.6-fold (5). Recent studies (22) indicate that the hepatotoxic and lethal effects of a single, nontoxic dose of $CHCl_3$ are also increased after dietary exposure of mice to 10 ppm CD. Neither mirex (M; 10 ppm) nor PB (225 ppm) in a similar dietary protocol increased the hepatotoxic or lethal effects of $CHCl_3$ (22; Table 1). Hepatotoxic and lethal effects of $BrCCl_3$ are also potentiated by CD (7,8), while the toxicities of $CHBr_3$ and CBr_4 are unaffected (9,10). Hepatotoxicity of neither trichloroethylene nor bromobenzene is potentiated by 10 ppm CD (11).

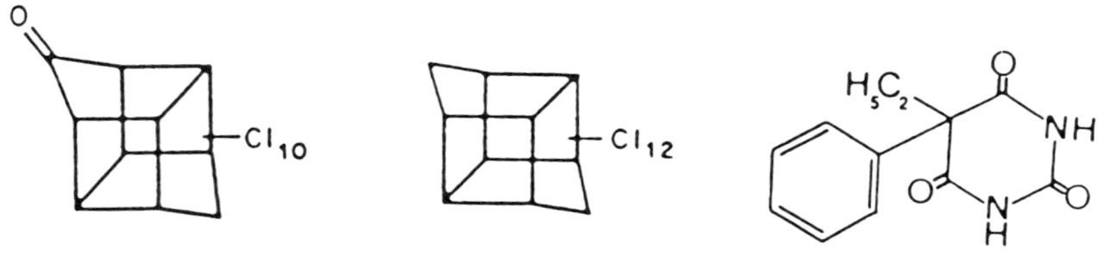

CHLORDECONE MIREX PHENOBARBITAL

Figure 1. *Structures of Mirex, Chlordecone, and Phenobarbital*

Table 1. *Amplification of Lethal Effects of Several Halomethanes by Subtoxic Levels of Chlordecone*

Dietary Pretreatment	Halomethane	48 hour LD_{50} ml/kg	95% Confidence Limits	Decrease LD_{50}
Female Rats				
Control	CCl_4	1.25	0.85 - 1.84	---
Chlordecone (10 ppm)	CCl_4	0.04*	0.03 - 0.7	26-fold
Male Rats				
Control	CCl_4	2.8	1.5 - 5.2	---
Chlordecone (10 ppm)	CCl_4	0.042*	was to low	67-fold
Phenobarbital (225 ppm)	CCl_4	1.7	1.2 - 2.3	1.6 fold
Control	$BrCCl_3$	0.110	0.056-0.250	---
Chlordecone (10 ppm)	$BrCCl_3$	0.027*	0.018-0.038	4.5-fold
Male Mice				
Control	$CHCl_3$	0.067	0.65 - 0.70	---
Chlordecone (10 ppm)	$CHCl_3$	0.16*	0.12 - 0.23	4.2-fold
Mirex (10 ppm)	$CHCl_3$	0.70	0.46 - 1.22	No change
Phenobarbital (225 ppm)	$CHCl_3$	0.70	0.41 - 1.18	No change

Adapted from reference 81 with permission. *Asterisks indicate significant change from the respective controls, $p < 0.05$.

The mechanism of this highly powerful and unusual interaction with halomethanes has eluded many investigations (1-28). Table 2 lists several candidate mechanisms that have been experimentally tested and the experimental evidence for or against them. Since exposure to nontoxic levels of CD results in the potentiation of the hepatotoxic and lethal effects of several halomethanes at single exposure to a nontoxic or subtoxic dose, the interaction represents a very attractive model for potential adverse health effects consequent to the interactions of environmental toxic chemicals at levels in the neighborhood of environmental significance.

Mechanism of Potentiation of Halomethane Hepatotoxicity by Other Chemicals

The capacity to enhance CCl_4 toxicity and hepatic necrosis has been attributed to a variety of chemicals: PB (29-32), aliphatic alcohols (33,34), ketones (35,36), DDT (37), and polychlorinated biphenyls (38) are representative example. 3-Methylcholanthrene was shown to be protective (39), while 3,4-benzo(a)pyrene was similar to PB in enhancing CCl_4 hepatotoxicity (40). Treatment either with

Table 2. *Candidate Mechanisms for Chlordecone Amplification of Carbontetrachloride Toxicity*

Candidate Mechanisms	Experimental Evidence
1. Enhanced Bioactivation of CCl_4	In part
2. Increased Lipid Peroxidation	No
3. Estrogenic Property of CD	No
4. Increased Ca^{2+} Accumulation	Yes
5. Suppressed Hepatocellular Regeneration	Yes

cysteamine, cysteine (41) or SKF-525A (41,42) afford protection against CCl_4 hepatotoxicity. Hepato- and nephrotoxicity of $CHCl_3$ is potentiated by aliphatic alcohols (43,44), polybrominated biphenyls (45), ketones (44), and PB (46).

The widely accepted theory in vogue for the mechanism of xenobiotic-induced enhancement of hepatocellular injury caused by CCl_4 is that its bioactivation to $\cdot CCl_3$ and the formation of $CCl_3O_2\cdot$ free-radical (29-32) is increased. There is also evidence for increased covalent binding of CCl_4 to liver tissue upon bioactivation, although the concept of a direct-relationship between the extent of covalent binding and liver injury has been found not to be tenable (3). However, there is a general acceptance of the concept that increased production of $\cdot CCl_3$ or $CCl_3O_2\cdot$ radicals leads to increased lipid peroxidation culminating in greater hepatotoxicity.

Mitigated Hepatocellular Regeneration and Tissue Repair in Chlordecone-Amplification of Halomethane Toxicity

A close perusal of the liver sections indicated that hepatocytes were greatly stimulated to divide at 6 h after the administration of a low dose of CCl_4 alone. This hepatocellular regeneration was invariably in the areas vicinal to necrotic cells within the hepatolobular region. This stimulation of cell division was entirely suppressed in animals consuming food contaminated with 10 ppm CD. These studies indicate a strong and progressive amplification of liver injury after the combination treatment, while the liver recovers from the effects of CCl_4 alone with time. In fact, the lethality experiments indicate that all animals in CD+CCl_4 (100 μl/kg) combination group die, while those receiving either compound alone survive without any discernible ill effects (5,6). These studies (51,52) allowed us to conceptualize a novel mechanism based on the suppression of hepatocellular regeneration and tissue repair in the livers of animals treated with the CD+CCl_4 combination (16). Extracellular entry of Ca^{2+} is associated with these events (16), although if in fact excessive hepatocellular Ca^{2+} is responsible for arrested hepatocellular regeneration, or the mechanism by which excessive Ca^{2+} might precipitate such an event, remain to be investigated.

Most recent studies (53-56) were designed to test the hypothesis concerning suppressed hepatocellular regeneration. If suppressed hepatocellular regeneration is in fact the reason for the highly amplified liver injury after CD + CCl_4 treatments then the toxicity of this combination treatment should be mitigated in a liver stimulated to regenerate. Partial hepatectomy is known to stimulate hepatocellular regeneration in a remarkable fashion and this experimental model is suitable for testing the hypothesis (53). Histomorphometric analysis established that hepatocellular regeneration measured as mitotic index after partial hepatectomy was identical in normal and CD treated rats during a time-course of 1 through 7 days. Regeneration was maximal at 1 and 2 days after partial hepatectomy and decreased after 4 days. Since by 7 days the original liver weight is recovered, hepatocellular regeneration also returns to the normal level of mitotic activity by day 7. Neither sham operation nor exposure to CD changed the basal level of minimal mitotic activity. Hepatocellular regeneration was unaltered by the CD treatment (51). These findings established the validity of using the partial hepatectomy model to test the regeneration hypothesis.

The second phase of this study involved testing the effect of CD+CCl_4 combination treatment in partial hepatectomized animals (53). Hepatotoxicity of CCl_4 administered 1,2,4, or 7 days after the surgical manipulation was measured by way of the serum enzymes (alanine aminotransferase, aspartate aminotransferase and isocitric dehydrogenase), in CD treated as well as in all of the necessary control groups. Liver injury was decreased significantly after partial hepatectomy in comparison to the corresponding sham hepatectomy. The protection by partial hepatectomy was maximal on day 2, at the time of maximal hepatocyte regeneration and was minimal on day 7 at the time of minimal regeneration. These findings showed that stimulation of hepatocellular regeneration by partial hepatectomy reults in protection against unpotentiated or chlordecone-potentiated CCl_4 hepatotoxicity.

Another group of rats was similarly treated 1 day after surgery for lethality studies (53). In these experiments, even though the liver size in the partial hepatectomized animals was less than 4 g (70% of the liver removed by partial hepatectomy, and after 1 day only a very small fraction of that liver weight is regained), the full 100 μl CCl_4/kg dose was administered to the partial hepatectomized animals in order to simplify the experiments. Despite not adjusting the dose of CCl_4 downward (and therefore, really a larger dose in partial hepatectomized rats in comparison to sham or control groups), the results showed a 70% protection in the survival of the rats (53).

Interest in the role of liver microsomal cytochrome P-450 was rekindled with the concern about the slightly decreased cytochrome P-450 content in partial hepatectomized (20%) livers (53-56). A reasonable criticism of the model could be leveled on the basis of decreased cytochrome P-450 in regenerating livers. Studies with $CoCl_2$ treated rats have shown that even a 62% decrease in microsomal cytochrome P-450 content did not affect the liver injury associated with CD + CCl_4 treatment (56). More recent studies indicate that neither the $^{14}CCl_4$ disposition, nor bioactivation were decreased in partially hepatectomized rats (57). Similar results were obtained in animals receiving prior treatment of $CoCl_2$ to decrease the normal content of hepatic microsomal cytochrome P-450 by roughly 60% (53,57). Thus regardless of whether cytochrome P-450 was decreased by 20% (by partial hepatectomy) or by 60% (by $CoCl_2$), the bioactivation of CCl_4 (100 μl/kg) was unaffected *in vivo*. These *in vivo* studies provide compelling exacting experimental evidence that for such small doses of CCl_4, even a significant decrease in hepatic microsomal cytochrome P-450 is inconsequential for bioactivation of CCl_4.

Mechanism of Suppressed Cell Division

The mechanism by which hepatocellular regeneration, ordinarily stimulated by a low dose of CCl_4, is suppressed by CD is of significant interest. This mechanism is of particular interest since stimulation of hepatocellular regeneration by partial hepatectomy is not affected by CD (53). Since a precipitous and irreversible drop in hepatic glycogen was observed in animals treated with CD + CCl_4 combination (51-53, 58), we have hypothesized that hepatocellular bankruptcy in energy leads to incapacitation and failure of the cells to divide (59-61).

In view of the previously demonstrated ability of CD at higher doses to inhibit mitochrondrial energy metabolism (62-64), it was of interest to examine the effect of 10 ppm CD on the energy status of the hepatocytes. Oligomycin-sensitive ATPase activity of hepato-

mitochondria isolated from rats maintained either on a normal, powdered diet or on a 10 ppm CD diet for 15 days was not affected by 10 ppm CD (59). ATP levels were not affected by the dietary exposure to CD (59). Therefore, it appears that while exposure to 5 to 10-fold higher levels of CD is known to interfere with hepatocellular energy metabolism (62), the low, nontoxic level used for the halomethane interaction studies does not interfere with this aspect of hepatocellular biochemistry. In this connection it should be noted that the findings concerning the effect of CD treatment on hepatic Ca^{2+} levels (65-68) are consistent with the above data. Treatment with CD even up to 10 times the level used in the interaction studies does not affect hepatic Ca^{2+} (65-68). Furthermore, the failure of 10 ppm CD exposure to adversely affect the stimulation of hepatocellular regeneration by partial hepatectomy is also consistent with the above observations.

The possibility that the CD + CCl_4 combination treatment may result in severely impaired energy metabolism in the liver was examined (59). Hepatic ATP levels are depleted within 1 h after the administration of CCl_4 to rats preexposed to 10 ppm CD, while the same dose of CCl_4 alone has no effect on the energy metabolism in the liver (59). If bankruptcy in cellular energy is the mechanism underlying the failure of the hepatocytes to divide during the critical early period of progressive injury, then providing an external source of cellular energy during that time should result in protection from liver injury. Administration of fructose 1,6-bisphosphate immediately following CCl_4 liver injury (60). Likewise, protection against a large dose of CCl_4 administered to normal animals was also observed (61). In both cases, the protection was accompanied by increased ATP levels and modifications in hepatic polyamine levels, conducive for hepatocellular regeneration (60,61,69).

Proposed Mechanism for Chlordecone Amplification of Halomethane Hepatotoxicity and Lethality

Many attempts to explain the remarkable amplification of halomethane toxicity based on currently accepted putative mechanisms failed (1-28). It seemed advantageous to hypothesize the mechanism for 'why an ordinarily nontoxic dose of CCl_4 is nontoxic?'

Figure 2 illustrates the mechanism of recovery from the limited liver injury observed after the administration of a low dose of CCl_4 alone. Within 6 h after the administration of a low dose of CCl_4, limited hepatocellular necrosis occurs. This limited liver injury is inflicted by the same widely accepted mechanism of CCl_4 bioactivation followed by lipid peroxidation. This limited hepatolobular injury is evident, as centrilobular necrosis, balloon cells, lipid accumulation, and lipid peroxidation are initiated as a consequence of the bioactivation of CCl_4. By mechanisms hitherto unexplored, simultaneously the liver tissue responds by stimulating hepatocellular regeneration (51,52). Most interestingly, this hepatocellular division is maximal at 6 h, even though the limited injury evident as centrilobular

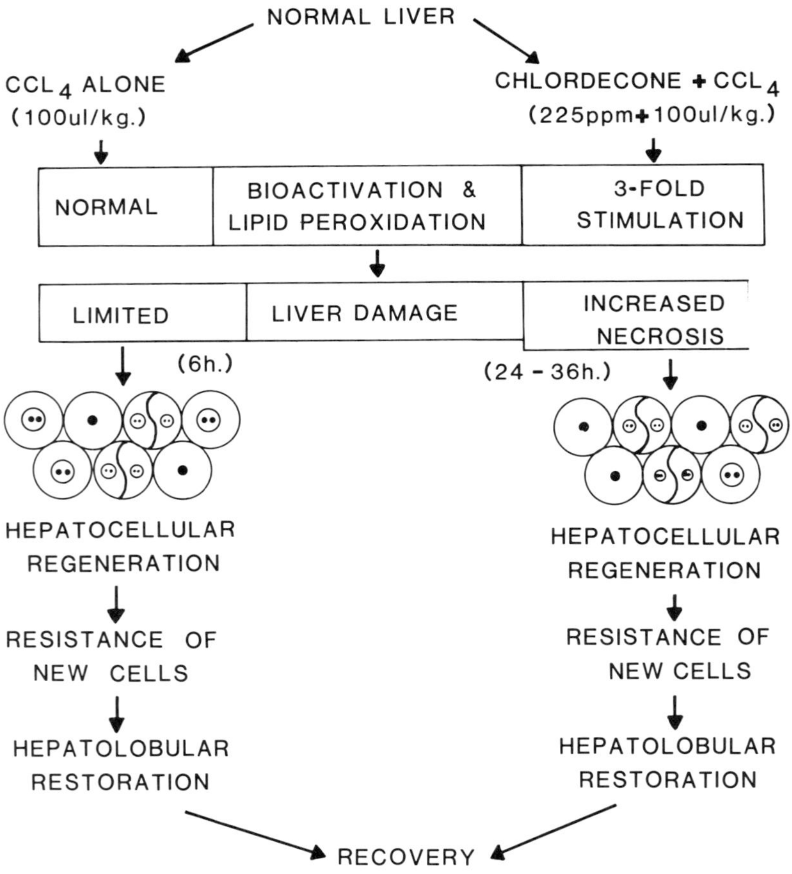

Figure 2. *Proposed Mechanism for the Highly Amplified Potentiation of CCl_4 Toxicity by CD.* The scheme depicts the concept of suppressed hepatocellular regeneration, simply permitting what is normally limited liver injury by a subtoxic dose of CCl_4 to progress in the absence of hepatolobular repair. This progression of toxicity, unchecked due to the absence of hepatolobular restoration, is evident after the CD + CCl_4 combination treatment. These events lead to complete hepatic failure, culminating in animal death.

necrosis is only beginning to manifest. Although the molecular events responsible for the stimulation of hepatocellular division have not been explored, we have observed mobilization of glycogen, the principal form of hepatic energy resource, prior to the cell division (51,52). Glycogen levels are restored after cell division has been adequately stimulated. The limited hepatocellular necrosis enters the progressive phase between 6 and 12 h (51,52) while the hepatocellular regeneration and tissue repair continue. By 24 h, no significant liver injury is evident. These observations allow one to propose that stimulation of hepatocellular regeneration is a protective response of the liver, occurs very early after the administration of a low dose of CCl_4 and leads to replacement of dead cells, thereby restoring the hepatolobular architecture.

Furthermore, it can be proposed that this remarkable biological event results in another important protective action. It is known that newly divided liver cells are relatively resistant to toxic chemicals (70-73). Therefore, in addition to the restoration of the hepatolobular architecture by cell division, by virtue of the relatively greater resistance of the new cells, the liver tissue is able to overcome the imminence of greater injury during the progressive phase (6 to 12 h), speeding up the process of overall recovery (Figure 2). Maintaining the integrity of hepatolobular structure, especially with cells of greater resistance during the progressive phase of liver injury, is particularly significant. By 6 h over 75% of the administered CCl_4 would have been eliminated in the expired air (26) and only less than 25% would be available for continued injury. Relative resistance of the newly divided cells at this critical time frame, as the animal eliminates most of the remainder of the CCl_4 from the body would be an added critical defense mechanism. At later time points (12 h and onwards), most of the CCl_4 would have been eliminated by the animal and hence continued cellular regeneration during this time period and at later time points is critical to restore hepatolobular architecture during and after the progressive phase of injury.

Administration of the same low dose of CCl_4 to animals maintained on food contaminated with low dose of CD results in initial injury by the same mechanisms of bioactivation of CCl_4 and lipid peroxidation. The liver injury in this case is slightly greater by virtue of approximately doubled rate of bioactivation of CCl_4 in livers of animals preexposed to CD (13,16,26,50). The liver injury thus initiated, enters the progressive phase between 6 to 12 h after the administration of CCl_4 and this phase appears to be accelerated in the absence of tissue repair mechanism (51,52). The highly unusual amplification of CCl_4 toxicity relates to the suppression of the initial hepatocellular regeneration, otherwise ordinarily stimulated by CCl_4 within 6 h (Figure 2).

At this juncture, although speculative, experimental observations permit invoking a role of hepatocellular bankruptcy in cellular energy. Under conditions of increased hepatocellular injury, mobilization of hepatic glycogen is initiated in order to stimulate hepatocellular division (74). Under these conditions of increased demand for cellular energy (extrusion of extracellular Ca^{2+} from the cells, protection against free-radical mediated injury etc.), cells are incapacitated due to bankruptcy in cellular energy. Stimulation of cell division, which should have normally occurred, cannot occur. The failure of cell division, which should have normally occurred, cannot occur. The failure of cell division has two important implications: first, the integrity of hepatolobular structure cannot be restored; second, unavailability of newly divided, relatively resistant cells predisposes the liver to continuation of liver injury during the progressive phase (6 or 12 h and beyond). The progressive and irreversible injury results as a consequence of the abolished tissue repair mechanisms, leading to complete hepatic failure (1-16), followed by animal death (5-16).

Many studies have shown a biphasic increase in hepatocellular Ca^{2+} levels in CCl_4 toxicity (65-68,76-78). The unusual aspect of excessive Ca^{2+} accumulation observed in livers treated with the CD + CCl_4 combination is that it occurs at a dose of CCl_4 not ordinarily associated with the causation of increased hepatic Ca^{2+}. Furthermore, CD alone, even at 10 times the higher dose than used in the interaction studies does not increase hepatocellular Ca^{2+} (65). Although *in vitro* studies with cellular organelles have been employed to speculate that the failure of organelle Ca^{2+} pumps leads to increased cytosolic Ca^{2+} levels (76,79), our studies indicate that at no time point these organelles contain decreased Ca^{2+} (65-68). Indeed, only significant change observed with regard to organelle Ca^{2+} is increased Ca^{2+} in the organelles in association with increased liver injury (65-68). Therefore, there is no *in vivo* evidence for decreased Ca^{2+} content in the organelles, which is in contradiction to the *in vitro* findings (76,79). Figure 3 illustrates a proposal on how Ca^{2+} might be involved in the progressive phase of the amplification of CCl_4 hepatotoxicity by CD. The primary mechanism leading to a highly amplified toxicity is a failure on the part of the biological events leading to hepatocellular division. Increased accumulation of extracellular Ca^{2+} (68) during the progressive phase of liver injury would be consistent with the significant loss of biochemical homeostasis in hepatocytes (Figure 3). Earlier histomorphometric (51,52) as well as biochemical studies (58) have shown that glycogen levels drop very rapidly after CCl_4 administration to CD treated animals. Increased cytosolic Ca^{2+} (65-68) would be expected to result in activation of phosphorylase *b* to phosphorylase *a*, the enzyme responsible for glycogenolysis. Phosphorylase *a* activity is progressively elevated in hepatocytes (59). The rise in Ca^{2+}, elevated phosphorylase *a* activity

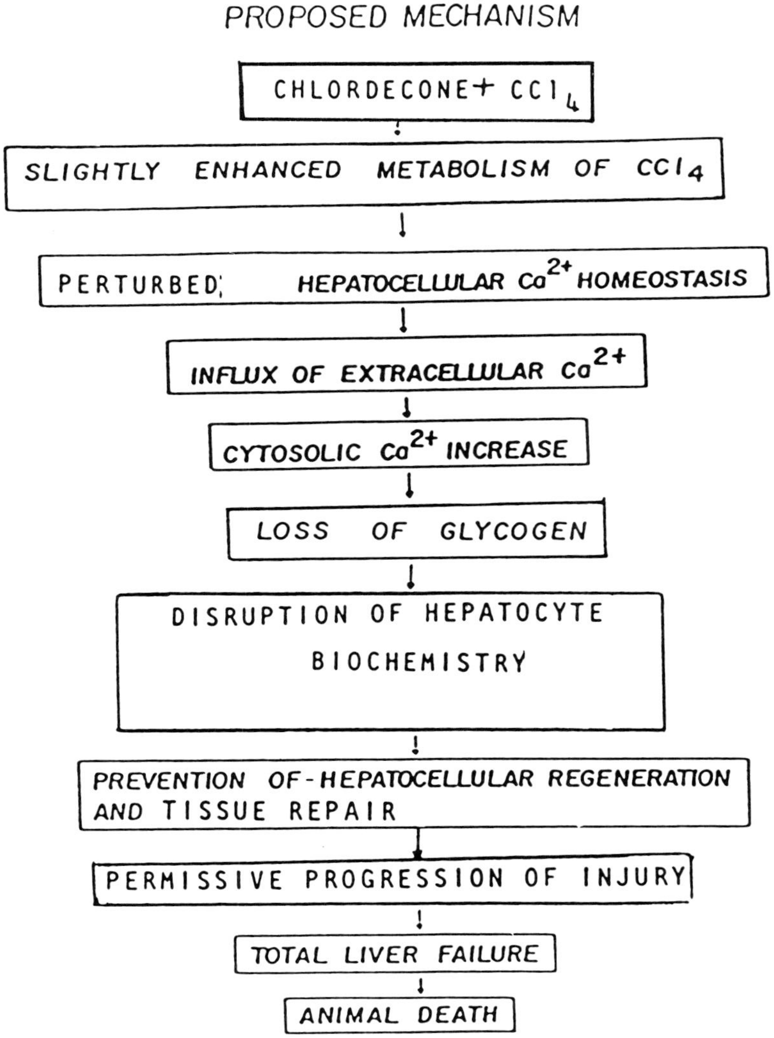

Figure 3. *Proposed Mechanism for Chlordecone-Induced Amplification of CCl_4 Toxicity Indicting the Biochemical Events Leading to the Failure of the Hepatocytes to Divide.* Suppression of otherwise stimulated hepatocellular division leads to amplification of otherwise limited injury.

(59) and precipitous glycogenolysis (51,52,58) are experimental observations consistent with the rapid depletion of cellular energy (59) on the one hand and irreversible increase in cytosolic Ca^{2+} (65-68) on the other. Fig. 3 illustrates the working hypothesis that integrates the biochemical events pursuant to accumulation of Ca^{2+} and the suppressed hepatocellular regeneration and failed tissue repair, leading to uncontrolled amplification of liver injury.

Discussion

An intriguing aspect of the experimental framework leading to the proposed mechanism is the observation that PB, even at significantly higher doses (225 ppm in the diet for 15 days) does not potentiate the hepatotoxic and lethal effects of CCl_4 nearly to the same extent as CD (1-3,5,12,13,16; Table 1). Although certain histopathological parameters of liver injury such as hepatocellular necrosis and ballooned cell response are indicative of significantly enhanced hepatotoxicity by PB (13), if the animals are left alone, this injury does not progress to significantly increased lethality (Table 1; 5). Hepatic microsomal cytochrome P-450 is approximately doubled by prior dietary exposure to 225 ppm PB (3,5) and the bioactivation of CCl_4 is tripled (13,26), and these parameters are consistent with the enhanced initiation of liver injury measured by histopathology, elevation of serum transaminases and other hepatic enzymes, or by hepatic function (1-3,5,12,13,26-28). Nevertheless, the liver injury neither progresses in an accelerated fashion nor is irreversible, as indicated by the reversal of liver injury and animal recovery (80).

How can we surmise the lack of enhanced lethality of CCl_4 in PB exposed rats, despite enhanced liver injury at 24 hours after CCl_4 administration? Induction of hepatomicrosomal cytochrome P-450 results in approximate tripling of CCl_4 bioactivation and increased lipid peroxidation (13, 16, 26-28). Thus the putative mechanisms responsible for enhanced liver injury are present. Enhanced liver injury (3,5,12,13,16,26-28) is consistent with these observations (81). It should be recalled that the liver is normally able to respond by stimulation of hepatocellular regeneration after a low dose of CCl_4 within 6 h (Fig. 2). While PB exposure results in greater injury, the liver's ability to respond by stimulated cell division is not completely compromised, as evidenced by the stimulation of hepatocellular regeneration starting at 6 h and continuing through 72 h. Therefore, hepatocellular regeneration is stimulated, thereby counteracting the enhanced liver injury, which leads to recovery from increased liver injury. In view of the enhanced liver injury, restoration of normal hepatolobular architecture takes longer than the approximate 24 h required upon administration of a low dose of CCl_4 alone. However, although the hepatocellular regeneration is somewhat delayed (80), it is enhanced, apparently tempered by the demand for more extensive restoration of hepatolobular architecture as a consequence of greater injury (74). Hence, the overall effect of PB-induced potentiation of CCl_4 injury is merely to delay the stepped up hepatocellular regeneration, tissue repair, and restoration of hepatolobular architecture. The prolongation of these normal responses of the liver are a consequence of the enhanced liver injury, inflicted by the enhanced putative injurious mechanisms.

The hypothesis that suppression of hepatocellular regeneration and tissue repair leads to the progression of liver injury was tested in partial hepatectomy model employing histomorphometric analysis of liver sections (53). In recent studies (54), we have employed ^3H-thymidine pulse-labelling techniques followed by quantitation of radiolabeled DNA in hepatocellular nuclei. ^3H-Thymidine incorporation into nuclear DNA and the percentage of labelled cells as evidenced by autoradiography of liver sections were significantly elevated in rats maintained on normal diet at 1 to 2 hours after CCl_4 (100 μl/kg) administration and returned (100 μl/kg) administration and returned to basal level by 6 h. Serum transaminases in these rats undergoing sham operation and partial hepatectomy were not altered. However, these enzymes were significantly elevated in CD treated rats following CCl_4 (100 μl/kg) administration, CCl_4-induced serum enzyme elevations were significantly lower in 2 days post partial hepatectomy rats when compared to sham operated rats or 7 days post partial hepatectomy rats maintained on CD diet, indicating that CD-potentiated CCl_4 hepatotoxicity is significantly mitigated in livers stimulated for regenerative activity. CCl_4 decreased the DNA levels significantly in all groups of rats receiving CD diet, but not in 2 day post partial hepatectomized rats receiving CD diet. ^3H-Thymidine incorporation, percentage of labelled cells and number of grains per cell were significantly decreased at 2 h in 2 days post partial hepatectomy rats receiving the CD + CCl_4 combination treatment, reflecting the suppression of cell proliferation after CCl_4 administration to CD fed rats. These results indicate that partial hepatectomy affords protection against CD + CCl_4 interaction. The protection against hepatotoxic and lethal effects of CD + CCl_4 combination by previously stimulated hepatocellular regeneration might be explained by two consequences of stimulated cell division. First, the hepatocellular architecture is renovated by the newly divided cells. Second, by virtue of the well known resistance of the newly divided cells, the progressive phase of toxicity is inhibited. These findings are supportive of our previous findings that suppression of hepatocellular regeneration, aside from the bioactivation of CCl_4, is involved in CD + CCl_4 combination toxicity.

Since we have gathered significant and substantial information in support of the concept that progression of liver injury depends on the biology of hepatocellular regeneration and hepatolobular restoration (16, 51-53,55), we focused our attention to test these concepts after CCl_4 injury employing only CCl_4 at a dose sufficient to cause significant liver injury. In other words, can we view the injury of a large dose of CCl_4 as that caused by two events? First event being the widely accepted bioactivation followed by lipid peroxidation. The second event being a lack of hepatocellular regeneration and hepatolobular restoration. The progression of injury is dependent on the second event. If progression of massive liver injury after administration of a clearly toxic dose of CCl_4 (*e.g.*, 2.5 ml/kg) is due to impaired cell division and accordingly suppressed restoration of hepatolobular architecture, then stimulation of the liver cells to divide by partial hepatectomy, before hand, should result in a mitigation of the massive injury. If not, these animals should be just as sensitive to the hepatotoxic and lethal effects of CCl_4. Therefore, administration of CCl_4 (2.5 ml/kg) 2 days after partial hepatectomy should result in much, much less toxicity, if the premise for the hypothesis is correct. Furthermore, administration of CCl_4 7 days after partial hepatectomy should restore the original toxicity since stimulation of liver cells for division would have phased out by 7 days. The results of these recent studies (55) indicate that partial hepatectomy affords protection against CCl_4 hepatotoxicity and lethality if CCl_4 is administered at the time of stimulated hepatocellular regeneration.

Several reports indicating that isolated hepatocytes prepared from partially hepatectomized rats are more resistant to hepatotoxins like aflatoxin B_1, methylmethane sulfonate, adriamycin, methotrexate and $CHCl_3$ (70-72) have appeared. These studies suggest that new dividing cells are resistant to hepatotoxins. These findings imply that a single biological event resulting in stimulated hepatocellular regeneration may accomplish two objectives. First, a restoration of hepatolobular architecture by replacing necrotic cells resulting in repair and renovation (16). Second, in view of the reported resistance of newly dividing cells to several hepatotoxins, the newer architecture of the hepatolobules is inherently protected (53,55). In the meantime, the majority of the administered CCl_4 would be eliminated (26) allowing a more complete recovery from ordinarily anticipated hepatotoxic and lethal effects of a toxic dose of CCl_4.

These findings suggest that hepatocellular regeneration during the early time points after the administration of CCl_4 is an important determinant of the progression (or repair of it) of the liver injury and consequent destruction (or restoration) of the hepatolobular architecture and function. These studies also advance our understanding of the mechanism of hepatotoxic and lethal effects of CCl_4 as being primarily related to the accelerated progression of liver injury due to suppressed hepatocellular regeneration and hepatolobular restoration. This concept is in contrast to the widely accepted putative mechanism, which invokes only bioactivation followed by runaway lipid peroxidation as the events determining the course of the progressive phase of liver injury. The present concept being advanced, accepts bioactivation and lipid peroxidation as the initiating events of cell injury but does not ascribe a role as the determinants of the progressive phase of liver injury. The biological issue of whether the cells can regenerate or not is the determinant of the progression of liver injury, and therefore, the ultimate outcome of hepatotoxicity and lethality. Since a variety of experimental evidence indicates that the progression of degenerative injury is due primarily to suppressed hepatocellular regeneration leading to hepatolobular dysrepair (51-55).

The next major step in our understanding of the mechanism is the mechanistic basis for the failure of the hepatocytes to divide. Recently, we have demonstrated that significant decrease in the levels of polyamines and related enzymes involved in polyamine metabolism occurs in the livers of rats exposed to CD and CCl_4 combination (69). Polyamines have been implicated to have an important role in cell proliferation and tissue regeneration (82-84). Recovery from CD + CCl_4 toxicity depends upon the extent to which residual hepatocytes can increase their levels of polyamines consequently effecting an increase in liver regeneration (69). Since these critical events leading to hepatocellular regeneration might be expected to require significant cellular energy, the failure of hepatocytes to divide (51-53) might be a consequence of bankruptcy in cellular energy. Recent work indicates that levels of ATP decrease significantly within one hour in livers of rats treated with CD and CCl_4 combination (54,55), which would directly affect the induction of polyamine-related enzymes, polyamine synthesis, and consequently liver regeneration. Neither CD alone nor CCl_4 alone affected the ATP levels (55), nor the hepatic cellular regeneration (51-54). This led to a hypothesis that the failure of hepatocytes to divide is due to unavailability of cellular energy (60,61). If true, supplementation of a readily utilizable source of energy should result in protection. Fructose 1,6-bisphosphate, a metabolite of the Embden-Meyerhoff pathway, was chosen as a readily available source of ATP without the need for any ATP utilization for energy production. Results indicate that exogenously provided fructose 1,6-bisphosphate significantly decreases CD + CCl_4-potentiated hepatotoxicity (61). Furthermore, the levels of enzymes involved in polyamine synthesis or interconversion were stimulated in the sugar bisphosphate-treated rats exposed to CD + CCl_4. Increased ATP and the levels of enzymes involved in polyamine synthesis or interconversion were stimulated in the sugar bisphosphate-treated rats exposed to CD + CCl_4. Increased ATP levels observed in CD + CCl_4 + fructose 1,6-diphosphate treated rats appear sufficient to support hepatocellular regeneration and tissue repair. The results indicate that the energy status of liver is a critical factor important for the recovery of animals from CD + CCl_4 potentiated hepatotoxicity (61).

The most promising aspect of our contribution to this area is the uncovering of the mechanism underlying such interactions. The interaction represents a pesticide shown to be an animal carcinogen (85) potentiating the toxicity of known animal carcinogens such as CCl_4 (86) and $CHCl_3$ (87). While the mechanistic inquiry into this most unusual toxic interaction between two carcinogenic compounds resulting in most severe toxic response of biological significance continues, the question of whether the carcinogenic response of the halomethanes is altered by exposure to CD has not been investigated. Such an inquiry in future studies would be of interest since exposure to extremely low levels of the pesticide and the halomethanes would be required for animal survival long enough to permit detection of any carcinogenic responses.

The progress that has been made in understanding the mechanism indicates that when such life threatening interactions occur or are predictable on the basis of known or documentable exposure to combinations of chemicals, an understanding of the biology of stimulated (or suppressed) cellular regeneration is likely to offer avenues for therapeutic intervention. For instance, if availability of cellular energy is indeed the predominant reason behind failed biological response, therapeutic intervention by externally supplying cellular energy would be indicated. Future work may also indicate that cellular mechanisms involving more intricate derangements in the biochemical pathways or other aspects of cellular homeostasis might be involved. Persuing the search for the underlying mechanism(s) for such complex interactions of chemicals is likely to be rewarding in the advancement of our basic understanding of the life processes and how these homeostatic mechanisms might be altered by exposure to chemicals in our environment.

Acknowledgements

This effort was made possible by the 1988 Burroughs Wellcome Toxicology Scholar Award, grant support from the Department of the Air Force grant AFOSR-88-0009, and by the Harry G. Armstrong Aerospace Medical Research Laboratory a part of the U.S. Air Force installation, restoration program funded under the defense enviornmental restoration account through U.S. EPA Award CR814053.

References

1. Curtis, L.R., Williams, W.L., and Mehendale, H.M. (1979) *Toxicol. Appl. Pharmacol.* **51**, 283-293
2. Curtis, L.R., Williams, W.L., and Mehendale, H.M. (1979) *Toxicology* **13**, 77-90
3. Curtis, L.R., and Mehendale, H.M. (1980) *Drug Metab. Dispos.* **8**, 23-27
4. Curtis, L.R., Thureson-Klein, A.K., and Mehendale, H.M. (1980) *J. Toxicol. Environ. Health* **7**, 499-517.
5. Klingensmith, J.S., and Mehendale, H.M. (1982) *Toxicol. Lett.* **11**, 149-154
6. Agarwal, A.K., and Mehendale, H.M. (1983) *Toxicology* **26**, 231-242
7. Klingensmith, J.S., and Mehendale, H.M. (1981) *Toxicol. Appl. Pharmacol.* **61**, 378-384
8. Agarwal, A.K., and Mehendale, H.M. (1982) *Fundam. Appl. Toxicol.* **2**, 161-167
9. Agarwal, A.K., and Mehendale, H.M. (1983) *Toxicol. Lett.* **15**, 251-257
10. Agarwal, A.K., Mehendale, H.M., and Berndt, W.O. (1983) *Toxicol. Lett.* **17**, 57-62
11. Mehendale, H.M., and Lockard, V.G. (1982) *Toxicologist* **2**, 37
12. Klingensmith, J.S.Lockard, V.G., and Mehendale, H.M. (1983) *Exp. Mol. Pathol.* **39**, 1-10
13. Klingensmith, J.S., and Mehendale, H.M. (1983) *Drug Metab. Dispos.* **11**, 329-334
14. Agarwal, A.K., and Mehendale, H.M. (1984) *Toxicology* **29**, 325-323
15. Agarwal, A.K., and Mehendale, H.M. (1983) *Fundam. Appl. Toxicol.* **3**, 507-511
16. Mehendale, H.M. (1984) *Fundam. Appl. Toxicol.* **4**, 295-305
17. Hewitt, W.R., Miyajima, H., Cote, M.G., and Plaa, G.L. (1979) *Toxicol. Appl. Pharmacol.* **48**, 509-517
18. Cianflone, D.J., Hewitt, W.R., Villeneave, D.C., Plaa, G.L. (1980) *Toxicol. Appl. Pharmacol.* **53**, 140-149
19. Ijima, M., Cote, M.G., and Plaa, G.L. (1983) *Toxicol. Lett.* **17**, 307-314

20. Hewitt, L.A., Hewitt, W.R., and Plaa, G.L. (1983) *Fundam. Appl. Toxicol.* **3**, 489-495
21. Hewitt, L.A., Caille, G., and Plaa, G.L. (1986) *Can. J. Physiol. Pharmacol.* **64**, 477-482
22. Purushotham, K.R.Lockard, V.G., and Mehendale, H.M. (1988) *Toxicol. Pathol.* **16**, 27-34
23. Hewitt, W.R, Miyajima, H., Cote, M.G., and Plaa, G.L. (1980) *Fed. Proc.* **39**, 3118-3123
24. Charbonneau, M., Ijima, M., Cote, M.G., and Plaa, G.L. (1985) *Toxicology* **35**, 95-112
25. Glende, Jr.E.A., and Lee, P.Y. (1985) *Exp. Mol. Pathol.* **42**, 167-174
26. Mehendale, H.M., and Klingensmith, J.S. (1988) *Toxicol. Appl. Pharmacol.* 93, 247-256
27. Klingensmith, J.S., and Mehendale, H.M. (1984) *Life Sci.* **33**, 2339-2348
28. Davis, M.E., and Mehendale, H.M. (1980) *Toxicology* **15**, 91-103
29. Noguchi, T., Fong, K.L., Lai, E.K., Alexander, S.S., King, M.M., Olson, L., Poyer, J.L., and McCay P.P. (1982) *Biochem. Pharmacol.* **31**, 615-624
30. Frank, H., Haussman, H.J., and Remmer, H. (1982) *Chemico-Biol. Inter.* **40**, 193-208
31. DeGroot, H., and Haas, W. (1981) *Biochem. Pharmacol.* **30**, 2343-2347
32. Fander, U., Haas, W., and Kroner, H. (1982) *Exp. Mol. Pathol.* **36**, 34-43
33. Klaassen, C.D., and Plaa, G.L. (1966) *Toxicol. Appl. Pharmacol.* **9**, 139-151
34. Brabeck, M.J., Owens, J.B., Kenel, M., Sorscher, D., and Cornish, H.H. (1982) *Drug Chem. Toxicol.* **5**, 143-154
35. Pilon, D., Charbonneau, M., Brodeur, J., and Plaa, G.L. (1986) *Toxicol. Appl. Pharmcol.* **85**, 49-59
36. Jernigan, J.D., Pounds, J.G., and Harbison, R.D. (1983) *Fundam. Appl. Toxicol.* **3**, 22-26
37. McLean, A.E.M., and McLean, E.K. (1966) *Biochem. J.* **100**, 564-571
38. Carlson, G.P. (1975) *Toxicology* **5**, 69-77
39. Suarez, K.A., Carlson, G.P., Fuller, G.C., and Fausto, N. (1972) *Toxicol. Appl. Pharmacol.* **23**, 171-177
40. Pitchumoni, C.S., Stenger, R.J., Rosenthal, W.S., and Johnson, E.A. (1972) *J. Pharmacol. Exp. Ther.* **181**, 227-233
41. Ferreyra, E.C., DeFenos, V.M., Bernacchi, A.S., DeCastro, C.R., and Castro, J.A. (1983) *Toxicol. Appl. Pharmacol.* **42**, 513-521
42. Smuckler, E.A., and Hultin, T. (1966) *Exp. Mol. Pathol.* **5**, 504-515
43. Kutob, S.D., and Plaa, G.L. (1962) *J. Pharmacol. Exp. Ther.* **135**, 245-251
44. Branchflower, R.V., and Pohl L.R. (1981) *Toxicol. Appl. Pharmacol.* **61**, 407-413
45. Kluwe, W.M., and Hook, J.B. (1978) Toxicol. *Appl. Pharmacol.* **45**, 861-869
46. Pohl, L.R. (1979) *Rev. Biochem. Toxicol.* **1**, 79-107

47. Recknagel, R.O., and Glende, Jr.E.A. (1977) *Handbook of Physiology*. (Lee, D. H. K. ed) Section 9, p 591-601. American Physiological Society, Bethesda, MD. Williams and Wilkins, Baltimore, MD
48. Reynolds, E.S., Treinen, R.J., Farish, H.H., and Moslen, M.T. (1984) *Arch. Toxicol. (Suppl)* **7**, 303-306
49. Slater, T.F. (1984) *Biochem. J.* **222**, 1-15
50. Mehendale, H.M. (1989) *Rev. Biochem. Toxicol.* **10**, 91-138
51. Lockard, V.G., O'Neal, R.M., and Mehendale, H.M. (1983) *Exp. Mol. Pathol.* 39, 230-245
52. Lockard, V.G., O'Neal, R.M., and Mehendale, H.M. (1983) *Exp. Mol. Pathol.* **39**, 246-255
53. Bell, A.N., Young, R.A., Lockard, V.G., and Mehendale, H.M. (1988) *Arch. Toxicol.* **61**, 392-405
54. Kodavanti, P.R.S., Joshi, U.M., Young, R.A., Bell, A.N., and Mehendale, H.M. (1989). *Arch. Toxicol.* **63**, 367-375
55. Kodavanti, P.R.S., Joshi, U.M., Lockard, V.G., and Mehendale, H.M. (1989) *J. Appl. Toxicol.* **9**, 367-375.
56. Prasada Rao, K.S., Joshi, U.M., Young, R.A., and Mehendale, H.M. (1989) *Toxicol. Pathol.* **17**, 496-504
57. Young, R.A., and Mehendale, H.M. (1989) *J. Biochem. Toxicol.* **4**, 211-219
58. Bell, A.N., and Mehendale, H.M. (1987) *Toxicol. Lett.* **35**, 191-200
59. Kodavanti, P.R.S., Kodavanti, U.P., and Mehendale, H.M. (1990) *Biochem. Pharmacol.* **39**, In Press
60. Rao, S.B., and Mehendale, H.M. (1989) *Biochem. J.* **262**, 721-724
61. Rao, S.B., and Mehendale, H.M. (1989) *Intern. J. Biochem.* **21**, 949-954
62. Desaiah, D., Ho, I.K., and Mehendale, H.M. (1977) *Toxicol. Appl. Pharmacol.* **39**, 219-228
63. Desaiah, D., Ho, I.K., and Mehendale, H.M. (1977) *Biochem. Pharmacol.* **26**, 1115-1159
64. Carmines E.L., Carchman, R.A., and Borzelleca, J.F. (1979) *Toxicol. Appl. Pharmacol.* **49**, 543-550
65. Agarwal, A.K., and Mehendale, H.M. (1984) *Toxicology* **30**, 17-24
66. Agarwal, A.K., and Mehendale, H.M. (1984) *Life Sci.* **34**, 141-148
67. Agarwal, A.K., and Mehendale, H.M. (1984) *Environ. Health Persp.* **57**, 289-291
68. Agarwal, A.K., and Mehendale, H.M. (1986) *Toxicol. Appl. Pharmacol.* **83**, 342-348
69. Rao, S.B., Young, R.A., and Mehendale, H.M. (1989) *J. Biochem. Toxicol.* **4**, 55-63
70. Chang, L.W., Periera, M.A., and Klaunig, J.E. (1985) *Toxicol. Appl. Pharmacol.* **80**, 274-283
71. Roberts, E., Ahluwalia, M.B., Lee, G., Chan, C., Sharma, D.S.R., and Farber, E. (1983) *Cancer Res.* **43**, 28-34

72. Ruch, R.J., Klaunig, J.E., and Pereira, M.A. (1985) *Cancer Lett.* **26**, 295-301
73. Ahmadizadeh, M., Jui, C.,Echt, R., and Hook, J.B. (1984) *Toxicology* **31**, 342-352
74. Mehendale, H.M., Purushotham, K.R., and Lockard, V.G. (1989) *Exp. Mol. Pathol.* **51**, 31-47
75. Mehendale, H.M. (1989) *Toxicol. Lett.* **49**, 211-241
76. Moore, L. (1980) *Biochem. Pharmacol.* **29**, 2505-2511
77. Kroner, H. (1982) *Biochem. Pharmacol.* **31**, 1069-1073
78. Schanne, F.A.X., Kane, A.B., Young, E.E., and Farber, J.L. (1979) Science 206, 700-702
79. Lowery, K., Glende, Jr.E.A., and Recknagel, R.O. (1981) *Biochem. Pharmacol.* **30**, 135-140
80. Kodavanti, P.R.S., and Mehendale, H.M. (1988) *Fed. Am. Soc. Exp. Biol.* **2**, A408
81. Mehendale, H.M. (1990) *Med. Hypoth.* **32**, In Press
82. Janne, J., Poso, H., and Raina, A. (1978) *Biochim. Biophys. Acta* **4**, 241-293
83. Pegg, A.E., and McCann, P.P. (1982) *Am. J. Physiol.* **243**, C 212-221
84. Tabor, C.W., and Tabor, H. (1984) *Ann. Rev. Biochem.* **53**, 749-790
85. Cueto, Jr.C., Page, N.P., and Saffiotti, U. (1979) *Report on carcinogenesis bioassay of technical grade CD (Kepone^R).* National Cancer Institute, DHEW Publication, NIH-76-1278, 1-27
86. Della-Porta, G., Terracini, B., and Shubik, P. (1961) *J. Natl. Cancer Inst.* 26, 855-862
87. National Cancer Institute (1976) *Report on Carcinogenesis bioassay of chloroform.* NIC, Bethesda, MD

Structure Activity Correlations of 1,4-Dihydropyridines as Reducing Agents

Serjinder Singh, Minakshi Chanana, Harjit Singh, Sarita Bhalla, and Navneet Sahota
Department of Chemistry, Guru Nanak Dev University, Amritsar 143 005 India

Abstract

Kinetic investigations of reductions with *sym*-3,5-disubstituted-1,4-dihydropyridines indicate a linear Hammette type correlation between the nature of the substituent and the reducing power of the 1,4-dihydropyridine. Extrapolation of these plots predicted a very high reactivity for 1,4-dihydropyridine-3,5-dicarboxylate dianion moiety. This was confirmed by *in situ* generation of the latter by alkaline hydrolysis of Hantzsch Ester in the presence of substrates and the facile reduction of these substrates normally not reduced by Hantzsch Ester as such.

The 1,4-dihydropyridine moiety is of immense importance as a constituent of biologically important coenzymes NADH and NADPH (1). The tendency of this moiety to release elements of a "hydride ion" and aromatise to pyridinium ion makes it a useful reducing agent in biochemical processes. The unsubstituted 1,4-dihydropyridine is a highly reactive species towards not only oxidants but also electrophiles and protic solvents (2) and hence is not convenient if to be employed as reducing agent in the laboratory. However, nature employs the reasonably stabilized 1,4-dihydronicotinamide moiety instead. The carboxamide group at the 3-position decreases the lone pair electron density on the ring nitrogen thereby lessening its tendency to eject a hydride ion from C-4. Even so, the N-alkyl-1,4-dihydronicotinamides although are analogous to NADH still are susceptible to hydration, dimerization, alkylation, oxidation and disproportionation under ordinary experimental conditions. Furthermore, common functional groups such as aldehydes, ketones, imines, and enones are as such unaffected by these dihydropyridines. An attempt to employ severe experimental conditions tend to destroy the latter. It was therefore pertinent to examine the effects of various substituents at different positions of the 1,4-dihydropyridine ring on its reactivity towards substrates to be reduced. A great deal of work has gone on (3) to investigate the use of various 1,4-dihydropyridine

Biological Oxidation Systems, Volume 2

derivatives as reducing agents for a variety of functional groups in organic compounds. Analogous to the mechanism of NADH mediated enzymatic reductions the nonenzymatic reductions using 1,4-dihydropyridines as NADH models also require activation of the otherwise inactive substrates. Activation is usually achieved by making the hydride acceptor atom in the substrate molecule more electron deficient by protonation or coordination to metal cations of heteroatoms in multiple bonds as in carboxyl compounds, imines etc. Alternatively, such bonds are linked to electron withdrawing group such as halomethyl, nitro etc. to achieve reduction. It has been observed that in general, extensive substitution of the ring sterically protects the dihydropyridines against side reactions with the solvent. Thus, decahydroacridine (1, R_1R_2= $-(CH_2)_2-$) is less susceptible to reaction with acidic solvents and side reactions than 1,4-dihydropyridine. But octahydroacridinedione (1, R_1R_2 = $-CO(CH_2)_3-$) is highly stable towards even dilute mineral acids still being able to reduce substrates. Symmetrically substituted 1,4-dihydropyridines of the Hantzsch Ester type (1, R_1R_2 = $-CO(CH_2)_3-$) are not only stable to solvents but also have reducing ability comparable to that of N-alkyl-1,4-dihydronicotinamides.

The substituents at positions 3- and 5- have a direct effect on the reducing power of the ring. Especially if the groups are able to get conjugated with the lone pair on the nitrogen atom the ability to eject hydride from C-4 is reduced. Similarly, the substituents at C-2 and C-6 enhance the reducing ability of the ring if the former can undergo hyperconjugation with the ring as in the case of 2,6-dimethyl-derivatives. Thus, the Hantzsch Ester derivatives are better reducing agents and less prone to solvent reactions than the corresponding compounds without 2,6-dimethyl substituents. The latter also are protected by these methyl groups sterically against enamine type reactions. Furthermore such compounds are very easily prepared compared with nicotinamides.

Results and Discussion

We have carried out an investigation of effects of different groups X on the reducing abilities of compounds 2. Several compounds of general structure 2 and differing in substituents X were prepared and kinetics of their reactions with common substrates investigated. Since carbenium ions are known hydride abstractors (4) these were chosen as the substrates to avoid mechanistic ambiguity. Malachite green is a stable carbenium ion and a convenient substrate whose reduction could be easily monitored spectrophotometrically and exists in ionic form thereby avoiding any problems of preforming the substrate. Pseudo first order rate constants were calculated (Table 1) by investigating the reactions of different 1,4-dihydropyridines with Malachite green in acetonitrile as solvent. The plot of the kinetic data gave a Hammette type linear relationship (Fig. 1) between the reactivity and the electronic nature of the substituents X at positions 3 and 5 of the dihydropyridines (2). The reactivity had the expected order COOR>$COCH_3$>CN.

Table 1. *Effect of 3,5-substituent on Reduction of Malachite Green with Various NADH Models*

NADH Model	2a	2b	2c	2d	2e	2f
X	CN	COC_6H_5	$COCH_3$	$COOCH_3$	COOEt	COO^-
$K_{obs} \times 10^2$	0.20	2.00	3.33	20.00	23.00	12,589 (predicted)
Relative Rate	1	10	17	100	115	62,955

The above behavior of these compounds is also in line with their general reactions such as reaction with aqueous acids and air. Whereas, the compounds with ester groups are rapidly destroyed in aqueous mineral acids or trifluoroacetic acid, the corresponding nitrile or diacetyl compounds are fairly stable (5). Similar order of reactivity is observed in NADH model studies where ketones, imines etc., are reduced.

An extrapolation of the Hammette plot to compound 2 (X = COO^-) indicated that the rate constant for this compound should be more than five hundred times higher than that for Hantzsch Ester (2, X = COOEt). Enhancement of reducing power by this magnitude implied that substrates normally not reduced by Hantzsch ester unless activated should get reduced with 2, (X = COO^-) unassisted by activating factors such as metal chelation, irradiation , protonation etc.

Indeed, 2 (X = COO^-) generated *in situ* by alkaline hydrolysis of Hantzsch ester (2, X = COOEt) under inert atmosphere reduced a variety of substrates (Table 2.) such as ketones, imines, enones, quinones etc., in fairly good yields. It may be pointed out that these substrates as such are not reduced by Hantzsch ester in the dark. The Hantzsch ester as such shows absorption maximum at 372 nm but

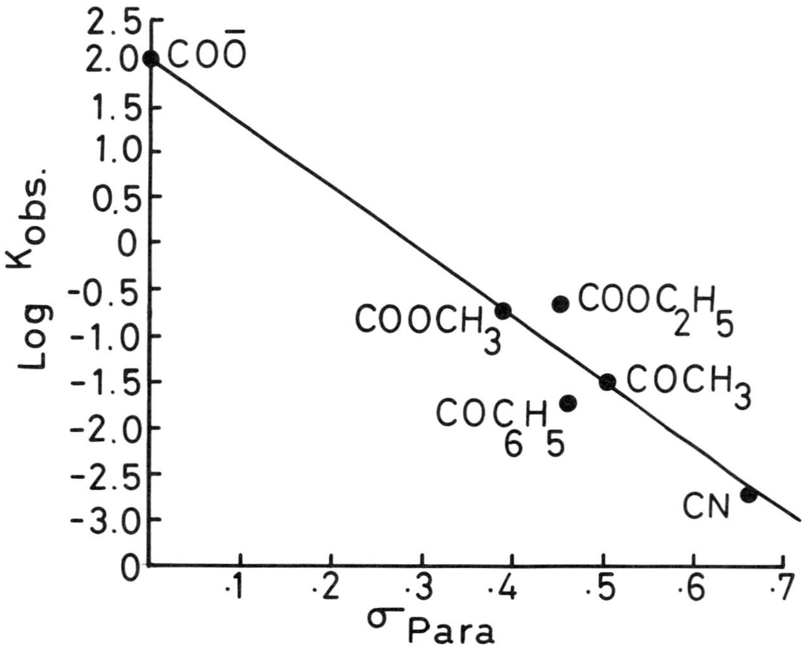

Figure 1. *Hammette Correlation Plot for Reduction of Malachite Green with Different 3,5-disubstituted-1,4-dihydropyridines*

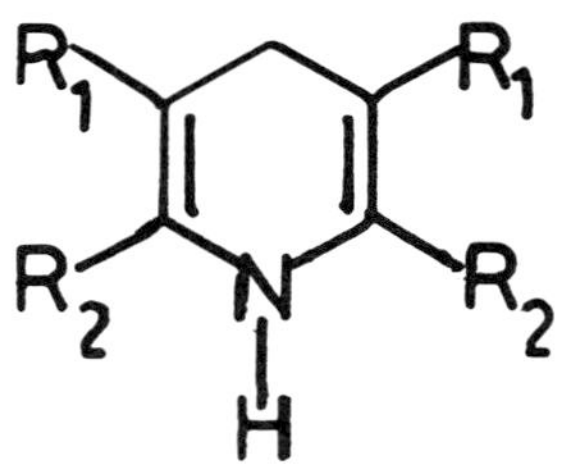

(1)

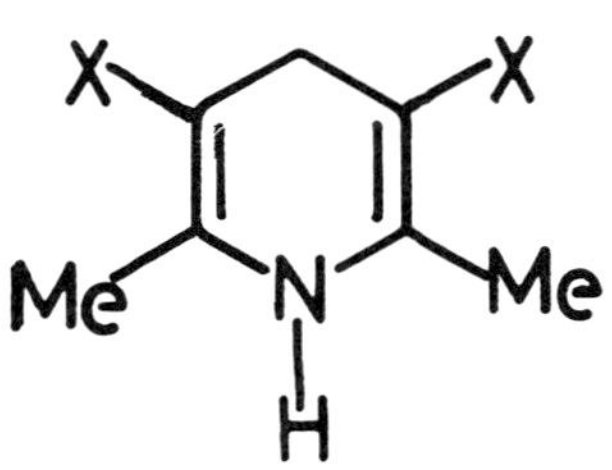

(2)

during hydrolysis first shows an absorption maximum at 350 nm followed by one at 330 nm corresponding presumably to the mono- and dicarboxylate anion species respectively. Attempts to isolate these species failed, only the corresponding pyridinecarboxylic acids could be isolated due to the air sensitivity of these highly reactive reducing agents. For good yields of the reduced products, 3-4 fold molar excess of the reducing agent was preferred. Aqueous alcohol was a suitable solvent but tert-butanol was preferred as the corresponding pyridinecarboxylate salts separated out and the reduced substrate remained in solution making isolation of the products convenient.

Table 2. *Reduction of Substrates with Dipotassium Salt of 3,5-dicarboxy-1,4-dihydropyridine*

Substrate	Products	Yield(%)
PhCHO	$PhCH_2OH$	95
$C_6H_5COC_6H_5$	C_6H_5-$CH(OH)C_6H_5$	89
$C_6H_5COCH_3$	C_6H_5-$CH(OH)CH_3$	72
$C_6H_5CH{=}CHCOC_6H_5$	$C_6H_5(CH_2)_2COC_6H_5$	35
	$C_6H_5(CH_2)_2CH(OH)C_6H_5$	18
	$C_6H_5CH{=}CHCH(OH)C_6H_5$	25
$C_6H_5CH{=}NC_6H_5$	$C_6H_5CH_2NHC_6H_5$	60
Cyclohexanone	Cyclohexanol	62
Phenanthraquinone	9,10-Phenanthrenediol	80
Anthraquinone	9,10-Dihydroyanthracane	84

We believe that the important reducing compound is 2, (X = COO^-) and not the half hydrolyzed Hantzsch ester . The latter should have approximately the same reducing power as that of ethyl N-alkyl-1,4-dihydronicotinate not known to reduce the above mentioned substrates. The products formed by reduction of enones indicate reduction of the double bond due to activation by the carbonyl group followed to a certain extent by the reduction of the carbonyl. Partially, the reduction of the carbonyl takes place by 1,2-hydride addition to the enone but main product is due to 1,4-hydride addition. This behavior is analogous to that shown by the enones towards other nucleophiles thereby indicating that the mechanism of hydrogen transfer by this NADH model is through hydride transfer and not one electron transfer. In conclusion, the systematic investigation of reactivities of substances with gradual variations in structure helps not only to understand the underlying factors but also to predict the reactivity and hence design a useful reagent.

Acknowledgements

We are grateful to the Council of Scientific and Industrial Research (CSIR), India, and Department of Science and Technology, Govt of India (DST), for financial support.

References

1. Kosower, E.M., (1962) in *Molecular Boichemistry*, pp 166, McGraw-Hill; Kaplan, N.O (1960) Enzymes, Vol 3, pp 105, Sund, H., (1968) in *Biological Oxidations* (Singer,T.P. ed.) pp 603
2. Johnson, C.C., Gardner, J.L., Suelter, C.H., and Meltzer, D.E., (1963) *Biochemistry* **2**, 689
3. Kill, R.J. and Widdowson, D.A. (1978) in *Bioorganic Chemistry* (vanTamelan, E.E. ed.) Vol. 4 pp 237-275, Academic Press, New York
4. Nenitzescn, C.D., (1973) in *Carbonium Ions* (Olah, G.A and Schleyer, P.R.) Vol. *4*, pp 463
5. Singh, S., Gill, S., Sharma, V.K., and Nagrath, S., (1986) *J. Chem. Soc., Perkin* **I**, 1273

Modulation of Tubulin Tyrosinolation in Human Neutrophils by the NADPH Oxidase-Mediated Respiratory Burst

Jayasree Nath
Department of Hematology, Walter Reed Army Institute of Research
Washington, DC 20307-5100, USA

Abstract

Microtubules, a major component of the neutrophil cytoskeleton, are believed to be functionally important in neutrophil inflammatory responses and host-defense mechanisms. Tubulin, a heterodimer consisting of α- and β- chains, is the major subunit protein of microtubules. The α-chain of tubulin can be ezymatically modified by a reversible addition of a single tyrosine residue to its carboxy-terminal glutamate (tubulin tyrosinolation). Although the functional role of tubulin tyrosinolation still remains to be defined, a 2-3 fold stimulation of tubulin tyrosinolation is induced when neutrophils are activated by the peptide chemoattractant fmet-leu-phe. Our studies show that the observed stimulation of tubulin tyrosinolation is closely associated with activation of the superoxide-generating NADPH oxidase and associated cellular redox changes in stimulated PMNs. The fmet-leu-phe-induced tubulin tyrosinolation is not observed under anaerobic conditions and is markedly inhibited (50-100%) in the presence of a variety of reducing agents and scavengers of oxygen radicals. A strong correlation between neutrophil oxidative metabolism and the modulation of tubulin tyrosinolation is further substantiated by the following observations: (1) The fmet-leu-phe-stimulated tubulin tyrosinolation is absent in neutrophils of patients with chronic granulomatous disease that have a defective NADPH oxidase; (2) Expression of fmet-leu-phe-induced tubulin tyrosinolation is coincident with induced differentiation and acquisition of the NADPH oxidase-mediated respiratory burst activity in differentiated HL-60 cells; (3) By direct demonstration of inhibition of fmet-leu-phe-stimulated tubulin tyrosinolation and superoxide generation in parallel, both in human neutrophils and in differentiated HL-60 cells.

Introduction

The ability of neutrophils (PMNs) to orient and migrate in response to chemotactic stimuli and to phagocytize foreign particles, depends on the dynamic properties of the cell membrane (1) and the cytoskeletal components (1-4), *i.e.* microtubules and the microfilaments. Microtubules are believed to be essential for the initial orientation and subsequent morphological stability during sustained locomotion (1-4), while microfilaments are considered to be the actual motile machinery of phagocytic cells (5). Furthermore, microtubules are functionally important component of the PMN cytoskeleton with respect to its inflammatory responses such as granule mobilization and degranulation (6,7). High-voltage electron microscopic studies have revealed that upon activation of PMNs, not only is there an increase in the total number of centriole-associated microtubules, there is also a substantial increase in the average length of the microtubules (4). Although much remains to be learned about the precise role of microtubules in PMN inflammatory responses, their integrity and cellular organization are important functional components of PMN host-defense mechanisms.

Tubulin, a heterodimer consisting of α- and β-chains, is the major subunit protein of microtubules (8). The α-chain of tubulin is subject to a unique biochemical modification, whereby the carboxyl terminus can be enzymatically modified by a reversible addition of a single tyrosine residue to its carboxy-terminal glutamate (9-11). A tubulin-tyrosine ligase (ligase) catalyzes the ATP-dependent addition of tyrosine to the carboxyl terminus of tubulin α-chain (tubulin tyrosinolation). A distinctly different carboxypeptidase-type enzyme which releases the tyrosine (detyrosinolation), has also been identified and characterized (12). The latter appears to be the initial modification, since tyrosine also is the coded terminal residue (13,14). Although the presence or absence of the carboxy-terminal tyrosine has not been found to affect assembly or disassembly of microtubules *in vitro* (12) or *in vivo* (15), in our earlier studies with a varietry of neuronal and non-neuronal cells, we have observed significant changes in the state of tyrosinolation of tubulin in cells undergoing cytoskeletal reorganization, e.g. during cellular differentiation (16) or in mitosis (17). During the course of these studies we observed a 2-3 fold stimulation of tubulin tyrosinolation in rabbit peritoneal neutrophils (18) and also in human peripheral blood PMNs (19), when these cells were activated by the synthetic peptide chemoattractant N-formyl methionyl-leucyl-phenylalanine (fmet-leu-phe). Subsequent studies demonstrated that the observed stimulation of tubulin tyrosinolation was closely correlated with activation of the NADPH oxidase-mediated respiratory burst and associated redox changes in stimulated PMNs (20).

Phagocytic cells (neutrophils, macrophages) react to a variety of stimulants by the production of superoxide anions (O_2^-) and provide an excellent example of a biological oxidation system. The enzymatic activity responsible for this "respiratory burst" resides in a membrane-bound NADPH oxidase, which reduces molecular oxygen to O_2^- (21). In subsequent reactions, O_2^- is transformed to bactericidal products (21). It is generally believed that the O_2^- forming NADPH oxidase contains a characteristic b-type cytochrome (cytochrome b_{558}) and a FAD-containing flavoprotein (22,23). The biochemical sequence of the activation pathway of the NADPH oxidase involves several different post-receptor steps depending on the agonist used. Although the exact sequence of events is not fully understood, recent reports from a number of laboratories indicate the critical involvement of a protein kinase C (PK-C) and some essential cytosolic activating factor(s), in the stimulus-response coupling for the activation of the NADPH oxidase (24).

As mentioned above, during studies designed to understand the possible cellular mechanism of modulation of tubulin tyrosinolation in human PMNs, a strong correlation between activation of the O_2^--generating NADPH oxidase and stimulation of tubulin tyrosinolation, was observed (20,25,26). The experimental approaches that led to this conclusion included manipulation of the oxidative-reductive status of stimulated PMNs (19,20,25) and exploitation of PMNs obtained from patients whose oxidative metabolism were known to be either exaggerated (19) or depressed (20). Some of these studies that are relevant to the theme of the present symposium, are briefly discussed in this review.

Experimental Procedures

For all experiments described in the present review, freshly prepared human peripheral blood PMNs were used (19,20). Because tubulin tyrosinolation is a posttranslational modification (9-11), the reaction can be studied in intact functional cells by measuring the incorporation of radiolabelled tyrosine into trichloroacetic acid-insoluble fraction, in the complete absence of protein synthesis (16). The details of this method have been published elsewhere (16,18-20). The experimental conditions utilized to study tubulin tyrosinolation in intact functional PMNs (18,19) or in other neuronal (16,27) and non-neuronal cells (15,28), allow exclusive incorporation of [^{14}C]-tyrosine into tubulin α-chains (18,19). SDS-urea polyacrylamide gel electrophoresis and subsequent fluorography reveals a single radioactive protein band which co-migrates with authentic [^{14}C]-tyrosinolated tubulin (as shown in Fig. 4). Kinetics of O_2^- release by stimulated PMNs were assessed by continuous monitoring of the superoxide dismutase-inhibitable reduction of ferricytochrome C at 550 nm (29). The details of the procedure have been recently described (26,30).

Results

Stimulation of PMN Tubulin Tyrosinolation

As shown in Fig. 1, a 2-3 fold stimulation of tubulin tyrosinolation is observed when PMNs are activated by the synthetic peptide chemoattractant fmet-leu-phe. Furthermore, similar to other fmet-leu-phe receptor-mediated functional responses in PMNs (e.g. chemotaxis, O_2^- generation, degranulation), the stimulation of tubulin tyrosinolation, as induced by fmet-leu-phe, is also significantly inhibited by prior incubation of PMNs with pertussis toxin (PT). As shown in Fig. 2, PT inhibits fmet-leu-phe-induced stimulation of

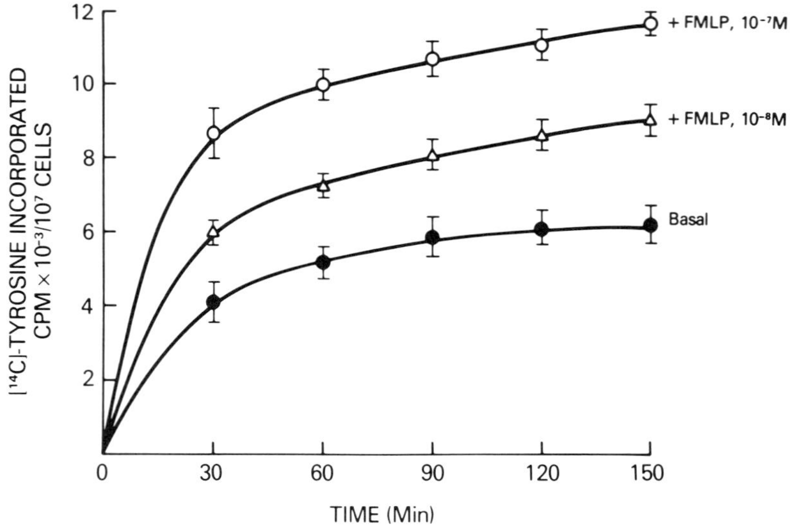

Figure 1. *Stimulation of Posttranslational Incorporation of Tyrosine in PMNs Activated with fmet-leu-phe (FMLP).* PMNs at 10^7/ml were preincubated for 30 min at 37° C with protein synthesis inhibitors. FMLP and [^{14}C]-tyrosine (5 μCi = 0.01 μmole) were added at 0 min. TCA-insoluble radioactivity was measured at indicated times. The results are means ± SEM of three separate experiments. Reproduced from (19).

tubulin tyrosinolation in a dose-dependent manner. These results are consistent with data published by other investigators which demonstrate that the signal-transduction pathways via fmet-leu-phe receptor is coupled to a PT-sensitive GTP-binding protein (31-34).

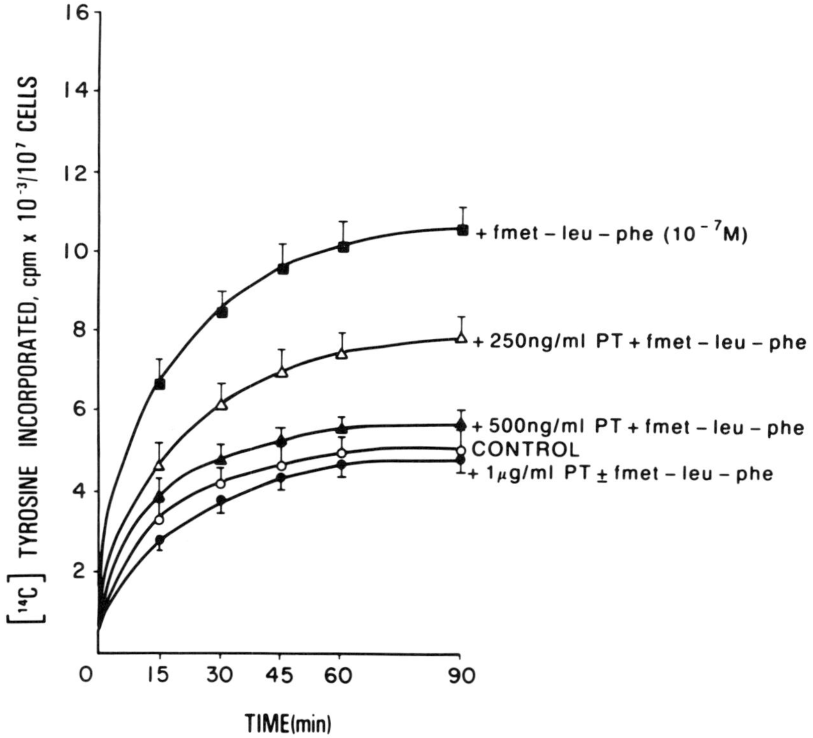

Figure 2. *Dose-Dependent Inhibition of fmet-leu-phe-Stimulated Tubulin Tyrosinolation by Pertussis Toxin (PT).* PMNs at 10^7/ml were incubated for 90 min with indicated concentrations of PT prior to studying posttranslational incorporation of $[^{14}C]$-tyrosine. All other experimental conditions were identical to those described for Fig. 1. The results are means ± SEM of three different experiments.

Effects of Ascorbic Acid and Other Reducing Agents on PMN Tubulin Tyrosinolation

One of the earliest experimental results which suggested a possible association between changes in PMN redox state and modulation of tubulin tyrosinolation, came from the experimental data shown in Fig. 3. Ascorbic acid was found to have a profound inhibitory effect both on the rate and the final extent of posttranslational incorporation of $[^{14}C]$-tyrosine in fmet-leu-phe-stimulated

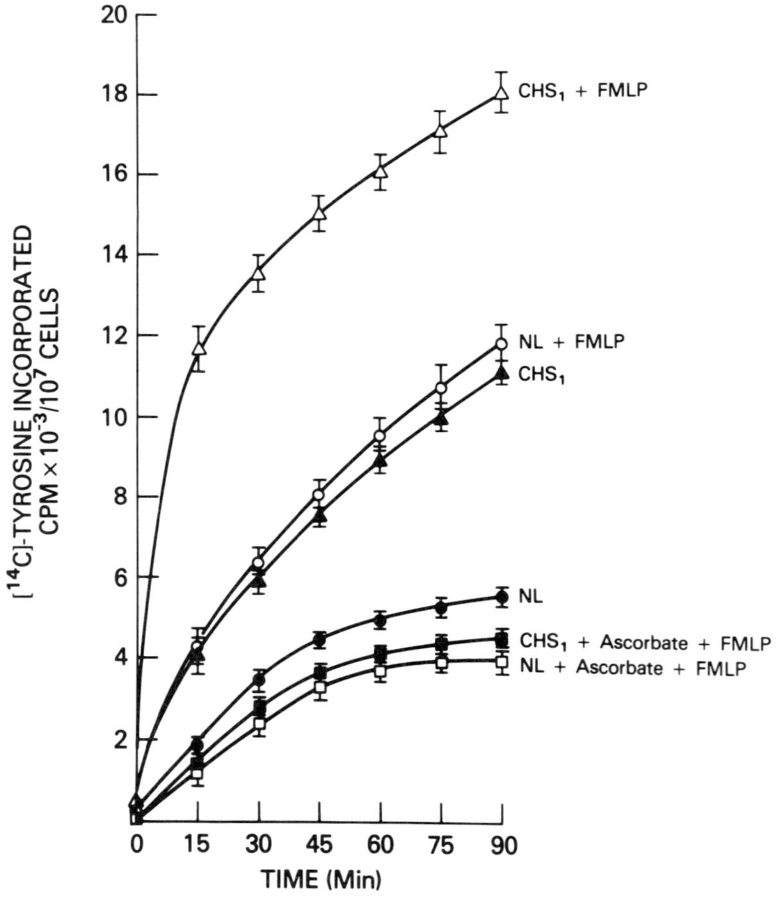

Figure 3. *Effect of in vitro Addition of Ascorbate on the Resting and fmet-leu-phe (FMLP)-Stimulated Incorporation of Tyrosine in Normal and Chediak-Higashi Syndrome (CHS_1) PMNs.* Where indicated, 10^{-4} M ascorbate was added during incubation with protein synthesis inhibitors prior to addition of 10^{-7} M FMLP and [^{14}C]-tyrosine at 0 min. Other experimental conditions were the same as in Figure 1. The results are means ± SEM for three separate experiments (reproduced from Ref. 41).

PMNs. This was found to be true in normal PMNs and also in PMNs from patients with Chediak Higashi Syndrome (CHS) that have an exaggerated oxidative metabolism with an abnormally high level of tubulin tyrosinolation (19). The highly specific nature of the observed posttranslational incorporation of [^{14}C]-tyrosine in PMNs is shown in

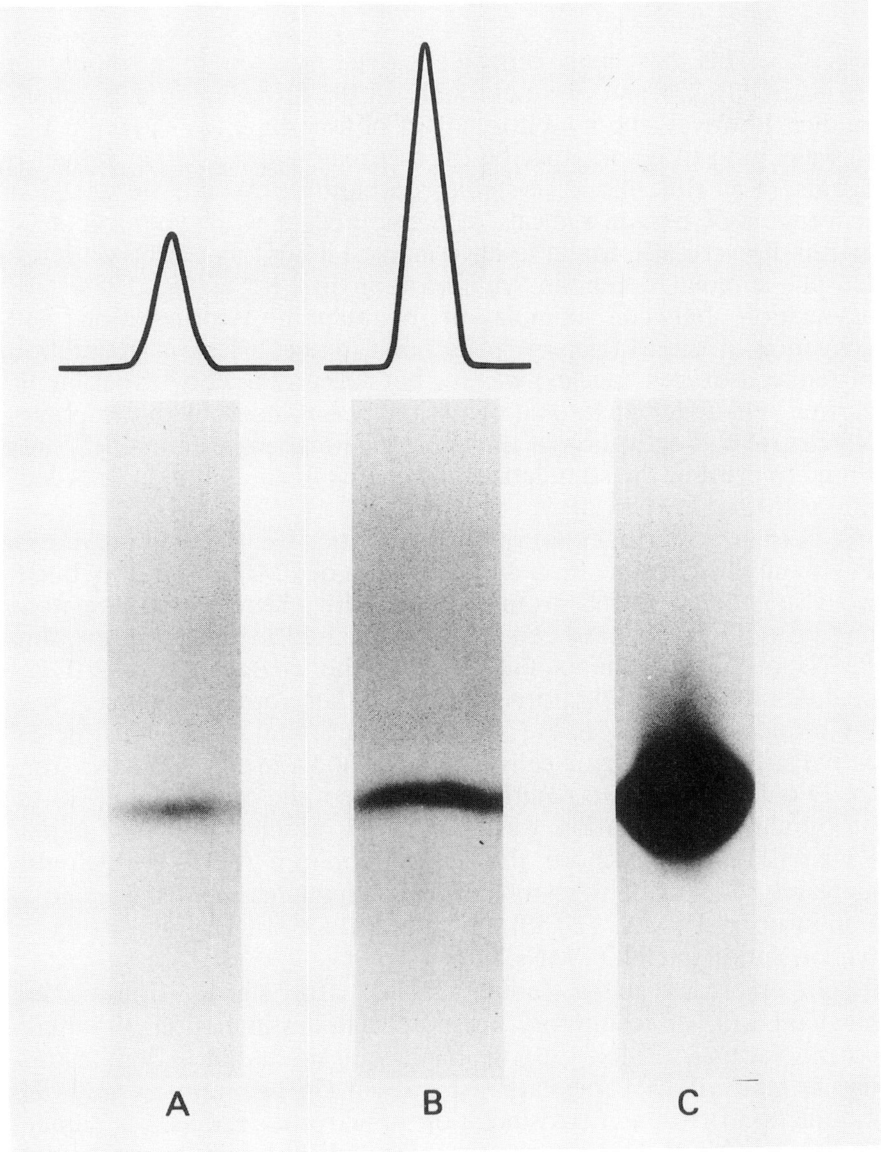

Figure 4. *Quantitative Demonstration of the Specific Posttranslational Incorporation of [^{14}C]-Tyrosine into Tubulin α-chains of Normal and Chediak-Higashi Syndrome (CHS$_1$) PMNs.* Equal amounts of [^{14}C]-tyrosinolated PMN samples were subjected to SDS-urea polyacrylamide gel electrophoresis and the gel was processed for fluorography. Details of the procedure are reported in Refs. 18 and 19. Lane A: normal PMNs; Lane B: CHS$_1$ PMNs; Lane C: an overexposed sample of authentic [^{14}C]-tyrosinolated tubulin preparation. The upper panel (in lanes A and B) represents the densitometric scans of the corresponding radioactive bands in the lower panel (reproduced from Ref. 19).

Fig. 4. This has been repeatedly demonstrated in a variety of neuronal (16,27) and non-neuronal (18,19,28) cells. In subsequent studies, however, the inhibitory effect of ascorbate (as shown in Fig. 3) was found to be not specific for ascorbate alone. Other reducing agents (e.g. glutathione, cysteine, dethiothreitol) and intracellular scavengers of oxygen radicals (e.g. thiourea, dihydroxybenzoic acid, mannitol) were also found to have marked inhibitory effects on fmet-leu-phe-stimulated tubulin tyrosinolation in PMNs (19,20,25). An association between stimulation of tubulin tyrosinolation and activation of the PMN respiratory burst (and subsequent generation of reduced oxygen species) was further substantiated by experiments carried out under anaerobic conditions. An anaerobic atmosphere, which prevents activation of the neutrophil respiratory burst (35), was found to prevent the stimulation of tubulin tyrosinolation in fmet-leu-phe-activated PMNs (20).

In order to further confirm the association between stimulation of PMN tubulin tyrosinolation and activation of PMN respiratory burst, we also studied PMNs from patients with chronic granulomatous disease (CGD). It has been well-documented in the literature that PMNs of CGD patients fail to elicit the characteristic NADPH oxidase-mediated respiratory burst, and therefore, cannot generate the reduced oxygen species (36) that are essential for the bactericidal activity of the phagocytic cells (36). As shown in Fig. 5, PMNs from CGD patients failed to respond to fmet-leu-phe and the stimulation of tubulin tyrosinolation was completely absent in these patient PMNs (20). Furthermore, the Ca^{2+} ionophore A23187, which also activates PMN oxidative burst (37) and stimulates tubulin tyrosinolation in normal PMNs (20,38), failed to induce stimulation of tubulin tyrosinolation in CGD PMNs.

In more recent studies related to signal transduction mechanism(s) of respiratory burst-associated stimulation of tubulin tyrosinolation in PMNs (26), we have been successful in demonstrating a more direct correlation between O_2^- generation and the modulation of tubulin tyrosinolation in activated PMNs. A parallel inhibition of both O_2^- generation and tubulin tyrosinolation in fmet-leu-phe-stimulated PMNs is shown in Fig. 6. During the course of these studies we were also able to demonstrate that in the human myeloid leukemia cell line HL-60, expression of the fmet-leu-phe-stimulated tubulin tyrosinolation coincided with the acquisition of the NADPH oxidase-mediated respiratory burst activity (26), which occur during induced cellular differentiation of these cells. The details of these studies have recently been published (26) and therefore, are not further discussed in the present review.

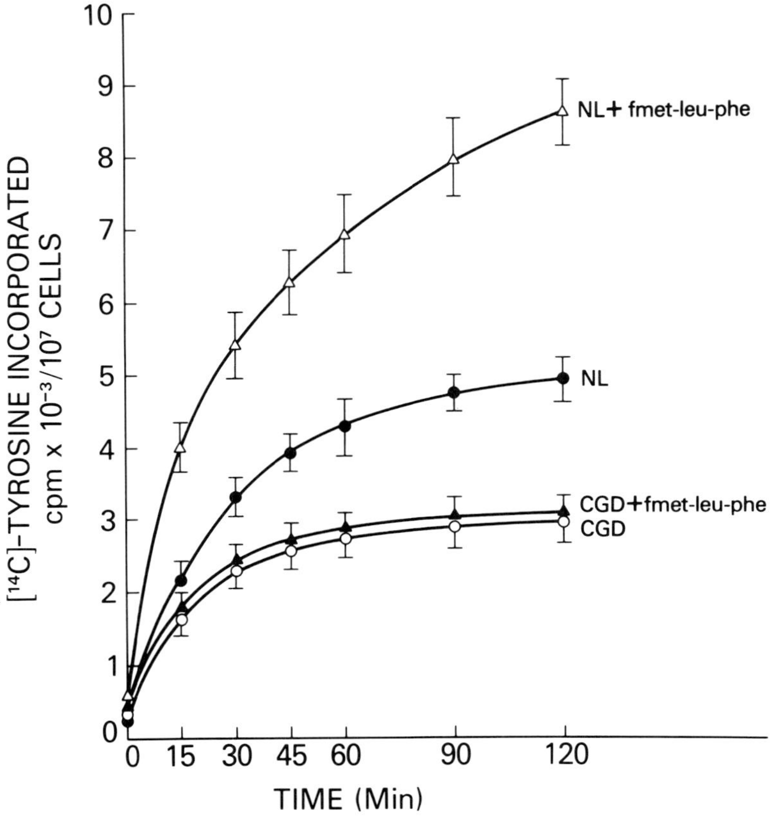

Figure 5. *Resting and fmet-leu-phe-Stimulated Posttranslational Incorporation of Tyrosine in Normal (NL) and Chronic Granulomatous Disease (CGD) PMNs.* Experimental conditions were identical to those described in Figure 1. Data are means ± SEM of studies in PMNs of seven different normal individuals and in seven different patients with CGD (reproduced from Ref. 20).

Discussion

Although the cellular functional role of tubulin tyrosinolation still remains to be understood, its rapid stimulation in activated PMNs provides the only example of a biological system where such modulation of tubulin tyrosinolation has been observed. It is also worthwhile to emphasize that the observed stimulation of PMN tubulin tyrosinolation, that appears to be closely associated with the

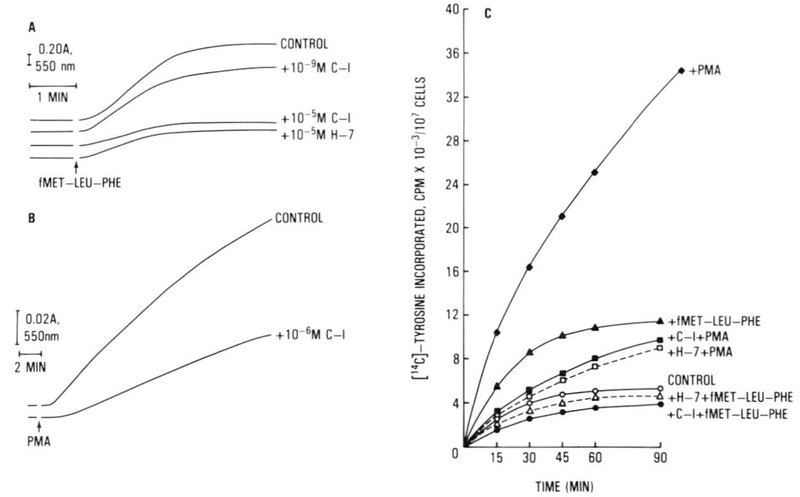

Figure 6. *Parallel Inhibition of O_2^- Production, fmet-leu-phe-stimulated Tubulin Tyrosinolation and PMA-Induced Posttranslational Incorporation of Tyrosine into Multiple Proteins by PK-C inhibitor C-I or H-7 in human PMNs.* Panel A, inhibition of fmet-leu-phe-induced O_2^- generation; panel B, inhibition of PMA-induced O_2- generation; panel c, inhibition of tubulin tyrosinolation and random posttranslational incorporation of tyrosine. Where indicated, 10^{-5} M C-I or H-7 was added at 37°C, 10 min prior to stimulation of PMNs by 10^{-7} M fmet-leu-phe or 20 ng/ml PMA (reproduced from Ref. 26).

changes in the redox state of activated PMNs, occurs only in intact functional cells and not in broken cell preparations. As our earlier studies have demonstrated (18,19), the stimulation is not due to changes in the specific activity of the ligase or the amount of cellular tubulin (18,19) in activated PMNs. Furthermore, in *in vitro* experiments carried out in a cell-free system, the ligase-catalyzed tyrosinolation of PMN tubulin was not found to be affected by any of the experimental manipulations that modulated tubulin tyrosinolation in intact functional PMNs (J. Nath, unpublished results). Moreover, microtubule-related cytoskeletal reorganization that occur upon activation of PMNs (1-4), was not found to be a necessary prerequisite for the induction of tubulin tyrosinolation in these cells (18,19). One is also impressed by the strictly intracellular nature of the observed modulation of tubulin tyrosinolation in stimulated PMNs. Although the activation of the NADPH oxidase-mediated respiratory burst appears to be an essential requirement for the observed stimulation to occur in activated PMNs (20,26), extracellularly generated O_2^- anions (e.g. by a xanthine/xanthine oxidase system) were not found to increase the posttranslational incorporation of tyrosine in unstimulated PMNs and conversely, extracellular scavengers of oxygen radicals (e.g. superoxide dismutase) did not inhibit the stimulation of tubulin tyrosinolation in intact PMNs (J. Nath, unpublished results).

The biochemical characterization of the modulation of PMN tubulin tyrosinolation has not yet been achieved. This is primarily due to the small amount of tubulin (1% of the total PMN protein) present in these cells. Because tubulin is known to have critical sulfhydryl groups that are essential for its functional integrity (39,40), it is not unreasonable to speculate that the dramatic changes in the redox state of activated PMNs could possibly influence the properties of tubulin as a substrate, that in turn, may significantly alter the rates of tyrosinolation or detyrosinolation, in stimulated PMNs. The studies related to the effects of ascorbate and other reducing agents on PMN tubulin tyrosinolation (19,25,41) are also consistent with this idea. Understanding the biochemical mechanism(s) of modulation of PMN tubulin tyrosinolation, and the reaction's possible functional role in stimulated PMNs, still remain as challenging questions in neutrophil biology.

Acknowledgement

The author wishes to thank Drs. Martin Flavin, Elliot Schiffmann, John Gallin and Daniel Wright for their support and encouragement during various phases of this study, and Mary West for typing the manuscript.

References

1. Oliver, J.M., and Berlin, R.D. (1979) in *Soc. Exptl. Biol. Symposia,* **33**, 277-298
2. Malech, H.L., Root, R.K., and Gallin, J.I. (1977) *J. Cell Biol* **75**, 666-693
3. Oliver, J.M. (1978) *Am. J. Pathol.* **93**, 221-278
4. Schliwa, M., Pryzwansky, K.B., and Euteneur, U. (1982) *Cell,* **31**, 705-717
5. Stossel, T.P. (1978) in *Leukocyte Chemotaxis.* Raven Press, N.Y. 143-160
6. Hirsch, J.G., and Cohn, Z.A. (1960) *J. Exp. Med.* **112**, 1005-1015
7. Wright, D.G. (1982) *Adv. Host Def. Mech.* **1**, 75-110
8. Cleveland, D.W., and Sullivan K.F. (1985), *Ann. Rev. Biochem.* **54**, 331-365
9. Raybin, D., and Flavin, M. (1977) *Biochemistry.* **16**, 2189-2194
10. Arce, C.A., Hallak, M.E., Rodriquez, J.A., Barra, H.S., and Caputto, R. (1978) *J. Neurochem.* **31**, 205-210
11. Raybin, D., and Flavin, M. (1977) *J. Cell Biol.* **73**, 492-504

12. Kumar, N., and Flavin, M. (1981) *J. Biol. Chem.* **256**, 7678-7686
13. Lemischka, I.R., Farmer, S., Racaniello, V.R., Sharp, P.A. (1981) *J. Mol. Biol.* **151**, 101-120
14. Valenzuela, P., Quiroga, M. Zaldiver, J., Rutter, W.J., Kirschner, M.W., and Cleveland, D.W. (1981) *Nature,* **289**, 650-655
15. Kobayashi, T., and Flavin, M. (1981) *Comp. Biochem. Physiol. B.* **69**, 387-392
16. Nath, J., and Flavin, M. (1978) *J. Biol. Chem.* **254**, 11505-11510
17. Nath, J., Whitlock, J., and Flavin, M. (1978) *J. Cell Biol.* **79** (2), 294a (abstr.)
18. Nath, J., Flavin, M., and Schiffmann, E. (1981) *J. Cell Biol.* **91**, 232-239
19. Nath, J., Flavin, M., and Gallin, J.I. (1982) *J. Cell Biol.* **95**, 519-526
20. Nath, J., and Gallin, J.I. (1983) *J. Clin. Invest.* **71**, 1273-1281
21. Klebanoff, S.J., and Clark, R.A. (1978) in *The Neutrophil: Function and Clinical Disorders*, North-Holland Publishing Co., Amsterdam, 283-408
22. Curnutte, J.T., and Babior, B.M. (1987) *In Adv. in Human Genetics, Plenum Publishing Corp., N.Y.,* **16**, 229-297
23. Bellavite, P. (1988) *Free Radical Biol. Med.* **4**, 225-261
24. Babior, B.M., (1988) *Arch. Biochem. Biophys.* **264**, 361-367
25. Nath, J. and Gallin, J.I. (1984) in *White Cell Mechanics: Basic Science and Clinical Aspects*, Alan R. Liss, Inc., N.Y., 95-110
26. Nath, J., Powledge, A., and Wright, D.G., (1989) *J. Biol. Chem.* **264**, 848-855
27. Nath, J., and Flavin, M. (1980) *J. Neurochem.* **35**, 693-706
28. Nath, J., and Flavin, M. (1984) *Biochim. Biophys. Acta.* **803**, 314-322
29. Cohen, H.J., and Chovaniec, M.E. (1978) *J. Clin. Invest.* **61**, 1081-1087
30. Nath, J., and Powledge, A. (1988) *Biochem. Biophys. Res. Commun.* **156**, 1376-1382
31. Verghese, M.W., Smith, C.D., and Snyderman, R. (1985) *Biochem. Biophys. Res. Commun.* **127**, 450-457
32. Berridge, M.J. (1984) *Biochem. J.* **220**, 345-360
33. Kikuchi, A., Kozawa, O., Kaibuchi, K., Katada, T., Ui, M., and Takai, Y. (1986) *J. Biol. Chem.* **261**, 11558-11562
34. Lad, P.M., Olson, C.V., Grewal, I.S., and Scott, S.J. (1985) *Proc. Natl. Acad. Sci. U.S.A.* **82**, 8643-8647
35. Baehner, R.L., Boxer, L.A., and Davis, J. (1976) *Blood.* **48**, 309-328
36. Klebanoff, S.J., and Clark, R.A. (1978) in *The Neutrophil: Functional and Clinical Disorders*. North-Holland Publishing Co. The Netherlands. 641-709
37. Becker, E.L., Siguran, M., and Oliver, J.M. (1979) *Am. J. Pathol.* **95**, 81-97
38. Nath, J., and Gallin, J.I. (1986) *J. Immunol.* **136**, 628-635
39. Mellon, M.G. and Rebhun, L.I. (1976) *J. Cell Biol.* **70**, 226-238
40. Nath, J., and Rebhun, L.I. (1976) *J. Cell Biol.* **68**, 440-450
41. Nath, J., and Gallin, J.I. (1987) *Annals. N.Y. Acad. Sci.* **498**, 216-228

Isolation and Characterization of a Mammalian Hydroxyacid-Oxoacid Transhydrogenase

E.E. Kaufman, T. Nelson and H.M. Fales
Laboratory of Cerebral Metabolism, National Institute of Mental Health
National Institutes of Health, Bethesda, MD 20892 USA

Abstract

A transhydrogenase that catalyzes the oxoacid-dependent oxidation of specific hydroxyacids has been isolated from the soluble fraction of rat kidney mitochondria (1). The hydroxyacids that have been shown to be good substrates for this enzyme are γ-hydroxybutyrate, L-β-hydroxybutyrate, and D-α-hydroxyglutarate; the oxoacids that can function as hydrogen acceptors include α-ketoglutarate, succinic semialdehyde, oxalacetate and α-ketoadipate.

Introduction

A mitochondrial transhydrogenase that catalyzes the oxoacid-dependent oxidation of the hydroxyacids, γ-hydroxybutyrate, L-β-hydroxybutyrate, and D-α-hydroxyglutarate has been isolated and partially purified from rat kidney (1).

This enzyme was discovered in the course of our investigation of the initial step in catabolic pathway of one of these hydroxyacids, γ-hydroxybutyrate. It has been established that the overall catabolic pathway for this neuroactive compound proceeds by the following series of steps (2,3).

1) γ-hydroxybutyrate → succinic semialdehyde

2) succinic semialdehyde → succinate

3) succinate → → → → $CO_2 + H_2O$.

The second step in this pathway is catalyzed by the mitochondrial enzyme, succinic semialdehyde dehydrogenase, and the third by the mitochondrial enzymes of the citric acid cycle. An $NADP^+$-dependent oxidoreductase that can catalyze the first step in the reaction has been isolated from liver cytosol (4,5).

The preparation of a polyclonal antibody (6) allowed us to ask whether this cytosolic dehydrogenase is the only enzyme that catalyzes the initial step in γ-hydroxybutyrate catabolism. Failure to block more than approximately 50% of the conversion of [1-^{14}C]γ-hydroxybutyrate to $^{14}CO_2$ with the antibody in a crude homogenate of rat kidney suggested the presence of at least one additional enzyme that can catalyze the first reaction.

Since we had previously shown that the cytosolic $NADP^+$-dependent oxidoreductase was not present in either the mitochondrial or the microsomal fraction (4), we could use the ability of the mitochondrial fraction to catalyze the complete catabolic pathway, *i.e.*, the conversion of [1-^{14}C]-γ-hydroxybutyrate to $^{14}CO_2$, as test of whether or not the mitochondrial fraction contains an alternate enzyme capable of catalyzing the initial step in the pathway. When this was tried, we found that the washed mitochondrial fraction could catalyze the complete oxidation of GHB to CO_2 and, when compared to the activity in the unfractionated homogenate, the activity in the mitochondrial fraction ranged from 31% for kidney to 54% for liver and 44% for brain (6). These results suggested the presence of an alternate mitochondrial enzyme that could catalyze the oxidation of γ-hydroxybutyrate to succinic semialdehyde.

Characterization of the Pyridine Nucleotide-Independent Oxidation of γ-Hydroxybutyrate by a Mitochondrial Enzyme.

A frozen and thawed kidney mitochondrial preparation was found to catalyze the oxidation of [1-^{14}C]γ-hydroxybutyrate to [^{14}C]succinic semialdehyde in the presence of either added NADH or NADPH with glutamate and glutamate dehydrogenase added as a reduced pyridine nucleotide generating system. Little or no activity was observed when NAD^+ or $NADP^+$ were substituted for the reduced cofactor. This was consistent with our previous failure to detect either NAD^+- or $NADP^+$-dependent γ-hydroxybutyrate dehydrogenase activity in the mitochondrial fraction (4).

The results of experiments in which the requirements of the mitochondrial enzyme were examined are shown in Table 1. Again, the addition of NADPH with glutamate and glutamate dehydrogenase produced good activity; however, the substitution of glucose-6-phosphate and glucose-6-phosphate dehydrogenase as the generating system reduced the rate from 160 nmol/30 min to 15 nmol/30 min. To test the possibility that the α-ketoglutarate produced by the glutamate/glutamate dehydrogenase reduced pyridine nucleotide-generating system might be the required substance, α-ketoglutarate was added without NADPH or a generating system. As can be seen from Table 1, the activity with α-ketoglutarate alone was slightly better, 207 nmol/30 min, than that obtained with NADPH, glutamate and glutamate dehydrogenase. This result suggested that the mitochondrial enzyme might be an α-ketoglutarate dioxygenase.

Table 1. *Requirements for the Conversion of γ-Hydroxybutyrate (GHB) to Succinic Semialdehyde (SSA) by Dialyzed Rat Kidney Mitochondria*[a]

Additions	Enzyme	Incubation time	nmol of SSA/ flask
		min at 37°C	
None	N	30	0
Glutamate + NADPH + glutamate dehydrogenase	N	30	160
Glucose-6-P + NADPH + glucose-6-P dehydrogenase	N	30	15
α-Ketoglutarate	N	30	207
α-Ketoglutarate	B	30	2
α-Ketoglutarate	N	0	1

[a] Dialyzed rat kidney mitochondria (4 x frozen-thawed) were prepared as previously described (1). The incubated reaction mixture was deproteinized, neutralized and assayed for SSA using an enzymatic assay (4). The reaction mixture contained 100 μmol of Hepes, pH 7.0, 3 μmol of GHB, four times frozen-thawed mitochondria and additions as indicated. Final volume was 1 ml. The abbreviations used are: N, native; B, boiled.

If this were so, one would expect that the products derived from α-ketoglutarate would be succinate and CO_2 (7) and that CO_2 formation would be stoichiometric with the formation of succinic semialdehyde from γ-hydroxybutyrate. These predictions were examined in an experiment shown in Fig. 1A and B in which γ-hydroxybutyrate and [1-^{14}C]α-ketoglutarate were incubated with either native enzyme or boiled enzyme. The incubated reaction mixture was assayed both for succinic semialdehyde formation (Fig. 1A) and for $^{14}CO_2$ formation (Fig. 1B). From Fig. 1A it is apparent that succinic semialdehyde formation is dependent on the presence of both substrates and native enzyme. The dotted line in Fig. 1B represents the $^{14}CO_2$, stoichiometric with succinic semialdehyde formation, that would have been expected had the enzyme been an α-ketoglutarate dioxygenase. The solid line shows that the actual amount of $^{14}CO_2$ formed was no greater in the flasks containing the native enzyme than in those containing boiled enzyme. It was clear from this experiment that the carboxyl group of α-ketoglutarate adjacent to the keto group was not eliminated as CO_2 in this experiment.

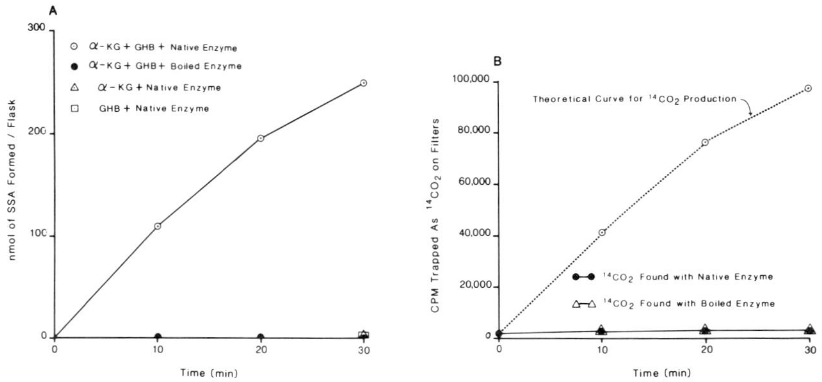

Figure 1. *Time Course of Succinic Semialdehyde Formation (A) and $^{14}CO_2$ Formation (B) in a Reaction Mixture Containing γ-Hydroxybutyrate (GHB) and α-Keto[1-^{14}C] glutarate(αKG)*: Measurement of succinic semialdehyde and $^{14}CO_2$ were carried out as previously described (1). Additions were as indicated in the Figure. Flasks contained native or boiled enzyme as indicated.

When the contents of the flask that had been incubated with γ-hydroxybutyrate, α-keto-[1-^{14}C]glutarate and native enzyme were subjected to paper chromatography, a new peak containing the [1-^{14}C] label appeared. From the total counts found in the new peak it could be calculated that the new [1-^{14}C] containing product was formed in amounts stoichiometric with succinic semialdehyde formation. This peak was absent in the reaction mixture incubated with boiled enzyme or without γ-hydroxybutyrate (1). This finding was consistent with our failure to find $^{14}CO_2$ as a product and also with our failure to find any inhibition when the reaction was carried out under anaerobic conditions.

When a combination of gas chromatography and mass spectroscopy were performed on an incubated reaction mixture containing γ-hydroxybutyrate, α-ketoglutarate, and native enzyme and the results compared with those obtained with a boiled enzyme control and a zero time control, a new peak appeared in the tracing from the gas chromatographic analysis of the TMS derivatives made from an incubated sample containing native enzyme (Fig. 2A). This peak was identified by mass spectrometric techniques as the TMS derivative of α-hydroxyglutarate (Fig. 2B).

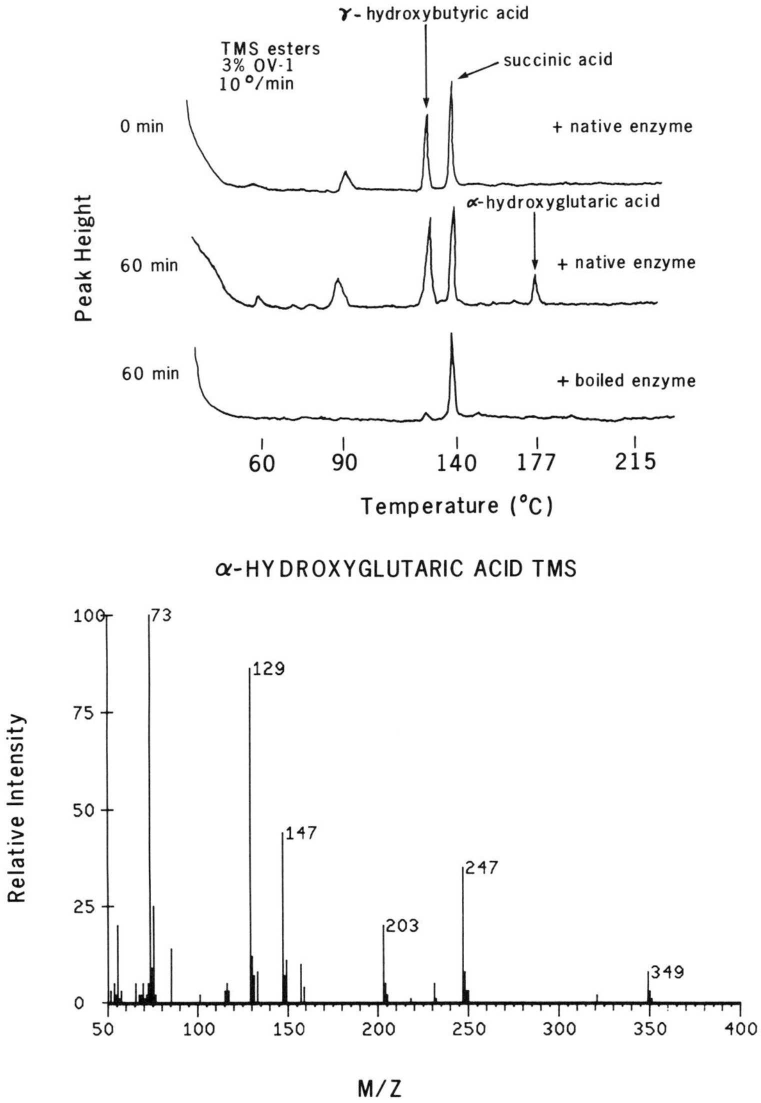

Figure 2. *Gas Chromatographic and Mass Spectral Analysis of the Products of the Reaction of γ-Hydroxybutyrate (GHB) and α-Ketoglutarate(α-KG) Catalyzed by a Mitochondrial Oxido-Reductase.* Gas chromatographic (**A**) and mass spectral (**B**) analyses were performed on three deproteinized reaction mixtures. **A, top** panel: a zero time reaction containing α-ketoglutarate, γ-hydroxybutyrate and native enzyme; **center panel**: a reaction mixture with α-ketoglutarate, γ-hydroxybutyrate, and native enzyme incubated for 60 min; **bottom panel**: the same except boiled enzyme was used. **B** is the mass spectral analysis of the new peak (α-hydroxyglutaric acid) found in the complete reaction mixture incubated for 60 min with native enzyme (**center panel, A**).

The identification of the second product of the reaction as α-hydroxyglutarate was consistent with 1) the failure to find $^{14}CO_2$ formed from α-keto-[1-^{14}C]glutarate and 2) with the appearance of a product which retained the ^{14}C label and was formed in stoichiometric amounts with succinic semialdehyde.

The reaction catalyzed by mitochondrial enzyme therefore was proposed to be:

γ-hydroxybutyrate + α-ketoglutarate ⇄ succinic semialdehyde + α-hydroxyglutarate

This reaction was shown to be reversible. With succinic semialdehyde and α-hydroxyglutarate as the substrates, α-ketoglutarate and γ-hydroxybutyrate were formed in approximately stoichiometric amounts (1).

These results suggested that the mitochondrial enzyme was a transhydrogenase which catalyzed the transfer of a hydrogen attached to the hydroxyl-bearing carbon of γ-hydroxybutyrate to the ketone-bearing carbon of α-ketoglutarate. To examine this possibility, DL-γ-deutero-γ-hydroxy-butyrate was incubated with α-ketoglutarate in the presence of partially purified transhydrogenase. The reaction products, after being derivatized with N,0-Bis (trimethylsilyl)-trifluoracetamide, BSTFA, were subjected to gas chromatographic-mass spectral analysis (Table 2). The substrate, γ-deutero-γ-hydroxybutyrate, was found to be 80.8% deuterated. Examination of two ion pairs ($M^{+\bullet}$-COOTMS and $M^{+\bullet}$-CH_3) derived from α-hydroxyglutarate showed that one, m/z 248/247 had a 38% excess of deuterium, whereas the other, m/z 350/349, had a 42% excess of deuterium. The average of the two was 40%. This is nearly the theoretical incorporation (1/2 of 80.8) since only one of the two optical isomers would be expected to be utilized by the enzyme. In the transfer of deuterium or hydrogen from the hydroxyl-bearing carbon of the hydroxyacid to the ketone-bearing carbon of the oxoacid, presumably there is an intermediate step in which the deuterium or hydrogen is transferred to some group on the enzyme prior to its transfer to the oxoacid. Conversion of only half of the DL-γ-deutero-γ-hydroxybutyrate suggests that a chiral enzyme-bound intermediate is involved. Furthermore, the fact that the expected amount of deuterium was found in the oxoacid indicates that the hydrogen or deuterium on the prosthetic group is probably not exchangeable, making it unlikely that a flavin prosthetic group is involved in the reaction.

Table 2. *Demonstration of the Transfer of Deuterium from DL-γ-Deutero-γ-Hydroxybutyrate to α-Ketoglutarate to Form α-Deutero-α-Hydroxyglutarate Catalyzed by the Mitochondrial Transhydrogenase*[a]

Substrate or Product (Tris-TMS derivative)	m/z pair (considered)	% Deuterated as calculated from m/z pair
Substrate		
γ-hydroxybutyrate	234/233	80.8%
Product		
α-hydroxyglutarate	248/247 ($M^{+\bullet}$-COOTMS)	38.0%
	350/349 ($M^{+\bullet}$-CH_3)	42.0%

[a] The reaction mixture contained 5 μmol of γ-deutero-γ-hydroxybutyrate, 5 μmol of α-ketoglutarate, 100 μmol of Hepes buffer, pH 7.0, 350 μl of partially purified transhydrogenase, and sufficient H_2O to bring the volume to 1.0 ml. This reaction mixture was incubated for 60 min at 37°C, deproteinized, and neutralized. The resulting solution was used for gas chromatographic mass spectral analysis

Properties of the Mitochondrial Transhydrogenase

The substrate specificity of the transhydrogenase is shown in Table 3. Of the hydroxyacids tested, γ-hydroxybutyrate and D-α-hydroxyglutarate were the most active; L-β-hydroxybutyrate had approximately 45% of the activity of α-hydroxyglutarate. D-β-Hydroxybutyrate, DL-α-hydroxy-butyrate, D- and L-malate and D- and L-lactate all had little or no activity as substrates for this enzyme. Other naturally occurring oxoacids namely, pyruvate, α-ketobutyrate, acetoacetate, oxalactate, β-ketobutyrate, α-ketoadipate and β-ketoadipate were tried as hydrogen acceptors in place of α-ketoglutarate. The four and six-carbon α-ketodicarboxylic acids, oxalacetate and α-ketoadipate had 19 and 24%, respectively, of the activity observed with the five-carbon dicarboxylic acid, α-ketoglutarate. The α-keto-monocarboxylic acids, pyruvate and α-ketobutyrate had small amounts of activity, whereas the β-ketoacids had little or no activity. Both chain length and the position of the keto group appear to be important factors in determining the activity of the oxoacid substrate. In addition to these ketoacids, succinic semialdehyde has been found to be at least as good a substrate as α-ketoglutarate when assayed as the hydrogen acceptor for L-β-hydroxybutyrate.

Table 3. *The Relative Rates of the Transhydrogenase Reaction with Other Hydroxyacids and Oxoacids*

Hydroxyacid (with SSA* as co-substrate)	Relative Rate	Oxoacid (with GHB[a] as co-substrate)	Relative Rate
D-α-Hydroxyglutarate	100	α-Ketoglutarate	100
L-α-Hydroxyglutarate	0	α-Ketoadipate	24
DL-β-Hydroxybutyrate	43	Oxalacetate	18.5
L-β-Hydroxybutyrate	45	Pyruvate	8.0
D-β-Hydroxybutyrate	0	α-Ketobutyrate	4.3
DL-α-Hydroxybutyrate	0	Acetoacetate	0
D-Lactate	0	β-Ketoglutarate	0.5
L-Lactate	1.2	β-Ketoadipate	0
D-Malate	0		
L-Malate	0		

[a] SSA = Succinic semialdehyde, GHB = γ-hydroxybutyrate

The kinetic constants for some of the substrates of this transhydrogenase have been determined and are presented in Table 4.

At this time the transhydrogenase has been only partially purified. Starting with a soluble mitochondrial fraction from rat kidney, ammonium sulfate fractionation, chromatography on Sephadex G-150, and DEAE cellulose column chromatography have yielded a 26 fold purification with a 12 % recovery.

Table 4. *Kinetic Constants for the Partially Purified Hydroxyacid-Oxoacid Transhydrogenase*

Substrate	Km_{app}
	M
γ-hydroxybutyrate	3.0×10^{-4}
L-β-hydroxybutyrate	3.0×10^{-3}
α-Hydroxyglutarate	4.2×10^{-4}
α-Ketoglutarate	1.8×10^{-5}
Succinic semialdehyde	4.6×10^{-6}

Discussion

The mitochondrial transhydrogenase described in this paper catalyzes at least three reactions that may be of physiological significance. One of these is the reaction in which γ-hydroxybutyrate and α-ketoglutarate are converted to succinic semialdehyde and α-hydroxyglutarate. Since the only other enzyme known to be capable of catalyzing the oxidation of γ-hydroxybutyrate to succinic semialdehyde is the cytosolic γ-hydroxybutyrate dehydrogenase, the reaction catalyzed by this transhydrogenase would provide an alternate mitochondrial pathway for the oxidation of γ-hydroxybutyrate to succinic semialdehyde, the first reaction in the catabolism of γ-hydroxybutyrate to CO_2 and H_2O. A comparison of the rates of the two enzymes catalyzing the initial step in the catabolic pathway for γ-hydroxybutyrate is shown in Table 5.

This enzyme also catalyzes the reverse reaction in which D-α-hydroxyglutarate and succinic semialdehyde are stoichiometrically converted to α-ketoglutarate and γ-hydroxybutyrate, respectively. Whether or not this reaction is of importance in the mammalian biosynthesis of γ-hydroxybutyrate is as yet to be determined. A third reaction catalyzed by this enzyme is the succinic semialdehyde-dependent or α-ketoglutarate-dependent oxidation of L-β-hydroxybutyrate to acetoacetate. In one case, the enzyme catalyzes the conversion of L-β-hydroxybutyrate and succinic semialdehyde to acetoacetate and γ-hydroxybutyrate, respectively, and in the other, the conversion of L-β-hydroxybutyrate and α-ketoglutarate to form acetoacetate and α-hydroxyglutarate.

Table 5. *A Comparison of the Capacity of the Cytosolic and Mitochondrial Fractions of Brain and Kidney to Oxidize γ-Hydroxybutyrate (GHB)*

Tissue	Enzyme	nmol/min/gram Tissue
BRAIN	GHB dehydrogenase[a] (cytosolic)	10.1
	Transhydrogenase[a] (mitochondrial)	16.6
KIDNEY	GHB dehydrogenase[a] (cytosolic)	324
	Transhydrogenase[a] (mitochondrial)	172

[a] All enzyme activities were measured under V_{max} conditions.

Pinson *et al.* (8) have presented evidence that heart cells in culture produce substantial amounts of free L-β-hydroxybutyrate. Furthermore, Weber and Edmond (9), Hunter and Sadler (10) and others have demonstrated that L-β-hydroxy-[^{14}C]butyrate can be converted to $^{14}CO_2$ *in vivo*. It has generally been believed that the β-hydroxyacyl-CoA-dehydrogenase (EC 1.1.1.35) which is specific for the acyl-CoA derivatives of the L-β-hydroxyacid is the only enzyme capable of catalyzing the oxidation of L-β-hydroxybutyrate (13). This reaction would require free L-β-hydroxybutyrate to be activated to form the acyl-CoA derivative before it could be oxidized. The discovery of a hydroxyacid-oxoacid transhydrogenase raises the possibility that *in vivo*, free L-β-hydroxybutyrate is converted directly to acetoacetate and subsequently metabolized to CO_2 and H_2O. In combination with D-β-hydroxybutyrate dehydrogenase (EC 1.1.1.30), also a mitochondrial enzyme, L-β-hydroxybutyrate could be converted to D-β-hydroxybutyrate with acetoacetate as an intermediate.

The hydroxyacid-oxoacid transhydrogenase described in this paper can be distinguished from the other known mammalian transhydrogenases located in the mitochondrion on the basis of substrate specificity and location within the mitochondria. Probably the most extensively studied transhydrogenase is the pyridine nucleotide transhydrogenase (E.C.1.6.1.1) (12) located in the mitochondrial membrane where reduction of $NADP^+$ by NADH catalyzed by this enzyme is coupled to the translocation of protons from the cytosolic side to the matrix side of the mitochondrial inner membrane (13). A second transhydrogenase, also with very different substrate specificity, is the glutathione-insulin transhydrogenase (E.C.1.8.4.2) which catalyzes a sulfhydryl-disulfide interchange with insulin and other proteins. This enzyme has been found not only in mitochondria, but also in microsomes, cytosol and other subcellular structures (14).

A transhydrogenase (EC 1.1.99.7) which catalyzes a similar reaction to those described in this report, *i.e.*,

$$\text{L-lactate} + \text{oxalacetate} \rightleftarrows \text{pyruvate} + \text{L-malate}$$

has been isolated from *Micrococcus lactilyticus* (15). Although both enzymes are hydroxyacid-oxoacid transhydrogenases, the substrate specificities of the bacterial and the mammalian enzymes are quite different since the mammalian mitochondrial enzyme has little or no activity with either L-malate or L-lactate (Table 3).

The bacterial enzyme has been found to contain a tightly bound NAD^+/NADH prosthetic group (15). We have not yet purified the mitochondrial enzyme sufficiently to determine whether or not it contains a bound prosthetic group and, if so, to determine the nature of the prosthetic group. Work on purification of this enzyme, as well as on the possible involvement and identity of a prosthetic group, is now in progress.

References

1. Kaufman, E.E., Nelson, T., Fales, H.M., and Levin, D.M. (1988) *J. Biol. Chem.* **263**, 16872-16879
2. Doherty, J.D., Stout, R.W., and Roth, R.H. (1975) *Biochem. Pharmacol.* **24**, 469-474
3. Möhler, H., Patel, A.J., and Balázs, R. (1976) *J. Neurochem.* **27**, 253-258
4. Kaufman, E.E., Nelson, T., Goochee, C., and Sokoloff, L. (1979) *J. Neurochem.* **32**, 699-712
5. Kaufman, E.E., and Nelson, T. (1981) *J. Biol. Chem.* **256**, 6890-6894
6. Kaufman, E.E., Nelson, T., Miller, D., and Stadlan, N. (1988) *J. Neurochem.* **51**, 1079-1084
7. Abbott, M.T., and Udenfriend, S. (1974) in Molecular Mechanisms of Oxygen Activation (Hayaishi, O., ed) pp. 168-214, Academic Press, New York.
8. Pinson, A., Desgres, J., and Heller, M. (1979) *J. Biol. Chem.* **254**, 8331-8335
9. Webber, R.J., and Edmond, J. (1977) *J. Biol. Chem.* **252**, 5222-5226
10. Hunter, E.S., III, and Sadler, T.W. (1987) *Metabolism* **36**, 558-561
11. Lehninger, A.L., and Greville, G.D. (1953) *Biophys. Acta* **12**, 188-207
12. Kaplan, N.O., Colowick, S.P., and Neufeld, E.F. (1953) *J. Biol. Chem.* **205**, 1-15
13. Hoek, J.B., and Rydström, J. (1988) *Biochem. J.* **254**, 1-10
14. Taylor, C.A., and Varandani, P.T. (1985) *Eur. J. Cell Biol.* **39**, 97-104
15. Allen, S.H.G., and Patil, J.R. (1972) *J. Biol. Chem.* **247**, 909-916

INDEXES

The following index is composed of key words submitted by the authors of the individual contributions. It is not intended as a comprehensive subject index. Page citations refer to first pages of articles for which a term serves as a key word.

D

Dr. Deepa Agarwal
ITRC, Post Box No. 80
M.G. Marg, Lucknow-226 001, India
Phone: 40106, 40107

Dr. Deepak K. Agarwal
ITRC, Post Box No. 80
M.G. Marg, Lucknow-226 001, India
Phone: 47586

Mr. C.V. Anand
Assistant Professor
Department of Biochemistry
M.S.R. Medical College
Bangalore, India

Ms. Usha Anand
Assistant Professor
Department of Biochemistry
M.S.R. Medical College
Bangalore, India

Dr. Robert F. Anderson
Mount Vernon Hospital
CRC Gray Laboratory
Northwood, Middlesex HA6 2JR U.K.
Phone: 09274 28611
Fax: 0923 835210

V.B. Angadi
Department of Organic Chemistry
Indian Institute of Science
Bangalore 560 012 India

Prof. A.R. Aroor
Dept. of Clinical Biochem. & Nutrition
Kasturba Medical College
Manipal - 576 119, Karnataka, India
Phone: 20060, Ext. 326

Ms. Usha Arunachalam
Department of Biochemistry
Indian Institute of Science
Bangalore 560 012, India
Phone: 344411 Ext. 309
Fax: 0812-341683

Prof. K.T. Augusti
Department of Biochemistry
University of Kerala
Kariavattom Trivandrum
695 581, Kerala, India
Phone: 8078

Dr. Ann Aust
Dept. of Chemistry and Biochemistry
Utah State University
Logan, UT 84322, USA
Phone: (801) 750-1629
Fax: (801) 750-3390

Dr. Steve D. Aust
Biotechnology Center
Utah State University
Logan, UT 84322, USA
Phone: (802) 750-2730
Fax: (801) 750-2755

Prof. June E. Ayling
Pharmacology Dept., 3130 MSB
University of South Alabama
Mobile, Alabama 36688, USA
Phone: (205) 460-6128
Fax: (205) 460-6464

Dr. Manashi Bagchi
Dept. of Sugery; Room L-1087
University of CT Health Center
School of Medicine
Farmington, CT 06032, USA
Phone: 46-3491

Dr. Steven W. Bailey
Pharmacology Dept., 3130 MSB
University of South Alabama
Mobile, Alabama 36688, USA
Phone: (205) 460-6787
Fax: (205) 460-6464

Prof. D. Balasubramanian
Center for Cellular and
Molecular Biology
Hyderabad, 500 007 India
Phone: 22102 Ext. 52

Prof. Kunissery A. Balasubramanian
Dept. of Gastroenterology
Christian Medical College & Hospital
Vellore 632 004, India

S.N. Balasubrahmanyam
Indian Institute of Science
Dept. of Organic Chemistry
Bangalore 560 012 India

Dr. Ranajit K. Banerjee
Dept. of Physiology
Indian Institute of Chemical Biology
4, Raja S.C. Mullick Road
Calcutta, India
Phone: 46-3493 Ext. 359

Savithra A. Banerji
Institute of Science
Dept. of Biochemistry
15, Madam Cama Road
Bombay, 400 032 India

Mr. Anil K. Bansal
Dept. of Biochemistry
GGS Medical College
Faridkot, India

Prof. P. Mahaboob Basha
Dept. of Studies in Zoology
Bangalore University
Bangalore, India
Phone: 601626 Ext. 214

Mr. Sanjay Batra
Division of Biochemistry
Central Drug Research Institute
P.O. Box No. 173
Lucknow, India

Dr. H.S. Bevinakatti
Alchemie Research Centre
PO Box 155
Thane-Belapur Road
Thane 400 601, Maharashtra, India

Dr. K. Seetharam Bhat
Indian Council of Medical Research
National Institute of Nutrition
Jamai Osmania PO
Hyderbad, India
Phone: 868909

K. Vikram Bhat
P.O. Kungipettu
Udupi 576 102 D.K. India

Mr. Manzoor A. Bhat
Dept. of Biochemistry
Indian Institute of Science
Bangalore 560 012 India

Dr. Surendra Prasad Bhatnagar
Chemicals and Plastics India, Ltd.
A-203, Blue Cross Chambers
Infantory Road Cross
Bangalore, India

Prof. P.K. Bhattacharyya
Indian Institute of Chemical Biology
4 Raja S C Mullick Road
Jadavpur, Calcutta 700 032, India

Anil K. Bhutey
17-A15, Govt. Colony
Haji-ali, Bombay 400 034 India

Dr. Elizabeth Blèe
Lab. d'Enzymologie cellulaire
et Molec.
Institut de Botanique
28 rue Goethe
Strasbourg, Cedex 67083, France

Dr. Sekhar S. Boddupalli
Dept. of Biochemistry
University of Texas
SW Medical Center at Dallas
5323 Harry Hines Blvd.
Dallas, TX 75235, USA

Dr. Alan R. Brash
Dept. of Pharmacology
Vanderbilt Univ. Medical School
Nashville, TN 37232, USA
Phone: (615) 322-3015
Fax: (605) 322-4707

Dr. Jay R. Burgess
Dept. of Biological and
Marine Sciences
University of California
Building 569
Santa Barbara, CA 93106, USA
Phone: 805-961-4007

Dr. Jorge H. Capdevila
Dept. of Medicine
Vanderbilt University
Medical Center North S-3223
Nashville, TN 37205, USA
Phone: (615) 322-4794
Fax: (615) 343-7156

Dr. Thandavan Chandra
PSG Inst. of Medical
Sciences & Research
Coimbatore, India
Phone: 33121

S. Chandrasekhar
Dept. of Organic Chemistry
Indian Institute of Science
Bangalore 560 012 India

S. Chandrasekharan
Dept. of Organic Chemistry
Indian Institute of Science
Bangalore 560 012 India

J. Chandraseknar
Dept. of Organic Chemistry
Indian Institute of Science
Bangalore 560 012 India

Mr. N.CH. Varada Charyulu
Dept. of Biochemistry
S.K. University
Anantapur, India

Prof. Indu B. Chatterjee
Dept. of Biochemistry
University College of Science
35 Ballygunge Circular Road
Calcutta, India
Phone: 47-3681

Dr. Jayasri Chaudhuri
Dept. of Biochemistry
Calcutta University
35 Ballygunge Circular Road
Calcutta, India
Phone: 47-3681

Dr. Shive M.S. Chauhan
Dept. of Chemistry
University of Delhi
Delhi, India
Phone: 2521521/331

Dr. Carmine J. Coscia
Dept. of Biochemistry
St. Louis Univ. School of Medicine
1402 S. Grand St.
St. Louis, MO 63104 USA
Phone: (314) 577-8160
Fax: (314) 772-1307

Sumati Jos D'Souza
Bhabha Atomic Research Centre
Biochemistry Division
Trombay, Bombay 400 085 India

Prof. Dipak K. Das
Cardiovascular Division
University of CT School of Medicine
Farmington, CT 06032, USA

Mr. Mukul Das
IRTC, P.O. Box No. 80
M.G. Marg, Lucknow-226 001, India
Phone: 47586, 48228

Dr. Undurti N. Das
Dept. of Medicine
Nizam's Inst. of Medical Sciences
Punjagutta, Hyderabad 500 482, India

Dr. Dipak Dasgupta
Crystallography & Molecular
Biology Division
Saha Institute of Nuclear Physics
1/AF, Bidhan Nagar
Calcutta - 70064, India

Prof. Kelvin J.A. Davies
Institute for Toxicology
Dept. of Biochemistry
University of Southern California
1985 Zonal Ave., HSC-PSC 614-616
Los Angeles, CA 90033, USA
Phone: 213-224-7452(7895)

Mrs. Anuradha S. Deshpande
Food Technology &
Enzyme Engineering Division.
Bhabha Atomic Research Centre
Bombay, India
Phone: 551 4910 Ext. 2544

K. Devaprakash
Alur (Mandal)
Molagapalli (P.O.)
Kurnool (D.T.) A.P. India

Mr. K. Dhanajaya
Dept. of Biochemistry
S.K. University
Anantapur, India

Prof. Bhola Nath Dhawan
C.D.R.I.
Lucknow, India

V. Dhananjayanath
S.K. University
Biochemistry Division
Anantapur 515 003 India

Prof. A. T. Diplock
Division of Biochemistry
Guy's and Saint Thomas's Hospitals
University of London
London Bridge, United Kingdom
Phone: 1-955-4521
Fax: 1-403-7195

Mr. P.D. Dwivedi
ITRC, PO Box 80
M.G. Marg, Lucknow
226 001 India

Prof. John W. Eaton
DIGHT Laboratories
University of Minnesota
400 Church Street SE
Minneapolis, MN 55455, USA
Phone: (612) 626-1951
Fax: (602) 378-3601

Dr. Robert W. Egan
Allergy Dept.
Schering Corporation
60 Orange St.
Bloomfield, NJ 07003, USA
Phone: (201) 429-4552
Fax: (201) 429-4039

Dr. Thomas E. Eling
NIEHS
Laboratory of Molecular Biophysics
PO Box 12233
Research Triangle, NC 27709, USA
Phone: (919) 541-3425

Prof. Ronald W. Estabrook
Dept. of Biochemistry
Univ. TX Southwestern Medical Ctr.
5323 Harry Hines Blvd.
Dallas, TX 75235, USA
Phone: (214) 688-3456
Fax: (214) 688-8855

Mr. Pervez F. Firozi
Biochemistry Division
Bhabha Atomic Research Centre
Bombay, India
Phone: 551 4910 Ext. 2335

Dr. Jurgen Fuchs
Zentrum der Dermatologie
Universitasklinikum
Theodor Stern Kai 7
6000 Frankfurt/M 70, FRG
Phone: 049-69-6301-6835
Fax: 049-69-6301-6301

Amos S. Gaikwad
Indian Institute of Science
Dept. of Biochemistry
Bangalore 560 012 India

Prof. Hiranmoy Gangopadhyay
Dept. of Sugery
Univ. of CT School of Medicine
Farmington, CT 06032, USA

Jacob G. Ghazarian
Dept. of Biochemistry
Medical College of Wisconsin
8701 Watertown Plank Road
Milwaukee, Wisconsin 53226, USA
Phone: (414) 257-8447
Fax: (414) 257-2008

Dr. Sudhir K. Goel
ITRC, Post Box 80
M.G. Marg, Lucknow 226 001, India
Phone: 47586

R. Gopalar
St. Thomas Mount
Bush Boake Allen (India) Ltd.
Madras 600 016 India

Prof. John W. Gorrod
Chelsea Dept. of Pharmacy
University of London
Manresa Rd., London, U.K.
Phone: 01-351 2488, Fax: 01-351 5307

Dr. Frederick P. Guengerich
Dept. of Biochemistry
Vanderbilt University
638 Medical Research Building
Nashville, TN 37232-0146, USA
Phone: (615) 322-2261
Fax: (615) 322-3141

Prof. Irwin C. Gunsalus
Dept. of Biochemistry
University of Illinois
1209 W. California St.
Urbana, IL 61801, USA

Mr. T.L. Gururaja
Indian Institute of Science
Bangalore 560 012 India

Prof. Barry Halliwell
Dept. of Biochemistry
Medical Faculty of King's College
The Strand, London, England

Dr. Gordon A. Hamilton
Dept. of Chemistry
Penn State University
152 Davey Lab
University Park, PA 16802, USA

Prof. Sven Hammarström
Dept. of Cell Biology
Faculty of Health Sciences
S-581 85, Linköping, Sweden
Phone: +46 13 194277
Fax: +46 13 194149

Dr. George Hammons
NCTR, HFT-110
Jefferson, AR 72079, USA
Phone: (501) 541-4552

Ms. Tapas K. Hazra
Dept. of Biochemistry
Calcutta University
35, Ballgunge Circular Road
Calcutta-19, India
Phone: 47-3681

Dr. George Hildenbrandt
Penn State University
232 Fenske Lab
University Park, PA 16802, USA

V.R. Holla
Indian Institute of Science
Dept. of Organic Chemistry
Bangalore 560 012 India

Mr. M. Mahaboob Hussain
Indian Council of Medical Research
National Institute of Nutrition
Jamai Osmania PO
Hyderabad, India
Phone: 868909

Dr. Keith U. Ingold
Division of Chemistry
National Res. Council of Canada
Ottawa, Ontario K1A 0R6 Canada
Phone: (613) 990-0938

Monika Jain
Indian Institute of Science
Dept. of Organic Chemistry
Bangalore 560 012 India

Ms. Shailu K. Jagtiani
Biochemistry Division
Bhabha Atomic Research Centre
Trombay, Bombay 400 085, India

Dr. D. R. Janero
Research Dept.
Pharmaceuticals Division
CIBA-GEIGY Corporation
Summit, New Jersey 07901 USA

Ms. A. Jayadeep
Dept. of Biochemistry
University of Kerala
Kariavattom, Trivandrum
695 581, India

Dr. N. Jayakumari
Assistant Professor of Biochemistry
Sree Chitra Tirunal Institute
for the Medical Sciences
Trivandrum-11, India

Dr. Bollineni Jayasundaramma
Dept. of Zoology
S.V. University
Tirupati, India
Phone: 21472

Dr. Donald M. Jerina
Secondary Oxidation Mechanisms
LBC, NIDDK, NIH
Rockville Pike
Bethesda, MD 20892, USA

T.V. John
St. Thomas Mount
Bush Boake Allen (India) Ltd.
Madras 600 016 India

Theresa Joseph
Indian Institute of Science
Dept. of Organic Chemistry
Bangalore 560 012 India

Dinesh K. Joshi
Indian Institute of Science
Dept. of Organic Chemistry
Bangalore 560 012 India

Dr. Prakash C. Joshi
Phototoxicology Laboratory
ITRC, P.O. Box 80
M.G. Marg, Lucknow 226 001, India

Dr. Valerian E. Kagan
Institute of Physiology
Bulgarian Academy of Sciences
Sofia, Bulgaria
Phone: 70-52-59

Dr. Poonam Kakkar
Ecotoxicolgy Section
Industrial Toxicolgy Res. Centre
Post Box No. 80
M.G. Marg, Lucknow 226 001, India
Phone: 33786

Jayashree Prabnakar Kamat
Bhabha Atomic Research Centre
Biochemistry Division
Bombay 400 085 India

T.B. Karegouder
Gulbarga University
Dept. of Biochemistry
Gulbarga 585 106 India

T.R. Kasturi
Dept. of Organic Chemistry
Indian Institute of Science
Bangalore 560 012 India

Dr. Elaine E. Kaufman
Laboratory of Cerebral Metabolism
National Institute of Mental Health
9000 Rockville Pike
Bethesda, MD 20892, USA
Phone: (301) 496-1371
Fax: (301) 480-1668

Dr. Seymour Kaufman
Laboratory of Neurochemistry
Natl. Institute for Mental Health
NIH, Bldg. 36, Rm. 3D30
Bethesda, MD 20892, USA
Phone: (301) 496-3579
Fax: (3010 496-9935

A.R. Kidwai
196, Zakir Bagh
Okhla Road
New Delhi 110 025 India

Prof. Judith P. Klinman
Dept. of Chemistry
University of CA at Berkeley
Berkeley, CA 94720, USA
Phone: (415) 642-2668
Fax: (415) 642-8369

K.K. Kohli
PGIMER, Chandigarh
Dept. of Biochemistry
U.P. 160 012 India

Dr. P.R. Krishnaswamy
Vittal Mallya Scientific
Research Foundation
K.R. Road, Bangalore, India

Mr. Chitta Suresh Kumar
Dept. of Biochemistry
S.K. University
Anantapur, India

Dr. Pradeep G. Kumar
School of Life Sciences
Devi Ahilya Vishwavidyalaya
Vigyan Bhawan, Khandwa Road
Indore, India

Prof. Raj Kumar
Dept. of Dermatology, M/C 624
University of Illinois, Chicago
808 S. Wood St.
Chicago, IL 60612, USA
Phone: (312) 996-1194

Dr. Rajiv Kumar
Division of Physiology
Central Drug Research Institute
Lucknow, India

Dr. Sunil Kumar
Dept. of Biochemistry
PGIMER
Chandigarh, India

Dr. Sushil Kumar
Laboratory Environ. Carcinogenesis
ITRC, P.B. No. 80
M.G. Marg, Lucknow 226 001, India
Phone: 40107

Dr. Ramadasan Kuttan
Amala Cancer Research Centre
Amala Nagar
Trichur - 680 553, Kerala India
Phone: 20950

Ms. Malini Laloraya
Endocrinology Lab
School of Life Science
Devi Ahilya Vishwavidyalaya
Vigyan Bhawan, Khandwa Road,
Indore 452 001 India

Prof. William E.M. Lands
Dept. of Biological Chemistry
University of Illinois
College of Medicine at Chicago
1853 W. Polk St.
Chicago, IL 60612 USA
Phone: (312) 996-7670
Fax: (312) 413-0364

Dr. John D. Lipscomb
Department of Biochemistry
University of Minnesota
4-225 MIllard Hall
Minneapolis, Minnesota 55455, USA
Phone: 612-625-6564, Fax: 612-626-5657

Dr. Bilur R. Lokesh
Dept. of Food Chemistry
CFTRI
Mysore, India

Dr. George H. Lorimer
Central R&D Department
E. I. DuPont DeNemours & Co.
Wilmington, DE 19880-0402
Phone: 302-695-4584, Fax: 302-695-4296

Dr. Jammada P. Machaiah
Food Tech. & Enzyme Res. Div.
Bhabha Atomic Research Centre
Bombay, India
Phone: 5514910 Ext. 2767

Xavier Macnado
Indian Institute of Science
Dept. of Organic Chemistry
Bangalore 560 012 India

Dr. Jane F. Maddox
Penn State University
115 Henning Building
University Park, PA 16802 USA

K.M. Madyastha
Chairman
Dept. of Organic Chemistry
Indian Institute of Science
Bangalore 560 012 India

Mr. Milind C. Mahajan
Dept. of Biochemistry
Indian Institute of Science
Bangalore 560 012 India
Phone: 344411 Ext. 2309

U. Maitra
Dept. of Organic Chemistry
Indian Institute of Science
Bangalore 560 012 India

Dr. S.J. Makhija
Biochemistry Unit, Chemistry Dept.
S.B.E.S. College of Science
Aurangabad, India

Nandita B. Mangalore
Institute of Science
Dept. of Biochemistry
15, Madame Cama Road
Bombay 400 032 India

Dr. Bengt Mannervik
Dept. of Biochemistry
Biomedical Center, University of Uppsala
PO Box 576
Uppsala, Sweden

Dr. Sara Rani Marcus
Dept. of Biochemistry
M.S. Ramaiah Medical College
Gokula, Bangalore 560 054, India
Phone: 369190 Ext. 11

Prof. Lawrence J. Marnett
Dept. of Biochemistry & Chemistry
Center in Molecular Toxicology
Vanderbilt University
Nashville, TN 37232, USA

Prof. Vincent Massey
Dept. of Biological Chemistry
University of Michigan Medical School
1301 Catherine Road
Ann Arbor, MI 48109, USA
Phone: (313) 764-7196

Dr. Harihara M. Mehendale
Dept. of Pharmacology and Toxicol.
University of Miss. Medical Center
2500 North State Street.
Jackson, MS 39216, USA
Phone: (601) 984-1618
Fax: (601) 984-1262

Mr. Sudhir Mehrotra
Ecotoxicology Section
ITRC, P.O. Box 80
M.G. Marg, Lucknow 226 001, India
Phone: 33786

Dr. B. Haresh Mehta
Dept. of Neurochemistry
NIHMANS
P.B. No. 2900
Bangalore 560 029, India

Dr. Hersch B. Mehta
School of Medicine and VAMC
Univ. of California San Diego
San Diego, CA 92103, USA

Prashant P. Mehta
Indian Institute of Science
Dept. of Biochemistry
Bangalore 560 012 India

Seema V. Mishra
The Institute of Science
Dept. of Biochemistry
15, Madam Cama Road
Bombay 400 032 India

Dr. Hara P. Misra
Dept. of Biomedical Sciences
College of Veterinary Medicine
Virginia Tech
Blacksburg, VA 24061, USA

S.C. Misra
A-203, Blue Cross Chambers
Chemicals & Plastics India Ltd.
Infantry Cross Road
Bangalore 560 001 India

Dr. Thomas J. Moorehead
Norwich Eaton Pharmaceuticals, Inc.
PO Box 191
Norwich, NY 13815, USA

Dr. Bhagavatula Moorthy
Dept. of Biochemistry
Medical College of Wisconsin
8701 Watertown Plank Road
Milwaukee, Wisconsin 53226, USA
Phone: (414) 257-8447
Fax: (414) 257-2008

Dr. Jason Morrow
Division of Clinical Pharmacology
Vanderbilt University
Nashville, TN 37232-6601 USA

Mr. Jayati Mullick
Dept. of Biochemistry
Vallabhbhai Patel Chest Institute
University of Delhi
Delhi, India
Phone: 2928994

Dr. B. Nagarajan
Dept. of Microbiology & Tumour
Biochemistry
Cancer Institute
Madras, India

S. Nagini
Annamalai University
Department of Chemistry
Annamalainagar 608 002 India

Dr. Afshan Naheed
Dept. of Biochemistry
Faculty of Life Sciences
Aligrah Muslim University
Aligrah, India

Prof. Madhavan G. Nair
Dept. of Biochemistry
College of Medicine
University of South Alabama
Mobile, AL 36688 USA

Dr. Lalit M. Narurkar
Biochemistry Division
Bhabha Atomic Research Centre
Trombay, Bombay 400 085, India

Dr. Jayasree Nath
Dept. Hematology
Division of Medicine
Walter Reed Army Inst. of Research
Washington, D.C. 20307, USA
Phone: (202) 576-3557
Fax: (202) 576-0703

Dr. Daniel W. Nebert
Lab of Developmental Pharm.
NICHD, NIH,Building 6, Room 416
Bethesda, MD 20892 USA
Phone: (301) 496-5128
Fax: (301) 480-9354

Dr. Santosh Nigam
Dept. Gynecolocigal Endocrinology
Clinicum University
Steglitz Hindenberg Damm #30
D1000 Berlin FRG
Phone: (030) 798-2467
Fax: (030) 798-4141

Prof. Mitsuhiro Nozaki
Dept. of Biochemistry
Shiga University of Medical Science
Seta, Ohtsu, Shiga 520-21, Japan
Phone: 0775-48-2156
Fax: 0775-48-2048

Prof. Peter J. O'Brien
Faculty of Pharmacy
University of Toronto
19 Russell St.
Toronto, Ontario M5S 1A1, Canada

Prof. Sten Orrenius
Dept. of Toxicology
Karolinska Institute
PO Box 60400
Stockholm, Sweden
Phone: 46-8-33 58 74
Fax: 46-8-32 90 41

Prof. C. R. Pace-Asciak
Dept. of Pediatrics & Pharmacology
Research Institute
Hospital for Sick Children
555 University Ave.
Toronto, Ontario M5G 1X8, Canada
Phone: (416) 598-5755
Fax: (416) 598-6897

Prof. Lester Packer
Dept. of Molecular and Cell Biology
University of California
251A Life Science Annex
Berkeley, CA 94720, USA

Potturi Padma
12-2-419/A/7 Alapatinagar
c/o P.L.N. Rao, Mehdipatnam
Hyderabad 500028 India

Prof. Govindarajan Padmanaban
Dept. of Biochemistry
Indian Institute of Science
Bangalore 560 012 India

Dr. Krishna U. Pai
Molecular Bio. & Agricultural Div.
Bhabha Atomic Research Centre
Bombay, India

Dr. Mallayan Palaniandavar
Dept. of Chemistry
Bharthidasan University
Tiruchirapally, India

Dr. Vaithianathan Palaniappan
P.O. Box 2006
Virginia Commonwealth University
1001 West Main Street
Richmond, VA 23284 USA
Phone: (804) 367-1298

Mr. Vikas C. Pandey
Division of Biochemistry
Central Drug Research Institute
Lucknow, India

Mr. Kumudray P. Pandya
Industrial Toxicolgy Research Centre
P.O. Box 80
M.G. Marg, Lucknow 226 001 India
Phone: 47586

Prof. Sampath Parthasarathy
Dept. of Medicine, M-013D
Div. of Endocrinology and Metabolism
Basic Science Building, Room 1080
La Jolla, CA 92093 USA
Phone: (619) 534-4401 or 4402
Fax: (619) 534-6774

Dr. Jawaharlal W. Patel
Division of Pulmonary Medicine
University of Florida
Health Center Box J-225
Gainesville, FL 32610 USA

Mr. Rajesh Patel
Dept. of Microbiology
and Biotechnology Centre
M.S. University
Baroda, Gujarat 391 750 India
Phone: 327796

Mr. Umesh D. Patel
Dept. of Biochemistry
M.S. University of Baroda
Baroda, India
Phone: 327796

C. Paulraj
Indian Institute of Science
Dept. of Organic Chemistry
Bangalore 560 012 India

Dr. Cecil B. Pickett
Merck Frosst Canada Inc.
C.P./P.O. 1005
Pointe Claire, Dorval
Quebec H9R 4P8 Canada
Phone: (514) 630-2683
Fax: (514) 630-2624

Prof. Ned Porter
Dept. of Chemistry
Duke University
Durham, NC 27706 USA
Phone: (919) 684-3420

Dr. Panguluri Surya Prabha
Dept. of Medicine
Nizam's Institute of Medical Sciences
Punjagutta, Hyderbad 500 482 India
Phone: 22889 Ext. 50

Dr. T. Rajendra Prakash
Dept. of Botany
S.V. University
Tirupati - 517 502, (A.P.) India

Mr. Alur R.K. Prasad
Dept. of Biochemistry
S.V. University
Tirupathi - 517 502, (A.P.) India

Dr. Renuka Prasad
Dept. of Surgery
University of CT Health Center
Farmington, CT 06032 USA

Dr. Hullahalli R. Prasanna
N.C.T.R./FDA
Dept. of Health & Human Services
Jefferson, AR 72079 USA
Phone: (501) 541-4546
Fax: (501) 541-4576

Dr. Basayya G. Pujar
Dept. of Chemistry
Karnatak University
Dharwad, India

Dr. Narayan S. Punekar
B.S.E. Group, Chemistry Dept.
Indian Institute of Technology
Powai, Bombay 400 076, India

Ms. Purnima
Dept. of Biochemistry
Panjab University
Chandigarh, India

Mr. Venugopal Radjendirane
Dept. of Biochemistry
Indian Institute of Science
Bangalore 560 012 India
Phone: 344411 Ext.2309
Fax: 011-91-0812-341683

Mr. Arshad Rahman
Dept. of Biochemistry
Aligarh Muslim University
Aligarh, India
Phone: 5741

Dr. K.V. Rajagopalan
Duke University Medical Center
234 Nanaline Duke Bldg.
Durham, NC 27710 USA
Phone: (919) 684-3120
Fax: (919) 684-8885

Dr. Candadai S. Ramadoss
Vittal Mallya Scientific
Research Foundation
K.R. Road, P.O. Box 406
Bangalore, India
Phone: 623951/606967

Golla Ramakrishna
S.K. University
Dept. of Microbiology
Anantapur 515 003 Inida

Santha Ramakrishnan
Indian Institute of Science
Dept. of Biochemistry
Bangalore 560 012 India

Prof. Rallapalli Ramamurthi
Professor of Zoology
S.V. University
Tirupati, India
Phone: 21472

Dr. T. Ramasarma
Indian Institute of Science
Department of Biochemistry
Bangalore 560 012, India

A.K.S. Bhujanga Rao
Senior Scientist, R&D Centre
SIPCOT Industrial Complex
Reckitt and Coleman of India Ltd.
Hosur 635 126 India

Dr. C. Gundu Rao
Research Manager, R&D Centre
Reckitt & Colman of India, Ltd.
176, Sipcot
HOSUR, TN 635 126, India

Dr. Gondi S. Rao
Industrial Research Toxicol. Centre
P.O. Box 80
M.G. Marg, Lucknow 226 001, India
Phone: 47586

G.S.R. Subba Rao
Dept. of Organic Chemistry
Indian Institute of Science
Bangalore 560 012 India

Dr. Gundu H.R. Rao
Dept. of Laboratory Medicine
and Pathology
P.B.198 UMHC
University of Minnesota
Minneapolis, MN 55455, USA
Phone: (612) 626-2717
Fax: (612) 626-3824

P. Ramakrishna Rao
S.K. University
Dept. of Biochemistry
Anantapur 515 003, India

Dr. Srinivas B. Rao
Dept. of Neurochemistry
NIMHANS, PB2900
Bangalore, India

Dr. Sudha R. Rao
Molecular Bio. & Agricultural Div.
Bhabha Atomic Research Centre
Tromboy, Bombay 400 085, India

Thara Rao
Indian Institute of Science
Dept. of Organic Chemistry
Bangalore 560 012 India

Dr. U. Mallikarjuna Rao
Nizam's Inst. of Medical Sciences
Punjagutta, Hyderbad 500 482, India
Phone: 001-402-559-4090
Fax: 001-402-559-4651

Dr. Venkateshwar C. Rao
Div. of Nutrition & Endocrinology
American Health Foundation
1 Dana Rd.
Valhalla, NY 10595, USA

Prof. Samuel M. Rapoport
Humboldt University
Institute of Biochemistry
Hessische Str. 3-4
Berlin, GDR
Phone: 28 386 12

Mr. Veerasamy Ravichandran
Dept. of Medical Biochemistry
University of Madras
Taramani, Madras 600 113 India
Phone: 413849

M. Ravindranath
Indian Institute of Science
Dept. of Organic Chemistry
Bangalore 560 012 India

S.D. Ravindranath
Scientist
CSIR Complex
Palampur 176 061 India

Dr. Vihayalakshmi Ravindranath
Dept. of Neurochemistry
NIHMANS
Bangalore, India
Phone: 642121 Ext. 211

Dr. Pradip K. Ray
Dept. of Biochemistry and Biophysics
University of Kalyani
Kalyani 741 235, West Bengal

Dr. P. Reddanna
S. V. University
Department of Zoology
Tirupati, India

Dr. Bandaru S. Reddy
Dept. of Nutrition & Endocrinology
Naylor Dana Institute for
Disease Prevention
American Health Foundation
Valhalla, NY 10595, USA

Dr. C. Channa Reddy
Penn State University
226 Fenske Lab
University Park, PA 16802, USA
Phone: (814) 863-1625
Fax: (814) 863-1696

Dr. Golla R. Reddy
Div. of Biochemistry & Food Science
Indian Veterinary Research Institute
Izatnagar, India
Phone: 74638

Ms. G. Panduranga Reddy
Research Scholar in Zoology
S.V. University
Tirupathi - 517 502, (A.P.) India
Phone: 20166 Ext. 304

Prof. Janardan K. Reddy
Dept. of Pathology
Northwestern Univ. Medical School
303 E. Chicago Ave.
Chicago, IL 60611, USA
Phone: (312) 908-7948
Fax: (312) 908-0365

Mr. K.K. Reddy
Department of Physical
Anthropology & Biochemistry
S.V. University
Tirupati - 517 502, (A.P.), India
Phone: 20166 Ext. 263

Dr. P.S. Reddy
Dept. of Zoology
S.V. University
Tirupati, India

Prof. Donald J. Reed
Dept. of Biochemistry & Biophysics
Oregon State University
Weniger 535
Corvallis, OR 97331, USA

Dr. L. Jackson Roberts
Division of Clinical Pharmacology
Vanderbilt Univ. School of Medicine
Nashville, TN 37232, USA
Phone: (615) 343-3511
Fax: (615) 322-4707

C.D. Roy
Indian Institute of Science
Dept. of Organic Chemistry
Bangalore 560 012 India

Dr. Bharathi P. Salimath
Dept. of Biochemistry
University of Mysore
Manasagangothri
Mysore 570 006, India
Phone: 22525 Ext. 52

Dr. Timir B. Samanta
Dept. of Biochemistry
Bose Institute
93/1 A.P.C. Road
Calcutta, India
Phone: 35-2402

Dr. Rajinder S. Sangwan
Central Institute of Medicinal &
Aromatic Plants
Post Bag No. 1, P.O. RSM Nagar
Lucknow, India
Phone: 74043 Ext. 32

Ameeta C. Satyaprakash
The Institute of Science
Dept. of Biochemistry
15, Madam Cama Road
Bombay 400 032 India

Dr. Richard W. Scholz
115 Henning Bldg.
Univeristy Park, PA 16802, USA
Phone: (814) 863-2163

Prof. Ramasamy Selvam
Dept. of Medical Biochemistry
University of Madras
Taramani, Madras 600 113, India
Phone: 413849

Mrs Sharmila Sengupta
Crystallography & Molecular
Biology Division
Saha Institute of Nuclear Physics
I/AF Bidhan Nagar
Calcutta 700 064, India
Phone: 37-0659

S. Senray
A 203 Blue Cross Chambers
Chemicals & Plastics (India) Ltd.
Infantry Cross Road
Bangalore 560 001 India

Dr. Charles Serhan
Hematology Division
Brigham & Women's Hospital
75 Francis Street
Boston, MA 02115, USA
Phone: (617) 732-5875
Fax: (617) 732-4144

Dr. Prahlad K. Seth
ITRC, Post Box 80
M.G. Marg, Lucknow 226 001, India
Phone: 47287, 40106

Mr. Arun K. Sethi
Dept. of Experimental Medicine
P.G.I.M.E.R.
Research Block B
Chandigarh, India

Y.I. Sethna
University of Bombay
Dept. of Life Sciences
Vidyanagar, Bombay 400 098 India

V.K. Shalini
CFTRI
Dept. of Nutrition & Food Safety
Mysore 570 013 India

V.N. Shankar
Indian Institute of Science
Dept. of Organic Chemistry
Bangalore 560 012 India

Dr. Sukesh C. Sharma
School of Life Sciences
Guru Nanak Dev University
Amritsar, India
Phone: 65265

Uma H. Shinde
The Institute of Sciences
Dept. of Biochemistry
15, Madam Cama Road
Bombay 400 032 India

Dr. Vidya ShivaSwamy
Dept. of Biochemistry
Indian Institute of Science
Bangalore 560 012 India

Mr. B.R. ShivaKumar
Dept. of Neurochemistry
NIHMANS
Bangalore, India
Phone: 642121 Ext. 211

Mr. Onkar P. Shukla
Division of Biochemistry
Central Drug Research Institute
P.O. Box 173
Lucknow, India

Dr. K.S. Sidhu
Dept. of Zoology
Punjab Agricultural University
Ludhiana - 141 004, Punjab, India
Phone: 51960 Ext. 382

Prof. Helmut Sies
Physiological Chemistry I
University of Dusseldorf
Moorenstr. 5 Dusseldorf, FRG
Phone: 0211/311 27 07
Fax: 011-211-311 30 29

Dr. Jaswant Singh
Biochemistry Section
Regional Research Laboratory
Jammu-Tawi, India
Phone: 5450, 47493

H.P. Singh
CSIR Complex
Scientist
Palampur 176 061 India

Ms. Neelam Singh
CSIR Complex
Palam Pur
Distt. Kangra, H.P. 176 061, India
Phone: 340

Prof. Serjinder Singh
Chemistry Dept.
Guru Nannak Dev University
Amritsar, India
Phone: 62219

Dr. O.P. Singhal
Division of Dairy Chemistry
National Dairy Research Institute
Karnal, India
Phone: 2840
Phale P. Sitakant
Indian Institute of Science
Dept. of Biochemistry
Bangalore 560 012 India

Dr. Roy J. Soberman
Harvard Medical School
Seeley J.Mudd Building
250 Longwood Ave.
Boston, MA 02115, USA
Phone: (617) 732-1422
Fax: (617) 732-0979

B.V. Somashekar
S.V. University
Dept. of Biochemistry
Tirupati 517 502 India

Dr. Girdhari Lal Soni
Dept. of Microbiology
Punjab Agricultural University
Ludhiana, India
Phone: 51960/330

Mr. A. Sreenivas
Dept. of Biochemistry
S.K. University
Anantapur, India

A. Srikrishna
Dept. of Organic Chemistry
Indian Institute of Science
Bangalore 560 012 India

Prof. Birendra N. Srimani
Dept. of Sugery
Univ. of CT School of Medicine
Farmington, CT 06032 USA

Dr. Leela Srinivas
Dept. of Nutrition & Food Safety
C.F.T.R.I.
Mysore, India
Phone: 37451/22660

Dr. P. Srinivasarao
National Institute of Nutrition
Indian Council of Medical Research
Hyderabad 500 007 India
Phone: 868909

Dr. Earl R. Stadtman
Laboratory of Biochemistry
NHLBI, NIH, Building 3, Room 222
Bethesda, MD, 20892, USA

Dr. Thressa C. Stadtman
National Heart, Lung & Blood Institute
NIH, Building 3, Room 108
Bethesda, MD, 20892, USA

Dr. Steen Steenken
Max -Planck-Institut fur Strahlenchemie
Stiftstr. 34-36
433 Mulheim/Ruhr, FRG

Dr. Manickam Sugumaran
Dept. of Biology
University of Massachusetts
Harbor Campus
Boston, MA 02125, USA
Phone: (617) 929-8429
Fax: (617) 265-7173

A. Suresh
S.K. University
Zoology Dept.
Anantapur 515 003 India

Dr. Mohiuddin M. Taher
Liver Care Unit
Osmania General Hospital
Hyderabad, India
Phone: 40121 Ext. 277

Mr. B.L. Tekwani
Division of Biochemistry
Central Drug Research Institute
P.O. Box No. 173
Lucknow, India
Phone: 45932

James Terner
P.O. Box 2006
Virginia Commonwealth University
1001 West Main Street
Richmond, VA 23284 USA
Phone: (804) 367-1298

Dr. Indu Shekhar Thakur
Dept. of Environmental Sciences
College of Basic Sciences & Humanities
G.B. Pant University of Agriculture & Technology
Pantnagar, Nainital, India

Ms. Mini Thomas
Biochemistry Division
Bhabha Atomic Research Centre
Bombay, India
Phone: 551 4910 Ext. 2335

Dr. Kedam Thyagataju
Dept. of Biochemistry
S.V. University
Tirupati, India

Dr. Ming Tien
Dept. of Molecular and Cell Biology
Penn State University
303 Althouse Lab
University Park, PA 16802, USA

Mr. Nitin S. Udar
Dept. of Microbiology & Biotechnology Centre
M.S. University
Baroda - 390 001, Gujarat India
Phone: 554204

Dr. Heinz Ulrich
Asta Pharma AG
Weismullerstr.45
D-600 Frankfurt/Main 1, FRG

William S. Utley
Dept. of Pharmacology and Toxicol.
University of Miss. Medical Center
2500 North State Street.
Jackson, MS 39216, USA
Phone: (601) 984-1618
Fax: (601) 984-1262

Dr. C. S. Vaidyanathan
Indian Institute of Science
Department of Biochemistry
Bangalore 560 012, India

B. Mahendra Varma
S.K. University
Dept. of Zoology
Anantapur, PIN 414 003 India

Mr. B. Venkaiah
Dept. of Biochemistry
S. V. University
Tirupati, India
Phone: 22022

Prof. N. Venkatasubramanian
INBRI
Post Bag 397
Bangalore, India

K. Venkatesan
Dept. of Organic Chemistry
Indian Institute of Science
Bangalore 560 012 India

V. VenkatesanIndian
Institute of Science
Dept. of Organic Chemistry
Bangalore 560 012 India

Aruna Venkatesh
C.F.T.R.I.
Dept. of Food Chemistry
Mysore 570 013 India

Mr. Venkataraman Venkateshwar
Dept. of Biochemistry
Indian Institute of Science
Bangalore 560 012 India

Dr. Joseph J. Villafranca
Dept. of Chemistry
Penn State University
152 Davey Lab
University Park, PA 16802, USA
Phone: (814) 865-4221
Fax: (814) 865-3314

Dr. P.N. Viswanathan
Ecotoxicology Section
ITRC, Post Box 80
M.G. Marg, Lucknow 226 001, India
Phone: 33786

Mr. Bharat Kumar R. Vyas
Dept. of Microbiology
Gujarat University School of Science
Ahmedabad, India
Phone: 443225

Khursheed K. Wadia
Govt. Staff Quarters
Panchasheel
Bldg. 5/10, J.J. Hospital Compound
Byculla, Bombay 400 008 India

Dr. Jay Whelan
Dept. of Food Science
Cornell University
Stocking Hall
Ithaca, NY 14853, USA
Phone: (607) 255-7655
Fax: (607) 254-4868

Dr. Kunio Yagi
Institute of Applied Biochemsitry
Yagi Memorial Park
Mitake, Gifu 505-01, Japan
Phone: 0574-67-5500
Fax: 0574-67-5310

Mr. Bhupesh N. Yagnik
Dept. of Microbiology
Gujarat University School of Science
Ahmedabad, India
Phone: 443225

Dr. Shozo Yamamoto
Dept. of Biochemistry
Tokushima University
School of Medicine
Kuramotocho, Tokushima 770, Japan
Fax: 0886-33-0771

Prof. Guido Zimmer
Gustav-Embden-Zentrum
der Bio. Chemie
Universitat Frankfurt
Theodor-Stern-Kai 7
6 Frankfurt am Main, FRG
Phone: 69-6301-5652
Fax: 6301-6301